CHAPMAN & HALL/CRC
Monographs and Surveys in
Pure and Applied Mathematics   139

SPECTRAL THEORY FOR

RANDOM AND

NONAUTONOMOUS

PARABOLIC EQUATIONS

AND APPLICATIONS

CHAPMAN & HALL/CRC
Monographs and Surveys in Pure and Applied Mathematics

Main Editors

H. Brezis, *Université de Paris*
R.G. Douglas, *Texas A&M University*
A. Jeffrey, *University of Newcastle upon Tyne (Founding Editor)*

Editorial Board

R. Aris, *University of Minnesota*
G.I. Barenblatt, *University of California at Berkeley*
H. Begehr, *Freie Universität Berlin*
P. Bullen, *University of British Columbia*
R.J. Elliott, *University of Alberta*
R.P. Gilbert, *University of Delaware*
R. Glowinski, *University of Houston*
D. Jerison, *Massachusetts Institute of Technology*
K. Kirchgässner, *Universität Stuttgart*
B. Lawson, *State University of New York*
B. Moodie, *University of Alberta*
L.E. Payne, *Cornell University*
D.B. Pearson, *University of Hull*
G.F. Roach, *University of Strathclyde*
I. Stakgold, *University of Delaware*
W.A. Strauss, *Brown University*
J. van der Hoek, *University of Adelaide*

CRC Press
Taylor & Francis Group
6000 Broken Sound Parkway NW, Suite 300
Boca Raton, FL 33487-2742

First issued in paperback 2019

© 2008 by Taylor & Francis Group, LLC
CRC Press is an imprint of Taylor & Francis Group, an Informa business

No claim to original U.S. Government works

ISBN-13: 978-1-58488-895-6 (hbk)
ISBN-13: 978-0-367-38759-4 (pbk)

This book contains information obtained from authentic and highly regarded sources Reasonable efforts have been made to publish reliable data and information, but the author and publisher cannot assume responsibility for the validity of all materials or the consequences of their use. The Authors and Publishers have attempted to trace the copyright holders of all material reproduced in this publication and apologize to copyright holders if permission to publish in this form has not been obtained. If any copyright material has not been acknowledged please write and let us know so we may rectify in any future reprint

Except as permitted under U.S. Copyright Law, no part of this book may be reprinted, reproduced, transmitted, or utilized in any form by any electronic, mechanical, or other means, now known or hereafter invented, including photocopying, microfilming, and recording, or in any information storage or retrieval system, without written permission from the publishers.

For permission to photocopy or use material electronically from this work, please access www.copyright.com (http://www.copyright.com/) or contact the Copyright Clearance Center, Inc. (CCC) 222 Rosewood Drive, Danvers, MA 01923, 978-750-8400. CCC is a not-for-profit organization that provides licenses and registration for a variety of users. For organizations that have been granted a photocopy license by the CCC, a separate system of payment has been arranged.

Trademark Notice: Product or corporate names may be trademarks or registered trademarks, and are used only for identification and explanation without intent to infringe.

CV 09.05.2019 1408

Library of Congress Cataloging-in-Publication Data

Mierczynski, Janusz.
  Spectral theory for random and nonautonomous parabolic equations and applications / Janusz Mierczynski and Wenxian Shen.
      p. cm. -- (Monographs and surveys in pure and applied mathematics)
  Includes bibliographical references and index.
  ISBN 978-1-58488-895-6 (alk. paper)
  1. Differential equations, Parabolic. 2. Evolution equations. 3. Spectral theory (Mathematics) I. Shen, Wenxian, 1961- II. Title. III. Series.

QA377.M514 2008
515'.3534--dc22   2007048063

Visit the Taylor & Francis Web site at
http://www.taylorandfrancis.com

and the CRC Press Web site at
http://www.crcpress.com

CHAPMAN & HALL/CRC
Monographs and Surveys in
Pure and Applied Mathematics 139

SPECTRAL THEORY FOR

RANDOM AND

NONAUTONOMOUS

PARABOLIC EQUATIONS

AND APPLICATIONS

Janusz Mierczyński

Wenxian Shen

CRC Press
Taylor & Francis Group
Boca Raton London New York

CRC Press is an imprint of the
Taylor & Francis Group, an **informa** business

A CHAPMAN & HALL BOOK

To Janusz's Mother and Edyta,

and

Ruijun, Bonny, Charles

Contents

Preface

Spectral theory for linear parabolic equations plays a fundamental role in the study of nonlinear parabolic problems. It is well developed for smooth linear elliptic and periodic parabolic equations. It is also quite well understood for general linear elliptic and periodic parabolic equations. In recent years, much attention has been paid to the extension of spectral theory for linear elliptic and periodic parabolic equations to general time dependent and random linear parabolic equations.

The goal of this monograph is to give a clear and essentially self-contained account of the spectral theory, in particular, principal spectral theory for general time dependent and random linear parabolic equations and systems of such equations. We establish a unified approach to the study of the principal spectral theory: we start to develop the abstract general theory, in the framework of weak solutions, and then specialize to the cases of random and nonautonomous equations. Among others, fundamental properties of the principal spectrum and principal Lyapunov exponents for nonautonomous and random linear parabolic equations are investigated and applications of the developed principal spectral theory to uniform persistence for competitive Kolmogorov systems of nonautonomous and random nonlinear parabolic equations are discussed. The monograph contains many new results, and puts already known results in a new perspective.

The works by H. Amann ([3]–[6]), the works by D. Daners ([29]–[34]), the book by R. Dautray and J.-L. Lions ([36]), the book by D. Henry ([48]), the book by O. A. Ladyzhenskaya [O. A. Ladyženskaja], V. A. Solonnikov and N. N. Ural'tseva [N. N. Ural'ceva] ([70]), and the book by G. M. Lieberman ([73]) are the main sources for the fundamentals (mainly existence, uniqueness, continuous dependence of solutions and Harnack inequalities for positive solutions) for the development of the spectral theory and applications in this monograph.

We have benefited a lot by reading the works by Z. Lian and K. Lu ([72]), J. Húska ([59]), J. Húska and P. Poláčik ([60]), and J. Húska, P. Poláčik and M. V. Safonov ([61]) on Lyapunov exponents for general random dynamical systems and on principal Floquet bundle and exponential separation for general time dependent parabolic equations. We are also indebted to many other people whose works provide the basics for the monograph. The second author has benefited greatly from the collaborations with V. Hutson and G. T.

Vickers on spectral theory for parabolic as well as other types of evolution operators.

This monograph was written under the partial support of NSF grants INT-0341754 and DMS-0504166. The first author was also supported by the research funds for 2005–2008 (grant MENII 1 PO3A 021 29, Poland).

During the preparation of this monograph, the first author visited Auburn University in the summers of 2004 and 2005 and in the spring of 2007. He thanks the faculty of the Department of Mathematics and Statistics for their hospitality. Both of the authors would like to thank Professors Tomasz Dłotko, Georg Hetzer, and Kening Lu for helpful discussions and references.

We are very grateful to the people in Chapman & Hall/CRC, in particular, Sunil Nair, Marsha Pronin, Sarah Morris, Ari Silver, and Tom Skipp for their assistance and cooperation.

The monograph is prepared in LaTeX. Our thanks go to Shashi Kumar for invaluable technical assistance.

Janusz Mierczyński
Institute of Mathematics and Computer Science
Wrocław University of Technology
mierczyn@pwr.wroc.pl

Wenxian Shen
Department of Mathematics and Statistics
Auburn University
wenxish@auburn.edu

Symbol Description

D	Bounded domain in $\mathbb{R}^N$	X	Phase space for scalar parabolic equations
∂D	Boundary of D		
$\mathcal{B}$	Boundary operator	$\mathbf{X}$	Phase space for systems of parabolic equations
$\boldsymbol{\nu}$	The outer unit normal on ∂D		
		Π	Topological or random linear skew-product semiflow
$(\Omega, \mathfrak{F}, \mathbb{P})$	Probability space		
θ	Metric flow on $(\Omega, \mathfrak{F}, \mathbb{P})$	Φ	Topological or random (nonlinear) skew-product semiflow
$\mathfrak{B}$	Family of Borel sets		
μ	Invariant measure		
α_0	Ellipticity constant	ρ_-	Lower principal resolvent
Y	Parameter space for a single linear equation or a general metric space	ρ_+	Upper principal resolvent
		Σ	Principal spectrum
		$\lambda_{\min}$	Minimum of the principal spectrum
$\mathbf{Y}$	Parameter space for a system of linear equations	$\lambda_{\max}$	Maximum of the principal spectrum
σ	Translation flow on Y (or on $\mathbf{Y}$) or general flow or semiflow on Y		
		$\lambda(\mu)$	Principal Lyapunov exponent
Z	Parameter space for a single nonlinear equation	λ_{princ}	Principal eigenvalue
$\mathbf{Z}$	Parameter space for a system of nonlinear equations	φ_{princ}	Principal eigenfunction of an elliptic equation
ζ	Translation flow on Z (or on $\mathbf{Z}$)	γ_0	Separating exponent
		ψ	test function

Chapter 1

Introduction

Reaction–diffusion equations or systems in bounded domains have been used to model many evolution processes in science and engineering, for example, Lotka–Volterra competitive and predator–prey systems, color pattern formation in butterflies and sea shells, tumor growth, just to mention a few in biology. Regardless of the details of the model, one of the common requirements is to investigate the spectral problem for an associated linear evolution problem. This is often required as a tool for nonlinear problems, for example when considering stability or invasion (in the ecological context).

Traditionally most evolution processes are considered in time independent and spatially homogeneous environments. However, in nature, many evolution processes are subject to various variations of the external environments, and the media of the processes are also heterogeneous. General time dependent and random parabolic equations and systems of such equations are therefore of great interest since they can take the above facts into account in modeling evolution processes. A vast amount of research has been carried out toward various dynamical aspects of nonautonomous and random parabolic equations (see, for a few examples, [7], [8], [17], [23], [24], [25], [26], [27], [37], [38], [54], [98], [110], [111]).

As a basic tool for nonlinear problems, it is of great significance to investigate spectral theory for general nonautonomous linear parabolic equations of the form

$$\frac{\partial u}{\partial t} = \sum_{i=1}^{N} \frac{\partial}{\partial x_i} \left(\sum_{j=1}^{N} a_{ij}(t,x) \frac{\partial u}{\partial x_j} + a_i(t,x)u \right)$$

$$+ \sum_{i=1}^{N} b_i(t,x) \frac{\partial u}{\partial x_i} + c_0(t,x)u, \qquad x \in D, \tag{1.0.1}$$

complemented with the boundary conditions

$$\mathcal{B}(t)u = 0, \qquad \text{on } \partial D, \tag{1.0.2}$$

where $D \subset \mathbb{R}^N$ is a bounded domain and $\mathcal{B}$ is a boundary operator of either

1

the Dirichlet or Neumann or Robin type, that is,

$$
\mathcal{B}(t)u = \begin{cases}
u & \text{(Dirichlet)} \\[2mm]
\displaystyle\sum_{i=1}^{N}\Big(\sum_{j=1}^{N} a_{ij}(t,x)\frac{\partial u}{\partial x_j} + a_i(t,x)u\Big)\nu_i & \text{(Neumann)} \\[4mm]
\displaystyle\sum_{i=1}^{N}\Big(\sum_{j=1}^{N} a_{ij}(t,x)\frac{\partial u}{\partial x_j} + a_i(t,x)u\Big)\nu_i \\[4mm]
\qquad + d_0(t,x)u. & \text{(Robin)}
\end{cases}
\tag{1.0.3}
$$

($\boldsymbol{\nu} = (\nu_1, \nu_2, \ldots, \nu_N)$ denotes the unit normal on the boundary ∂D pointing out of D, interpreted in a certain weak sense (in the regular sense if ∂D is sufficiently smooth)).

It is also of great importance to investigate spectral theory for general random linear parabolic equations of the form

$$
\frac{\partial u}{\partial t} = \sum_{i=1}^{N}\frac{\partial}{\partial x_i}\Big(\sum_{j=1}^{N} a_{ij}(\theta_t\omega, x)\frac{\partial u}{\partial x_j} + a_i(\theta_t\omega, x)u\Big)
$$
$$
+ \sum_{i=1}^{N} b_i(\theta_t\omega, x)\frac{\partial u}{\partial x_i} + c_0(\theta_t\omega, x)u, \quad x \in D,
\tag{1.0.4}
$$

complemented with the boundary conditions

$$
\mathcal{B}(\theta_t\omega)u = 0, \qquad \text{on } \partial D,
\tag{1.0.5}
$$

where $\mathcal{B}$ is a boundary operator of either the Dirichlet or Neumann or Robin type, that is, $\mathcal{B}(\theta_t\omega)$ is of the same form as $\mathcal{B}(t)$ in (1.0.3) with $a_{ij}(t,x) = a_{ij}(\theta_t\omega, x)$, $a_i(t,x) = a_i(\theta_t\omega, x)$, and $d_0(t,x) = d_0(\theta_t\omega, x)$, $\omega \in \Omega$ and $((\Omega, \mathfrak{F}, \mathbb{P})$, $\{\theta_t\}_{t\in\mathbb{R}})$ is an ergodic metric dynamical system. It is important as well to investigate spectral theory of nonautonomous and random linear parabolic equations in nondivergence form and coupled systems of nonautonomous and random linear parabolic equations.

Spectral theory is well understood for smooth elliptic or time periodic parabolic equations. For example, it is well known that the eigenvalue λ_{princ} to the eigenvalue problem

$$
\begin{cases}
\Delta u(x) + h(x)u(x) = \lambda u(x), & x \in D, \\[2mm]
\dfrac{\partial u}{\partial \boldsymbol{\nu}}(x) = 0, & x \in \partial D,
\end{cases}
$$

where both the domain D and the function $h\colon \bar{D} \to \mathbb{R}$ are sufficiently smooth, having the largest real part (called the *principal eigenvalue*) is real, simple, and an eigenfunction φ_{princ} corresponding to it (called *principal eigenfunction*) can be chosen so that $\varphi_{\text{princ}}(x) > 0$ for $x \in D$. Hence all positive solutions of a time independent linear parabolic equation are attracted in the direction toward the one-dimensional space spanned by a principal eigenfunction of the associated elliptic eigenvalue problem (*principal eigenspace*) and the solutions lying in the complementary space of the principal eigenspace decay exponentially faster than positive solutions (which is referred to as an *exponential separation* property).

The principal eigenvalue and principal eigenfunction theory is of special interest in applications since it provides necessary and/or sufficient conditions for exponential stability and/or instability in nonlinear problems. The concepts of principal eigenvalue and principal eigenfunction and their properties were extended in [50] to time-periodic parabolic equations (see also [35]). The extension to general time dependent and random parabolic equations is of great difficulty since many approaches which can be successfully applied to time-periodic problems fail to be useful for general time dependent and random problems. Nevertheless, quite a lot of linear theories for time almost-periodic, and general nonautonomous, or even random parabolic problems have been established in various publications, see for example [19], [20], [21], [22], [29], [30], [59], [60], [61], [62], [79], [81], [82], [84], [92], [94], and [97]. The established theories have also found great applications (see [51], [64], [83], [85], [93], [103]).

In the past several years we studied nonautonomous/random linear parabolic equations. A general theory of principal spectrum for nonautonomous linear parabolic equations with certain smoothness (both the coefficients and the domain are sufficiently smooth) has been established, serving as a generalization of the well-known theory of principal eigenvalues and principal eigenfunctions for elliptic equations ([81], [82]). As a counterpart, a theory of principal Lyapunov exponents for random linear parabolic equations with certain smoothness has also been established ([81], [82]). For general nonautonomous linear parabolic equations, many fundamental results about existence, uniqueness, continuous dependence on coefficients of solutions are established in [29], [30], [36], [70], and various versions of Harnack inequalities are developed in [10], [40], [59], [60], [61], [69], [73], [88]. Recently a spectral theory for such equations has also been obtained in [59], [60], [61], mostly for the Dirichlet boundary condition case. There is surely a need to develop an adequate spectral theory for general nonautonomous and random parabolic equations with Neumann or Robin boundary conditions. As a basic tool for the study of nonlinear parabolic problems, it is also of great importance to collect existing as well as newly developed linear theories for general and smooth nonautonomous and random parabolic equations in a monograph.

The objective of this monograph is to give a hopefully clear and essentially self-contained account of the spectral theory, in particular, principal spectral theory for general time dependent and random linear parabolic equations and systems. We follow the following unified approach for the investigation of the spectral theory: we start to develop the abstract general theory, in the framework of weak solutions (mild solutions in the case of systems of parabolic equations), and then specialize to the cases of random and nonautonomous equations. We treat all types of boundary conditions in the same manner. Our exposition focuses on equations in the divergence form, however we provide remarks on corresponding theories for equations in the nondivergence form. On the regularity of the coefficients, we assume the boundedness and measurability, that is, in (1.0.1)+(1.0.2), we assume that a_{ij}, a_i, b_i, and c_0 are bounded and measurable on $(-\infty, \infty) \times D$ and that d_0 is nonnegative bounded and measurable on $\partial D \times (-\infty, \infty)$. In (1.0.4)+(1.0.5), we assume that a_{ij}^ω, a_i^ω, b_i^ω, and c^ω are uniformly bounded in $\omega \in \Omega$ and are measurable on $(-\infty, \infty) \times D$ for each $\omega \in \Omega$ and that d_0^ω is nonnegative uniformly bounded in $\omega \in \Omega$ and is measurable on $\partial D \times (-\infty, \infty)$ for each $\omega \in \Omega$, where $a_{ij}^\omega(t, x) = a_{ij}(\theta_t \omega, x)$, etc. As for the regularity of the domain D, no assumption is needed in the Dirichlet boundary condition case and it is assumed that D is Lipschitz in the Neumann or Robin boundary condition case. We prove various additional properties when the coefficients and the domain turn out to be smooth.

1.1   Outline of the Monograph

First of all, in Chapter 2 we establish fundamental theories in a general setting, i.e., for a general family of nonautonomous equations.

To be more specific, let Y be a (norm-)bounded subset of $L_\infty(\mathbb{R} \times D, \mathbb{R}^{N^2+2N+1}) \times L_\infty(\mathbb{R} \times \partial D, \mathbb{R})$ that is closed (hence, compact) in the weak-* topology of that space and is translation invariant (see Section 1.3).

We then consider (1.0.1)+(1.0.2) for a whole family of coefficients $a \in Y$. The reason for considering (1.0.1)+(1.0.2) for a whole family of coefficients $a \in Y$ is at least fourfold. First, even when we start with only one nonautonomous equation, in many proofs one has to use the procedure of passing to a limit of a sequence of time-translated equations, which can be most easily put in the context of linear skew-product semidynamical systems on a bundle whose base space consists of the closure of all the time translates of the coefficients of the original equation. Second, when considering random equations, their coefficients belong to some family. Third, sometimes we have to compare the properties of the principal spectrum for two equations or even investigate the continuity of the principal spectrum with respect to parameters. And fourth, to study the stability of an invariant set of a nonlinear equation, we need to consider the linearized equations along all the solutions in the invariant set. To emphasize the coefficients and the boundary terms in the problem (1.0.1)+(1.0.2), we will write $(1.0.1)_a + (1.0.2)_a$.

We list in Chapter 2 some basic assumptions including the uniform ellipticity of the time dependent parabolic equations in the general setting introduced above and the Lipschitz continuity of the underlying domain of the equations in the Neumann or Robin boundary condition case. We introduce the concept of weak solutions of the equations in the general setting in the space $L_2(D)$ and collect basic properties of weak solutions which will be needed in later chapters, including local regularity, Harnack inequalities, comparison properties, compactness, continuity with respect to initial data as well as the coefficients of the equations. The solutions of the equations in the general setting are shown to form a skew-product semiflow with fibre or phase space $L_2(D)$. Additional properties are proved when both the domain and the coefficients are smooth. Several remarks are provided for the parabolic problems in nondivergence form.

In Chapter 3, the concepts of principal spectrum and principal Lyapunov exponents and exponential separation of the skew-product semiflow induced from a family of equations in the general setting are introduced. Various basic properties of principal spectrum and principal Lyapunov exponents are presented. Existence of exponential separation and existence and uniqueness of entire positive solutions are shown under quite general assumptions, namely, that positive solutions satisfy appropriate Harnack inequalities. In addition, we present a multiplicative ergodic theorem for a family of equations in the general setting. We also collect several properties of parabolic equations on one-dimensional space domain in an appendix.

Chapter 4 concerns principal spectrum and principal Lyapunov exponents of nonautonomous and random parabolic equations. First the concepts of principal spectrum and principal Lyapunov exponents of nonautonomous and random parabolic equations are introduced in terms of proper family of parabolic equations associated to the given nonautonomous and random equations, which extend the classical concept of principal eigenvalue of elliptic and periodic parabolic equations. Applying the theories developed in Chapters 2 and 3, fundamental properties are then proved, including continuity with respect to the perturbation of coefficients and monotonicity with respect to zero order terms.

In Chapter 5, we investigate the effect of time (space) dependence and randomness of zero order terms on principal spectrum and principal Lyapunov exponents of nonautonomous and random parabolic equations. It is shown that neither time (space) dependence nor randomness will reduce principal spectrum and principal Lyapunov exponents and they are indeed increased except in degenerate cases. More precisely, we show that in the general case the principal spectrum (principal Lyapunov exponent) of a nonautonomous (random) parabolic equation is always greater than or equal to that of the corresponding time-averaged equation. We also show that in the smooth case, the principal spectrum (principal Lyapunov exponent) of a nonautonomous (random) parabolic equation is strictly greater than that of the time-averaged equation except in the case that the coefficient can be decomposed as the sum of a spatially dependent term and a time dependent term. Similar results are proved about the effect of space dependence of zero order terms on principal spectrum and principal Lyapunov exponents of nonautonomous and random parabolic equations. In the biological context these results mean that invasion by a new species is always easier in the time and space dependent case. In addition, we explore the effect of the shape of the domain on principal spectrum and principal Lyapunov exponents and extend the well-known Faber–Krahn inequalities for elliptic and time-periodic parabolic problems to general time dependent and random problems.

Chapter 6 is to extend the linear theory for scalar nonautonomous and random parabolic equations to cooperative systems of such equations. More precisely, we consider the following cooperative systems of nonautonomous parabolic equations

$$
\begin{cases}
\dfrac{\partial u_k}{\partial t} = \displaystyle\sum_{i=1}^{N} \dfrac{\partial}{\partial x_i}\left(\sum_{j=1}^{N} a_{ij}^k(t,x)\dfrac{\partial u_k}{\partial x_j} + a_i^k(t,x)u_k\right) \\[3mm]
\qquad + \displaystyle\sum_{i=1}^{N} b_i^k(t,x)\dfrac{\partial u_k}{\partial x_i} + \sum_{l=1}^{K} c_l^k(t,x)u_l, \qquad x \in D, \\[3mm]
\mathcal{B}^k(t)u_k = 0, \qquad\qquad\qquad\qquad\qquad\qquad x \in \partial D,
\end{cases} \tag{1.1.1}
$$

where $\mathcal{B}^k$ is a boundary operator of either the Dirichlet or Neumann or Robin type, that is, $\mathcal{B}^k(t) = \mathcal{B}(t)$ ($\mathcal{B}(t)$ is as in (1.0.3)) with $a_{ij}(t,x) = a_{ij}^k(t,x)$, $a_i(t,x) = a_i^k(t,x)$, and $d_0(t,x) = d_0^k(t,x)$, $k = 1,2,\ldots,K$, and the following cooperative systems of random parabolic equations

$$
\begin{cases}
\dfrac{\partial u_k}{\partial t} = \displaystyle\sum_{i=1}^{N} \dfrac{\partial}{\partial x_i}\left(\sum_{j=1}^{N} a_{ij}^k(\theta_t\omega,x)\dfrac{\partial u_k}{\partial x_j} + a_i^k(\theta_t\omega,x)u_k\right) \\[3mm]
\qquad + \displaystyle\sum_{i=1}^{N} b_i^k(\theta_t\omega,x)\dfrac{\partial u_k}{\partial x_i} + \sum_{l=1}^{K} c_l^k(\theta_t\omega,x)u_l, \qquad x \in D, \\[3mm]
\mathcal{B}^k(\theta_t\omega)u_k = 0, \qquad\qquad\qquad\qquad\qquad\qquad x \in \partial D,
\end{cases} \tag{1.1.2}
$$

where $k = 1,2,\ldots,K$, $\omega \in \Omega$, $((\Omega,\mathfrak{F},\mathbb{P}),\{\theta_t\}_{t\in\mathbb{R}})$ is an ergodic metric dynamical system, and for each $\omega \in \Omega$, $\mathcal{B}^k(\theta_t\omega) = \mathcal{B}(t)$ with $a_{ij}(t,x) = a_{ij}^k(\theta_t\omega,x)$, $a_i(t,x) = a_i^k(\theta_t\omega,x)$, and $d_0(t,x) = d_0^k(\theta_t\omega,x)$ ($\mathcal{B}(t)$ is as in (1.0.3)). We extend the theories developed in Chapters 2–5 for nonautonomous and random parabolic equations to the above cooperative systems of nonautonomous and random parabolic equations. While doing so, a linear theory is first established for a general family of cooperative systems of parabolic equations, that is, (1.1.1) for all $\mathbf{a} = (a_{ij}^k, a_i^k, b_i^k, c_l^k, d_0^k)$ in a subset $\mathbf{Y}$ of $L_\infty(\mathbb{R} \times D, \mathbb{R}^{K(N^2+2N+K)}) \times L_\infty(\mathbb{R} \times \partial D, \mathbb{R}^K)$.

In the last chapter we consider the applications of the linear theory developed in Chapters 2 to 5 to the uniform persistence issue in systems of random and nonautonomous nonlinear parabolic equations of Kolmogorov type. We focus on the uniform persistence of the following two species competitive Kolmogorov systems of random partial differential equations:

$$
\begin{cases}
\dfrac{\partial u_1}{\partial t} = \Delta u_1 + f_1(\theta_t\omega, x, u_1, u_2)u_1, & x \in D, \\[2mm]
\dfrac{\partial u_2}{\partial t} = \Delta u_2 + f_2(\theta_t\omega, x, u_1, u_2)u_2, & x \in D, \\[2mm]
\mathcal{B}u_1 = 0, & x \in \partial D, \\[2mm]
\mathcal{B}u_2 = 0, & x \in \partial D,
\end{cases}
\tag{1.1.3}
$$

where $\omega \in \Omega$, $((\Omega, \mathfrak{F}, \mathbb{P}), \{\theta_t\}_{t\in\mathbb{R}})$ is an ergodic metric dynamical system, $\mathbf{f} = (f_1, f_2)\colon \Omega \times \bar{D} \times [0, \infty) \times [0, \infty) \to \mathbb{R}^2$, and $\mathcal{B}$ is either the Dirichlet or Neumann boundary operator, i.e.,

$$
\mathcal{B} := \begin{cases}
\mathrm{Id} & \text{(Dirichlet)} \\[2mm]
\dfrac{\partial}{\partial \boldsymbol{\nu}} & \text{(Neumann)}
\end{cases}
\tag{1.1.4}
$$

and the uniform persistence of the following two species competitive Kolmogorov systems of nonautonomous partial differential equations :

$$
\begin{cases}
\dfrac{\partial u_1}{\partial t} = \Delta u_1 + f_1(t, x, u_1, u_2)u_1, & x \in D, \\[2mm]
\dfrac{\partial u_2}{\partial t} = \Delta u_2 + f_2(t, x, u_1, u_2)u_2, & x \in D, \\[2mm]
\mathcal{B}u_1 = 0, & x \in \partial D, \\[2mm]
\mathcal{B}u_2 = 0, & x \in \partial D,
\end{cases}
\tag{1.1.5}
$$

where $\mathbf{f} = (f_1, f_2)\colon \mathbb{R} \times \bar{D} \times [0, \infty) \times [0, \infty) \to \mathbb{R}^2$, and $\mathcal{B}$ is as in (1.1.4). To this end, by applying the linear theory in Chapters 1–5, we first establish uniform persistence theorems for the following random parabolic equation of Kolmogorov type:

$$
\begin{cases}
\dfrac{\partial u}{\partial t} = \Delta u + f(\theta_t\omega, x, u)u, & x \in D, \\[2mm]
\mathcal{B}u = 0, & x \in \partial D,
\end{cases}
\tag{1.1.6}
$$

where $f\colon \Omega \times \bar{D} \times [0, \infty) \mapsto \mathbb{R}$ and $\mathcal{B}$ is as in (1.1.4), and the following nonautonomous parabolic equation of Kolmogorov type:

$$
\begin{cases}
\dfrac{\partial u}{\partial t} = \Delta u + f(t, x, u)u, & x \in D, \\[2mm]
\mathcal{B}u = 0, & x \in \partial D,
\end{cases}
\tag{1.1.7}
$$

where $f\colon \mathbb{R} \times \bar{D} \times [0, \infty) \mapsto \mathbb{R}$ and $\mathcal{B}$ is as in (1.1.4). Uniform persistence theorems are then established for (1.1.3) and (1.1.5). While doing all the above, global attracting dynamics and uniform persistence theories are first established for a general family of nonlinear parabolic equations of Kolmogorov type (i.e., (1.1.7) for all f in a set Z of certain admissible functions) and for a general family of competitive Kolmogorov systems of parabolic equations (i.e., (1.1.5) for all $\mathbf{f}$ in a set $\mathbf{Z}$ of certain admissible functions).

We have chosen to provide the fundamentals (mainly existence, uniqueness, continuous dependence of solutions and Harnack inequalities for positive solutions) for the introduction of spectral theory rather than to actually carry out such analysis, and we supply appropriate references where specific results are quoted. For the exposition to be self-contained, we provide proofs for some (already known) spectral results in existing publications whenever we feel it would be more helpful for the reader.

1.2   General Notations and Concepts

$\mathbb{R}$ denotes the set of reals, $\mathbb{Z}$ denotes the set of integers, and $\mathbb{N}$ denotes the set of positive integers. For $t \in \mathbb{R}$, $\lfloor t \rfloor$ stands for the greatest integer smaller than or equal to t.

If $E \subset \mathbb{R}^m$, it is always considered with the topology induced from the standard topology on $\mathbb{R}^m$.

For a measurable subset $E \subset \mathbb{R}^m$, $|E|$ denotes the m-dimensional Lebesgue measure of E.

All Banach spaces are assumed to be real. For a Banach space B let $\|\cdot\|_B$ denote its norm.

Let B_1, B_2 be Banach spaces. The symbol $\mathcal{L}(B_1, B_2)$ denotes the space of all bounded linear operators from B_1 into B_2 endowed with the norm topology. The norm in $\mathcal{L}(B_1, B_2)$ is denoted by $\|\cdot\|_{B_1, B_2}$.

The symbol $\mathcal{L}_s(B_1, B_2)$ denotes the vector space of all bounded linear operators from B_1 into B_2, but endowed with the strong operator topology.

Instead of $\mathcal{L}(B, B)$ ($\mathcal{L}_s(B, B)$, resp.) we write $\mathcal{L}(B)$ ($\mathcal{L}_s(B)$, resp.).

For B a Banach space and B^* its dual, we denote by $\langle \cdot, \cdot \rangle_{B, B^*}$ the duality pairing between them. For H a Hilbert space, $\langle \cdot, \cdot \rangle_{H, H}$ stands for the inner product in H.

For $E \subset \mathbb{R}^m$, B a Banach space and $k = 0, 1, 2, \ldots$, we write $C^k(E, B)$ for the Banach space of k-times continuously differentiable B-valued functions defined on E whose derivatives up to order k are bounded on B. $C^k(E, B)$ is assumed to be endowed with the standard C^k-norm. Instead of $C^0(E, B)$ we write $C(E, B)$, and instead of $C^k(E, \mathbb{R})$ we write $C^k(E)$.

For a compact $E \subset \mathbb{R}^m$, we denote by $\overset{\circ}{C}(E, B)$ the (closed) linear subspace of the Banach space $C(E, B)$ consisting of functions taking value 0 on the boundary of E.

For $E \subset \mathbb{R}^m$, $k = 0, 1, 2, \ldots$ and $\alpha \in (0, 1)$, we denote by $C^{k+\alpha}(E)$ the Banach space of functions in $C^k(E)$ whose derivatives up to order k are Hölder continuous with exponent α, uniformly in $x \in E$. $C^{k+\alpha}(E)$ is assumed to be given the standard norm. Instead of $C^{0+\alpha}(E)$ we write $C^{\alpha}(E)$.

For $E \subset \mathbb{R}^m$ and $k = 1, 2, \ldots$ we denote by $C^{k-}(E)$ the Banach space of functions in $C^{k-1}(E)$ whose derivatives of up to order $k - 1$ are Lipschitz continuous, uniformly in $x \in E$. $C^{k-}(E)$ is assumed to be given the standard norm.

For $E_1 \subset \mathbb{R}^{m_1}$, $E_2 \subset \mathbb{R}^{m_2}$, $k = 0, 1, 2, \ldots$ and $l = 0, 1, 2, \ldots$, we denote by $C^{k,l}(E_1 \times E_2)$ the Banach space of real-valued functions $u = u(x_1, x_2)$ such that all the derivatives $\partial^{i+j} u / \partial x_1^i \partial x_2^j$ with $0 \leq i \leq k$, $0 \leq j \leq l$, are continuous and bounded on $E_1 \times E_2$. $C^{k,l}(E_1 \times E_2)$ is assumed to be endowed with the standard norm.

For $E_1 \subset \mathbb{R}^{m_1}$, $E_2 \subset \mathbb{R}^{m_2}$, $k = 0, 1, 2, \ldots$, $l = 0, 1, 2, \ldots$, $\alpha \in (0, 1)$ and $\beta \in (0, 1)$, we denote by $C^{k+\alpha, l+\beta}(E_1 \times E_2)$ the Banach space of functions from $C^{k,l}(E_1 \times E_2)$ such that all the derivatives $\partial^{i+j} u / \partial x_1^i \partial x_2^j$ with $0 \leq i \leq k$, $0 \leq j \leq l$, are Hölder continuous in x_1 with exponent α and in x_2 with exponent β, uniformly in $(x_1, x_2) \in E_1 \times E_2$. $C^{k+\alpha, l+\beta}(E_1 \times E_2)$ is assumed to be endowed with the standard norm.

For $E_1 \subset \mathbb{R}^{m_1}$, $E_2 \subset \mathbb{R}^{m_2}$, $k = 1, 2, \ldots$ and $l = 1, 2, \ldots$, we denote by $C^{k-,l-}(E_1 \times E_2)$ the Banach space of functions from $C^{k-1,l-1}(E_1 \times E_2)$ such that all the derivatives $\partial^{i+j} u / \partial x_1^i \partial x_2^j$ with $0 \leq i \leq k - 1$, $0 \leq j \leq l - 1$, are Lipschitz continuous in x_1 and in x_2, uniformly in $(x_1, x_2) \in E_1 \times E_2$. $C^{k-,l-}(E_1 \times E_2)$ is assumed to be endowed with the standard norm.

For $E \subset \mathbb{R}^m$, let $\mathcal{D}(E)$ stand for the vector space of (real-valued) C^{∞} functions with compact supports in E (test functions), and let $\mathcal{D}'(E)$ stand for the corresponding vector space of distributions.

We collect now some facts about measurable functions defined on a Lebesgue-measurable subset E of $\mathbb{R}^m$ and taking values in a Banach space B (for a reference, see [74] or [99]). We will assume B to be separable. To start with, recall that a function $u \colon E \to B$ is called *simple* if there are Lebesgue-measurable pairwise disjoint sets $E_1, \ldots, E_m \subset E$, $E_1 \cup \cdots \cup E_m = E$, and elements $u_1, \ldots, u_m \in B$ such that $u(x) = u_j$ for any $x \in E_j$ ($1 \leq j \leq m$).

DEFINITION 1.2.1 *A function $u \colon E \to B$, where B is a separable Banach space, is called* measurable *if one of the following (mutually equivalent) conditions is satisfied:*

(i) *There is a sequence $(u_n)_{n=1}^{\infty}$ of simple functions such that $\|u_n(x) - u(x)\|_B$ converges to 0 as $n \to \infty$, for Lebesgue-a.e. $x \in E$,*

(ii) *For any open subset $A \subset B$ (or for any closed subset $A \subset B$), $u^{-1}(A)$ is Lebesgue-measurable,*

(iii) *For any $v^* \in B^*$ the function $[E \ni x \mapsto \langle u(x), v^* \rangle_{B, B^*} \in \mathbb{R}]$ is Lebesgue-measurable.*

A function satisfying (i) is usually referred to as *strongly measurable*, whereas a function satisfying (iii) is usually called *weakly measurable*. The equivalence of (i) and (iii) is a consequence of the Pettis theorem.

The following two lemmas can be easily proved.

LEMMA 1.2.1
A continuous function $u \colon E \to B$ is measurable.

LEMMA 1.2.2
If $u \colon E \to B$ is measurable and $f \colon B \to B_1$ is continuous, where B and B_1 are separable Banach spaces, then the composition $f \circ u \colon E \to B_1$ is measurable.

For $1 \le p < \infty$, a measurable function $u \colon E \to B$ belongs to $L_p(E, B)$ if the (measurable) function $[\, E \ni x \mapsto (\|u(x)\|_B)^p \in \mathbb{R}\,]$ belongs to $L_1(E, \mathbb{R})$. The norm in $L_p(E, B)$ is defined as

$$\|u\|_{L_p(E,B)} := \left(\int_E (\|u(x)\|_B)^p \, dx \right)^{1/p}.$$

A measurable function $u \colon E \to B$ belongs to $L_\infty(E, B)$ if the (measurable) function $[\, E \ni x \mapsto \|u(x)\|_B \in \mathbb{R}\,]$ belongs to $L_\infty(E, \mathbb{R})$. The norm in $L_\infty(E, B)$ is defined as

$$\|u\|_{L_\infty(E,B)} := \operatorname{ess\,sup} \{\, \|u(x)\|_B : x \in E \,\}.$$

LEMMA 1.2.3
Let $1 \le p \le \infty$. If $u \in L_p(E, B)$, where the Banach space B is separable, then there is a sequence $(u_n)_{n=1}^{\infty}$ of simple functions such that $\|u_1\|_{L_p(E,B)} \le \|u_2\|_{L_p(E,B)} \le \cdots \le \|u\|_{L_p(E,B)}$, $\|u_n\|_{L_p(E,B)} \to \|u\|_{L_p(E,B)}$ as $n \to \infty$, and $u_n(x)$ converges in B to $u(x)$ as $n \to \infty$, for Lebesgue-a.e. $x \in E$.

PROOF   See [74, Lemma 21-2.5].   □

Instead of $L_p(E, \mathbb{R})$ we write $L_p(E)$. (But for the notation $L_p(\mathbb{R} \times \partial D)$ see Section 1.3.)

For $1 \le p \le \infty$, E a Lebesgue-measurable subset of $\mathbb{R}^m$ and $k = 1, 2, \dots$ we denote by $W_p^k(E)$ the Banach space of real-valued functions whose generalized derivatives up to order k belong to $L_p(E)$.

For $1 \le p \le \infty$, E_1 a Lebesgue-measurable subset of $\mathbb{R}^{m_1}$, E_2 a Lebesgue-measurable subset of $\mathbb{R}^{m_2}$, and $l = 0, 1, 2, \dots$, we denote by $W_p^{l,2l}(E_1 \times E_2)$ $(W_p^{l,0}(E_1 \times E_2)$, $W_p^{0,l}(E_1 \times E_2))$ the Banach space of real-valued functions $u = u(x_1, x_2)$ such that all generalized derivatives $\partial^{i+j} u / \partial x_1^i \partial x_2^j$ $(\partial^i u / \partial x_1^i,$ $\partial^j u / \partial x_2^j)$ with $0 \le 2i + j \le 2l$ $(0 \le i \le l,\ 0 \le j \le l)$ belong to $L_p(E_1 \times E_2)$.

For $1 \leq p \leq \infty$ and E a Lebesgue-measurable subset of $\mathbb{R}^m$, $L_{p,\mathrm{loc}}(E)$ stands for the Fréchet space consisting of real-valued functions such that for any compact subset $E_1 \subset E$ the restriction $u|_{E_1}$ belongs to $L_p(E_1)$.

For a metric space S, by $\mathfrak{B}(S)$ we denote the countably additive algebra of all Borel subsets of S, and by $C(S)$ we denote the Banach space of all bounded continuous real functions on S with the supremum norm.

A *topological flow* (or a *topological dynamical system*) on a metric space Y is a continuous mapping

$$\sigma \colon \mathbb{R} \times Y \to Y$$

satisfying the following properties (where $\sigma_t(\cdot)$ stands for $\sigma(t, \cdot)$):

(TF1) $\sigma_0 = \mathrm{Id}_Y$,

(TF2) $\sigma_{s+t} = \sigma_s \circ \sigma_t$ for any $s, t \in \mathbb{R}$.

It follows from (TF1) and (TF2) that

(TF3) $(\sigma_t)^{-1} = \sigma_{-t}$ for any $t \in \mathbb{R}$.

Sometimes for a topological flow σ we write $\sigma = \{\sigma_t\}_{t \in \mathbb{R}}$. Also, we can write $(Y, \sigma) = (Y, \{\sigma_t\}_{t \in \mathbb{R}})$. If Y is compact, we call (Y, σ) a *compact flow*.

A *topological semiflow* (or a *topological semidynamical system*) on a metric space Y is a mapping

$$\sigma \colon [0, \infty) \times Y \to Y$$

satisfying the following properties (where $\sigma_t(\cdot)$ stands for $\sigma(t, \cdot)$):

(TSF0) σ restricted to $(0, \infty) \times Y$ is continuous; moreover, for each $y \in Y$ the mapping $[\, [0, \infty) \ni t \mapsto \sigma_t y \in Y \,]$ is continuous,

(TSF1) $\sigma_0 = \mathrm{Id}_Y$,

(TSF2) $\sigma_{s+t} = \sigma_s \circ \sigma_t$ for any $s, t \geq 0$.

Sometimes for a topological semiflow σ we write $\sigma = \{\sigma_t\}_{t \geq 0}$. Also, we can write $(Y, \sigma) = (Y, \{\sigma_t\}_{t \geq 0})$.

From now on until revoking (Y, σ) denotes a topological semiflow. Let $d(\cdot, \cdot)$ stand for the metric on Y.

A set $A \subset Y$ is *invariant* if $\sigma_t(A) = A$ for any $t \geq 0$. Sometimes we say that A is *invariant under σ* (or σ-*invariant*).

A closed invariant set $A \subset Y$ is called an *isolated invariant set* if there is a neighborhood of U of A such that A is the largest invariant set contained in U.

A set $A \subset Y$ is *forward invariant* if $\sigma_t(A) \subset A$ for any $t \geq 0$.

For $y \in Y$ the *forward orbit* of y is defined as $O^+(y) := \{\, \sigma_t y : t \geq 0 \,\}$. For $A \subset Y$ the *forward orbit* of A is defined as $O^+(A) := \{\, \sigma_t(A) : t \geq 0 \,\}$.

It should be remarked that in the literature forward invariant sets, forward orbits, etc., are usually called positively invariant sets, positive orbits, etc. But in the present monograph "positive" is reserved for "belonging (or related) to the cone of nonnegative functions (or its interior)."

We say that $A \subset Y$ *attracts* $B \subset Y$ if for each $\epsilon > 0$ there is $T = T(\epsilon) \geq 0$ such that $\sigma_t(B)$ is contained in the ϵ-neighborhood of A, for any $t \geq T$.

The ω-*limit set* of $y \in Y$ is defined as

$$\omega(y) := \bigcap_{t \geq 0} \operatorname{cl} O^+(\sigma_t y).$$

There holds: $z \in Y$ belongs to $\omega(y)$ if and only if there is a sequence $(t_n)_{n=1}^{\infty} \subset \mathbb{R}$ such that $\lim_{n \to \infty} t_n = \infty$ and $\lim_{n \to \infty} \sigma(t_n, y) = z$. $\omega(y)$ is closed and invariant. Moreover, $\omega(y) = \omega(\sigma_t y)$ for any $t \geq 0$.

The following lemma follows from general theory of topological semiflows (see [46]).

LEMMA 1.2.4
Assume that for some $t \geq 0$ the forward orbit $O^+(\sigma_t y)$ has compact closure. Then $\omega(y)$ is nonempty, compact and connected, and attracts y.

The ω-*limit set* of $A \subset Y$ is defined as

$$\omega(A) := \bigcap_{t \geq 0} \operatorname{cl} O^+(\sigma_t(A)).$$

There holds: $z \in Y$ belongs to $\omega(A)$ if and only if there are sequences $(t_n)_{n=1}^{\infty} \subset \mathbb{R}$ and $(y_n)_{n=1}^{\infty} \subset A$ such that $t_n \to \infty$ and $\sigma(t_n, y_n) \to z$ as $n \to \infty$. $\omega(A)$ is closed and invariant. Moreover, $\omega(A) = \omega(\sigma_t(A))$ for any $t \geq 0$.

Similar to Lemma 1.2.4, there holds (see [46])

LEMMA 1.2.5
Assume that for some $t \geq 0$ the forward orbit $O^+(\sigma_t(A))$ has compact closure. Then $\omega(A)$ is nonempty and compact, and attracts A. Moreover, if A is connected then $\omega(A)$ is connected, too.

A nonempty compact invariant set $A \subset Y$ is an *attractor* if there is a neighborhood C of A such that A attracts C, or, equivalently, $\omega(C) = A$.

A nonempty compact invariant set $\Gamma \subset Y$ is said to be the *global attractor* if it attracts each bounded $B \subset Y$.

Observe that if (Y, σ) possesses a global attractor Γ then for each bounded $B \subset Y$ there holds $\emptyset \neq \omega(B) \subset \Gamma$. Further, Γ is the maximal compact invariant set in the sense that if $A \subset Y$ is compact invariant then necessarily $A \subset \Gamma$. Consequently, a global attractor is uniquely determined.

Let $A \subset Y$ be a forward invariant set. Then the restriction $\sigma|_{[0,\infty) \times A}$ satisfies all the properties (TSF0)–(TSF2) of a semiflow, and is referred to as the *restriction* of the semiflow σ to A (usually denoted simply by $\sigma|_A$).

From now on, again until revoking, (Y, σ) will denote a compact flow.

A set $A \subset Y$ is *invariant* if $\sigma_t(A) = A$ for each $t \in \mathbb{R}$.

For $y \in Y$ the *backward orbit* of y is defined as $O^-(y) := \{\sigma_t y : t \leq 0\}$. For $A \subset Y$ the *backward orbit* of A is defined as $O^-(A) := \{\sigma_t(A) : t \leq 0\}$.

For $y \in Y$ the *orbit* of y is defined as $O(y) := \{\sigma_t y : t \in \mathbb{R}\}$. For $A \subset Y$ the *orbit* of A is defined as $O(A) := \{\sigma_t(A) : t \in \mathbb{R}\}$.

The definitions of the ω-limit set of a point and of a set are the same as in the case of semiflows. We have that $\omega(y) = \omega(\sigma_t y)$ for any $t \in \mathbb{R}$, and $\omega(A) = \omega(\sigma_t(A))$ for any $t \in \mathbb{R}$.

The α-*limit set* of $y \in Y$ is defined as

$$\alpha(y) := \bigcap_{t \leq 0} \operatorname{cl} O^-(\sigma_t y).$$

There holds: $z \in Y$ belongs to $\alpha(y)$ if and only if there is a sequence $(t_n)_{n=1}^{\infty} \subset \mathbb{R}$ such that $\lim_{n \to \infty} t_n = -\infty$ and $\lim_{n \to \infty} \sigma(t_n, y) = z$ as $n \to \infty$. $\alpha(y)$ is closed and invariant. Moreover, $\alpha(y) = \alpha(\sigma_t y)$ for any $t \in \mathbb{R}$.

The α-*limit set* of $A \subset Y$ is defined as

$$\alpha(A) := \bigcap_{t \leq 0} \operatorname{cl} O^-(\sigma_t(A)).$$

There holds: $z \in Y$ belongs to $\alpha(A)$ if and only if there are sequences $(t_n)_{n=1}^{\infty} \subset \mathbb{R}$ and $(y_n)_{n=1}^{\infty} \subset A$ such that $t_n \to -\infty$ and $\sigma(t_n, y_n) \to z$. $\alpha(A)$ is closed and invariant. Moreover, $\alpha(A) = \alpha(\sigma_t(A))$ for any $t \in \mathbb{R}$.

A set $B \subset Y$ is a *repeller* if there is a neighborhood C of B such that $\alpha(C) = B$. Consequently, B is compact and invariant.

Let A be an attractor. We say that $y \in Y$ belongs to the *attraction basin* of A if $\omega(y) \subset A$. The complement in Y of the attraction basin of A is a repeller (called the *repeller dual to A*).

Let B be a repeller. We say that $y \in Y$ belongs to the *repulsion basin* of B if $\alpha(y) \subset B$. The complement in Y of the repulsion basin of B is an attractor (called the *attractor dual to B*).

A finite ordered family $\{A_1, \dots, A_k\}$ of nonempty pairwise disjoint compact invariant sets is called a *Morse decomposition* of Y if for each $y \in Y \setminus \bigcup_{i=1}^{k} A_i$ there are $1 \leq i_1 < i_2 \leq k$ such that $\alpha(y) \subset A_{i_1}$ and $\omega(y) \subset A_{i_2}$.

Let $A \subset Y$ be an invariant set. Then the restriction $\sigma|_{\mathbb{R} \times A}$ satisfies the properties (TF1), (TF2) of a flow, and is referred to as the *restriction* of the flow σ to A (usually denoted simply by $\sigma|_A$).

A compact flow (Y, σ) is called *topologically transitive* if there is $y_0 \in Y$ such that $Y = \operatorname{cl} O(y_0)$.

A compact flow (Y, σ) is called *minimal* if the only closed invariant nonempty set is Y itself.

From now on until further notice, (Y, σ) has no specific meaning.

Let $(\Omega, \mathfrak{F})$ be a measurable space (that is, Ω is a set and $\mathfrak{F}$ is a countably additive algebra of subsets of Ω). A $(\mathfrak{B}(\mathbb{R}) \times \mathfrak{F}, \mathfrak{F})$-measurable mapping $\theta \colon \mathbb{R} \times$

$\Omega \to \Omega$ is a *measurable flow* (or a *measurable dynamical system*) on $(\Omega, \mathfrak{F})$ if it satisfies the following properties (where $\theta_t(\cdot)$ stands for $\theta(t, \cdot)$).

(MF1) $\theta_0 = \mathrm{Id}_\Omega$,

(MF2) $\theta_{s+t} = \theta_s \circ \theta_t$ for any $s, t \in \mathbb{R}$.

As a section of a measurable mapping, $\theta_t \colon \Omega \to \Omega$ is $(\mathfrak{F}, \mathfrak{F})$-measurable, for each $t \in \mathbb{R}$. Further, from (MF1) and (MF2) it follows that $(\theta_t)^{-1}$ is $(\mathfrak{F}, \mathfrak{F})$-measurable, and

(MF3) $(\theta_t)^{-1} = \theta_{-t}$ for any $t \in \mathbb{R}$.

Sometimes for a measurable flow θ we write $\theta = \{\theta_t\}_{t \in \mathbb{R}}$. Also, we can write $((\Omega, \mathfrak{F}), \theta) = ((\Omega, \mathfrak{F}), \{\theta_t\}_{t \in \mathbb{R}})$.

For $((\Omega, \mathfrak{F}), \theta)$ a measurable flow, a set $A \subset \Omega$ is *invariant* if $\theta_t(A) = A$ for any $t \in \mathbb{R}$. Sometimes we say that A is *invariant under θ* (or θ-*invariant*).

We say that a triple $(\Omega, \mathfrak{F}, \mathbb{P})$ is a *probability space* if $(\Omega, \mathfrak{F})$ is a measurable space and $\mathbb{P}$ is a probability measure on $\mathfrak{F}$.

For a probability space $(\Omega, \mathfrak{F}, \mathbb{P})$ denote by $L_1((\Omega, \mathfrak{F}, \mathbb{P}))$ the Banach space of all real-valued $(\mathfrak{F}, \mathfrak{B}(\mathbb{R}))$-measurable functions that are integrable with respect to $\mathbb{P}$, with the standard norm.

Let $(\Omega_1, \mathfrak{F}_1)$ and $(\Omega_2, \mathfrak{F}_2)$ be measurable spaces, and let $\mathbb{P}_1$ be a probability measure on $\mathfrak{F}_1$. For a $(\mathfrak{F}_1, \mathfrak{F}_2)$-measurable mapping $F \colon \Omega_1 \to \Omega_2$, we define $F\mathbb{P}_1$, the *image of $\mathbb{P}_1$ with respect to F*, by

$$F\mathbb{P}_1(A) := \mathbb{P}_1(F^{-1}(A)) \quad \text{for any} \quad A \in \mathfrak{F}_2.$$

$F\mathbb{P}_1$ so defined is a probability measure on $\mathfrak{F}_2$.

For a measurable flow $((\Omega, \mathfrak{F}), \theta)$ we say that a probability measure $\mathbb{P}$ on $\mathfrak{F}$ is θ-*invariant* if $\theta_t \mathbb{P} = \mathbb{P}$ for each $t \in \mathbb{R}$. In such a case we will call $((\Omega, \mathfrak{F}, \mathbb{P}), \theta)$ a *metric dynamical system* (or a *metric flow*). Sometimes we write $((\Omega, \mathfrak{F}, \mathbb{P}), \{\theta_t\}_{t \in \mathbb{R}})$.

For a metric dynamical system $((\Omega, \mathfrak{F}, \mathbb{P}), \theta)$ we say that the invariant measure $\mathbb{P}$ is *ergodic* if for any θ-invariant set $A \in \mathfrak{F}$ one has either $\mathbb{P}(A) = 0$ or $\mathbb{P}(A) = 1$.

A metric dynamical system $((\Omega, \mathfrak{F}, \mathbb{P}), \theta)$ is *ergodic* if the invariant measure $\mathbb{P}$ is ergodic. The following Birkhoff Ergodic Theorem will be often utilized throughout the monograph.

LEMMA 1.2.6 (Birkhoff's Ergodic Theorem)
Let $((\Omega, \mathfrak{F}, \mathbb{P}), \theta)$ be an ergodic metric dynamical system and $h \in L_1((\Omega, \mathfrak{F}, \mathbb{P}))$. Then there is a θ-invariant set $\tilde{\Omega} \in \mathfrak{F}$ such that $\mathbb{P}(\tilde{\Omega}) = 1$ and

$$\lim_{t \to \infty} \frac{1}{t} \int_0^t h(\theta_s \omega)\, ds = \int_\Omega h(\cdot)\, d\mathbb{P}(\cdot)$$

for any $\omega \in \tilde{\Omega}$.

PROOF   See [7] or references therein.   ☐

For Y a compact metric space, by a *measure* on Y we mean a probability measure on $\mathfrak{B}(Y)$. For a compact flow $(Y, \sigma) = (Y, \{\sigma_t\}_{t \in \mathbb{R}})$, a probability measure μ on Y is said to be an *invariant measure* of (Y, σ) if $\mu(\sigma_t(A)) = \mu(A)$ for any $t \in \mathbb{R}$ and $A \in \mathfrak{B}(Y)$ (hence $((Y, \mathfrak{B}(Y), \mu), \sigma)$ is a metric dynamical system). An invariant measure μ of (Y, σ) is *ergodic* if for any σ-invariant set $A \in \mathfrak{B}(Y)$ one has either $\mu(A) = 0$ or $\mu(A) = 1$. A compact flow (Y, σ) is said to be *uniquely ergodic* if there is a unique (necessarily ergodic) invariant measure for σ.

Regarding the existence of (ergodic) invariant measure of a compact flow (Y, σ), we have

LEMMA 1.2.7 (Krylov–Bogolyubov Theorem)
If σ is a topological flow on a compact metric space Y then there exist (ergodic) invariant measures of (Y, σ).

PROOF   See [107, Theorem 6.10].   ☐

Consider a *product bundle* $S \times Y$, where S and Y are metric spaces (notice that Y is the base and S is a model fiber). Let σ be a topological semiflow on Y. We say that

$$\Phi \colon [0, \infty) \times S \times Y \to S \times Y$$

is a *topological skew-product semiflow* on $S \times Y$ covering σ if it can be written as

$$\Phi(t; u, a) = (\phi_t(a, u), \sigma_t a), \quad t \geq 0, \ u \in S, \ a \in Y,$$

and has the following properties (where $\Phi_t(\cdot, \cdot)$ stands for $\Phi(t; \cdot, \cdot)$):

(TSP0) Φ restricted to $(0, \infty) \times S \times Y$ is continuous; moreover, for each $(z, a) \in S \times Y$ the mapping $[\,[0, \infty) \ni t \mapsto (\phi_t(z, a), \sigma_t a) \in S \times Y\,]$ is continuous,

(TSP1) $\Phi_0 = \mathrm{Id}_{S \times Y}$,

(TSP2) $\Phi_{t+s} = \Phi_t \circ \Phi_s$ for any $t, s \geq 0$.

When B is a Banach space, a *topological linear skew-product semiflow*

$$\Pi \colon [0, \infty) \times B \times Y \to B \times Y,$$

$$\Pi(t; u, a) = (U(t, a)u, \sigma_t a)$$

on the product Banach bundle $B \times Y$ covering a topological semiflow σ on Y is a topological skew-product semiflow with the property that for each $t \geq 0$ and each $a \in Y$ the mapping $[\, B \ni u \mapsto U(t, a)u \in B \,]$ belongs to $\mathcal{L}(B)$. We write that mapping as $U_a(t, 0)$.

When S is a subset of a Banach space, a *topological C^1 skew-product semiflow*

$$\Phi \colon [0, \infty) \times S \times Y \to S \times Y$$

on the product bundle $S \times Y$ covering a topological semiflow σ on Y is a topological skew-product semiflow with the property that for each $t \geq 0$ and each $a \in Y$ the mapping $[\, S \ni u \mapsto \phi_t(u, a) \in S \,]$ is of class C^1, and, moreover, the derivatives in u depend continuously on $(t, u, a) \in (0, \infty) \times S \times Y$.

Consider a *measurable bundle* $S \times \Omega$, where S is a metric space and $(\Omega, \mathfrak{F})$ is a measurable space (notice that Ω is the base and S is a model fiber). Let θ be a metric flow on $(\Omega, \mathfrak{F}, \mathbb{P})$. We say that a mapping

$$\Phi \colon [0, \infty) \times S \times \Omega \to S \times \Omega$$

is a *continuous random skew-product semiflow* on $S \times \Omega$ covering θ if it can be written as

$$\Phi(t; u, \omega) = (\phi_t(\omega, u), \theta_t \omega), \quad t \geq 0, \ u \in S, \ \omega \in \Omega,$$

and has the following properties (where $\Phi_t(\cdot, \cdot)$ stands for $\Phi(t; \cdot, \cdot)$):

(RSP0) Φ is $(\mathfrak{B}([0, \infty)) \times \mathfrak{B}(S) \times \mathfrak{F}, \mathfrak{B}(S) \times \mathfrak{F})$-measurable,

(RSP1) $\Phi_0 = \mathrm{Id}_{X \times Y}$,

(RSP2) $\Phi_{t+s} = \Phi_t \circ \Phi_s$ for any $t, s \geq 0$,

(RSP3) For any $t \geq 0$ and $\omega \in \Omega$ the mapping $\Phi_t(\omega, \cdot)$ is continuous.

When B is a Banach space, a *random linear skew-product semiflow*

$$\Pi \colon [0, \infty) \times B \times \Omega \to B \times \Omega,$$

$$\Pi(t; u, \omega) = (U(t, \omega)u, \theta_t \omega)$$

on the measurable Banach bundle $B \times \Omega$ covering a metric flow θ on $(\Omega, \mathfrak{F}, \mathbb{P})$ is a random skew-product semiflow with the property that for each $t \geq 0$ and each $\omega \in \Omega$ the mapping $[\, B \ni u \mapsto U(t, \omega)u \in B \,]$ belongs to $\mathcal{L}(B)$. We write that mapping as $U_\omega(t, 0)$.

1.3   Standing Assumptions

We assume that $D \subset \mathbb{R}^N$ is a bounded domain (that is, an open connected set), with boundary ∂D.

Subsets of $\mathbb{R}$ are always considered with the (one-dimensional) Lebesgue measure.

In all the notations of the form "$L_p(\mathbb{R} \times D, \cdot)$," or "a.e. on D," etc., the domain D is assumed to be endowed with the N-dimensional Lebesgue measure, whereas in all the notations of the form "$L_p(\mathbb{R} \times \partial D, \cdot)$," or "a.e. on ∂D," etc., the boundary ∂D is assumed to be endowed with the $(N-1)$-dimensional Hausdorff measure H_{N-1}. When the boundary ∂D is (at least) Lipschitz, then the $(N-1)$-dimensional Hausdorff measure on ∂D equals the ordinary surface measure on ∂D.

We denote the operator of differentiation in t (acting on functions of $N+1$ variables $u = u(t, x_1, \ldots, x_N)$) by $\frac{\partial}{\partial t}$, and we denote the operator of differentiation in x_i ($1 = 1, \ldots, N$) by $\frac{\partial}{\partial x_i}$. Sometimes instead of $\frac{\partial}{\partial t}$ we write ∂_t, and instead of $\frac{\partial}{\partial x_i}$ we write ∂_{x_i}.

Throughout Chapters 2 to 5 of the monograph we will write

$$a = ((a_{ij})_{i,j=1}^N, (a_i)_{i=1}^N, (b_i)_{i=1}^N, c_0, d_0),$$

where a_{ij}, a_i, b_i and c_0 are the coefficients of the equation (1.0.1), and d_0 is the coefficient in the boundary condition (in the Dirichlet or Neumann case d_0 is set to be equal to zero).

For any $a = ((a_{ij})_{i,j=1}^N, (a_i)_{i=1}^N, (b_i)_{i=1}^N, c_0, d_0)$ and any $t \in \mathbb{R}$ we define the *time-translate* $a \cdot t$ of a by

$$a \cdot t := ((a_{ij} \cdot t)_{i,j=1}^N, (a_i \cdot t)_{i=1}^N, (b_i \cdot t)_{i=1}^N, c_0 \cdot t, d_0 \cdot t),$$

where $a_{ij} \cdot t(\tau, x) := a_{ij}(\tau + t, x)$ for $s, \tau \in \mathbb{R}$, $x \in D$, etc.

We fix a countable dense subset $\{g_1, g_2, \ldots\}$ of the unit ball in $L_1(\mathbb{R} \times D, \mathbb{R}^{N^2+2N+1}) \times L_1(\mathbb{R} \times \partial D, \mathbb{R})$ such that for each $k \in \mathbb{N}$ there exists $K = K(k) > 0$ with the property that $g_k(t, \cdot) = 0$ for a.e. $t \in \mathbb{R} \setminus [-K, K]$.

For any $a^{(1)}, a^{(2)} \in L_\infty(\mathbb{R} \times D, \mathbb{R}^{N^2+2N+1}) \times L_\infty(\mathbb{R} \times \partial D, \mathbb{R})$ put

$$d(a^{(1)}, a^{(2)}) := \sum_{k=1}^{\infty} \frac{1}{2^k} |\langle g_k, (a^{(1)} - a^{(2)}) \rangle_{L_1, L_\infty}|. \tag{1.3.1}$$

We make the following assumptions on Y, the family of admissible coefficients.

(A1-1) Y is a (norm-)bounded subset of $L_\infty(\mathbb{R} \times D, \mathbb{R}^{N^2+2N+1}) \times L_\infty(\mathbb{R} \times \partial D, \mathbb{R})$ that is closed (hence, compact) in the weak-* topology of that space.

(A1-2) Y is *translation invariant*: If $a \in Y$ then $a \cdot t \in Y$, for each $t \in \mathbb{R}$.

(A1-3) $d_0 = 0$ for all $a = ((a_{ij})_{i,j=1}^N, (a_i)_{i=1}^N, (b_i)_{i=1}^N, c_0, d_0) \in Y$ (in the Dirichlet or Neumann cases) or $d_0 \geq 0$ for all $a = ((a_{ij})_{i,j=1}^N, (a_i)_{i=1}^N, (b_i)_{i=1}^N, c_0, d_0) \in Y$ (in the Robin case).

(Y, d) is a compact metric space. When speaking of convergence of sequences in Y, continuity of mappings from and/or to Y, etc., that space is assumed to be endowed with the weak-* topology, or, which is equivalent, with the topology generated by the metric d.

The mapping $[\, Y \times \mathbb{R} \ni (a, t) \mapsto a \cdot t \in Y \,]$ is continuous. For $t \in \mathbb{R}$ put $\sigma_t \colon Y \to Y$ to be $\sigma_t a := a \cdot t$. The family $\sigma = \{\sigma_t\}_{t \in \mathbb{R}}$ is a flow on the compact metrizable space Y.

We will write simply $\|\cdot\|$ for the standard norm in $L_2(D)$. Similarly, $\langle \cdot, \cdot \rangle$ will stand for the standard inner product in the Hilbert space $L_2(D)$.

Let B_1, B_2 be Banach spaces whose members are (equivalence classes of) real functions defined on D. We write

$$B_1 \hookrightarrow B_2$$

if B_1 continuously embeds into B_2, and

$$B_1 \overset{c}{\hookrightarrow} B_2$$

if $B_1 \hookrightarrow B_2$ and the embedding is compact (completely continuous).

Let $X \hookrightarrow L_1(D)$. In X we define the *nonnegative cone* X^+ as

$$X^+ := \{\, u \in X : u(x) \geq 0 \quad \text{for a.e.} \quad x \in D \,\}.$$

If $X \hookrightarrow C(\bar{D})$ then

$$X^+ = \{\, u \in X : u(x) \geq 0 \quad \text{for all} \quad x \in \bar{D} \,\}.$$

The nonnegative cone X^+ is a closed and convex subset of X, having the following properties:

- If $u \in X^+$ and $r \geq 0$ then $ru \in X^+$.

- If $u \in X^+$ and $-u \in X^+$ then $u = 0$.

For any $u_1, u_2 \in X$ we write

$$u_1 \leq u_2 \quad \text{if} \quad u_2 - u_1 \in X^+,$$
$$u_1 < u_2 \quad \text{if} \quad u_2 - u_1 \in X^+ \setminus \{0\}.$$

The reversed symbols $\geq$ and $>$ are used in the usual way.

In the Banach spaces $X = L_p(D)$, $1 \leq p \leq \infty$, or $X = C(\bar{D})$, the norm has the following *monotonicity* property: For any $u_1, u_2 \in X$, if $0 \leq u_1 \leq u_2$ then $\|u_1\|_X \leq \|u_2\|_X$.

Sometimes it happens that the interior of the nonnegative cone X^+ is nonempty. We denote then that interior by X^{++}. Also, for any $u_1, u_2 \in X$ we write

$$u_1 \ll u_2 \quad \text{if} \quad u_2 - u_1 \in X^{++}.$$

The reversed symbol $\gg$ is used in the usual way.

If the boundary ∂D of D is of class C^1, we denote by $\mathring{C}^1(\bar{D})$ the Banach space consisting of $u \in C^1(\bar{D})$ such that $u(x) = 0$ for all $x \in \partial D$. Recall that $\boldsymbol{\nu} = (\nu_1, \nu_2, \ldots, \nu_N)$ denotes the outer unit normal on the boundary ∂D pointing out of D.

LEMMA 1.3.1

Assume additionally that D has a boundary ∂D of class $C^{2+\alpha}$, for some $0 < \alpha < 1$.

(1) *The interior $C^1(\bar{D})^{++}$ of the nonnegative cone $C^1(\bar{D})^+$ is nonempty, and is characterized by*

$$C^1(\bar{D})^{++} = \{\, u \in C^1(\bar{D})^+ : u(x) > 0 \quad \text{for all} \quad x \in \bar{D}\,\}. \qquad (1.3.2)$$

(2) *The interior $\mathring{C}^1(\bar{D})^{++}$ of the nonnegative cone $\mathring{C}^1(\bar{D})^+$ is nonempty, and is characterized by*

$$\mathring{C}^1(\bar{D})^{++} = \{\, u \in \mathring{C}^1(\bar{D})^+ : u(x) > 0 \quad \text{for all} \quad x \in D$$
$$\text{and} \quad (\partial u / \partial \boldsymbol{\nu})(x) < 0 \quad \text{for all} \quad x \in \partial D\,\}.$$
$$(1.3.3)$$

PROOF We prove only (2), as the main idea of the proof of (1) is the same (but the details are much simpler).

We apply a construction of a *collar* of the boundary ∂D of the manifold $\bar{D}$ (see, e.g., [55]). We extend the C^1 vector field $\boldsymbol{\nu} \colon \partial D \to \mathbb{R}^N$ to a C^1 vector field $\tilde{\boldsymbol{\nu}}$ defined on some compact (relative) neighborhood V of ∂D in $\bar{D}$. Denote by $\rho = \rho(t, x)$ the local flow of the vector field $\tilde{\boldsymbol{\nu}}$. The standard theorem on the C^1 dependence of solutions of systems of ordinary differential equations on initial values guarantees the existence of $\eta > 0$ such that the restriction of ρ to $[-\eta, 0] \times \partial D$ is a C^1 diffeomorphism into some compact (relative) neighborhood V_1 of ∂D in $\bar{D}$.

Fix u belonging to the right-hand side of (1.3.3). For $x \in V_1$, $x = \rho(s, \tilde{x})$, $s \in [-\eta, 0]$, $\tilde{x} \in \partial D$, we write $\tilde{u}(s, \tilde{x}) := u(x)$ (in other words, $\tilde{u}$ is the representation of the restriction $u|_{V_1}$ in the $(s, \tilde{x})$-coordinates). The second condition in (1.3.3) translates into $(\partial \tilde{u}/\partial s)(0, \tilde{x}) < 0$ for all $\tilde{x} \in \partial D$. By the compactness of ∂D, there exist $\delta > 0$ and $\epsilon_1 > 0$ such that for any $v \in X$ with $\|v - u\|_{\mathring{C}^1(\bar{D})} < \epsilon_1$ one has $(\partial \tilde{v}/\partial s)(s, \tilde{x}) < 0$ for all $s \in [-\delta, 0]$ and all $\tilde{x} \in \partial D$. Denote $V_2 := \rho([-\delta, 0] \times \partial D)$. V_2 is a compact (relative) neighborhood of ∂D in $\bar{D}$. We conclude that $v(x) > 0$ for any $v \in X$ with $\|v - u\|_{\mathring{C}^1(\bar{D})} < \epsilon_1$ and any $x \in V_2 \setminus \partial D$. Denote by V_2' the closure of $\bar{D} \setminus V_2$. Since u is positive on the compact set V_2', it is bounded away from zero on V_2'. There is $\epsilon_2 > 0$ such that for any $v \in X$, if $\|v - u\|_{\mathring{C}^1(\bar{D})} < \epsilon_2$ then $v|_{V_2'}$ is positive and bounded away from zero. Consequently, if $\|v - u\|_{\overset{\circ}{C}{}^1(\bar{D})} < \epsilon$, where $\epsilon := \min\{\epsilon_1, \epsilon_2\}$, then $v \in \overset{\circ}{C}{}^1(\bar{D})^+$. This proves the "⊃" inclusion.

Let φ_{princ} be some (nonnegative) principal eigenvalue of the elliptic equation $\Delta u = 0$ on D with the Dirichlet boundary conditions. The standard regularity theory and maximum principles (see, e.g., [43]) guarantee that φ_{princ} belongs to the right-hand side of (1.3.3), hence to $\overset{\circ}{C}{}^1(\bar{D})^{++}$. Finally, let $u \in \overset{\circ}{C}{}^1(\bar{D})^{++}$. There is $\epsilon > 0$ such that $u - \epsilon\varphi_{\text{princ}} \in \overset{\circ}{C}{}^1(\bar{D})^+$, therefore $u(x) \geq \epsilon\varphi_{\text{princ}} > 0$ for all $x \in D$, which gives further that $\frac{\partial u}{\partial \boldsymbol{\nu}}(x) \leq \epsilon \frac{\partial \varphi_{\text{princ}}}{\partial \boldsymbol{\nu}}(x) < 0$ for all $x \in \partial D$. ☐

Throughout Chapter 6 of the monograph, $K \geq 1$ is a fixed integer. We write

$$\mathbf{a} = (a_{ij}^k, a_i^k, b_i^k, c_l^k, d_0^k)$$

and

$$\sigma_t \mathbf{a} \equiv \mathbf{a} \cdot t := (a_{ij}^k \cdot t, a_i^k \cdot t, b_i^k \cdot t, c_l^k \cdot t, d_0^k \cdot t),$$

where $i, j = 1, 2, \ldots, N$, $k, l = 1, 2, \ldots, K$, $a_{ij}^k \cdot t(\tau, x) := a_{ij}^k(t + \tau, x)$ for $t, \tau \in \mathbb{R}$, $x \in D$, etc.

We assume $\mathbf{Y} \subset L_\infty(\mathbb{R} \times D, \mathbb{R}^{K(N^2+2N+K)}) \times L_\infty(\mathbb{R} \times \partial D, \mathbb{R}^K)$ satisfies

(A1-4) $\mathbf{Y}$ is a (norm-)bounded subset of $L_\infty(\mathbb{R} \times D, \mathbb{R}^{K(N^2+2N+K)}) \times L_\infty(\mathbb{R} \times \partial D, \mathbb{R}^K)$ that is closed (hence, compact) in the weak-* topology of that space.

(A1-5) $\mathbf{Y}$ is *translation invariant*: If $\mathbf{a} \in \mathbf{Y}$ then $\mathbf{a} \cdot t \in \mathbf{Y}$, for each $t \in \mathbb{R}$.

(A1-6) $d_0^k = 0$ for all $\mathbf{a} = (a_{ij}^k, a_i^k, (b_i)^k, c_l^k, d_0^k) \in \mathbf{Y}$ (in the Dirichlet or Neumann cases) or $d_0^k \geq 0$ for all $\mathbf{a} = (a_{ij}^k, a_i^k, (b_i)^k, c_l^k, d_0^k) \in \mathbf{Y}$ (in the Robin case).

We assume that $\mathbf{Y}$ is endowed with the weak-* topology. Thus $(\mathbf{Y}, \sigma)$ is a compact flow, where $\sigma = \{\sigma_t\}_{t \in \mathbb{R}}$ and $\sigma_t \mathbf{a} = \mathbf{a} \cdot t$.

For a given $\mathbf{a} = (a_{ij}^k, a_i^k, b_i^k, c_l^k, d_0^k) \in \mathbf{Y}$, let $a^k := (a_{ij}^k, a_i^k, b_i^k, 0, d_0^k) := ((a_{ij}^k)_{i,j=1}^N, (a_j^k)_{i=1}^N, (b_i^k)_{i=1}^N, 0, d_0^k)$ and $\mathbf{C_a} := (c_l^k)_{l,k=1,2,\ldots,K}$. Let $P^k(\mathbf{a})$ be defined by $P^k(\mathbf{a}) := a^k$ and $Y^k := \{ P^k(\mathbf{a}) : \mathbf{a} \in \mathbf{Y} \}$ $(k = 1, 2, \ldots, K)$. Hence $Y^k \subset L_\infty(\mathbb{R} \times D, \mathbb{R}^{N^2+2N+1}) \times L_\infty(\mathbb{R} \times \partial D, \mathbb{R})$ satisfies (A1-1)–(A1-3) for $k = 1, 2, \ldots, K$.

For $1 \leq p \leq \infty$ we denote

$$\mathbf{L}_p(D) := (L_p(D))^K.$$

If $1 \leq p < \infty$ we define the norm in $\mathbf{L}_p(D)$ by

$$\|\mathbf{u}\|_p := \left(\sum_{k=1}^K \int_D |u_k(x)|^p \, dx \right)^{1/p}.$$

We define the norm in $\mathbf{L}_\infty(D)$ by

$$\|\mathbf{u}\|_\infty := \max_{1 \leq k \leq K} \operatorname{ess\,sup}\{\,|u_k(x)| : x \in D\,\}.$$

(In both cases, $\mathbf{u} = (u_1, \ldots, u_K)$.) We write simply $\|\cdot\|$ for the standard norm in the Hilbert space $\mathbf{L}_2(D)$.

For $\mathbf{u}, \mathbf{v} \in \mathbb{R}^K$, $\mathbf{u} = (u_1, \ldots, u_K)$, $\mathbf{v} = (v_1, \ldots, v_K)$, we write

$$\mathbf{u} \leq \mathbf{v} \quad \text{if} \quad u_k \leq v_k \quad \text{for } 1 \leq k \leq K,$$

$$\mathbf{u} < \mathbf{v} \quad \text{if} \quad \mathbf{u} \leq \mathbf{v} \text{ and } \mathbf{u} \neq \mathbf{v}.$$

The reversed symbols $\geq$ and $>$ are used in the usual way.

Let $\mathbf{X} \hookrightarrow \mathbf{L}_1(D)$. In $\mathbf{X}$ we define the *nonnegative cone* $\mathbf{X}^+$ as

$$\mathbf{X}^+ := \{\,\mathbf{u} \in \mathbf{X} : \mathbf{u}(x) \geq \mathbf{0} \quad \text{for a.e.} \quad x \in D\,\}.$$

If $\mathbf{X} \hookrightarrow C(\bar{D}, \mathbb{R}^K)$ then

$$\mathbf{X}^+ = \{\,\mathbf{u} \in \mathbf{X} : \mathbf{u}(x) \geq \mathbf{0} \quad \text{for all} \quad x \in \bar{D}\,\}.$$

The nonnegative cone $\mathbf{X}^+$ is a closed and convex subset of $\mathbf{X}$, having the following properties:

- If $\mathbf{u} \in \mathbf{X}^+$ and $r \geq 0$ then $r\mathbf{u} \in \mathbf{X}^+$.

- If $\mathbf{u} \in \mathbf{X}^+$ and $-\mathbf{u} \in \mathbf{X}^+$ then $\mathbf{u} = \mathbf{0}$.

For any $\mathbf{u}, \mathbf{v} \in \mathbf{X}$ we write

$$\mathbf{u} \leq \mathbf{v} \quad \text{if} \quad \mathbf{v} - \mathbf{u} \in \mathbf{X}^+,$$
$$\mathbf{u} < \mathbf{v} \quad \text{if} \quad \mathbf{v} - \mathbf{u} \in \mathbf{X}^+ \setminus \{\mathbf{0}\}.$$

The reversed symbols $\geq$ and $>$ are used in the usual way.

Sometimes it happens that the interior of the nonnegative cone $\mathbf{X}^+$ is nonempty. We denote then that interior by $\mathbf{X}^{++}$. Also, for any $\mathbf{u}, \mathbf{v} \in \mathbf{X}$ we write

$$\mathbf{u} \ll \mathbf{v} \quad \text{if} \quad \mathbf{v} - \mathbf{u} \in \mathbf{X}^{++}.$$

The reversed symbol $\gg$ is used in the usual way.

Notice that, if $\mathbf{X} = (X)^K$ then $\mathbf{X}^+ = (X^+)^K$.

If the boundary ∂D of D is of class C^1, we denote by $\mathring{C}^1(\bar{D}, \mathbb{R}^K)$ the Banach space consisting of $\mathbf{u} \in C^1(\bar{D}, \mathbb{R}^K)$ such that $\mathbf{u}(x) = \mathbf{0}$ for all $x \in \partial D$.

We have the following corollary of Lemma 1.3.1.

LEMMA 1.3.2

Assume additionally that the boundary ∂D of D is of class $C^{2+\alpha}$, for some $0 < \alpha < 1$.

(1) *The interior $C^1(\bar{D}, \mathbb{R}^K)^{++}$ of $C^1(\bar{D}, \mathbb{R}^K)^+$ is nonempty, and is characterized by*

$$C^1(\bar{D}, \mathbb{R}^K)^{++} = \{\, \mathbf{u} = (u_1, \ldots, u_K) \in C^1(\bar{D}, \mathbb{R}^K)^+ : u_k(x) > 0 \\ \text{for all } x \in \bar{D} \text{ and all } 1 \le k \le K \,\}.$$

(1.3.4)

(2) *The interior $\mathring{C}^1(\bar{D}, \mathbb{R}^K)^{++}$ of $\mathring{C}^1(\bar{D}, \mathbb{R}^K)^+$ is nonempty, and is characterized by*

$$\mathring{C}^1(\bar{D}, \mathbb{R}^K)^{++} = \{\, \mathbf{u} = (u_1, \ldots, u_K) \in \mathring{C}^1(\bar{D}, \mathbb{R}^K)^+ : u_k(x) > 0 \\ \text{for all } x \in D \text{ and all } 1 \le k \le K, \\ \text{and } (\partial u_k / \partial \boldsymbol{\nu})(x) < 0 \text{ for all } x \in \partial D, \ 1 \le k \le K \,\}.$$

(1.3.5)

Throughout Chapter 7 of the monograph, D is assumed to be $C^{3+\alpha}$.

We denote by Z the set of admissible functions for (1.1.7), $Z = \{\, g \colon \mathbb{R} \times \bar{D} \times [0, \infty) \to \mathbb{R} : g$ satisfies certain conditions $\}$, and by $\mathbf{Z}$ the set of admissible functions for (1.1.5), $\mathbf{Z} = \{\, \mathbf{g} = (g_1, g_2) \colon \mathbb{R} \times \bar{D} \times [0, \infty) \times [0, \infty) \to \mathbb{R} \times \mathbb{R} : \mathbf{g}$ satisfies certain conditions $\}$.

X denotes some fractional power space of the Laplacian operator Δ in $L_p(D)$ with the corresponding boundary condition such that

$$X \hookrightarrow C^1(\bar{D}).$$

$$\mathbf{X} := X \times X.$$

Chapter 2

Fundamental Properties in the General Setting

Introduction

In the present chapter we establish some fundamental properties for a general family of parabolic equations.

Let Y be a subset of $L_\infty(\mathbb{R} \times D, \mathbb{R}^{N^2+2N+1}) \times L_\infty(\mathbb{R} \times \partial D, \mathbb{R})$ satisfying (A1-1)-(A1-3) (see in Section 1.3). We may write $a = (a_{ij}, a_i, b_i, c_0, d_0)$ for $a = ((a_{ij})_{i,j=1}^N, (a_i)_{i=1}^N, (b_i)_{i=1}^N, c_0, d_0) \in Y$ if no confusion occurs. For a given $a = (a_{ij}, a_i, b_i, c_0, d_0)$, we may assume that $a_{ij}(t,x)$, $a_i(t,x)$, $b_i(t,x)$, and $c_0(t,x)$ are defined and bounded for all $(t,x) \in \mathbb{R} \times D$, and $d_0(t,x)$ is defined and bounded for all $(t,x) \in \mathbb{R} \times \partial D$.

For each $a = (a_{ij}, a_i, b_i, c_0, d_0) \in Y$ we consider

$$\frac{\partial u}{\partial t} = \sum_{i=1}^N \frac{\partial}{\partial x_i} \left(\sum_{j=1}^N a_{ij}(t,x) \frac{\partial u}{\partial x_j} + a_i(t,x)u \right)$$

$$+ \sum_{i=1}^N b_i(t,x) \frac{\partial u}{\partial x_i} + c_0(t,x)u, \quad t > s, \; x \in D, \qquad (2.0.1)$$

complemented with the boundary conditions

$$\mathcal{B}_a(t)u = 0, \quad t > s, \; x \in \partial D, \qquad (2.0.2)$$

where $s \in \mathbb{R}$ is an initial time and $\mathcal{B}_a$ is a boundary operator of either the Dirichlet, or Neumann, or Robin type, that is,

$$\mathcal{B}_a(t)u = \begin{cases} u & \text{(Dirichlet)} \\ \sum_{i=1}^N \left(\sum_{j=1}^N a_{ij}(t,x)\partial_{x_j}u + a_i(t,x)u \right)\nu_i & \text{(Neumann)} \\ \sum_{i=1}^N \left(\sum_{j=1}^N a_{ij}(t,x)\partial_{x_j}u + a_i(t,x)u \right)\nu_i \\ \qquad + d_0(t,x)u & \text{(Robin).} \end{cases} \qquad (2.0.3)$$

To emphasize the coefficients and the boundary terms in the problem (2.0.1)+ (2.0.2), we will write (2.0.1)$_a$+(2.0.2)$_a$. We use standard notion of solutions of (2.0.1)+(2.0.2), i.e., weak solutions (see [10], [30], [70], [73]).

First, in Section 2.1 we list basic assumptions, i.e., (A2-1) (the uniform ellipticity), (A2-2) (a very weak condition on the regularity of ∂D), and (A2-3) (a condition on perturbation of coefficients), and introduce the definition of weak solutions. Then, in Section 2.2 we collect basic properties of weak solutions of (2.0.1)+(2.0.2), including regularity, Harnack inequalities, monotonicity, joint continuity, compactness, etc., which are needed in the following chapters. It is shown that (2.0.1)+(2.0.2) generates a skew-product semiflow on the product bundle $L_2(D) \times Y$. The adjoint problem of (2.0.1)+(2.0.2) is considered in Section 2.3. In Section 2.4 we discuss the satisfaction of the assumption (A2-3), and show that (A2-3) is fulfilled under some very general condition. We study in Section 2.5 the case that the coefficients and the domain of (2.0.1)+(2.0.2) are sufficiently smooth. This chapter is ended up with some remarks in Section 2.6 about the solutions of nonautonomous equations in nondivergence form.

2.1   Assumptions and Weak Solutions

In this section we list our basic assumptions (A2-1)–(A2-3) and introduce the concept of weak solutions.

Consider (2.0.1)$_a$+(2.0.2)$_a$, $a \in Y$. First of all, we assume that the principal parts of all the elements of Y are uniformly elliptic:

(A2-1) (Uniform ellipticity) *There exists $\alpha_0 > 0$ such that for all $a \in Y$ there holds*

$$\sum_{i,j=1}^{N} a_{ij}(t,x)\,\xi_i\,\xi_j \geq \alpha_0 \sum_{i=1}^{N} \xi_i^2 \quad \text{for a.e. } (t,x) \in \mathbb{R} \times D \text{ and all } \xi \in \mathbb{R}^N,$$

(2.1.1)

$$a_{ij}(t,x) = a_{ji}(t,x) \quad \text{for a.e. } (t,x) \in \mathbb{R} \times D, \quad i,j = 1,2,\dots,N.$$

Let V be defined as follows

$$V := \begin{cases} \mathring{W}_2^1(D) & \text{(Dirichlet)} \\ W_2^1(D) & \text{(Neumann)} \\ W_{2,2}^1(D,\partial D) & \text{(Robin)} \end{cases}$$

(2.1.2)

where $\mathring{W}^1_2(D)$ is the closure of $\mathcal{D}(D)$ in $W^1_2(D)$ and $W^1_{2,2}(D, \partial D)$ is the completion of

$$V_0 := \{\, v \in W^1_2(D) \cap C(\bar{D}) : v \text{ is } C^\infty \text{ on } D \text{ and } \|v\|_V < \infty \,\}$$

with respect to the norm $\|v\|_V := (\|\nabla v\|^2_2 + \|v\|^2_{2, \partial D})^{1/2}$.

If no confusion occurs, we will write $\langle u, u^* \rangle$ for the duality between V and V^*, where $u \in V$ and $u^* \in V^*$.

LEMMA 2.1.1
If D is Lipschitz then

(1) $W^1_{2,2}(D, \partial D) = W^1_2(D)$ *up to norm equivalence;*

(2) $\mathring{W}^1_2(D) = \{\, u \in W^1_2(D) : u|_{\partial D} = 0 \,\}$.

PROOF   (1) See [30, (2.6)].
(2) See [45, Theorem 1.5].   □

For $s \le t$ let

$$W = W(s, t; V, V^*) := \{\, v \in L_2((s, t), V) : \dot{v} \in L_2((s, t), V^*) \,\} \qquad (2.1.3)$$

equipped with the norm

$$\|v\|_W := \left(\int_s^t \|v(\tau)\|^2_V \, d\tau + \int_s^t \|\dot{v}(\tau)\|^2_{V^*} \, d\tau \right)^{\frac{1}{2}},$$

where $\dot{v} := dv/d\tau$ is the time derivative in the sense of distributions taking values in V^* (see [36, Chapter XVIII] for definitions).

LEMMA 2.1.2
Let $s < t$. Then $W(s, t; V, V^*)$ embeds continuously into $C([s, t], L_2(D))$.

PROOF   See [36, Theorem 1, Chapter XVIII].   □

LEMMA 2.1.3
Let $s < t$ and $u \in W(s, t; V, V^*)$. Then $u \in L_2((s, t) \times D)$.

PROOF   For given $s < t$, $u \in W(s, t; V, V^*)$, and $n \in \mathbb{N}$, let

$$t_0 = s < t_1 = s + \frac{t-s}{n} < t_2 = s + \frac{2(t-s)}{n} < \cdots < t_n = t,$$

$$u_n(\tau, x) := \begin{cases} u(s, x), & \tau = s \\ u(t_i, x), & t_{i-1} < \tau \le t_i, \quad i = 1, 2, \ldots, n. \end{cases}$$

Then by Lemma 2.1.2,

$$\|u_n(\tau, \cdot) - u(\tau, \cdot)\| \to 0 \quad \text{as} \quad n \to \infty$$

for all $\tau \in (s, t)$.

Clearly, $u_n(\tau, x)$ is measurable in $(\tau, x) \in (s, t) \times D$. We claim that

$$u_n(\tau, x) \to u(\tau, x) \quad \text{for a.e.} \quad (\tau, x) \in (s, t).$$

Assume this is false. Then there is an $\epsilon_0 > 0$ such that $|E_{\epsilon_0}| > 0$, where

$$E_{\epsilon_0} := \{\, (\tau, x) : |u_n(\tau, x) - u(\tau, x)| \ge \epsilon_0 \quad \text{for infinitely many} \quad n \in \mathbb{N} \,\}.$$

Let $E_{\epsilon_0}(\tau) := \{\, x \in D : (\tau, x) \in E_{\epsilon_0} \,\}$. Then $[\, (s, t) \ni \tau \mapsto |E_{\epsilon_0}(\tau)| \in \mathbb{R} \,]$ is measurable, and

$$|E_{\epsilon_0}| = \int_s^t |E_{\epsilon_0}(\tau)| \, d\tau$$

(see [42]). Therefore there is $\tau_0 \in (s, t)$ such that $|E_{\epsilon_0}(\tau_0)| > 0$. This implies that

$$\|u_n(\tau_0, \cdot) - u(\tau_0, \cdot)\| \not\to 0 \quad \text{as} \quad n \to \infty.$$

This is a contradiction.

It then follows from the measurability of $u_n(\tau, x)$ that $u(\tau, x)$ is measurable in $(\tau, x) \in (s, t) \times D$. Therefore

$$\int_{(s,t) \times D} |u(\tau, x)|^2 \, d\tau \, dx = \int_s^t \int_D |u(\tau, x)|^2 \, dx \, d\tau < \infty$$

and $u \in L_2((s, t) \times D)$. $\qquad\qquad\qquad\qquad\qquad\qquad\qquad\qquad\qquad$ ☐

For $a \in Y$ denote by $B_a = B_a(t, \cdot, \cdot)$ the bilinear form on V associated with a,

$$B_a(t, u, v) := \int_D (a_{ij}(t, x) \partial_{x_j} u + a_i(t, x) u) \partial_{x_i} v \, dx$$
$$- \int_D (b_i(t, x) \partial_{x_i} u + c_0(t, x) u) v \, dx, \quad u, v \in V, \qquad (2.1.4)$$

in the Dirichlet and Neumann boundary condition cases, and

$$B_a(t, u, v) := \int_D (a_{ij}(t, x) \partial_{x_j} u + a_i(t, x) u) \partial_{x_i} v \, dx$$
$$- \int_D (b_i(t, x) \partial_{x_i} u + c_0(t, x) u) v \, dx$$
$$+ \int_{\partial D} d_0(t, x) u v \, dH_{N-1}, \quad u, v \in V, \qquad (2.1.5)$$

in the Robin boundary condition case, where H_{N-1} stands for the $(N-1)$-dimensional Hausdorff measure (we used the summation convention in the above).

LEMMA 2.1.4
Assume (A2-1). *Then the following holds.*

(i) *For any* $a \in Y$ *and for any* $u, v \in V$ *the function* $[\mathbb{R} \ni t \mapsto B_a(t, u, v) \in \mathbb{R}]$ *is* (*Lebesgue-*)*measurable.*

(ii) *There exists* $M_0 > 0$ *such that* $|B_a(t, u, v)| \le M_0 \|u\|_V \|v\|_V$ *for any* $a \in Y$, *a.e.* $t \in \mathbb{R}$ *and any* $u, v \in V$.

PROOF   See [36, Section XVIII.4.4].   ⬜

DEFINITION 2.1.1 (Weak solution)
A function $u \in L_2((s, t), V)$ *is a* weak solution *of* (2.0.1)+(2.0.2) *on* $[s, t] \times D$ $(s < t)$ *with initial condition* $u(s) = u_0$ *if*

$$-\int_s^t \langle u(\tau), v \rangle \, \dot{\psi}(\tau) \, d\tau + \int_s^t B_a(\tau, u(\tau), v) \psi(\tau) \, d\tau - \langle u_0, v \rangle \, \psi(s) = 0 \quad (2.1.6)$$

for all $v \in V$ *and* $\psi \in \mathcal{D}([s, t))$, *where* $\mathcal{D}([s, t))$ *is the space of all smooth real functions having compact support in* $[s, t)$.

PROPOSITION 2.1.1
If u *is a weak solution of* (2.0.1)+(2.0.2) *with initial condition* $u(s) = u_0$, *then* $u \in W(s, t; V, V^*)$.

PROOF   See [30, Theorem 2.4].   ⬜

Lemma 2.1.2 and Proposition 2.1.1 allow us to state the following.

PROPOSITION 2.1.2
Assume (A2-1). *If* u *is a weak solution of* (2.0.1)+(2.0.2) *on* $[s, t] \times D$ *with initial condition* $u(s) = u_0 \in L_2(D)$ *then* $u(s) = u_0$.

PROOF   See [36, Section XVIII.1.2].   ⬜

PROPOSITION 2.1.3 (Equivalence)
Assume (A2-1). *Let* $u \in L_2((s, t), V)$, $s < t$, $s \in \mathbb{R}$. u *is a weak solution of* (2.0.1)+(2.0.2) *on* $[s, t] \times D$ *with* $u(s) = u_0$ *if and only if* $u \in W(s, t; V, V^*)$

and for any $v \in W(s, t; V, V^*)$,

$$-\int_s^t \langle u(\tau), \dot{v}(\tau) \rangle \, d\tau + \int_s^t B_a(\tau, u(\tau), v(\tau)) \, d\tau$$
$$+ \langle u(t), v(t) \rangle - \langle u_0, v(s) \rangle = 0 \qquad (2.1.7)$$

PROOF   First note that $u \in L_2((s, t), V)$ is a weak solution of $(2.0.1)+$ $(2.0.2)$ on $[s, t] \times D$ with $u(s) = u_0$ if and only if $u \in W(s, t; V, V^*)$ and for all $v \in W(s, t; V, V^*)$ satisfying $v(t) = 0$,

$$-\int_s^t \langle u(\tau), \dot{v}(\tau) \rangle \, d\tau + \int_s^t B_a(\tau, u(\tau), v(\tau)) \, d\tau - \langle u_0, v(s) \rangle = 0 \qquad (2.1.8)$$

(see [30, Remark 2.3]). Hence we only need to prove that if $u \in W(s, t; V, V^*)$ is a weak solution of $(2.0.1)+(2.0.2)$ with $u(s) = u_0$, then for any $v \in W(s, t; V, V^*)$, $(2.1.7)$ holds. By Lemma 2.1.2 and Proposition 2.1.1, a weak solution u of $(2.0.1) +(2.0.2)$ on $[s, t] \times D$ belongs to $C([s, t], L_2(D))$.

Next, assume that $u \in W(s, t; V, V^*)$ is a weak solution of $(2.0.1)+(2.0.2)$ with $u(s) = u_0$. Let $v \in W(s, t; V, V^*)$ be such that $v(t) \in V$. Let $\tilde{v}(\tau) := v(\tau) - v(t)$. Then $\tilde{v} \in W(s, t; V, V^*)$ and $\tilde{v}(t) = 0$. Hence,

$$-\int_s^t \langle u(\tau), \dot{\tilde{v}}(\tau) \rangle \, d\tau + \int_s^t B_a(\tau, u(\tau), \tilde{v}(\tau)) \, d\tau - \langle u_0, \tilde{v}(s) \rangle = 0.$$

Note that

$$\int_s^t \langle u(\tau), \dot{\tilde{v}}(\tau) \rangle \, d\tau = \int_s^t \langle u(\tau), \dot{v}(\tau) \rangle \, d\tau,$$

$$\int_s^t B_a(\tau, u(\tau), \tilde{v}(\tau)) \, d\tau = \int_s^t B_a(\tau, u(\tau), v(\tau)) \, d\tau - \int_s^t B_a(\tau, u(\tau), v(t)) \, d\tau$$

and

$$\langle u_0, \tilde{v}(s) \rangle = \langle u_0, v(s) \rangle - \langle u_0, v(t) \rangle.$$

Note also that

$$\frac{d}{d\tau} \langle u(\tau), v(t) \rangle + B_a(\tau, u(\tau), v(t)) = 0$$

in the sense of distributions in $\mathcal{D}'((s, t))$. Since $[\, [s, t] \ni \tau \mapsto u(\tau) \in L_2(D)\,]$ is continuous (by Lemma 2.1.2), the function $[\, [s, t] \ni \tau \mapsto \langle u(\tau), v(t) \rangle \,]$ is continuous, and the function $[\, [s, t] \ni \tau \mapsto B_a(\tau, u(\tau), v(t)) \in \mathbb{R} \,]$ belongs to $L_1((s, t))$ (see Lemma 2.1.4). Consequently, we have

$$\int_s^t B_a(\tau, u(\tau), v(t)) \, d\tau = \langle u(s), v(t) \rangle - \langle u(t), v(t) \rangle.$$

Therefore $(2.1.7)$ holds for any $v \in W(s, t; V, V^*)$ with $v(t) \in V$.

Now for any $v \in W(s, t; V, V^*)$, $v(\tau) \in V$ for a.e. $\tau \in [s, t]$. Hence there is a sequence $(\tau_n)_{n=1}^{\infty} \subset [s, t]$ such that $\tau_n \to t$ and $v(\tau_n) \in V$. By the above arguments,

$$- \int_s^{\tau_n} \langle u(\tau), \dot{v}(\tau) \rangle \, d\tau + \int_s^{\tau_n} B_a(\tau, u(\tau), v(\tau)) \, d\tau$$
$$+ \langle u(\tau_n), v(\tau_n) \rangle - \langle u_0, v(s) \rangle = 0 \qquad (2.1.9)$$

for $n = 1, 2, \ldots$. Observe that, since the functions $[\, [s, t] \ni \tau \mapsto \langle u(\tau), \dot{v}(\tau) \rangle \in \mathbb{R}\,]$ and $[\, [s, t] \ni \tau \mapsto B_a(\tau, u(\tau), v(\tau)) \in \mathbb{R}\,]$ are in $L_1((s, t))$, there holds $\int_{\tau_n}^t \langle u(\tau), \dot{v}(\tau) \rangle \, d\tau \to 0$, $\int_{\tau_n}^t B_a(\tau, u(\tau), v(\tau)) \, d\tau \to 0$, and $\langle u(\tau_n), v(\tau_n) \rangle \to \langle u(t), v(t) \rangle$ as $n \to \infty$. It then follows from (2.1.9) that (2.1.7) holds for any $v \in W(s, t; V, V^*)$. $\quad\square$

PROPOSITION 2.1.4

Assume (A2-1). *Let u be a weak solution of* (2.0.1)+(2.0.2) *on* $[s, t] \times D$. *Then*

$$\|u(t)\|^2 - \|u(s)\|^2 = -2 \int_s^t B_a(\tau, u(\tau), u(\tau)) \, d\tau. \qquad (2.1.10)$$

PROOF   Observe that

$$\|u(t)\|^2 - \|u(s)\|^2 = 2 \int_s^t \langle u(\tau), \dot{u}(\tau) \rangle \, d\tau.$$

(see, e.g., [30, Lemma 3.3]), and apply Proposition 2.1.3 to $v = u$. $\quad\square$

DEFINITION 2.1.2 (Global weak solution)

A function $u \in L_{2,\text{loc}}((s, \infty), V)$ is a global weak solution *of* (2.0.1)+(2.0.2) *with initial condition $u(s) = u_0$, $s \in \mathbb{R}$, if for each $t > s$ its restriction $u|_{[s,t]}$ is a weak solution of* (2.0.1)+(2.0.2) *on* $[s, t] \times D$ *with initial condition $u(s) = u_0$.*

Global solutions of (2.0.1)+(2.0.2) exist under very general conditions. To be more specific, we make the following assumption.

(A2-2) (Boundary regularity) *For Dirichlet boundary conditions, D is a bounded domain. For Neumann or Robin boundary conditions, D is a bounded domain with Lipschitz boundary.*

Notice that under the assumption (A2-2), in the Robin boundary condition case the Hausdorff $(N-1)$-dimensional measure on H_{N-1} on ∂D reduces to the ordinary surface measure.

PROPOSITION 2.1.5 (Existence of global solution)
Assume (A2-1) and (A2-2). Then for any $s \in \mathbb{R}$ and any $u_0 \in L_2(D)$ there exists a unique global weak solution of (2.0.1)+(2.0.2) with initial condition $u(s) = u_0$.

PROOF   See [30, Theorem 2.4].   ▯

We write the unique global weak solution $[t \mapsto u(t)]$ of (2.0.1)+(2.0.2) with initial condition $u(s) = u_0$ as $U_a(t,s)u_0 := u(t)$, $t > s$. We write $U_a(s,s)u_0 = u_0$, for any $a \in Y$, $s \in \mathbb{R}$ and $u_0 \in L_2(D)$.

Important properties are given by the following results.

PROPOSITION 2.1.6
Assume (A2-1) and (A2-2). Then for any $a \in Y$ and $s \leq t$ one has

$$U_a(t,s) = U_{a \cdot s}(t-s,0). \tag{2.1.11}$$

PROOF   For $s = t$ there is nothing to prove. So assume $s < t$. Fix $u_0 \in L_2(D)$. Put $u_1(\tau) := U_a(\tau,s)u_0$ for $\tau \in [s,t]$, and $u_2(\tilde{\tau}) := U_{a \cdot s}(\tilde{\tau} - s, 0)u_0$ for $\tilde{\tau} \in [s,t]$. For any $v \in V$ and $\psi \in \mathcal{D}([s,t))$ the function u_1 satisfies the equation

$$-\int_s^t \langle u_1(\tau), v \rangle \dot{\psi}(\tau) \, d\tau + \int_s^t B_a(\tau, u_1(\tau), v)\psi(\tau) \, d\tau - \langle u_0, v \rangle \psi(s) = 0.$$

After the change of variables $\tilde{\tau} = \tau - s$ we obtain

$$-\int_0^{t-s} \langle u_1(\tilde{\tau} + s), v \rangle \dot{\psi}(\tilde{\tau} + s) \, d\tilde{\tau} + \int_0^{t-s} B_{a \cdot s}(\tilde{\tau}, u_1(\tilde{\tau} + s), v)\psi(\tilde{\tau} + s) \, d\tilde{\tau}$$
$$- \langle u_0, v \rangle \psi(s) = 0.$$

By the uniqueness of weak solutions, $u_2(\tau) = u_1(\tau)$ for $\tau \in [s,t]$.   ▯

PROPOSITION 2.1.7
Assume (A2-1) and (A2-2). Then for any $a \in Y$ and $s \leq t_1 \leq t_2$, one has

$$U_a(t_2, t_1) \circ U_a(t_1, s) = U_a(t_2, s). \tag{2.1.12}$$

PROOF   Assume $s < t_1 < t_2$. Fix $u_0 \in L_2(D)$, and put $u_1(t) := U_a(t,s)u_0$ for $t \in [s,t_1]$, $u_2(t) := U_a(t,t_1)u_1(t_1)$ for $t \in [t_1,t_2]$. Let $u(t)$ be defined by

$$u(t) := \begin{cases} u_1(t) & \text{for } t \in [s,t_1] \\ u_2(t) & \text{for } t \in [t_1,t_2]. \end{cases}$$

It is clear that $u \in L_2((s, t_2), V)$ and $u(t)$ satisfies (2.1.7) on $[s, t_2]$ for any $v \in W(s, t_2; V, V^*)$.

Note that for any $\psi \in \mathcal{D}([s, t_2))$ and $v \in V$, $\psi v \in W(s, t_2; V, V^*)$ and $\frac{d}{dt}(\psi v) = \frac{d\psi}{dt} v$. By the fact that $u(t)$ satisfies (2.1.7) on $[s, t_2]$, there holds

$$-\int_s^{t_2} \langle u(\tau), v \rangle \dot{\psi}(\tau) \, d\tau + \int_s^{t_2} B_a(\tau, u(\tau), v) \psi(\tau) \, d\tau - \langle u_0, v \rangle \psi(s) = 0.$$

It then follows from the definition of weak solution (Definition 2.1.1) that $u(t) = U_a(t, s)u_0$ for $t \in [s, t_2]$ and then $U_a(t_2, t_1) \circ U_a(t_1, s)u_0 = U_a(t_2, s)u_0$. □

As a consequence of Propositions 2.1.6 and 2.1.7 we obtain the following *cocycle equality*.

PROPOSITION 2.1.8
Assume (A2-1) *and* (A2-2). *Then for any* $a \in Y$ *and* $s, t \in [0, \infty)$ *we have*

$$U_{a \cdot s}(t, 0) U_a(s, 0) = U_a(s + t, 0). \tag{2.1.13}$$

As the set Y is assumed to have the property that $a \in Y$ and $t \in \mathbb{R}$ implies $a \cdot t \in Y$, the above results allow us to consider $U_a(t, 0)$ for $t > 0$ instead of $U_a(t, s)$ for $t > s$.

A third assumption imposed on Y regards the continuous dependence of solutions on parameters:

(A2-3) (Perturbation of coefficients) *For any sequence* $(a^{(n)})_{n=1}^\infty \subset Y$ *and any real sequence* $(t_n)_{n=1}^\infty$ *with* $t_n > 0$, *if*

$$\lim_{n \to \infty} a^{(n)} = a \quad \text{and} \quad \lim_{n \to \infty} t_n = t > 0,$$

then for any $u_0 \in L_2(D)$, $U_{a^{(n)}}(t_n, 0)u_0$ *converges to* $U_a(t, 0)u_0$ *in* $L_2(D)$.

We remark that (A2-3) is satisfied under some very general condition (see Section 2.4).

2.2   Basic Properties of Weak Solutions

In this section, we present some basic properties about weak solutions. We will assume (A2-1) and (A2-2) throughout this section and assume (A2-3) at some places.

First of all, the linear operator $U_a(t, 0)$ can be extended to $L_p(D)$. Indeed, we have:

PROPOSITION 2.2.1 (Extension in L_p)

Under the assumptions (A2-1) *and* (A2-2), *for any* $1 \le p \le \infty$, *any* $a \in Y$ *and any* $t > 0$ *there exists an operator* $U_{a,p}(t,0) \in \mathcal{L}(L_p(D))$ *such that* $U_{a,p}(t,0)u_0 = U_a(t,0)u_0$ *for all* $u_0 \in L_2(D) \cap L_p(D)$. *Moreover, for* $1 < p < \infty$ *and* $a \in Y$ *the mapping* $[\,[0,\infty) \ni t \mapsto U_{a,p}(t,0) \in \mathcal{L}_s(L_p(D))\,]$ *is continuous.*

PROOF   See [30, Theorem 5.1 and Corollary 5.3].   ☐

$U_{a,p}(t,s)$, for $s < t$, is understood as $U_{a \cdot s,p}(t-s,0)$. Further, $U_{a,p}(s,s)u_0 = u_0$. In the following we may write $U_a(t,s)$ instead of $U_{a,p}(t,s)$. We may also write $U(t,s)$ instead of $U_a(t,s)$, if no confusion occurs.

The next proposition gives us L_p–L_q estimates of $U_a(t,0)$.

PROPOSITION 2.2.2 (L_p–L_q estimates)

Assume (A2-1) *and* (A2-2). *Then there are constants* $M > 0$ *and* $\gamma > 0$ *such that*

$$\|U_a(t,0)\|_{L_p(D),L_q(D)} \le Mt^{-\frac{N}{2}(\frac{1}{p}-\frac{1}{q})}e^{\gamma t} \tag{2.2.1}$$

for $a \in Y$, $1 \le p \le q \le \infty$, $t > 0$.

PROOF   See [30, Corollary 7.2].   ☐

PROPOSITION 2.2.3 (Strong continuity in t)

Assume (A2-1) *and* (A2-2). *For any* $1 \le p < \infty$ *and any* $a \in Y$, *the mapping* $[\,(0,\infty) \ni t \mapsto U_{a,p}(t,0) \in \mathcal{L}_s(L_p(D))]$ *is continuous.*

PROOF   The continuity of $[\,(0,\infty) \ni t \mapsto U_{a,p}(t,0) \in \mathcal{L}_s(L_p(D))]$ for $1 < p < \infty$ follows from Proposition 2.2.1.

Let now $p = 1$. For any $(t_n)_{n=1}^{\infty} \subset (0,\infty)$ with $t_n \to t > 0$, let $\delta > 0$ be such that $t - \delta > 0$. Then by Proposition 2.2.2, for any $u_0 \in L_1(D)$, $U_{a,1}(\delta,0)u_0 \in L_2(D)$. Hence $U_{a,1}(t_n,0)u_0 = U_{a \cdot \delta,1}(t_n - \delta,0)U_{a,1}(\delta,0)u_0 = U_{a \cdot \delta,2}(t_n - \delta,0)U_{a,1}(\delta,0)u_0 \to U_{a \cdot \delta,2}(t - \delta,0)U_{a,1}(\delta,0)u_0 = U_{a,1}(t,0)u_0$.   ☐

PROPOSITION 2.2.4 (Local regularity)

Assume (A2-1) *and* (A2-2). *Let* $1 \le p \le \infty$. *Then for any* $0 < t_1 < t_2$ *there exists* $\alpha \in (0,1)$ *such that for any* $a \in Y$, *any* $u_0 \in L_p(D)$, *and any compact subset* $D_0 \subset D$ *the function* $[\,[t_1,t_2] \times D_0 \ni (t,x) \mapsto (U_a(t,0)u_0)(x)\,]$ *belongs to* $C^{\alpha/2,\alpha}([t_1,t_2] \times D_0)$. *Moreover, for fixed* t_1, t_2, *and* D_0, *the* $C^{\alpha/2,\alpha}([t_1,t_2] \times D_0)$-*norm of the above restriction is bounded above by a constant depending on* $\|u_0\|_p$ *only.*

PROOF It follows from Proposition 2.2.2 and from [70, Chapter III, Theorem 10.1]. ⬚

PROPOSITION 2.2.5 (Compactness)
Let (A2-1) through (A2-2) be satisfied and $1 \leq p \leq \infty$, $1 \leq q < \infty$. Then for any given $0 < t_1 \leq t_2$, if E is a bounded subset of $L_p(D)$ then $\{U_a(\tau,0)u_0 : a \in Y, \tau \in [t_1, t_2], u_0 \in E\}$ is relatively compact in $L_q(D)$.

PROOF Let $(\tau_n)_{n=1}^{\infty} \subset [t_1, t_2]$, $(a^{(n)})_{n=1}^{\infty} \subset Y$, and $(u_n)_{n=1}^{\infty} \subset E$. From Proposition 2.2.4 it follows that there are subsequences $(a^{(n_k)})_{k=1}^{\infty}$, $(\tau_{n_k})_{k=1}^{\infty}$, and $(u_{n_k})_{k=1}^{\infty}$ such that $U_{a^{(n_k)}}(\tau_{n_k},0)u_{n_k}$ converges to some $u_{\infty} \in L_{\infty}(D)$ uniformly on compact subsets of D. This together with Proposition 2.2.2 implies that $U_{a^{(n_k)}}(\tau_{n_k},0)u_{n_k}$ converges to u_{∞} in $L_q(D)$ for any $1 \leq q < \infty$.
⬚

PROPOSITION 2.2.6 (Joint continuity in t and u_0)
Assume (A2-1) and (A2-2). Let $1 \leq p \leq q < \infty$ and $a \in Y$. For any real sequence $(t_n)_{n=1}^{\infty}$ and any sequence $(u_n)_{n=1}^{\infty} \subset L_p(D)$, if $\lim_{n \to \infty} t_n = t$, where $t > 0$, and $\lim_{n \to \infty} u_n = u_0$ in $L_p(D)$, then $U_a(t_n,0)u_n$ converges in $L_q(D)$ to $U_a(t,0)u_0$.

PROOF First observe that
$$\|U_a(t_n,0)u_n - U_a(t,0)u_0\|_q \leq \|U_a(t_n,0)u_n - U_a(t_n,0)u_0\|_q$$
$$+ \|U_a(t_n,0)u_0 - U_a(t,0)u_0\|_q.$$

Next, by Proposition 2.2.2,
$$\|U_a(t_n,0)u_n - U_a(t_n,0)u_0\|_q \leq Mt_n^{-\frac{N}{2}(\frac{1}{p}-\frac{1}{q})} e^{\gamma t_n} \|u_n - u_0\|_p.$$

Hence $\|U_a(t_n,0)u_n - U_a(t_n,0)u_0\|_q \to 0$ as $n \to \infty$. Now, by Proposition 2.2.3, $\|U_a(t_n,0)u_0 - U_a(t,0)u_0\|_p \to 0$ as $n \to \infty$. We deduce from Proposition 2.2.5 that also $\|U_a(t_n,0)u_0 - U_a(t,0)u_0\|_q \to 0$. It then follows that
$$\|U_a(t_n,0)u_n - U_a(t,0)u_0\|_q \to 0 \quad \text{as} \quad n \to \infty.$$

⬚

PROPOSITION 2.2.7 (Positivity)
Assume (A2-1) and (A2-2). Let $1 \leq p \leq \infty$. For any $u_0 \in L_p(D)^+$ there holds $U_a(t,0)u_0 \in L_p(D)^+$ for all $a \in Y$ and $t \geq 0$.

PROOF For the case $1 < p < \infty$, the proposition follows from [30, Proposition 8.1]. The case $p = \infty$ then follows from the fact that $L_{\infty}(D)^+ \subset$

$L_p(D)^+$ for any $1 < p < \infty$. Now for the case $p = 1$, for any $u_0 \in L_1(D)^+$ there is $(u_n)_{n=1}^\infty \subset L_2(D)^+$ such that $\|u_n - u_0\|_{L_1(D)} \to 0$ as $n \to \infty$. Note that $U_a(t,0)u_n \in L_2(D)^+ \subset L_1(D)^+$ for $t > 0$ and $n = 1, 2, \ldots$. For any given $t > 0$, $U_a(t,0) \in \mathcal{L}(L_1(D))$, consequently $\|U_a(t,0)u_n - U_a(t,0)u_0\|_{L_1(D)} \to 0$ as $n \to \infty$ and there is subsequence $(u_{n_k})_{k=1}^\infty$ such that $(U_a(t,0)u_{n_k})(x) \to (U_a(t,0)u_0)(x)$ as $k \to \infty$ for a.e. $x \in D$. We then also have $U_a(t,0)u_0 \in L_1(D)^+$ for $t > 0$. ☐

In the following, $B(x_0; r)$ is defined by

$$B(x_0; r) := \{\, x \in \mathbb{R}^N : \|x - x_0\| \leq r \,\}.$$

PROPOSITION 2.2.8 (Interior Harnack inequality)
Assume (A2-1) and (A2-2). Let $1 \leq p \leq \infty$.

(1) *Given $r > 0$, there is $C_r > 0$ such that for any $x_0 \in D$ and $t_0 > 0$ satisfying $B(x_0; 2r) \subset D$ and $t_0 - 2r^2 > 0$, and any $a \in Y$ and $u_0 \in L_p(D)$ such that $(U_a(t,0)u_0)(x)$ is nonnegative in $[t_0 - 2r^2, t_0 + 2r^2] \times B(x_0; 2r)$, the following holds:*

$$\sup\{\, (U_a(t,0)u_0)(x) : t \in [t_0 - (29/16)r^2, t_0 - (7/4)r^2], \ x \in B(x_0; r/4) \,\}$$
$$\leq C_r \cdot \inf\{\, (U_a(t,0)u_0)(x) : t \in [t_0 + (31/16)r^2, t_0 + 2r^2], \ x \in B(x_0; r/4) \,\}.$$

(2) *For any $t_0 > 0$ there is $\delta_0 = \delta_0(t_0)$, $0 < \delta_0 < 1$, with the property that for any $0 < \delta < \delta_0$ there is $C_\delta > 0$ such that*

$$(U_a(t,0)u_0)(y) \leq C_\delta \cdot (U_a(t + \tau, 0)u_0)(x)$$

for any $a \in Y$, $t \geq \delta^2$, $\delta^2 \leq \tau \leq t_0$, $u_0 \in L_p(D)^+$, and any $x, y \in D^\delta :=$ $\{\xi \in D : d(\xi) > \delta\}$, where $d(\xi)$ denotes the distance of $\xi \in D$ from the boundary ∂D of D.

PROOF   (1) First, (1) with $p = 2$ follows from [63, Theorem 1] (see also [68], [73]). Next for any $p \geq 2$, $L_p(D) \subset L_2(D)$, hence (1) also holds. Now, for $1 \leq p < 2$, $U_a(t,0)u_0 = U_{a \cdot \delta}(t - \delta, 0)U_a(\delta, 0)u_0$ for any $\delta > 0$. Note that $u_1 = U_a(\delta, 0)u_0 \in L_2(D)$. Let $\delta > 0$ be so small that $t_0 - \delta - 2r^2 > 0$. Then $(U_{a \cdot \delta}(t,0)u_1)(x)$ is nonnegative in $[t_\delta - 2r^2, t_\delta + 2r^2] \times B(x_0; 2r)$, where $t_\delta = t_0 - \delta$. This implies that

$$\sup\{\, (U_{a \cdot \delta}(t,0)u_1)(x) : t \in [t_\delta - (29/16)r^2, t_\delta - \tfrac{7}{4}r^2], \ x \in B(x_0; r/4) \,\}$$
$$\leq C_r \cdot \inf\{\, (U_{a \cdot \delta}(t,0)u_1)(x) : [t_\delta + (31/16)r^2, t_\delta + 2r^2], \ x \in B(x_0; r/4) \,\},$$

which is equivalent to the desired result.

(2) By Proposition 2.2.7, $U_a(t,0)u_0 \geq 0$ for all $t > 0$. (2) then follows from (1). ☐

PROPOSITION 2.2.9 (Monotonicity on initial data)
Assume (A2-1) *and* (A2-2) *and* $1 \leq p \leq \infty$. *Let* $a \in Y$, $t > 0$ *and* $u_1, u_2 \in L_p(D)$.

(1) *If* $u_1 \leq u_2$ *then* $U_a(t, 0)u_1 \leq U_a(t, 0)u_2$.

(2) *If* $u_1 \leq u_2$, $u_1 \neq u_2$, *then* $(U_a(t, 0)u_1)(x) < (U_a(t, 0)u_2)(x)$ *for* $x \in D$.

PROOF (1) By Proposition 2.2.7, $U_a(t, 0)(u_2 - u_1) \geq 0$ for all $t > 0$. It then follows that $U_a(t, 0)u_1 \leq U_a(t, 0)u_2$ for all $t > 0$.
(2) Let $u_0 := u_2 - u_1$ and $\tilde{u}_0$ be given by

$$\tilde{u}_0(x) := \begin{cases} u_0(x) & \text{if } 0 \leq u_0(x) \leq 1 \\ 1 & \text{if } u_0(x) > 1. \end{cases}$$

Then $\tilde{u}_0 \in L_2(D)$ and $u_0 \geq \tilde{u}_0 > 0$. By Part (1), $U_a(t, 0)u_0 \geq U_a(t, 0)\tilde{u}_0 \geq 0$ for all $t \geq 0$. We claim that there is $\delta' > 0$ such that for any $\delta \in (0, \delta')$ there is $x_\delta \in D^\delta$ with the property that $(U_a(\delta^2, 0)\tilde{u}_0)(x_\delta) > 0$. Suppose not. Then there is a sequence $\delta_n \searrow 0$ such that $U_a((\delta_n)^2, 0)\tilde{u}_0 \equiv 0$ on D^{δ_n}. Consequently, for each compact subset $K \subset D$, $\int_K (U_a((\delta_n)^2, 0)\tilde{u}_0)(x)\, dx$ converge to 0 as $n \to \infty$. It follows from Proposition 2.2.1 that $\int_K (U_a((\delta_n)^2, 0)\tilde{u}_0)(x)\, dx$ converge in $L_2(D)$ to $\int_K \tilde{u}_0(x)\, dx$, which gives $\tilde{u}_0 = 0$. This is a contradiction.

Fix now $t_0 > 0$ and $x_0 \in D$. Take $\delta \in (0, \min\{\delta_0(t_0), d(x_0, \partial D), \sqrt{t_0/2}, \delta'\})$, where $\delta_0(t_0)$ is as in Proposition 2.2.8(2). We have thus $\delta^2 < t_0 - \delta^2 < t_0$. Take $y \in D^\delta$ such that $(U_a(\delta^2, 0)\tilde{u}_0)(y) > 0$. The interior Harnack inequality (Proposition 2.2.8(2)) implies $(U_a(t_0, 0)\tilde{u}_0)(x_0) > 0$, hence $(U_a(t_0, 0)u_0)(x_0) > 0$ and then $(U_a(t_0, 0)u_1)(x_0) < (U_a(t_0, 0)u_2)(x_0)$. □

In view of the above proposition, we say that a (global) weak solution u of (2.0.1)+(2.0.2) is a *positive weak solution* (on $[s, \infty) \times D$) if $u(t)(x) > 0$ for all $t > s$ and all $x \in D$.

PROPOSITION 2.2.10 (Monotonicity on coefficients)
Let (A2-1) *and* (A2-2) *be satisfied and* $1 \leq p \leq \infty$.

(1) *Assume the Dirichlet boundary condition. Let* $a^{(k)}$, $k = 1, 2$, *be such that* $a_{ij}^{(1)} = a_{ij}^{(2)}$, $a_i^{(1)} = a_i^{(2)}$, $b_i^{(1)} = b_i^{(2)}$, *but* $c_0^{(1)} \leq c_0^{(2)}$, *where equalities and inequalities are to be understood a.e. on* $\mathbb{R} \times D$. *Then*

$$U_{a^{(1)}}(t, 0)u_0 \leq U_{a^{(2)}}(t, 0)u_0$$

for any $t > 0$ *and any* $u_0 \in L_p(D)^+$.

(2) *Assume the Neumann or Robin boundary condition. Let* $a^{(k)}$, $k = 1, 2$, *be such that* $a_{ij}^{(1)} = a_{ij}^{(2)}$, $a_i^{(1)} = a_i^{(2)}$, $b_i^{(1)} = b_i^{(2)}$, *but* $c_0^{(1)} \leq c_0^{(2)}$, $d_0^{(1)} \geq$

$d_0^{(2)}$, where equalities and inequalities are to be understood a.e. on $\mathbb{R} \times D$ or a.e. on $\mathbb{R} \times \partial D$. Then

$$U_{a^{(1)}}(t, 0)u_0 \leq U_{a^{(2)}}(t, 0)u_0$$

for any $t > 0$ and any $u_0 \in L_p(D)^+$.

(3) Let $a^{(k)}$, $k = 1, 2$, be such that $a_{ij}^{(1)} = a_{ij}^{(2)}$, $a_i^{(1)} = a_i^{(2)}$, $b_i^{(1)} = b_i^{(2)}$, $c_0^{(1)} = c_0^{(2)}$, but $d_0^{(1)} \geq 0$, $d_0^{(2)} = 0$, where equalities and inequalities are to be understood a.e. on $\mathbb{R} \times D$ or a.e. on $\mathbb{R} \times \partial D$. Then

$$U_{a^{(1)}}^{\mathrm{R}}(t, 0)u_0 \leq U_{a^{(2)}}^{\mathrm{N}}(t, 0)u_0$$

for any $t > 0$ and any $u_0 \in L_p(D)^+$, where $U_a^{\mathrm{R}}(t, 0)u_0$ and $U_a^{\mathrm{N}}(t, 0)u_0$ denote the solutions of $(2.0.1)_a + (2.0.2)_a$ with Robin and Neumann boundary conditions, respectively.

(4) Let $a^{(k)}$, $k = 1, 2$, be such that $a_{ij}^{(1)} = a_{ij}^{(2)}$, $a_i^{(1)} = a_i^{(2)}$, $b_i^{(1)} = b_i^{(2)}$, $c_0^{(1)} = c_0^{(2)}$, but $d_0^{(2)} \geq 0$, where equalities and inequalities are to be understood a.e. on $\mathbb{R} \times D$ or a.e. on $\mathbb{R} \times \partial D$. Then

$$U_{a^{(1)}}^{\mathrm{D}}(t, 0)u_0 \leq U_{a^{(2)}}^{\mathrm{R}}(t, 0)u_0$$

for any $t > 0$ and any $u_0 \in L_p(D)^+$, where $U_a^{\mathrm{D}}(t, 0)u_0$ and $U_a^{\mathrm{R}}(t, 0)u_0$ denote the solutions of $(2.0.1)_a + (2.0.2)_a$ with Dirichlet and Robin boundary conditions, respectively.

PROOF   First of all, note that we only need to prove the proposition in the case that $u_0 \in L_2(D)^+$. In fact, if $u_0 \in L_p(D)^+$ with $p > 2$, we have $u_0 \in L_2(D)^+$. If $u_0 \in L_p(D)^+$ with $1 \leq p < 2$, there are $u_n \in L_2(D)^+$ such that $\|u_n - u_0\|_p \to 0$ as $n \to \infty$. For any given $a \in Y$ and $t > 0$, $U_a(t, 0) \in \mathcal{L}(L_p(D))$. Hence $U_a(t, 0)u_n \to U_a(t, 0)u_0$ in $L_p(D)$ as $n \to \infty$ and then there is a subsequence n_k such that $(U_a(t, 0)u_{n_k})(x) \to (U_a(t, 0)u_0)(x)$ for a.e. $x \in D$. Therefore we only need to prove the case that $u_0 \in L_2(D)^+$. In the rest of the proof, we assume $u_0 \in L_2(D)^+$.

(1) follows from the arguments of [30, Proposition 8.1]. For completeness, we provide a proof here.

Let $u_0 \in L_2(D)^+$ and $v(t) := U_{a^{(2)}}(t, 0)u_0 - U_{a^{(1)}}(t, 0)u_0$. It follows from Proposition 2.1.3 that $v(t)$ satisfies

$$-\int_0^t \langle v(\tau), \dot{\tilde{v}}(\tau) \rangle \, d\tau + \int_0^t B_{a^{(2)}}(\tau, v(\tau), \tilde{v}(\tau)) \, d\tau$$

$$-\int_0^t \langle (c_0^{(2)}(\tau, \cdot) - c_0^{(1)}(\tau, \cdot))(U_{a^{(1)}}(\tau, 0)u_0), \tilde{v}(\tau) \rangle \, d\tau + \langle v(t), \tilde{v}(t) \rangle = 0$$

for any $\tilde{v} \in W(0, t; V, V^*)$. Here and in the following, $V = \mathring{W}_2^1(D)$.

Put

$$\delta_0 := (\alpha_0)^{-1} \Big(\big(\sum_{i=1}^{N} \|a_i^{(2)}\|_\infty^2 \big)^{1/2} + \big(\sum_{i=1}^{N} \|b_i^{(2)}\|_\infty^2 \big)^{1/2} \Big) + \|(c_0^{(2)})^-\|_\infty,$$

where $(c_0^{(2)})^-$ is the negative part of $c_0^{(2)}$ and $\|\cdot\|_\infty$ denotes the $L_\infty(\mathbb{R} \times D)$-norm. Let $w(t) := e^{-\delta_0 t} v(t)$ and $w^-(t)$ be the negative part of $w(t)$.

Note that $\mathcal{D}([0,t];V)$ is dense in $W(0,t;V,V^*)$ (see [36, Lemma 1 in Section XVIII.1.2]). Choose $(w_m)_{m=1}^\infty \subset \mathcal{D}([0,t],V)$ such that $w_m \to w$ in the $W(0,t;V,V^*)$-norm. By Lemma 2.1.3, $w_m, w \in L_2((0,t) \times D)$ and hence $\|w_m - w\|_{L_2((0,t) \times D)} \to 0$ as $m \to \infty$. Without loss of generality, we may then assume that $w_m(\tau, x) \to w(\tau, x)$ as $m \to \infty$ for a.e. $(\tau, x) \in (0,t) \times D$.

For a given $\epsilon > 0$ let $f_\epsilon \colon \mathbb{R} \to \mathbb{R}$ be defined by

$$f_\epsilon(\xi) := \begin{cases} (\xi^2 + \epsilon^2)^{1/2} - \epsilon & \text{if } \xi < 0 \\ 0 & \text{if } \xi \geq 0. \end{cases} \tag{2.2.2}$$

Observe that $f_\epsilon \in C^1(\mathbb{R})$ and $f_\epsilon'(\xi)$ is bounded in $\xi \in \mathbb{R}$.

For $\tau \in [0,t]$ we define

$$f_\epsilon(w_m)(\tau)(x) := f_\epsilon(w_m(\tau)(x)), \quad f_\epsilon(w)(\tau)(x) := f_\epsilon(w(\tau)(x)),$$
$$f_\epsilon'(w_m)(\tau)(x) := f_\epsilon'(w_m(\tau)(x)), \quad f_\epsilon'(w)(\tau)(x) := f_\epsilon'(w(\tau)(x))$$

for a.e. $x \in D$. Then the function $[\, [0,t] \ni \tau \mapsto f_\epsilon(w_m)(\tau) \in V \,]$ is in $L_2((0,t),V)$ and the function $[\, [0,t] \ni \tau \mapsto \frac{d}{d\tau} f_\epsilon(w_m)(\tau) \in L_2(D) \,]$ is in $L_2((0,t), L_2(D))$, and, moreover, there holds

$$\int_0^t \Big\langle w(\tau), \frac{d}{d\tau} f_\epsilon(w_m)(\tau) \Big\rangle \, d\tau = \int_0^t \int_D w(\tau)(x) f_\epsilon'(w_m)(\tau)(x) \frac{dw_m}{dt}(\tau)(x) \, dx \, d\tau.$$

Then we have $f_\epsilon(w_m) \in W(0,t;V,V^*)$, and

$$-\int_0^t \int_D w(\tau)(x) f_\epsilon'(w_m)(\tau)(x) \frac{dw_m}{dt}(\tau)(x) \, dx \, d\tau$$
$$+ \int_0^t \Big(B_{a^{(2)}}(\tau, w(\tau), f_\epsilon(w_m)(\tau)) + \delta_0 \langle w(\tau), f_\epsilon(w_m)(\tau) \rangle \Big) \, d\tau$$
$$- \int_0^t \langle (c_0^{(2)}(\tau, \cdot) - c_0^{(1)}(\tau, \cdot))(U_{a^{(1)}}(\tau, 0) u_0), f_\epsilon(w_m)(\tau)) \rangle \, d\tau$$
$$+ \langle w(t), f_\epsilon(w_m)(t) \rangle - \langle w(0), f_\epsilon(w_m)(0) \rangle = 0.$$

Let $g_\epsilon \colon \mathbb{R} \to \mathbb{R}$ be given by

$$g_\epsilon(\xi) := \xi f_\epsilon'(\xi) = \begin{cases} \dfrac{\xi^2}{(\xi^2 + \epsilon^2)^{1/2}} & \text{if } \xi < 0 \\ 0 & \text{if } \xi \geq 0. \end{cases}$$

It is not difficult to see that g_ϵ and g'_ϵ are continuous and $|g_\epsilon(\xi)| \le |\xi|$ and $|g'_\epsilon(\xi)|$ is bounded in $\xi \in \mathbb{R}$. This implies that $g_\epsilon(w_m)(\cdot) := w_m(\cdot)f'_\epsilon(w_m)(\cdot)$, $g_\epsilon(w)(\cdot) := w(\cdot)f'_\epsilon(w)(\cdot) \in L_2((0,t),V)$. Moreover, by $w_m(\tau,x) \to w(\tau,x)$ as $m \to \infty$ for a.e. $(\tau,x) \in (0,t) \times D$, we have $w_m f'_\epsilon(w_m) \to w f'_\epsilon(w)$ as $m \to \infty$ in $L_2((0,t),V)$. This together with the fact that $\frac{dw_m}{dt} \to \frac{dw}{dt}$ as $m \to \infty$ in $L_2((0,t),V^*)$ implies

$$\int_0^t \int_D w_m(\tau)(x) f'_\epsilon(w_m)(\tau)(x) \frac{dw_m}{dt}(\tau)(x)\, dx\, d\tau$$

$$\to \int_0^t \left\langle w(\tau)f'_\epsilon(w)(\tau), \frac{dw}{dt}(\tau) \right\rangle d\tau \quad \text{as} \quad m \to \infty.$$

Observe that

$$\left| \int_0^t \int_D w(\tau)(x) f'_\epsilon(w_m)(\tau)(x) \frac{dw_m}{dt}(\tau)(x)\, dx\, d\tau \right.$$

$$\left. - \int_0^t \int_D w_m(\tau) f'_\epsilon(w_m)(\tau) \frac{dw_m}{dt}(\tau)\, dx\, d\tau \right|$$

$$\le \left(\int_0^t \int_D |w(\tau) - w_m(\tau)|^2\, dx\, d\tau \right)^{1/2} \left(\int_0^t \int_D \left| \frac{dw_m}{dt}(\tau) \right|^2 dx\, d\tau \right)^{1/2}$$

$$\to 0$$

as $m \to \infty$. We therefore have

$$\int_0^t \int_D w(\tau)(x) f'_\epsilon(w_m)(\tau)(x) \frac{dw_m}{dt}(\tau)(x)\, dx\, d\tau$$

$$\to \int_0^t \left\langle w(\tau)f'_\epsilon(w)(\tau), \frac{dw}{dt}(\tau) \right\rangle d\tau$$

as $m \to \infty$ and then

$$-\int_0^t \left\langle (wf'_\epsilon(w))(\tau), \frac{dw}{dt}(\tau) \right\rangle d\tau$$

$$+ \int_0^t \left(B_{a^{(2)}}(\tau, w(\tau), f_\epsilon(w)(\tau)) + \delta_0 \langle w(\tau), f_\epsilon(w)(\tau) \rangle \right) d\tau$$

$$- \int_0^t \langle (c_0^{(2)}(\tau,\cdot) - c_0^{(1)}(\tau,\cdot))(U_{a^{(1)}}(\tau,0)u_0), f_\epsilon(w)(\tau) \rangle\, d\tau$$

$$+ \langle w(t), f_\epsilon(w)(t) \rangle - \langle w(0), f_\epsilon(w)(0) \rangle = 0.$$

Note that $wf'_\epsilon(w) \to w^-$ in $L_2((0,t),V)$ as $\epsilon \to 0$. This together with

$f_\epsilon(w) \to w^-$ in $L_2((0,t), V)$ as $\epsilon \to 0$ implies that

$$- \int_0^t \left\langle w^-(\tau), \frac{dw}{dt}(\tau) \right\rangle d\tau$$
$$+ \int_0^t \left(B_{a^{(2)}}(\tau, w(\tau), w^-(\tau)) + \delta_0 \langle w(\tau), w^-(\tau) \rangle \right) d\tau$$
$$- \int_0^t \langle c_0^{(2)}(\tau, \cdot) - c_0^{(1)}(\tau, \cdot))(U_{a^{(1)}}(\tau, 0)u_0), w^-(\tau) \rangle d\tau$$
$$+ \langle w(t), w^-(t) \rangle - \langle w(0), w^-(0) \rangle = 0.$$

By [30, Lemma 3.3], there holds

$$\frac{1}{2} \left(\|w^-(t)\|^2 - \|w^-(0)\|^2 \right) = - \int_0^t \left\langle w^-(\tau), \frac{dw}{dt}(\tau) \right\rangle d\tau.$$

We then have

$$\|w^-(t)\|^2 = \|w^-(0)\|^2 - 2 \int_0^t \left(B_{a^{(2)}}(\tau, w^-(\tau), w^-(\tau)) + \delta_0 \|w^-(\tau)\|^2 \right) d\tau$$
$$- 2 \int_0^t \langle c_0^{(2)}(\tau, \cdot) - c_0^{(1)}(\tau, \cdot))(U_{a^{(1)}}(\tau, 0)u_0), w^-(\tau) \rangle d\tau.$$

As $c_0^{(2)} \geq c_0^{(1)}$, $\int_0^t \langle c_0^{(2)}(\tau, \cdot) - c_0^{(1)}(\tau, \cdot))(U_{a^{(1)}}(\tau, 0)u_0), w^-(\tau) \rangle d\tau \geq 0$. By [30, Lemma 3.1], $\int_0^t \left(B_a(\tau, w^-(\tau), w^-(\tau)) + \delta_0 \|w^-(\tau)\|^2 \right) d\tau \geq 0$. We then have $\|w^-(t)\| \leq \|w^-(0)\|$. But $w^-(0) = 0$. Hence $w^-(t) = 0$. It then follows that $w(t) \geq 0$ and then $v(t) \geq 0$, that is, $U_{a^{(2)}}(t, 0)u_0 \geq U_{a^{(1)}}(t, 0)u_0$.

(2) and (3) can be proved by the arguments similar to those in (1).

(4) First of all, let $d_0^{(2,n)}(t, x) := d_0^{(2)}(t, x) + n$ for $n = 1, 2, \ldots$. Then $(d_0^{(2,n)})_{n=1}^\infty \subset L_\infty(\mathbb{R} \times \partial D, \mathbb{R})$ and

$$d_0^{(2)}(t, x) \leq d_0^{(2,1)}(t, x) \leq d_0^{(2,2)}(t, x) \leq \cdots \leq d_0^{(2,n)}(t, x) \leq \cdots$$

for a.e. $(t, x) \in \mathbb{R} \times \partial D$. Let

$$a^{(2,n)} := (a_{ij}^{(2)}, a_i^{(2)}, b_i^{(2)}, c_0^{(2)}, d_0^{(2,n)})$$

for $n = 1, 2, \ldots$. Then by (2) we have

$$U_{a^{(2,1)}}^R(t, 0)u_0 \geq U_{a^{(2,2)}}^R(t, 0)u_0 \geq \cdots \geq U_{a^{(2,n)}}^R(t, 0)u_0 \geq \cdots \geq 0, \quad (2.2.3)$$

hence

$$\|U_{a^{(2,n)}}^R(t, 0)u_0\| \leq \|U_{a^{(2)}}^R(t, 0)u_0\| \quad (2.2.4)$$

for $n = 1, 2, \ldots$.

Let $u_n(\cdot) := U^{\mathrm{R}}_{a^{(2,n)}}(\cdot, 0)u_0$. By [30, Lemma 3.1],

$$\alpha_0 \int_0^t \|\nabla u_n(\tau)\|^2 \, d\tau$$

$$\leq 2 \int_0^t B^0_{a^{(2,n)}}(\tau, u_n(\tau), u_n(\tau)) \, d\tau + 2\delta_0 \int_0^t \|u_n(\tau)\|^2 \, d\tau$$

for $n = 1, 2, \ldots$, where α_0 is as in (2.1.1),

$$B^0_{a^{(2,n)}}(\tau, u_n(\tau), u_n(\tau)) := B_{a^{(2,n)}}(\tau, u_n(\tau), u_n(\tau))$$
$$- \int_{\partial D} d^{(2,n)}_0(\tau, x)(u_n(\tau)(x))^2 \, dH_{N-1},$$

and

$$\delta_0 := (\alpha_0)^{-1}\left(\left(\sum_{i=1}^N \|a_i^{(2)}\|_\infty^2\right)^{1/2} + \left(\sum_{i=1}^N \|b_i^{(2)}\|_\infty^2\right)^{1/2}\right) + \|(c_0^{(2)})^-\|_\infty.$$

Since $d_0^{(2,n)} \geq 0$, we have

$$B^0_{a^{(2,n)}}(\tau, u_n(\tau), u_n(\tau)) \leq B_{a^{(2,n)}}(\tau, u_n(\tau), u_n(\tau)).$$

Hence

$$\alpha_0 \int_0^t \|\nabla u_n(\tau)\|^2 \, d\tau \leq 2 \int_0^t B_{a^{(2,n)}}(\tau, u_n(\tau), u_n(\tau)) \, d\tau + 2\delta_0 \int_0^t \|u_n(\tau)\|^2 \, d\tau.$$

Because u_n is a weak solution, we obtain with the help of (2.1.10) that

$$\alpha_0 \int_0^t \|\nabla u_n(\tau)\|^2 \, d\tau \leq -\|u_n(t)\|^2 + \|u_0\|^2 + 2\delta_0 \int_0^t \|u_n(\tau)\|^2 \, d\tau. \quad (2.2.5)$$

By (2.2.4), $\int_0^t \|\nabla u_n(\tau)\|^2 \, d\tau$ is bounded uniformly in $n \in \mathbb{N}$. Hence $\{u_n|_{[0,t]} : n \in \mathbb{N}\}$ is bounded in $L_2((0,t), W_2^1(D))$. This makes sure that $(u_n|_{[0,t]})$ has a subsequence (denoted again by $(u_n|_{[0,t]})$) that converges weakly in $L_2((0,t), W_2^1(D))$ to some $u(\cdot)$.

Next, we show that $u(\tau) \in \mathring{W}_2^1(D)$ for a.e. $\tau \in [0,t]$. Note that, by Proposition 2.1.4,

$$\left|\int_0^t B_{a^{(2,n)}}(\tau, u_n(\tau), u_n(\tau)) \, d\tau\right| \leq \frac{1}{2}(\|u_n(t)\|^2 + \|u_n(0)\|^2)$$

and, by Lemma 2.1.4(ii),

$$\left|\int_0^t B^0_{a^{(2,n)}}(\tau, u_n(\tau), u_n(\tau)) \, d\tau\right| \leq M_0 \int_0^t \|u_n(\tau)\|^2_{W_2^1(D)} \, d\tau.$$

Hence

$$n \int_0^t \int_{\partial D} (u_n(\tau)(x))^2 \, dH_{N-1} \, d\tau$$

$$\leq \int_0^t \int_{\partial D} d_0^{(2,n)}(\tau, x)(u_n(\tau)(x))^2 \, dH_{N-1} \, d\tau$$

$$= \int_0^t B_{a^{(2,n)}}(\tau, u_n(\tau), u_n(\tau)) \, d\tau - \int_0^t B_{a^{(2,n)}}^0(\tau, u_n(\tau), u_n(\tau)) \, d\tau$$

$$\leq \frac{1}{2}(\|u_n(t)\|^2 + \|u_n(0)\|^2) + M_0 \int_0^t \|u_n(\tau)\|_{W_2^1(D)}^2 \, d\tau$$

for each $n \in \mathbb{N}$. But $\frac{1}{2}(\|u_n(t)\|^2 + \|u_n(0)\|^2) + M_0 \int_0^t \|u_n(\tau)\|_{W_2^1(D)}^2 \, d\tau$ is bounded uniformly in $n \in \mathbb{N}$, by (2.2.4) and (2.2.5). We then must have

$$\int_0^t \int_{\partial D} (u_n(\tau)(x))^2 \, dH_{N-1} \, d\tau \to 0$$

as $n \to \infty$, consequently

$$\int_0^t \int_{\partial D} u_n(\tau)(x) \, dH_{N-1} \, d\tau \to 0$$

as $n \to \infty$. Observe that

$$\langle u_n, 1 \rangle_{L_2((0,t), W_2^1(D))} = \int_0^t \int_{\partial D} u_n(\tau)(x) \, dH_{N-1} \, d\tau.$$

As u_n converge weakly in $L_2((0,t), W_2^1(D))$ to u, we have

$$\int_0^t \int_{\partial D} u(\tau)(x) \, dH_{N-1} \, d\tau = 0.$$

Further, since $u(\tau)$ is nonnegative, its trace $u(\tau)|_{\partial D}$ is nonnegative for a.e. $\tau \in [0, t]$. Therefore, $u(\tau)|_{\partial D} = 0$ for a.e. $\tau \in [0, t]$. By Lemma 2.1.1, we have $u(\tau) \in \mathring{W}_2^1(D)$ for a.e. $\tau \in [0, t]$.

We now prove that $u(t) = U_{a^{(1)}}^{\mathrm{D}}(t, 0)u_0$. Note that for any $n \in \mathbb{N}$ and for any $v \in W(0, t; V, V^*)$,

$$-\int_0^t \langle u_n(\tau), \dot{v}(\tau) \rangle \, d\tau + \int_0^t B_{a^{(2)}}^0(\tau, u_n(\tau), v(\tau)) \, d\tau$$

$$+ \int_0^t \int_{\partial D} d_0^{(2,n)}(\tau, x)(u_n(\tau)v(\tau))(x) \, dH_{N-1} \, d\tau + \langle u_n(t), v(t) \rangle - \langle u_0, v(0) \rangle$$

$$= -\int_0^t \langle u_n(\tau), \dot{v}(\tau) \rangle \, d\tau + \int_0^t B_{a^{(1)}}(\tau, u_n(\tau), v(\tau)) \, d\tau$$

$$+ \langle u_n(t), v(t) \rangle - \langle u_0, v(0) \rangle = 0,$$

where $V = \mathring{W}_2^1(D)$. Letting $n \to \infty$ and using again the fact that u_n converge weakly in $L_2((0,t), W_2^1(D))$ to u, we have

$$-\int_0^t \langle u(\tau), \dot{v}(\tau) \rangle \, d\tau + \int_0^t B_{a^{(1)}}(\tau, u(\tau), v(\tau)) \, d\tau + \langle u(t), v(t) \rangle - \langle u_0, v(0) \rangle = 0$$

for any $v \in W(0,t; V, V^*)$. This means that $u(t) = U_{a^{(1)}}^{\mathrm{D}}(t,0)u_0$ (see Proposition 2.1.3).

Finally, by (2.2.3), we have

$$U_{a^{(1)}}^{\mathrm{D}}(t,0)u_0 = u(t) \leq U_{a^{(2)}}^{\mathrm{R}}(t,0)u_0$$

for $t \geq 0$. (4) is thus proved.   □

PROPOSITION 2.2.11 (Joint measurability)
For any $a \in Y$, $u_0 \in L_2(D)$, and $T > 0$, $u(\cdot, \cdot) \in W_2^{0,1}((0,T) \times D)$, where $u(t,x) := (U_a(t,0)u_0)(x)$.

PROOF   First of all, by Lemma 2.1.3, $[\, (0,\infty) \times D \ni (t,x) \mapsto u(t,x) \in \mathbb{R} \,]$ is measurable and for any $T > 0$, $u \in L_2((0,T) \times D)$.

Next we prove that $u \in W_2^{0,1}((0,T) \times D)$. By the fact that $[\, (0,T) \ni t \mapsto u(t, \cdot) \,] \in L_2((0,T), V)$ and Lemma 1.2.3, there are simple functions $\phi_n \in L_2((0,T), V)$ such that

$$\|\phi_1\|_{L_2((0,T),V)} \leq \|\phi_2\|_{L_2((0,T),V)} \leq \cdots,$$

$\|\phi_n\|_{L_2((0,T),V)} \to \|u\|_{L_2((0,T),V)}$, and $\phi_n(t) \to u(t,\cdot)$ in V as $n \to \infty$ for a.e. $t \in (0,T)$. Let $\tilde{\phi}_n(t,x) := (\phi_n(t))(x)$, $t \in (0,T)$, $x \in D$. It is clear that $\tilde{\phi}_n \in W_2^{0,1}((0,T) \times D)$ and $\|\tilde{\phi}_n\|_{W_2^{0,1}((0,T) \times D)} = \|\phi_n\|_{L_2((0,T),V)} \leq \|u\|_{L_2((0,T),V)}$. Hence, without loss of generality, we may assume that $\tilde{\phi}_n$ weakly converges to $\tilde{\phi}$ in $W_2^{0,1}((0,T) \times D)$. Therefore, for any $\psi \in \mathcal{D}((0,T) \times D)$, we have

$$\int_{(0,T) \times D} \tilde{\phi}_n(t,x) \psi(t,x) \, dt \, dx \to \int_{(0,T) \times D} \tilde{\phi}(t,x) \psi(t,x) \, dt \, dx.$$

But

$$\int_{(0,T) \times D} \tilde{\phi}_n(t,x) \psi(t,x) \, dt \, dx = \int_0^T \left(\int_D \tilde{\phi}_n(t,x) \psi(t,x) \, dx \right) dt$$

$$\to \int_0^T \left(\int_D u(t,x) \psi(t,x) \, dx \right) dt = \int_{(0,T) \times D} u(t,x) \psi(t,x) \, dt \, dx.$$

It then follows that $u(t,x) = \tilde{\phi}(t,x)$ for a.e. $(t,x) \in (0,T) \times D$, hence $u \in W_2^{0,1}((0,T) \times D)$.   □

We proceed now to investigate consequences of (A2-3).

As the mapping $[\,Y \times \mathbb{R} \ni (a,t) \mapsto a \cdot t \in Y\,]$ is continuous, we have, in the light of Propositions 2.1.6 through 2.1.8, the following consequence of Proposition 2.2.6.

PROPOSITION 2.2.12

Let (A2-1)–(A2-3) *be satisfied. For any sequence* $(a^{(n)})_{n=1}^{\infty} \subset Y$, *any real sequences* $(s_n)_{n=1}^{\infty}$ *and* $(t_n)_{n=1}^{\infty}$, $s_n < t_n$, *if*

$$\lim_{n\to\infty} a^{(n)} = a, \quad \lim_{n\to\infty} s_n = s, \quad \lim_{n\to\infty} t_n = t, \ \text{where } s < t,$$

then for any $u_0 \in L_2(D)$, $U_{a^{(n)}}(t_n, s_n)u_0$ *converges to* $U_a(t,s)u_0$ *in* $L_2(D)$.

The next result is much more important.

PROPOSITION 2.2.13 (Joint continuity in t, u_0, and a)

Assume (A2-1) *through* (A2-3). *For any sequence* $(a^{(n)})_{n=1}^{\infty} \subset Y$, *any real sequence* $(t_n)_{n=1}^{\infty}$, *and any sequence* $(u_n)_{n=1}^{\infty} \subset L_p(D)$ $(2 \le p < \infty)$, *if* $\lim_{n\to\infty} a^{(n)} = a$, $\lim_{n\to\infty} t_n = t$, *where* $t > 0$, *and* $\lim_{n\to\infty} u_n = u_0$ *in* $L_p(D)$, *then* $U_{a^{(n)}}(t_n, 0)u_n$ *converges in* $L_p(D)$ *to* $U_a(t,0)u_0$.

PROOF   Since $L_p(D) \hookrightarrow L_2(D)$ for $2 \le p < \infty$, we assume $\lim_{n\to\infty} u_n = u_0$ in $L_2(D)$. Put $K := \sup\{\|U_{a^{(n)}}(t_n, 0)\| : n \in \mathbb{N}\}$. By Proposition 2.2.2, $K < \infty$. There holds

$$\|U_{a^{(n)}}(t_n, 0)u_n - U_a(t, 0)u_0\|$$
$$\le \|U_{a^{(n)}}(t_n, 0)u_n - U_{a^{(n)}}(t_n, 0)u_0\| + \|U_{a^{(n)}}(t_n, 0)u_0 - U_a(t, 0)u_0\|.$$

The first term on the right-hand side is bounded by $K\|u_n - u_0\|$, hence it converges to 0, and the second one converges to 0 by virtue of (A2-3). We deduce from Proposition 2.2.5 that $U_{a^{(n)}}(t_n, 0)u_n$ converges to $U_a(t, 0)u_0$ in $L_p(D)$, too.   □

We put

$$\Pi(t; u_0, a) = \Pi_t(u_0, a) := (U_a(t, 0)u_0, a \cdot t) \tag{2.2.6}$$

for $t \ge 0$, $a \in Y$ and $u_0 \in L_2(D)$.

Proposition 2.2.13, Lemma 2.1.2 and (2.1.13) guarantee that the mapping $\Pi = \{\Pi_t\}_{t\ge 0}$ so defined is a topological linear skew-product semiflow on the product Banach bundle $L_2(D) \times Y$ covering the topological (semi)flow $\sigma = \{\sigma_t\}_{t\in\mathbb{Z}}$, $\sigma_t a = a \cdot t$.

We shall refer to Π defined by (2.2.6) as the *(topological) linear skew-product semiflow on* $L_2(D) \times Y$ *generated by* (2.0.1)+(2.0.1).

By Proposition 2.2.9(2), the solution operator $U_a(t,0)$ ($a \in Y$, $t > 0$) has the property that, for any u_1, u_2 with $u_1 < u_2$ there holds $U_a(t,0)u_1 < U_a(t,0)u_2$. By adjusting the terminology used for semiflows on ordered metric spaces (see, e.g., [57]) to skew-product semiflows with ordered fibers we can say that the (topological) linear skew-product semiflow Π is *strictly monotone*.

2.3   The Adjoint Problem

We consider in this section the adjoint problem of (2.0.1)+(2.0.2), i.e., the backward parabolic equation

$$-\frac{\partial u}{\partial t} = \sum_{i=1}^{N} \frac{\partial}{\partial x_i} \left(\sum_{j=1}^{N} a_{ji}(t,x)\frac{\partial u}{\partial x_j} - b_i(t,x)u \right)$$

$$-\sum_{i=1}^{N} a_i(t,x)\frac{\partial u}{\partial x_i} + c_0(t,x)u, \quad t < s, \ x \in D, \tag{2.3.1}$$

where $s \in \mathbb{R}$ is a final time, complemented with the boundary conditions:

$$\mathcal{B}_a^*(t)u = 0, \quad t < s, \ x \in \partial D, \tag{2.3.2}$$

where $\mathcal{B}_a^*(t)u = \mathcal{B}_{a^*}(t)u$ with $a^* := ((a_{ji})_{i,j=1}^N, -(b_i)_{i=1}^N, -(a_i)_{i=1}^N, c_0, a_0)$ and $\mathcal{B}_{a^*}(t)$ is as in (2.0.3) with a being replaced by a^*.

Denote $U_a^*(t,s)$ ($t \leq s$) to be the (weak) solution operator of (2.3.1)+(2.3.2) (the weak solution of (2.3.1)+(2.3.2) is defined in a way similar to the weak solution of (2.0.1)+(2.0.2), see Definition 2.1.1).

Let $\tilde{U}_a(t,s)$ ($t \geq s$) be the (weak) solution operator of

$$\frac{\partial u}{\partial t} = \sum_{i=1}^{N} \frac{\partial}{\partial x_i} \left(\sum_{j=1}^{N} a_{ji}(-t,x)\frac{\partial u}{\partial x_j} - b_i(-t,x)u \right)$$

$$-\sum_{i=1}^{N} a_i(-t,x)\frac{\partial u}{\partial x_i} + c_0(-t,x)u, \quad t > s, \ x \in D, \tag{2.3.3}$$

where $s \in \mathbb{R}$ is an initial time, complemented with the boundary conditions:

$$\tilde{\mathcal{B}}_a^*(t)u = 0, \quad t > s, \ x \in \partial D, \tag{2.3.4}$$

where $\tilde{\mathcal{B}}_a^*(t)u = \mathcal{B}_{\tilde{a}^*}(t)u$ with $\tilde{a}^*(t,x) := a^*(-t,x)$ and $\mathcal{B}_{\tilde{a}^*}(t)$ is as in (2.0.3) with a being replaced by $\tilde{a}^*$.

Then we have

$$U_a^*(t,s) = \tilde{U}_a(-t,-s) \quad (t \leq s).$$

For $a \in Y$ denote by $\tilde{B}_a = \tilde{B}_a(t, \cdot, \cdot)$ the bilinear form on V associated with a,

$$\tilde{B}_a(t, u, v) := B_{\tilde{a}^*}(t, u, v) \tag{2.3.5}$$

where $B_{\tilde{a}^*}(t, u, v)$ is as in (2.1.4) with a being replaced by $\tilde{a}^*$ in the Dirichlet and Neumann boundary condition cases, and $B_{\tilde{a}^*}(t, u, v)$ is as in (2.1.5) with a being replaced by $\tilde{a}^*$ in the Robin boundary condition case. Similarly to Proposition 2.1.3, we have

PROPOSITION 2.3.1
Assume (A2-1). Let $u \in L_2((s, t), V)$. u is a weak solution of (2.3.3)+(2.3.4) on $[s, t] \times D$ ($t > s$) with $u(s) = u_0$ if and only if $u \in W(s, t; V, V^*)$ and for any $v \in W(s, t; V, V^*)$,

$$-\int_s^t \langle u(\tau), \dot{v}(\tau) \rangle \, d\tau + \int_s^t \tilde{B}_a(\tau, u(\tau), v(\tau)) \, d\tau$$
$$+ \langle u(t), v(t) \rangle - \langle u_0, v(s) \rangle = 0. \tag{2.3.6}$$

PROPOSITION 2.3.2
Assume (A2-1), (A2-2). If u and v are solutions of (2.0.1)+(2.0.2) and of (2.3.1)+(2.3.2) on $[s, t] \times D$, respectively, then $\langle u(\tau), v(\tau) \rangle$ is independent of τ for $\tau \in [s, t]$.

PROOF   First note that $v(\tau) = U_a^*(\tau, t)(v(t)) = \tilde{U}_a(-\tau, -t)(v(t))$ for $s \leq \tau \leq t$.

For any $s \leq \tau \leq t$, by Proposition 2.1.3,

$$\int_s^\tau \langle u(r), \dot{v}(r) \rangle \, dr = \int_s^\tau B_a(r, u(r), v(r)) \, dr$$
$$+ \langle u(\tau), v(\tau) \rangle - \langle u(s), v(s) \rangle. \tag{2.3.7}$$

By Proposition 2.3.1, we have

$$\int_s^\tau \langle v(r), \dot{u}(r) \rangle \, dr$$
$$= \int_{-\tau}^{-s} \langle \tilde{U}(r, -\tau) v(\tau), \dot{u}(-r) \rangle \, dr$$
$$= -\int_{-\tau}^{-s} \tilde{B}_a(r, \tilde{U}(r, -\tau) v(\tau), u(-r)) \, dr - \langle v(\tau), u(\tau) \rangle + \langle v(s), u(s) \rangle$$
$$= -\int_s^\tau \tilde{B}_a(-r, \tilde{U}(-r, -\tau) v(\tau), u(r)) \, dr - \langle v(\tau), u(\tau) \rangle + \langle v(s), u(s) \rangle$$
$$= -\int_s^\tau \tilde{B}_a(-r, v(r), u(r)) \, dr - \langle u(\tau), v(\tau) \rangle + \langle u(s), v(s) \rangle. \tag{2.3.8}$$

Note that

$$\int_s^\tau B_a(r, u(r), v(r)) \, dr - \int_s^\tau \tilde{B}_a(-r, v(r), u(r)) \, dr = 0$$

for any $s \leq \tau \leq t$. It then follows that

$$\int_s^\tau \langle v(r), \dot{u}(r) \rangle \, dr + \int_s^\tau \langle u(r), \dot{v}(r) \rangle \, dr = 0$$

for $s \leq \tau \leq t$. As both u and v are in $W(s, t; V, V^*)$, by [36, Section XVIII.1.2, Theorem 2] we have

$$\langle u(\tau), v(\tau) \rangle = \langle u(s), v(s) \rangle$$

for any $s \leq \tau \leq t$, so $\langle u(\tau), v(\tau) \rangle$ is independent of τ. □

PROPOSITION 2.3.3
Assume (A2-1) and (A2-2).

$$(U_a(t, s))^* = U_a^*(s, t) \quad \text{for any } a \in Y \text{ and any } s < t.$$

PROOF   First, recall that the dual $(U_a(t, s))^*$ $(a \in Y,\ s < t)$ of the linear operator $U_a(t, s)$ is defined by

$$\langle u_0, (U_a(t, s))^* v_0 \rangle = \langle U_a(t, s) u_0, v_0 \rangle, \quad u_0, v_0 \in L_2(D).$$

By Proposition 2.3.2,

$$\langle u(t), v(t) \rangle = \langle u(s), v(s) \rangle$$

where u and v are solutions of (2.0.1)+(2.0.2) and of (2.3.1)+(2.3.2) on $[s, t] \times D$, respectively. Let $u(\cdot) := U_a(\cdot, s) u_0$ and $v(\cdot) := U_a^*(\cdot, t) v_0$. Then $u(s) = u_0$, $v(t) = v_0$, and

$$\langle U_a(t, s) u_0, v_0 \rangle = \langle u(t), v(t) \rangle = \langle u(s), v(s) \rangle = \langle u_0, U_a^*(s, t) v_0 \rangle$$

for any $u_0, v_0 \in L_2(D)$. Consequently,

$$(U_a(t, s))^* = U_a^*(s, t) \quad \text{for any } a \in Y \text{ and any } s < t.$$

□

Observe that if (2.0.1)+(2.0.2) satisfies (A2-3), then (2.3.1)+(2.3.2) also satisfies (A2-3). In fact, let $(a^{(n)})_{n=1}^\infty \subset Y$ and $(t_n)_{n=1}^\infty \subset \mathbb{R}$ be such that $a^{(n)} \to a$ and $t_n \to t > 0$ as $n \to \infty$. Then for any $v \in L_2(D)$, $U_{a^{(n)}}(t_n, 0)v \to U_a(t, 0)v$ in $L_2(D)$ as $n \to \infty$. Now for any $u_0 \in L_2(D)$ and any $v \in L_2(D)$,

$$\langle v, U_{a^{(n)}}^*(0, -t_n) u_0 \rangle = \langle u_0, U_{a^{(n)}}(t_n, 0)v \rangle$$
$$\to \langle u_0, U_a(t, 0)v \rangle = \langle v, U_a^*(0, -t) u_0 \rangle$$

as $n \to \infty$. Therefore $U^*_{a^{(n)}}(0, -t_n)u_0 \to U^*_a(0, -t)u_0$ weakly in $L_2(D)$. By Proposition 2.2.5 and $U^*_a(t, s) = \tilde{U}_a(-t, -s)$ for any $a \in Y$ and $t < s$, without loss of generality, we may assume that $U^*_{a^{(n)}}(0, -t_n) \to u^*$ in $L_2(D)$. We then have $U^*_a(0, -t)u_0 = u^*$ and $U^*_{a^{(n)}}(0, -t_n)u_0 \to U^*_a(0, -t)u_0$ in $L_2(D)$. Hence having constructed a topological linear skew-product semiflow Π on the Banach bundle $L_2(D) \times Y$, we have the *dual* topological linear skew-product semiflow $\Pi^* = \{\Pi^*_t\}_{t \geq 0}$, defined by the formula:

$$\Pi^*(t, v_0, a) = \Pi^*_t(v_0, a) := ((U_{a \cdot (-t)}(t, 0))^* v_0, a \cdot (-t)) = (U^*_a(-t, 0)v_0, a \cdot (-t)),$$

where $t \geq 0$, $a \in Y$, and $v_0 \in L_2(D)$.

2.4 Perturbation of Coefficients

In this section we discuss the satisfaction of the assumptions (A2-1)–(A2-3) presented in the previous sections.

Consider (2.0.1)+(2.0.2). We first note that (A2-1) is a natural uniform ellipticity assumption and (A2-2) is a regularity condition of the domain D. We therefore assume throughout this section that (2.0.1)+(2.0.2) satisfies (A2-1) and (A2-2), and focus on the investigation on (A2-3) (perturbation property).

In this section we make also another standing assumption:

(A2-4) (Convergence almost everywhere)
In the Dirichlet or Neumann case:
For any sequence $(a^{(n)})$ converging in Y to a we have that $a^{(n)}_{ij} \to a_{ij}$, $a^{(n)}_i \to a_i$, $b^{(n)}_i \to b_i$ pointwise a.e. on $\mathbb{R} \times D$.
In the Robin case:
For any sequence $(a^{(n)})$ converging in Y to a we have that $a^{(n)}_{ij} \to a_{ij}$, $a^{(n)}_i \to a_i$, $b^{(n)}_i \to b_i$ pointwise a.e. on $\mathbb{R} \times D$, and $d^{(n)}_0 \to d_0$ pointwise a.e. on $\mathbb{R} \times \partial D$.

THEOREM 2.4.1
Consider (2.0.1)+(2.0.2). Let V be as in (2.1.2). Let $u_0 \in L_2(D)$ and $a^{(n)}$ be as in (A2-4). Then for each $t > 0$ the following holds:

(1) *The restrictions $U_{a^{(n)}}(\cdot, 0)u_0|_{[0,t]}$ converge weakly in $L_2((0, T), V)$ to $U_a(\cdot, 0)u_0|_{[0,t]}$.*

(2) *The functions $[\, [0, t] \times D \ni (\tau, x) \mapsto (U_{a^{(n)}}(\tau, 0)u_0)(x)\,]$ converge in the $L_2((0, t) \times D)$-norm to $[\, [0, t] \times D \ni (\tau, x) \mapsto (U_a(\tau, 0)u_0)(x)\,]$.*

(3) *For any $0 < t_0 < t$, the restrictions $U_{a^{(n)}}(\cdot, 0)u_0|_{[t_0,t]}$ converge in the $C([t_0, t], L_2(D))$-norm to $U_a(\cdot, 0)u_0|_{[t_0,t]}$.*

PROOF   We prove the theorem only for the Neumann or Robin boundary cases ($V = W_2^1(D)$ in these cases), the proof of the theorem for the Dirichlet case being similar, but simpler (cf. [30, Lemma 8.4]). Put $u_n(\cdot) := U_{a^{(n)}}(\cdot, 0)u_0$. First, by [30, Lemma 3.1],

$$\alpha_0 \int_0^t \|\nabla u_n\|^2 \, d\tau \le 2 \int_0^t B_{a^{(n)}}^0(\tau, u_n(\tau), u_n(\tau)) \, d\tau + 2\delta_0 \int_0^t \|u_n(\tau)\|^2 \, d\tau,$$

where α_0 is as in (2.1.1),

$$B_{a^{(n)}}^0(\tau, u_n(\tau), u_n(\tau)) := B_{a^{(n)}}(\tau, u_n(\tau), u_n(\tau)) - \int_{\partial D_n} d_0^{(n)} u_n(\tau) u_n(\tau) \, dH_{N-1},$$

and

$$\delta_0 := (\alpha_0)^{-1} \Big(\sup_{n \in \mathbb{N}} \big(\sum_{i=1}^N \|a_i^{(n)}\|_\infty^2 \big)^{1/2} + \sup_{n \in \mathbb{N}} \big(\sum_{i=1}^N \|b_i^{(n)}\|_\infty^2 \big)^{1/2} \Big) + \sup_{n \in \mathbb{N}} \big(\|c_0^{(n)}\|_\infty \big),$$

where $\|\cdot\|_\infty$ denotes the $L_\infty(\mathbb{R} \times D)$-norm. Since $d_0^{(n)} \ge 0$, we have

$$B_{a^{(n)}}^0(\tau, u_n(\tau), u_n(\tau)) \le B_{a^{(n)}}(\tau, u_n(\tau), u_n(\tau)).$$

Hence

$$\alpha_0 \int_0^t \|\nabla u_n\|^2 \, d\tau \le 2 \int_0^t B_{a^{(n)}}(\tau, u_n(\tau), u_n(\tau)) \, d\tau + 2\delta_0 \int_0^t \|u_n(\tau)\|^2 \, d\tau.$$

Since u_n is a weak solution of (2.0.1)+(2.0.2), we obtain with the help of (2.1.10) that

$$\alpha_0 \int_0^t \|\nabla u_n\|^2 \, d\tau \le \|u_0\|^2 - \|u_n(t)\|^2 + 2\delta_0 \int_0^t \|u_n(\tau)\|^2 \, d\tau.$$

Note that by Proposition 2.2.2 $\int_0^t \|u_n(\tau)\|^2 \, d\tau$ is bounded uniformly in $n \in \mathbb{N}$. Hence $\{ u_n|_{[0,t]} : n \in \mathbb{N} \}$ is bounded in $L_2((0,t), W_2^1(D))$. This makes sure that $(u_n|_{[0,t]})$ has a subsequence (denoted again by $(u_n|_{[0,t]})$) that converges weakly in $L_2((0,t), W_2^1(D))$ to some $u(\cdot)$. By Proposition 2.2.11, $u_n(t,x)$ is integrable on $[0,t] \times D$. It therefore follows that $[(0,t) \times D \ni (\tau, x) \mapsto u_n(\tau, x)]$ converge weakly in $L_2((0,t) \times D)$ to $[(0,t) \times D \ni (\tau, x) \mapsto u(\tau, x)]$, where we write $u(\tau, x) = u(\tau)(x)$.

An application of Proposition 2.2.4 allows us, after possibly taking a subsequence, to conclude that there exists a continuous function $v : (0,t) \times D \to \mathbb{R}$ such that for any $0 < t_0 < t$ and any compact $D_0 \subset D$ the functions $[[t_0, t] \times D_0 \ni (\tau, x) \mapsto u_n(\tau, x)]$ converge uniformly to the restriction $v|_{[t_0,t] \times D_0}$.

Fix for the moment $0 < t_0 < t$. From the L_2–L_∞ estimates in Proposition 2.2.2 it follows that there is $M_0 > 0$ such that $\|u_n(\tau)\|_\infty \le M_0$ for each $\tau \in [t_0, t]$ and each $n \in \mathbb{N}$. As a consequence, $\|v(\tau, \cdot)\|_\infty \le M_0$ for each

$\tau \in [t_0, t]$ (here again $\|\cdot\|_\infty$ means the $L_\infty(D)$-norm). Take $\epsilon > 0$. Let $D_0 \subset D$ be a compact set with $|D \setminus D_0| < \epsilon/8M_0^2$. Consequently,

$$\int_{D \setminus D_0} |u_n(\tau, x) - v(\tau, x)|^2 \, dx < \epsilon/2$$

for any $t_0 \leq \tau \leq t$ and any $n \in \mathbb{N}$. Now we take $n_0 \in \mathbb{N}$ so large that

$$\int_{D_0} |u_n(\tau, x) - v(\tau, x)|^2 \, dx < \epsilon/2$$

for any $t \leq \tau \leq t$ and any $n > n_0$. Therefore it follows that $u_n|_{[t_0,t]}$ ($\in C([t_0, t], L_2(D))$) converge uniformly, as functions from $[t_0, t]$ into $L_2(D)$, to $[\, [t_0, t] \ni \tau \to v(\tau, \cdot) \in L_2(D)\,]$ (which belongs to $C([t_0, t], L_2(D))$).

By the L_2–L_2 estimates in Proposition 2.2.2, there exists $M_1 > 0$ such that $\|u_n(\tau)\| \leq M_1$ for any $n \in \mathbb{N}$ and any $\tau \in [0, t]$. Let $\epsilon > 0$, and take $0 < t_0 < \min\{\epsilon/(8M_1^2), t\}$. One has

$$\int_0^{t_0} \int_D |u_n(\tau, x) - v(\tau, x)|^2 \, dx \, d\tau < \frac{\epsilon}{2}.$$

By the previous paragraph, there exists $n_1 \in \mathbb{N}$ such that

$$\int_{t_0}^t \int_D |u_n(\tau, x) - v(\tau, x)|^2 \, dx \, d\tau < \frac{\epsilon}{2}$$

for all $n > n_1$. Therefore the functions $[\, (0, t) \times D \ni (\tau, x) \mapsto u_n(\tau)(x)\,]$ converge in the $L_2((0, t) \times D)$-norm, hence weakly, to v. Consequently, $u(\tau) = v(\tau, \cdot)$ for each $\tau \in (0, t]$.

To conclude the proof it suffices to show that $u(\tau) = U_a(\tau, 0)u_0$ for any $\tau \in (0, t]$. For any $v \in W_2^1(D)$ and any $\psi \in \mathcal{D}([0, t))$,

$$-\int_0^t \langle u_n(\tau), v \rangle \dot\psi(\tau) \, d\tau + \int_0^t B_{a^{(n)}}(\tau, u_n(\tau), v)\psi(\tau) \, d\tau - \langle u_0, v \rangle \psi(0) = 0.$$

Observe that

$$\int_0^t \langle u_n(\tau), v \rangle \dot\psi(\tau) \, d\tau = \int_0^t \left(\int_D u_n(\tau, x)v(x) \, dx \right) \dot\psi(\tau) \, d\tau.$$

Hence

$$\int_0^t \langle u_n(\tau), v \rangle \dot\psi(\tau) \, d\tau \to \int_0^t \left(\int_D u(t, x)v(x) \, dx \right) \dot\psi(\tau) \, d\tau = \int_0^t \langle u(\tau), v \rangle \dot\psi(\tau) \, d\tau.$$

Observe also that

$$\int_0^t \left(\int_{\partial D} d_0^{(n)}(\tau, x) u_n(\tau, x) v(x) \, dH_{N-1} \right) \psi(\tau) \, d\tau$$

$$- \int_0^t \left(\int_{\partial D} d_0(\tau, x) u(\tau, x) v(x) \, dH_{N-1} \right) \psi(\tau) \, d\tau$$

$$= \int_0^t \left(\int_{\partial D} (d_0^{(n)}(\tau, x) - d_0(\tau, x)) u_n(\tau)(x) v(x) \, dH_{N-1} \right) \psi(\tau) \, d\tau$$

$$+ \int_0^t \left(\int_{\partial D} d_0(\tau, x)(u_n(\tau, x) - u(\tau)(x)) v(x) \, dH_{N-1} \right) \psi(\tau) \, d\tau.$$

Note that

$$\int_0^t \left(\int_{\partial D} d_0(\tau, x)(u_n(\tau, x) - u(\tau, x)) v(x) \, dH_{N-1} \right) \psi(\tau) \, d\tau \to 0$$

by the fact that $u_n \to u$ weakly in $L_2((0,t), W_2^1(D))$, hence the traces of u_n on ∂D converge weakly in $L_2((0,t), L_2(\partial D))$ to the trace of u on ∂D (see Lemma 2.1.1). Further, since $\{ u_n|_{[0,t]} : n \in \mathbb{N} \}$ is bounded in $L_2((0,t), W_2^1(D))$, $\{ \int_0^t \int_{\partial D} (u_n(\tau, x))^2 \, dH_{N-1} \, d\tau : n \in \mathbb{N} \}$ is a bounded sequence. We then have

$$\left| \int_0^t \left(\int_{\partial D} (d_0^{(n)}(\tau, x) - d_0(\tau, x)) u_n(\tau, x) v(x) \, dH_{N-1} \right) \psi(\tau) \, d\tau \right|^2$$

$$\leq \left(\int_0^t \left(\int_{\partial D} |d_0^{(n)}(\tau, x) - d_0(\tau, x)|^2 (v(x))^2 \, dH_{N-1} \right) (\psi(\tau))^2 \, d\tau \right)$$

$$\cdot \left(\int_0^t \left(\int_{\partial D} (u_n(\tau, x))^2 \, dH_{N-1} \right) d\tau \right)$$

$$\to 0$$

by the boundedness of $d_0^{(n)}$ in the $L_\infty(\mathbb{R} \times \partial D)$-norm and the convergence of $d_0^{(n)}$ to d_0 a.e. on $[0,t] \times \partial D$. Hence

$$\int_0^t \left(\int_{\partial D} d_0^{(n)}(\tau, x) u_n(\tau, x) v(x) \, dH_{N-1} \right) \psi(\tau) \, d\tau$$

$$\to \int_0^t \left(\int_{\partial D} d_0(\tau, x) u(\tau, x) v(x) \, dH_{N-1} \right) \psi(\tau) \, d\tau.$$

Similarly, we can prove that

$$\int_0^t B_{a^{(n)}}^0(\tau, u_n(\tau), v) \psi(\tau) \, d\tau + \int_0^t \left(\int_D c_0^{(n)}(\tau, x) u_n(\tau, x) v(x) \, dx \right) \psi(\tau) \, d\tau$$

$$\to \int_0^t B_a^0(\tau, u(\tau), v) \psi(\tau) \, d\tau + \int_0^t \left(\int_D c_0(\tau, x) u(\tau, x) v(x) \, dx \right) \psi(\tau) \, d\tau.$$

Now,

$$\int_0^t \int_D c_0^{(n)}(\tau, x) u_n(\tau, x) v(x) \psi(\tau)\, dx\, d\tau - \int_0^t \int_D c_0(\tau, x) u(\tau, x) v(x) \psi(\tau)\, dx\, d\tau$$

$$= \int_0^t \int_D c_0^{(n)}(\tau, x)(u_n(\tau, x) - u(\tau, x)) v(x) \psi(\tau)\, dx\, d\tau$$

$$+ \int_0^t \int_D (c_0^{(n)}(\tau, x) - c_0(\tau, x)) u(\tau, x) v(x) \psi(\tau)\, dx\, d\tau.$$

We estimate

$$\left| \int_0^t \int_D c_0^{(n)}(\tau, x)(u_n(\tau, x) - u(\tau, x)) v(x) \psi(\tau)\, dx\, d\tau \right|^2$$

$$\leq \left(\int_0^t \int_D |u_n(\tau, x) - u(\tau, x)|^2\, dx\, d\tau \right)$$

$$\left(\int_0^t \int_D |c_0^{(n)}(\tau, x)|^2 (v(x))^2 (\psi(\tau))^2\, dx\, d\tau \right)$$

$$\rightarrow 0.$$

By the weak-* convergence of $c_0^{(n)}$ to c_0,

$$\int_0^t \int_D (c_0^{(n)}(\tau, x) - c_0(\tau, x)) u(\tau, x) v(x) \psi(\tau)\, dx\, d\tau \rightarrow 0.$$

Consequently

$$\int_0^t \int_D c_0^{(n)}(\tau, x) u_n(\tau, x) v(x) \psi(\tau)\, dx\, d\tau \rightarrow \int_0^t \int_D c_0(\tau, x) u(\tau, x) v(x) \psi(\tau)\, dx\, d\tau.$$

It then follows that

$$- \int_0^t \langle u(\tau), v \rangle \dot{\psi}(\tau)\, d\tau + \int_0^t B_a(\tau, u(\tau), v) \psi(\tau)\, d\tau - \langle u_0, v \rangle \psi(0) = 0.$$

Therefore, $u(\cdot)$ is a weak solution of (2.0.1)+(2.0.2) on $[0, t] \times D$ with $u(0) = u_0$. In particular, the whole sequence (u_n) (and not only a subsequence) converges to u as required. □

2.5 The Smooth Case

In the present section we are considering the case when the domain and the coefficients are sufficiently regular for any solution to be a classical one.

We introduce the following assumption.

(A2-5) (Smoothness)
∂D *is an* $(N-1)$*-dimensional manifold of class* $C^{3+\alpha}$ *for some* $0 < \alpha <$
1*. Moreover, there is* $M > 0$ *such that for any* $a \in Y$*, the* $C^{2+\alpha,2+\alpha}(\mathbb{R} \times$
$\bar{D})$*-norms of* a_{ij} *and* a_i $(i, j = 1, 2, \ldots, N)$*, the* $C^{2+\alpha,1+\alpha}(\mathbb{R} \times \bar{D})$*-norms of* b_i
$(i = 1, 2, \cdots, N)$ *and* c_0*, and the* $C^{2+\alpha,2+\alpha}(\mathbb{R} \times \partial D)$*-norms of* d_0*, are bounded*
by M*.*

First of all, we have

LEMMA 2.5.1
Assume (A2-5)*. Then* $\lim_{n \to \infty} a^{(n)} = a$ *if and only if* $a_{ij}^{(n)}$ *converge to* a_{ij}*,*
$a_i^{(n)}$ *converge to* a_i*,* $b_i^{(n)}$ *converge to* b_i*,* $c_0^{(n)}$ *converge to* c_0*, all uniformly*
on compact subsets of $\mathbb{R} \times \bar{D}$*, and (in the Robin case)* $d_0^{(n)}$ *converge to* d_0
uniformly on compact subsets of $\mathbb{R} \times \partial D$*.*

PROOF   As the convergence in the open-compact topology implies convergence in the weak-* topology, the closure $\tilde{Y}$ of Y in the open-compact topology equals, as a set, Y. By the Ascoli–Arzelà theorem, $\tilde{Y}$ is compact. Compact topologies on the same set are identical. $\Box$

In the remainder of the present section we assume that (A2-1) and (A2-5) are satisfied. (A2-5) implies (A2-2). Also, by Lemma 2.5.1 (A2-5) implies (A2-4), from which it follows that (A2-3) is satisfied.

We will apply the theories developed in [3] to derive regularity properties, various estimates, and strong monotonicity of weak solutions of (2.0.1)+(2.0.2).

Observe that $(2.0.1)_a + (2.0.2)_a$ can be rewritten as

$$\frac{\partial u}{\partial t} = \sum_{i,j=1}^{N} a_{ij}(t, x) \frac{\partial^2 u}{\partial x_i \partial x_j} + \sum_{i=1}^{N} \tilde{b}_i(t, x) \frac{\partial u}{\partial x_i}$$
$$+ \tilde{c}_0(t, x) u, \quad t > s, \ x \in D, \tag{2.5.1}$$

complemented with the boundary conditions

$$\mathcal{B}(t) u = 0, \qquad t > s, \ x \in \partial D, \tag{2.5.2}$$

where

$$\mathcal{B}(t) u = \begin{cases} u & \text{(Dirichlet)} \\ \displaystyle\sum_{i,j=1}^{N} a_{ij}(t, x) \partial_{x_j} u \nu_i + \tilde{d}_0(t, x) u & \text{(Neumann)} \\ \displaystyle\sum_{i,j=1}^{N} a_{ij}(t, x) \partial_{x_j} u \nu_i + \tilde{d}_0(t, x) u & \text{(Robin)}, \end{cases}$$

with $\tilde{b}_i(t,x) := b_i(t,x) + a_i(t,x) + \sum_{j=1}^{N} \frac{\partial a_{ji}}{\partial x_j}(t,x)$, $\tilde{c}_0(t,x) := c(t,x) + \sum_{i=1}^{N} \frac{\partial a_i}{\partial x_i}(t,x)$, and $\tilde{d}_0(t,x) := \sum_{i=1}^{N} a_i(t,x)\nu_i$ in the Neumann case and $\tilde{d}_0(t,x) := d_0(t,x) + \sum_{i=1}^{N} a_i(t,x)\nu_i$ in the Robin case. Note that the boundary conditions in the Neumann and Robin cases are of the same form after rewriting. Note also that $\tilde{d}_0(t,x)$ may change sign. We point out that the theory presented in [3] applies to such a general case. More precisely, to apply that theory we only need the smoothness of the coefficients and the domain and the uniform ellipticity of (2.5.1)+(2.5.2) (see [3] for detail).

PROPOSITION 2.5.1 (Regularity up to boundary)
Let $1 \le p \le \infty$ and $u_0 \in L_p(D)$. Then for any $\alpha \in (0, 1/2)$, any $a \in Y$, and any $0 < t_1 < t_2$ the restriction $[[t_1, t_2] \ni t \mapsto U_a(t, 0)u_0]$ belongs to $C^1([t_1, t_2], C^\alpha(\bar{D})) \cap C([t_1, t_2], C^{2+\alpha}(\bar{D}))$. Moreover, there is $C = C(t_1, t_2, u_0) > 0$ such that

$$\|[[t_1, t_2] \ni t \mapsto U_a(t, 0)u_0]\|_{C^1([t_1,t_2], C^\alpha(\bar{D}))} \le C$$

and

$$\|[[t_1, t_2] \ni t \mapsto U_a(t, 0)u_0]\|_{C([t_1,t_2], C^{2+\alpha}(\bar{D}))} \le C$$

for all $a \in Y$.

PROOF For given $1 \le p \le \infty$ and $u_0 \in L_p(D)$, for any $t > 0$ and $1 < q < \infty$, one has $U_a(t, 0)u_0 \in L_q(D)$. The result then follows from [3, Corollary 15.3]. □

By Proposition 2.5.1, for any $u_0 \in L_p(D)$ $(1 \le p \le \infty)$, $U_a(t, 0)u_0$ turns out to be a *classical solution* of (2.5.1)+(2.5.2): for any $t > 0$ and $x \in D$ the equation (2.5.1) is satisfied pointwise, and for any $t > 0$ and $x \in \partial D$ the boundary condition (2.5.2) is satisfied pointwise.

Next, we derive other regularity properties and various estimates. To do so, for $a \in Y$ let $A(a)$ be the operator given by

$$A(a)u := \sum_{i=1}^{N} \frac{\partial}{\partial x_i}\left(\sum_{j=1}^{N} a_{ij}(0, x)\frac{\partial u}{\partial x_j} + a_i(0, x)u\right)$$
$$+ \sum_{i=1}^{N} b_i(0, x)\frac{\partial u}{\partial x_i} + c_0(0, x)u, \qquad x \in D,$$

and

$$\mathcal{B}(a) := \mathcal{B}_a(0),$$

where $\mathcal{B}_a(\cdot)$ is as in (2.0.3).

For given $1 < p < \infty$, $1 \leq q \leq \infty$ and $s > 0$, let $W_p^s(D)$, $H_p^s(D)$, and $B_{p,q}^s(D)$ be the Sobolev–Slobodetskiĭ spaces, the Bessel potential spaces, and the Besov spaces, respectively (see [3], [105] for definitions).

LEMMA 2.5.2
If $0 < s < \infty$, then, up to equivalent norms,

$$W_p^s(D) = \begin{cases} H_p^s(D) & \text{if } s \in \mathbb{N} \\ B_{p,p}^s(D) & \text{if } s \notin \mathbb{N}. \end{cases}$$

PROOF   See [105].   ⬚

For given $0 < \beta < 1$ and $1 < p < \infty$, let $(\cdot, \cdot)_{\beta,p}$ and $[\cdot, \cdot]_\beta$ be a real interpolation functor and a complex interpolation functor, respectively, (see [15], [105] for definitions), and let

$$V_p^\beta := (L_p(D), W_p^2(D))_{\beta,p} \tag{2.5.3}$$

and

$$\tilde{V}_p^\beta := [L_p(D), W_p^2(D)]_\beta. \tag{2.5.4}$$

LEMMA 2.5.3
Let $1 < p < \infty$ and $0 < \beta < 1$. Then the following holds.

(1) $V_p^\beta = B_{p,p}^{2\beta}(D)$;

(2) $\tilde{V}_p^\beta = H_p^{2\beta}(D)$.

PROOF   See [105, Theorem 2 in Section 4.3.1].   ⬚

For given $a \in Y$, $0 < \beta < 1$ and $1 < p < \infty$, let

$$V_p^\beta(a) := (L_p(D), V_p^1(a))_{\beta,p}, \tag{2.5.5}$$

and

$$\tilde{V}_p^\beta(a) := [L_p(D), \tilde{V}_p^1(a)]_\beta, \tag{2.5.6}$$

where $V_p^1(a) = \tilde{V}_p^1(a) := \{\, u \in W_p^2(D) : \mathcal{B}(a)u = 0 \,\}$.

LEMMA 2.5.4
Let $1 < p < \infty$ and $0 < \beta < 1$ with $2\beta - \frac{1}{p} \neq 0, 1$. Then the following holds.

(1) $V_p^\beta(a)$ *is a closed subspace of* V_p^β;

(2) $\tilde{V}_p^\beta(a)$ *is a closed subspace of* $\tilde{V}_p^\beta$.

PROOF   (1) follows from [3, Lemma 14.4].
(2) follows from [4, Lemma 5.1].   ⬜

Recall that, for given two Banach spaces X_1 and X_2 consisting of (equivalence classes of) real functions defined on D, $X_1 \hookrightarrow X_2$ means that X_1 is continuously embedded into X_2, and $X_1 \hookrightarrow\hookrightarrow X_2$ means that X_1 is compactly embedded into X_2. We have

LEMMA 2.5.5

(1) *For* $p > N/2$ *and* $\frac{N}{2p} < \beta \leq 1$ *there holds*

$$V_p^\beta \hookrightarrow\hookrightarrow C(\bar{D}) \tag{2.5.7}$$

and

$$\tilde{V}_p^\beta \hookrightarrow\hookrightarrow C(\bar{D}); \tag{2.5.8}$$

in particular

$$W_p^2(D) \hookrightarrow\hookrightarrow C(\bar{D}); \tag{2.5.9}$$

(2) *For* $p > N$ *and* $\frac{N}{2p} + \frac{1}{2} < \beta \leq 1$ *there holds*

$$V_p^\beta \hookrightarrow\hookrightarrow C^{1+\tilde{\beta}}(\bar{D}) \tag{2.5.10}$$

and

$$\tilde{V}_p^\beta \hookrightarrow\hookrightarrow C^{1+\tilde{\beta}}(\bar{D}), \tag{2.5.11}$$

where $0 < \tilde{\beta} < 2\beta - 1 - \frac{N}{p}$; *in particular*

$$W_p^2(D) \hookrightarrow\hookrightarrow C^{1+\tilde{\beta}}(\bar{D}), \tag{2.5.12}$$

where $0 < \tilde{\beta} < 1 - \frac{N}{p}$;

(3) *For given* $1 < p < \infty$ *and* $0 < \beta_1 < \beta_2 < 1$,

$$W_p^2(D) \hookrightarrow\hookrightarrow W_p^{2\beta_2}(D) \hookrightarrow\hookrightarrow W_p^{2\beta_1}(D) \hookrightarrow\hookrightarrow L_p(D), \tag{2.5.13}$$

$$W_p^2(D) \hookrightarrow\hookrightarrow V_p^{\beta_2} \hookrightarrow\hookrightarrow V_p^{\beta_1} \hookrightarrow\hookrightarrow L_p(D), \tag{2.5.14}$$

and

$$W_p^2(D) \hookrightarrow\hookrightarrow \tilde{V}_p^{\beta_2} \hookrightarrow\hookrightarrow \tilde{V}_p^{\beta_1} \hookrightarrow\hookrightarrow L_p(D); \tag{2.5.15}$$

(4) *For given $1 < p < \infty$ and $0 < \beta_1 < \beta_2 < 1$ with $2\beta_1 - \frac{1}{p}, 2\beta_2 - \frac{1}{p} \neq 0, 1$,*

$$V_p^1(a) \hookrightarrow V_p^{\beta_2}(a) \hookrightarrow V_p^{\beta_1}(a) \hookrightarrow L_p(D), \tag{2.5.16}$$

and

$$\tilde{V}_p^1(a) \hookrightarrow \tilde{V}_p^{\beta_2}(a) \hookrightarrow \tilde{V}_p^{\beta_1}(a) \hookrightarrow L_p(D). \tag{2.5.17}$$

PROOF　　(1) follows from the fact that if $mp > N$ then $W_p^{j+m}(D) \hookrightarrow C^j(\bar{D})$ (see [1, Theorem 6.2(7)]), together with Lemmas 2.5.2 and 2.5.3 (see also [3, Theorem 11.5]).

(2) follows from the fact that if $mp > N > (m-1)p$ and $0 < \tilde{\beta} < m - (N/p)$ then $W_p^{j+m}(D) \hookrightarrow C^{j+\tilde{\beta}}(\bar{D})$ (see [1, Theorem 6.2(8)]), together with Lemmas 2.5.2 and 2.5.3 (see also [3, Theorem 11.5]).

(3) (2.5.13) follows from [3, Theorem 11.5 and Corollary 15.2].

The continuity of the embeddings in (2.5.14) follows from [105, Theorem 4.6.1] and the compactness follows from (2.5.13) and Lemma 2.5.2.

The continuity of the embeddings in (2.5.15) also follows from [105, Theorem 4.1.1]. By [105, Theorem 4.6.2], for any $\epsilon > 0$ with $0 < \beta_1 - \epsilon < \beta_1 + \epsilon < \beta_2 - \epsilon < \beta_2 + \epsilon < 1$,

$$B_{p,p}^{\beta_2+\epsilon}(D) \hookrightarrow H_p^{\beta_2}(D) \hookrightarrow B_{p,p}^{\beta_2-\epsilon}(D) \hookrightarrow B_{p,p}^{\beta_1+\epsilon}(D) \hookrightarrow H_p^{\beta_1}(D) \hookrightarrow B_{p,p}^{\beta_1-\epsilon}(D).$$

This together with the compactness of the embeddings in (2.5.14) implies that the embeddings in (2.5.15) are also compact.

(4) It follows from (3) and Lemma 2.5.4.　　　　　　　　　　　　　　　□

By [3, Lemma 6.1 and Theorem 14.5] (see also [109]), we have

PROPOSITION 2.5.2

(1) *For any $1 < p < \infty$ and $u_0 \in L_p(D)$, $U_a(t,0)u_0 \in V_p^1(a \cdot t)$ for $t > 0$.*

(2) *For any $1 < p < \infty$ and $u_0 \in L_p(D)$, $U_a(t,0)u_0 \in \tilde{V}_p^1(a \cdot t)$ for $t > 0$.*

(3) *For any fixed $T > 0$ and $1 < p < \infty$ there is $C_p > 0$ such that*

$$\|U_a(t,0)\|_{L_p(D),W_p^2(D)} \leq \frac{C_p}{t}$$

for all $a \in Y$ and $0 < t \leq T$.

By [3, Theorems 7.1 and 14.5] we have

PROPOSITION 2.5.3
Suppose that $2\beta - 1/p \notin \mathbb{N}$. Then

(1) *for any $a \in Y$, $t \geq 0$ and $u_0 \in V_p^\beta(a)$ there holds $U_a(t,0)u_0 \in V_p^\beta(a \cdot t)$; moreover, the mapping $[\,[0,\infty) \ni t \mapsto U_a(t,0)u_0 \in V_p^\beta\,]$ is continuous;*

(2) *for any $a \in Y$, $t \geq 0$ and $u_0 \in \tilde{V}_p^\beta(a)$ there holds $U_a(t,0)u_0 \in \tilde{V}_p^\beta(a \cdot t)$; moreover, the mapping $[\,[0,\infty) \ni t \mapsto U_a(t,0)u_0 \in \tilde{V}_p^\beta\,]$ is continuous;*

(3) *for any $T > 0$ there is $C_{p,\beta} > 0$ such that*

$$\|U_a(t,0)u_0\|_{V_p^\beta} \leq C_{p,\beta}\|u_0\|_{V_p^\beta}$$

for any $a \in Y$, $0 \leq t \leq T$, and $u_0 \in V_p^\beta(a)$, and

$$\|U_a(t,0)u_0\|_{\tilde{V}_p^\beta} \leq C_{p,\beta}\|u_0\|_{\tilde{V}_p^\beta}$$

for any $a \in Y$, $0 \leq t \leq T$, and $u_0 \in \tilde{V}_p^\beta(a)$.

PROPOSITION 2.5.4 (Joint continuity in X)
Assume that X is a Banach space such that, for some $1 < p < \infty$,

$$W_p^2(D) \hookrightarrow X \hookrightarrow L_2(D).$$

For any sequence $(a^{(n)})_{n=1}^\infty \subset Y$, any real sequence $(t_n)_{n=1}^\infty$, and any sequence $(u_n)_{n=1}^\infty \subset L_2(D)$, if $\lim_{n\to\infty} a^{(n)} = a$, $\lim_{n\to\infty} t_n = t$, where $t > 0$, and $\lim_{n\to\infty} u_n = u_0$ in $L_2(D)$, then $U_{a^{(n)}}(t_n,0)u_n$ converges in X to $U_a(t,0)u_0$.

PROOF First of all, we have by Proposition 2.2.13 that $U_{a^{(n)}}(t_n,0)u_n$ converges in $L_2(D)$ to $U_a(t,0)u_0$. It follows from Propositions 2.2.2 and 2.5.2 and the assumption $W_p^2(D) \hookrightarrow X$ that there is a subsequence $(n_k)_{k=1}^\infty$ such that $U_{a^{(n_k)}}(t_{n_k},0)u_{n_k}$ converges in X to some u^*. We then must have $u^* = U_a(t,0)u_0$ and $U_{a^{(n_k)}}(t_{n_k},0)u_{n_k}$ converges in X to $U_a(t,0)u_0$. This implies that $U_{a^{(n)}}(t_n,0)u_n$ converges in X to $U_a(t,0)u_0$. □

Examples of Banach spaces X satisfying the assumptions of the above proposition include V_p^β, $\tilde{V}_p^\beta$ ($p \geq 2$, $0 < \beta < 1$), $C(\bar{D})$ (when $p > N/2$), $C^1(\bar{D})$ (when $p > N$), etc. (see Lemma 2.5.5).

PROPOSITION 2.5.5 (Norm continuity in X)
Assume that a Banach space X has the property that, for some $1 < p < \infty$,

$$W_p^2(D) \hookrightarrow X \hookrightarrow L_2(D).$$

For any sequence $(a^{(n)})_{n=1}^\infty \subset Y$ and any real sequence $(t_n)_{n=1}^\infty$, if $a^{(n)} \to a$ and $t_n \to t$ as $n \to \infty$, where $t > 0$, then $U_{a^{(n)}}(t_n,0)$ converges in $\mathcal{L}(X)$ to $U_a(t,0)$.

PROOF   Assume that $U_{a^{(n)}}(t_n, 0)$ does not converge in $\mathcal{L}(X)$ to $U_a(t, 0)$. Then we may assume that there are $\epsilon_0 > 0$ and $u_n \in X$ with $\|u_n\|_X = 1$ such that

$$\|U_{a^{(n)}}(t_n, 0)u_n - U_a(t, 0)u_n\|_X \geq \epsilon$$

for $n = 1, 2, \ldots$. By the assumption $X \hookrightarrow L_2(D)$, without loss of generality we may assume that there is $u_0 \in L_2(D)$ such that $\|u_n - u_0\| \to 0$ as $n \to \infty$. Then by Proposition 2.5.4 we have

$$\|U_{a^{(n)}}(t_n, 0)u_n - U_a(t, 0)u_0\|_X \to 0, \quad \|U_a(t, 0)u_n - U_a(t, 0)u_0\|_X \to 0$$

as $n \to \infty$. Hence

$$\|U_{a^{(n)}}(t_n, 0)u_n - U_a(t, 0)u_0\|_X \to 0,$$

which is a contradiction. Therefore, $U_{a^{(n)}}(t_n, 0)$ converges in $\mathcal{L}(X)$ to $U_a(t, 0)$.
□

Examples of Banach spaces satisfying the assumption of the above proposition include V_p^β, $\tilde{V}_p^\beta$ ($p \geq 2$, $0 < \beta < 1$), $C^1(\bar{D})$ (when $p > N$), etc. (see Lemma 2.5.5).

We proceed now to investigate the strong monotonicity property of the solution operators $U_a(t, 0)$. We will use the strong maximum principle and the Hopf boundary point principle for classical solutions. But before we do that we have to analyze whether the existing theory (as presented, e.g., in [44]) can be applied: notice that in the Robin case $\tilde{d}_0$ may change sign. We show that the zero-order coefficient can be made nonnegative by an appropriate change of variables.

Fix $a \in Y$ and $p > N$. Take u_0 to be a C^∞ real function whose support is a nonempty compact set contained in D (the existence of such a function follows by the C^∞ Urysohn lemma, see, e.g., [42, Lemma 8.18]). Let $u^*(\cdot, \cdot)$ be the solution of

$$\begin{cases} \dfrac{\partial u}{\partial t} = \displaystyle\sum_{i,j=1}^{N} a_{ij}(t, x)\dfrac{\partial^2 u}{\partial x_i \partial x_j}, & t > -1, \ x \in D, \\ \displaystyle\sum_{i,j=1}^{N} a_{ij}(t, x)\partial_{x_j} u \nu_i + u = 0, & t > -1, \ x \in \partial D, \end{cases}$$

with the initial condition $u(-1, \cdot) = u_0$. The initial function u_0 clearly belongs to V_p^1, and satisfies pointwise the boundary conditions at $t = -1$. Consequently, [4, Theorem 7.3(ii)] states that $[[-1, \infty) \ni t \mapsto u^*(t, \cdot) \in V_p^1]$ is continuous, from which it follows via (2.5.12) that $[[-1, \infty) \ni t \mapsto u^*(t, \cdot) \in C^1(\bar{D})]$ is continuous, too. In particular, u^* is continuous on $[-1, \infty) \times \bar{D}$. Further, u^* is a classical solution, so the strong maximum principle and the Hopf boundary point principle for parabolic equations imply that $u^*(t, x) > 0$ for $t > -1$ and $x \in \bar{D}$.

Now, let $v(t,x) := e^{M^* u^*(t,x)} u(t,x)$, where M^* is a positive constant (to be determined later). Then (2.5.1)+(2.5.2) with $s = 0$ becomes

$$\frac{\partial v}{\partial t} = \sum_{i,j=1}^{N} a_{ij}(t,x) \frac{\partial^2 v}{\partial x_i \partial x_j} + \sum_{i=1}^{N} \bar{b}_i(t,x) \frac{\partial v}{\partial x_i}$$

$$+ \bar{c}_0(t,x) v, \quad t > 0, \ x \in D, \tag{2.5.18}$$

complemented with the boundary conditions

$$\sum_{i,j=1}^{N} a_{ij}(t,x) \partial_{x_j} v \nu_i + \bar{d}_0(t,x) v = 0, \quad t > 0, \ x \in \partial D, \tag{2.5.19}$$

where

$$\bar{b}_i(t,x) := \tilde{b}_i(t,x) - M^* \left(\sum_{j=1}^{N} \left(a_{ij} \frac{\partial u^*}{\partial x_j} + a_{ji} \frac{\partial u^*}{\partial x_j} \right) \right),$$

$$\bar{c}_0(t,x) := \tilde{c}_0(t,x) - M^* \sum_{i=1}^{N} \tilde{b}_i(t,x) \frac{\partial u^*}{\partial x_i} + (M^*)^2 \sum_{i,j=1}^{N} a_{ij}(t,x) \frac{\partial u^*}{\partial x_i} \frac{\partial u^*}{\partial x_j},$$

$$\bar{d}_0(t,x) := \tilde{d}_0(t,x) + M^* e^{M^* u^*(t,x)} u^*(t,x).$$

We see that for any $T > 0$, there is $M^* = M^*(T) > 0$ such that $\bar{d}_0(t,x) > 0$ for $t \in [0,T]$ and $x \in \bar{D}$. Also, the coefficients $\bar{b}_i$, $\bar{c}_0$, and $\bar{d}_0$ are continuous, hence bounded on $[0,T] \times \bar{D}$. So the existing theory for classical solutions of parabolic equations can be applied to (2.5.18)+(2.5.19) and then to (2.5.1)+(2.5.2). In particular, we have the following result.

PROPOSITION 2.5.6 (Strong monotonicity on initial data)
Let $1 \le p \le \infty$ and $u_1, u_2 \in L_p(D)$. If $u_1 < u_2$, then

(i)
$$(U_a(t,0)u_1)(x) < (U_a(t,0)u_2)(x) \quad \text{for } a \in Y, \ t > 0, \ x \in D$$

and

$$\frac{\partial}{\partial \boldsymbol{\nu}} (U_a(t,0)u_1)(x) > \frac{\partial}{\partial \boldsymbol{\nu}} (U_a(t,0)u_2)(x) \quad \text{for } a \in Y, \ t > 0, \ x \in \partial D$$

in the Dirichlet case,

(ii)
$$(U_a(t,0)u_1)(x) < (U_a(t,0)u_2)(x) \quad \text{for } a \in Y, \ t > 0, \ x \in \bar{D}$$

in the Neumann or Robin case.

Recall that the nonnegative cone $V_p^\beta(a)^+$ of $V_p^\beta(a)$ is defined by

$$V_p^\beta(a)^+ := \{\, u \in V_p^\beta(a) : u(x) \ge 0 \quad \text{for a.e. } x \in D \,\}.$$

Similarly

$$\tilde{V}_p^\beta(a)^+ := \{\, u \in \tilde{V}_p^\beta(a) : u(x) \ge 0 \quad \text{for a.e. } x \in D \,\}.$$

If p and β are such that $V_p^\beta(a) \hookrightarrow C(\bar{D})$, then

$$V_p^\beta(a)^+ = \{\, u \in V_p^\beta(a) : u(x) \ge 0 \quad \text{for all } x \in \bar{D} \,\}.$$

Similarly, if p and β are such that $\tilde{V}_p^\beta(a) \hookrightarrow C(\bar{D})$, then

$$\tilde{V}_p^\beta(a)^+ = \{\, u \in \tilde{V}_p^\beta(a) : u(x) \ge 0 \quad \text{for all } x \in \bar{D} \,\}.$$

LEMMA 2.5.6
Assume that $p > N$ and $\frac{N}{2p} + \frac{1}{2} < \beta \le 1$. Let $a \in Y$.

(1) *In the case of the Dirichlet boundary conditions the interior $V_p^\beta(a)^{++}$ of the nonnegative cone $V_p^\beta(a)^+$ is nonempty, and is characterized by*

$$\begin{aligned}
V_p^\beta(a)^{++} = \{\, u \in V_p^\beta(a)^+ : u(x) &> 0 \quad \text{for all} \quad x \in D \\
\text{and} \quad (\partial u/\partial\boldsymbol{\nu})(x) &< 0 \quad \text{for all} \quad x \in \partial D \,\}.
\end{aligned}$$
(2.5.20)

(2) *In the case of the Neumann or Robin boundary conditions the interior $V_p^\beta(a)^{++}$ of the nonnegative cone $V_p^\beta(a)^+$ is nonempty, and is characterized by*

$$V_p^\beta(a)^{++} = \{\, u \in V_p^\beta(a)^+ : u(x) > 0 \quad \text{for all} \quad x \in \bar{D} \,\}. \qquad (2.5.21)$$

Analogous results hold for the complex interpolation spaces $\tilde{V}_p^\beta(a)$.

PROOF   We prove the lemma only for the real interpolation spaces $V_p^\beta(a)$. Fix $a \in Y$. It follows from Lemmas 2.5.4 and 2.5.5(2) that

$$V_p^\beta(a) \hookrightarrow C^1(\bar{D}).$$

(1) In the Dirichlet case $V_p^1(a)$ consists precisely of those elements of $W_p^2(D)$ whose trace on ∂D is zero. Since $V_p^1(a) \hookrightarrow C^1(\bar{D})$, any $u \in V_p^1(a)$ is a C^1 function vanishing on ∂D. By [3, Section 7], the image of the embedding $V_p^1(a) \hookrightarrow V_p^\beta(a)$ is dense. Because $V_p^\beta(a) \hookrightarrow C^1(\bar{D})$, we conclude that $V_p^\beta(a) \hookrightarrow \mathring{C}^1(\bar{D})$.

Denote by I the embedding $V_p^\beta(a) \hookrightarrow \mathring{C}^1(\bar{D})$. It follows from Lemma 1.3.1(2) that the right-hand side of (2.5.20) equals $I^{-1}(\mathring{C}^1(\bar{D})^{++})$, where $\mathring{C}^1(\bar{D})^{++}$ is an open subset of $\mathring{C}^1(\bar{D})$. This proves the "$\supset$" inclusion. Denote by φ_{princ} some (nonnegative) principal eigenfunction of the elliptic equation

$$0 = \sum_{i=1}^N \frac{\partial}{\partial x_i}\left(\sum_{j=1}^N a_{ij}(0,x)\frac{\partial u}{\partial x_j} + a_i(0,x)u\right)$$
$$+ \sum_{i=1}^N b_i(0,x)\frac{\partial u}{\partial x_i} + c_0(0,x)u, \quad x \in D,$$

with the Dirichlet boundary conditions. We have that $\varphi_{\text{princ}} \in V_p^1(a) \hookrightarrow V_p^\beta(a)$ and that it belongs to the right-hand side of (2.5.20), consequently to $V_p^\beta(a)^{++}$. Finally, let $u \in V_p^\beta(a)^{++}$. There is $\epsilon > 0$ such that $u - \epsilon\varphi_{\text{princ}} \in V_p^\beta(a)^+$, therefore $u(x) \geq \epsilon\varphi_{\text{princ}}(x) > 0$ for all $x \in D$, which gives further that $\frac{\partial u}{\partial \boldsymbol{\nu}}(x) \leq \epsilon\frac{\partial\varphi_{\text{princ}}}{\partial \boldsymbol{\nu}}(x) < 0$ for all $x \in \partial D$.

(2) In the Neumann or Robin cases, denote by I the embedding $V_p^\beta(a) \hookrightarrow C^1(\bar{D})$. It follows from Lemma 1.3.1(1) that the right-hand side of (2.5.21) equals $I^{-1}(C^1(\bar{D})^{++})$, where $C^1(\bar{D})^{++}$ is an open subset of $C^1(\bar{D})$. This proves the "$\supset$" inclusion. Denote by φ_{princ} some (nonnegative) principal eigenfunction of the elliptic equation

$$0 = \sum_{i=1}^N \frac{\partial}{\partial x_i}\left(\sum_{j=1}^N a_{ij}(0,x)\frac{\partial u}{\partial x_j} + a_i(0,x)u\right)$$
$$+ \sum_{i=1}^N b_i(0,x)\frac{\partial u}{\partial x_i} + c_0(0,x)u, \quad x \in D,$$

with the boundary conditions

$$0 = \begin{cases} \displaystyle\sum_{i=1}^N\left(\sum_{j=1}^N a_{ij}(0,x)\partial_{x_j}u + a_i(0,x)u\right)\nu_i, & x \in \partial D \quad \text{(Neumann)} \\ \displaystyle\sum_{i=1}^N\left(\sum_{j=1}^N a_{ij}(0,x)\partial_{x_j}u + a_i(0,x)u\right)\nu_i \\ \quad + d_0(0,x)u, & x \in \partial D \quad \text{(Robin)}. \end{cases}$$

We have that $\varphi_{\text{princ}} \in V_p^1(a) \hookrightarrow V_p^\beta(a)$ and that it belongs to the right-hand side of (2.5.21), consequently to $V_p^\beta(a)^{++}$. Finally, let $u \in V_p^\beta(a)^{++}$. There is $\epsilon > 0$ such that $u - \epsilon\varphi_{\text{princ}} \in V_p^\beta(a)^+$, therefore $u(x) \geq \epsilon\varphi_{\text{princ}}(x) > 0$ for all $x \in \bar{D}$. $\qquad\square$

In view of Lemma 2.5.6, the following result is a consequence of Propositions 2.5.6 and 2.5.3.

PROPOSITION 2.5.7

(1) *For any $1 \leq p \leq \infty$, any $q > N$ and any $\frac{N}{2q} + \frac{1}{2} < \beta \leq 1$ there holds*

$$U_a(t,0)(L_p(D)^+ \setminus \{0\}) \subset V_q^\beta(a \cdot t)^{++}$$

and

$$U_a(t,0)(L_p(D)^+ \setminus \{0\}) \subset \tilde{V}_q^\beta(a \cdot t)^{++},$$

for all $a \in Y$ and $t > 0$.

(2) *There holds*

$$U_a(t,0)(L_p(D)^+ \setminus \{0\}) \subset \mathring{C}^1(\bar{D})^{++}$$

in the Dirichlet case, or

$$U_a(t,0)(L_p(D)^+ \setminus \{0\}) \subset C^1(\bar{D})^{++}$$

in the Neumann or Robin cases, for all $a \in Y$ and $t > 0$.

Proposition 2.5.7(2) yields that

$$U_a(t,0)(\mathring{C}^1(\bar{D})^+ \setminus \{0\}) \subset \mathring{C}^1(\bar{D})^{++} \qquad (2.5.22)$$

for all $a \in Y$ and all $t > 0$ (in the Dirichlet case), and

$$U_a(t,0)(C^1(\bar{D})^+ \setminus \{0\}) \subset C^1(\bar{D})^{++} \qquad (2.5.23)$$

for all $a \in Y$ and all $t > 0$ (in the Neumann or Robin cases).

The property described in (2.5.22) (resp. (2.5.23)) can be written as: For each $a \in Y$ and $t > 0$, if $u_1, u_2 \in \mathring{C}^1(\bar{D})$ (resp. $u_1, u_2 \in C^1(\bar{D})$) and $u_1 < u_2$ then $u_1 \ll u_2$. In the existing terminology (see [57]) the linear operator $U_a(t,0) \colon \mathring{C}^1(\bar{D}) \to \mathring{C}^1(\bar{D})$ (resp. $U_a(t,0) \colon C^1(\bar{D}) \to C^1(\bar{D})$) is, for $a \in Y$ and $t > 0$, *strongly positive* (or *strongly monotone*).

PROPOSITION 2.5.8
For each $a \in Y$ and each $t > 0$ the linear operator $U_a(t,0)$ is injective.

PROOF   See [43, Chapter 6].   ∎

We finish the section with a remark on the adjoint problem. Observe that the adjoint equation $(2.3.1)_a$ with the corresponding boundary conditions $(2.3.2)_a$ can be rewritten as

$$-\frac{\partial u}{\partial t} = \sum_{i,j=1}^N a_{ji}^*(t,x)\frac{\partial^2 u}{\partial x_i \, \partial x_j}$$

$$+ \sum_{j=1}^N b_j^*(t,x)\frac{\partial u}{\partial x_j} + c_0^*(t,x)u, \quad t < s, \ x \in D, \qquad (2.5.24)$$

complemented with the boundary conditions

$$\mathcal{B}^*(t)u = 0, \quad t < s, \ x \in \partial D, \tag{2.5.25}$$

where

$$\mathcal{B}^*(t)u = \begin{cases} u & \text{(Dirichlet)} \\ \displaystyle\sum_{j,i=1}^{N} a_{ji}^*(t,x)\partial_{x_i}u\nu_j + d_0^*(t,x)u & \text{(Neumann)} \\ \displaystyle\sum_{j,i=1}^{N} a_{ji}^*(t,x)\partial_{x_i}u\nu_j + d_0^*(t,x)u & \text{(Robin)}, \end{cases}$$

with $a_{ji}^*(t,x) := a_{ij}(t,x)$, $b_j^*(t,x) := -b_j(t,x) - a_j(t,x) + \sum_{i=1}^{N}\frac{\partial a_{ji}}{\partial x_i}(t,x)$, $c_0^*(t,x) := c_0(t,x) - \sum_{i=1}^{N}\frac{\partial b_i}{\partial x_i}(t,x)$, $d_0^*(t,x) := -\sum_{j=1}^{N} b_j(t,x)\nu_j$ in the Neumann case and $d_0^*(t,x) := d_0(t,x) - \sum_{j=1}^{N} b_j(t,x)\nu_j$ in the Robin case.

All the results presented above in the present section carry over to the case of the adjoint problem.

2.6 Remarks on Equations in Nondivergence Form

In this section, we provide remarks on nonautonomous equations in nondivergence form. Consider

$$\frac{\partial u}{\partial t} = \sum_{i,j=1}^{N} a_{ij}(t,x)\frac{\partial^2 u}{\partial x_i \partial x_j} + \sum_{i=1}^{N} b_i(t,x)\frac{\partial u}{\partial x_i}$$
$$+ c_0(t,x)u, \quad t > s, \ x \in D, \tag{2.6.1}$$

complemented with the boundary conditions

$$\mathcal{B}(t)u = 0, \quad t > s, \ x \in \partial D, \tag{2.6.2}$$

where $D \subset \mathbb{R}^N$ is a bounded domain, $s \in \mathbb{R}$ is an initial time, and $\mathcal{B}$ is a boundary operator of either the Dirichlet or Neumann or Robin type, that is,

$$\mathcal{B}(t)u = \begin{cases} u & \text{(Dirichlet)} \\ \displaystyle\sum_{i=1}^{N} \partial_{x_i}u\bar{\nu}_i(t,x) & \text{(Neumann)} \\ \displaystyle\sum_{i=1}^{N} \partial_{x_i}u\bar{\nu}_i(t,x) + d_0(t,x)u, & \text{(Robin)} \end{cases}$$

where (in the Neumann or Robin cases) $(\bar{\nu}_1, \ldots, \bar{\nu}_N)$ is a (in general time dependent) vector field on ∂D pointing out of D.

First of all, if both the domain D and the coefficients are sufficiently smooth and, in the Neumann or Robin cases, $\bar{\nu}_i(t,x) = \sum_{j=1}^{N} a_{ji}(t,x)\nu_j(x)$, $1 \le i \le N$, (that is, the derivative is conormal), then (2.6.1)+(2.6.2) can be written in the divergence form and then the results in Section 2.5 apply.

In general, a proper notion of solutions of (2.6.1)+(2.6.2) is strong solutions. Roughly speaking, a function u is a *strong solution* of (2.6.1)+(2.6.2) on $(s,t) \times D$ if $u \in W_p^{1,2}((s,t) \times D) \cap C([s,t] \times \bar{D})$ is such that (2.6.1) holds (Lebesgue-) almost everywhere and (2.6.2) holds everywhere (see [59], [61], [73]).

For the Dirichlet case, under the additional assumption that the coefficients a_{ij} are continuous on $\mathbb{R} \times D$, it is proved in [61, Proposition 5.4] that for any $u_0 \in C(\bar{D})$ satisfying the boundary conditions, (2.6.1)+(2.6.2) has a unique strong solution with initial condition $u(s) = u_0$ (see also [73, Theorem 7.17] about the existence and uniqueness of solutions). We refer the reader to [61], [73], and references therein for various properties of strong solutions of (2.6.1)+(2.6.2), for example, maximum principle, a priori estimates, weak Harnack inequality, etc.

For the Neumann or Robin boundary condition case, we do not have all the results as in the Dirichlet case (see [59]). But many important properties of strong solutions in Dirichlet case still hold (see [59], [73], etc.).

Chapter 3

Spectral Theory in the General Setting

In this chapter, we introduce the definitions of principal spectrum and principal Lyapunov exponents and exponential separation for a family of general parabolic equations and present their basic properties. We also present a multiplicative ergodic theorem for a family of general parabolic equations. This chapter is organized as follows. In Section 3.1 we introduce the definitions of principal spectrum and Lyapunov exponents of (2.0.1)+(2.0.2) and study their basic properties. We introduce the definition of exponential separation and investigate relevant basic properties in Section 3.2. The existence of exponential separation is explored in Section 3.3. In Section 3.4 we present a multiplicative ergodic theorem. Special properties for a family of general smooth parabolic equations are discussed in Section 3.5. Some remarks on parabolic equations in nondivergence form are given in Section 3.6. This chapter ends up with an appendix on parabolic equations on one-dimensional space domain.

3.1 Principal Spectrum and Principal Lyapunov Exponents: Definitions and Properties

In the present section we introduce the definitions of principal spectrum and Lyapunov exponents of (2.0.1)+(2.0.2), or of Π (see (2.2.6)), and study their basic properties.

The standing assumption throughout the present section is that (2.0.1)+(2.0.2) satisfies (A2-1)–(A2-3). Π shall denote the topological linear skew-product semiflow generated on $L_2(D) \times Y$ by (2.0.1)+(2.0.2).

Recall that, for $a \in Y$ and $t > 0$, $\|U_a(t, 0)\|$ denotes the $\mathcal{L}(L_2(D))$-norm of the linear operator $U_a(t, 0)$. We introduce now a norm-like concept. For $a \in Y$ and $t > 0$ let

$$\|U_a(t, 0)\|^+ := \sup\{ \|U_a(t, 0)u_0\| : u_0 \in L_2(D)^+, \ \|u_0\| = 1 \}.$$

LEMMA 3.1.1
For any $a \in Y$ and any $t > 0$ one has

$$\|U_a(t,0)\|^+ = \|U_a(t,0)\|.$$

PROOF   The inequality $\|U_a(t,0)\|^+ \leq \|U_a(t,0)\|$ is obvious. To prove
the other inequality, notice that any $u_0 \in L_2(D)$ can be represented as
$u_0 = u_0^+ - u_0^-$, where $u_0^+(x) = \max\{u_0(x), 0\}$ for a.e. $x \in D$ and $u_0^-(x) = \max\{-u_0(x), 0\}$ for a.e. $x \in D$. Notice that for $|u_0| := u_0^+ + u_0^-$ one has
$\||u_0|\| = \|u_0\|$. The inequalities

$$
\begin{aligned}
|U_a(t,0)u_0| = |U_a(t,0)u_0^+ - U_a(t,0)u_0^-| &\leq |U_a(t,0)u_0^+| + |U_a(t,0)u_0^-| \\
&= U_a(t,0)u_0^+ + U_a(t,0)u_0^- = U_a(t,0)|u_0|
\end{aligned}
$$

give us, after imposing the norms, the desired inequality.   □

From now on until the end of Chapter 3 let Y_0 be a nonempty compact
connected invariant subset of Y.

DEFINITION 3.1.1 (Principal resolvent)   *A real number λ belongs
to the* principal resolvent *of Π over Y_0, denoted by $\rho(Y_0)$, if either of the
following conditions holds:*

- *There are $\epsilon > 0$ and $M \geq 1$ such that*

$$\|U_a(t,0)\| \leq Me^{(\lambda-\epsilon)t} \quad \text{for } t > 0 \text{ and } a \in Y_0$$

  (such λ are said to belong to the upper principal resolvent, *denoted by*
  $\rho_+(Y_0)$),

- *There are $\epsilon > 0$ and $M \in (0,1]$ such that*

$$\|U_a(t,0)\| \geq Me^{(\lambda+\epsilon)t} \quad \text{for } t > 0 \text{ and } a \in Y_0$$

  (such λ are said to belong to the lower principal resolvent, *denoted by*
  $\rho_-(Y_0)$).

In view of Lemma 3.1.1, in the above inequalities the $\|\cdot\|$-norms can be
replaced with $\|\cdot\|^+$-"norms," with the same M and ϵ.

DEFINITION 3.1.2 (Principal spectrum)   *The* principal spectrum
*of the topological linear skew-product semiflow Π over Y_0, denoted by $\Sigma(Y_0)$,
equals the complement in $\mathbb{R}$ of the principal resolvent of Π over Y_0.*

To study the basic properties of $\Sigma(Y_0)$, we first prove some auxiliary results.

LEMMA 3.1.2

(1) *For any $t_0 > 0$ there is $K_1 = K_1(t_0) \geq 1$ such that $\|U_a(t,0)\| \leq K_1$ for all $a \in Y_0$ and all $t \in [0,t_0]$.*

(2) *For any $t_0 > 0$ there is $K_2 = K_2(t_0) > 0$ such that $\|U_a(t,0)\| \geq K_2$ for all $a \in Y_0$ and all $t \in [0,t_0]$.*

PROOF   Part (1) is a consequence of the L_2–L_2 estimates (Proposition 2.2.2).

To prove (2), notice that by Proposition 2.2.9(2), $\|U_a(t,0)1\| > 0$ for all $a \in Y_0$ and $t > 0$, where 1 is identified with the function constantly equal to one. Since $Y_0 \times [t_0/2, t_0]$ is compact, Proposition 2.2.12 implies that the set $\{\|U_a(t,0)1\| : a \in Y_0,\ t \in [t_0/2, t_0]\}$ is bounded away from zero. Hence there is $M_1 > 0$ such that

$$\|U_a(t,0)\| \geq M_1 \quad \text{for} \quad t \in [t_0/2, t_0], \quad a \in Y_0.$$

Now, for any $0 < t < t_0/2$, $a \in Y_0$, and $u_0 \in L_2(D)$ with $\|u_0\| = 1$,

$$\begin{aligned}
&\|U_{a \cdot (-t_0/2)}(t + t_0/2, 0)u_0\| \\
&\leq \|U_{a \cdot (-t_0/2)}(t + t_0/2, t_0/2)\| \cdot \|U_{a \cdot (-t_0/2)}(t_0/2, 0)u_0\| \\
&\leq K_1 \|U_{a \cdot (-t_0/2)}(t + t_0/2, t_0/2)\| \qquad \text{(by Part (1))} \\
&= K_1 \|U_a(t,0)\|.
\end{aligned}$$

This implies that

$$M_1 \leq \|U_{a \cdot (-t_0/2)}(t + t_0/2, 0)\| \leq K_1 \|U_a(t,0)\|$$

for any $a \in Y_0$ and $0 < t < t_0/2$. Part (2) then follows with $K_2 = \min\{M_1, M_1/K_1\} = M_1/K_1$. $\square$

LEMMA 3.1.3
A real number λ belongs to the lower principal resolvent if and only if for any $\delta_0 > 0$ there are $\epsilon > 0$ and $\tilde{M} > 0$ such that

$$\|U_a(t,0)\| \geq \tilde{M}e^{(\lambda+\epsilon)t} \quad \text{for } t \geq \delta_0 \text{ and } a \in Y_0.$$

PROOF   The "only if" part follows from Definition 3.1.1 in a straightforward way. The "if" part follows from Lemma 3.1.2(2). $\square$

LEMMA 3.1.4
There exist $\delta_1 > 0$, $M_1 > 0$, and a real $\underline{\lambda}$ such that $\|U_a(t,0)\| \geq M_1 e^{\underline{\lambda} t}$ for all $a \in Y_0$ and all $t \geq \delta_1$.

PROOF   Pick $\delta' > 0$ sufficiently small that $D^{\delta'} := \{\, x \in D : \mathrm{dist}(x, \partial D) > \delta' \,\}$ is a nonempty bounded domain. Further, put $t_0 := (\delta')^2$ and $\delta := \delta'/\sqrt{2}$. It follows from the interior Harnack inequality (Proposition 2.2.8(2)) that there is $C_\delta > 0$ such that

$$(U_a(\delta^2, 0)1)(y) \le C_\delta \cdot (U_a(t, 0)1)(x)$$

for any $a \in Y_0$, any $t \in [2\delta^2, 3\delta^2]$, and any $x, y \in D^{\delta'}$. Without loss of generality, we assume that $C_\delta > 1$. By Proposition 2.2.9(2),

$$\sup\{\, (U_a(\delta^2, 0)1)(x) : x \in D^{\delta'} \,\} =: m(a) > 0.$$

Then by the arguments of Proposition 2.2.5,

$$\inf\{\, m(a) : a \in Y_0 \,\} := m > 0.$$

Consequently

$$\inf\{\, (U_a(t, 0)1)(x) : x \in D^{\delta'} \,\} \ge \frac{m(a)}{C_\delta} \ge \frac{m}{C_\delta}$$

for any $a \in Y_0$ and any $t \in [2\delta^2, 3\delta^2]$. Repeating the application of the interior Harnack inequality (Proposition 2.2.8(2)), we obtain that

$$C_\delta^{k-1}(U_a(t, 0)1)(x) \ge (U_a(\delta^2, 0)1)(y)$$

for any $a \in Y$, $t \in [k\delta^2, (k+1)\delta^2]$, $k = 2, 3, \ldots$, and $x, y \in D^{\delta'}$. It then follows that

$$\inf\{\, (U_a(t, 0)1)(x) : x \in D^{\delta'} \,\} \ge \frac{m}{(C_\delta)^{k-1}}$$

for any $a \in Y_0$ and any $t \in [k\delta^2, (k+1)\delta^2]$, $k = 2, 3, \ldots$. Thus the statement holds with $\delta_1 = 2\delta^2$, $M_1 = mC_\delta|D^{\delta'}|^{1/2}$, and $\underline{\lambda} = -\ln(C_\delta)/\delta^2$.   ▯

We start now to investigate the properties of the principal spectrum $\Sigma(Y_0)$. First of all, we have

THEOREM 3.1.1
The principal spectrum of Π over Y_0 is a compact nonempty interval $[\lambda_{\min}, \lambda_{\max}]$.

PROOF   We prove first that the upper principal resolvent $\rho_+(Y_0)$ is nonempty. Indeed, by the L_2–L_2 estimates (Proposition 2.2.2), there are $M > 0$ and $\gamma > 0$ such that $\|U_a(t, 0)\| \le Me^{\gamma t}$ for all $a \in Y_0$ and $t > 0$, hence $\gamma + 1 \in \rho_+(Y_0)$. Further, $\rho_+(Y_0)$ is a right-unbounded open interval $(\lambda_{\max}, \infty)$.

The lower principal resolvent $\rho_-(Y_0)$ is nonempty, too, since it contains, by Lemma 3.1.4, the real number $\lambda - 1$.

Consequently, as $\rho_-(Y_0) \cup \rho_+(Y_0) = \rho(Y_0)$ and $\rho_-(Y_0) \cap \rho_+(Y_0) = \emptyset$, one has $\Sigma(Y_0) = \mathbb{R} \setminus \rho(Y_0) = [\lambda_{\min}, \lambda_{\max}]$. $\quad\square$

The next theorem gives a characterization of the principal spectrum of Π.

THEOREM 3.1.2

(1) *For any sequence* $(a^{(n)})_{n=1}^\infty \subset Y_0$ *and any real sequences* $(t_n)_{n=1}^\infty$, $(s_n)_{n=1}^\infty$ *such that* $t_n - s_n \to \infty$ *as* $n \to \infty$ *there holds*

$$\lambda_{\min} \le \liminf_{n\to\infty} \frac{\ln \|U_{a^{(n)}}(t_n, s_n)\|}{t_n - s_n} \le \limsup_{n\to\infty} \frac{\ln \|U_{a^{(n)}}(t_n, s_n)\|}{t_n - s_n} \le \lambda_{\max}.$$

(2A) *There exist a sequence* $(a^{(n,1)})_{n=1}^\infty \subset Y_0$ *and a sequence* $(t_{n,1})_{n=1}^\infty \subset (0, \infty)$ *such that* $t_{n,1} \to \infty$ *as* $n \to \infty$, *and*

$$\lim_{n\to\infty} \frac{\ln \|U_{a^{(n,1)}}(t_{n,1}, 0)\|}{t_{n,1}} = \lambda_{\min}.$$

(2B) *There exist a sequence* $(a^{(n,2)})_{n=1}^\infty \subset Y_0$ *and a sequence* $(t_{n,2})_{n=1}^\infty \subset (0, \infty)$ *such that* $t_{n,2} \to \infty$ *as* $n \to \infty$, *and*

$$\lim_{n\to\infty} \frac{\ln \|U_{a^{(n,2)}}(t_{n,2}, 0)\|}{t_{n,2}} = \lambda_{\max}.$$

PROOF Part (1) is a direct consequence of the definition of the principal spectrum.

To prove (2A), notice that, since $\lambda_{\min} \notin \rho_-(Y_0)$, it follows from Lemma 3.1.3 (with $\delta_0 = 1$) that for each $n \in \mathbb{N}$ there are $a^{(n,1)} \in Y_0$ and $t_{n,1} > 0$ such that

$$\|U_{a^{(n,1)}}(t_{n,1}, 0)\| < \tfrac{1}{n} \exp \left((\lambda_{\min} + \tfrac{1}{n})t_{n,1}\right).$$

We claim that $\lim_{n\to\infty} t_{n,1} = \infty$. If not, there is a bounded subsequence $(t_{n_k,1})_{k=1}^\infty$, $n_k \to \infty$ as $k \to \infty$. It follows that $\|U_{a^{(n_k,1)}}(t_{n_k,1}, 0)\| \to 0$ as $k \to \infty$, which contradicts Lemma 3.1.2(2). Thus we have

$$\limsup_{n\to\infty} \frac{\ln \|U_{a^{(n,1)}}(t_{n,1}, 0)\|}{t_{n,1}} \le \lambda_{\min},$$

which together with Part (1) gives the desired result.

To prove (2B), notice that, since $\lambda_{\max} \notin \rho_+(Y_0)$, it follows from Definition 3.1.1 that for each $n \in \mathbb{N}$ there are $a^{(n,2)} \in Y_0$ and $t_{n,2} > 0$ such that

$$\|U_{a^{(n,2)}}(t_{n,2}, 0)\| > n \exp \left((\lambda_{\max} - \tfrac{1}{n})t_{n,2}\right).$$

We claim that $\lim_{n\to\infty} t_{n,2} = \infty$. If not, there is a bounded subsequence $(t_{n_k,2})_{k=1}^{\infty}$, $n_k \to \infty$ as $k \to \infty$. It follows that $\|U_{a^{(n_k,2)}}(t_{n_k,2},0)\| \to \infty$ as $k \to \infty$, which contradicts Lemma 3.1.2(1). Thus we have

$$\liminf_{n\to\infty} \frac{\ln \|U_{a^{(n,2)}}(t_{n,2},0)\|}{t_{n,2}} \geq \lambda_{\max},$$

which together with Part (1) gives the desired result.   □

Recall that for any $a \in Y$ we write $a = (a_{ij}, a_i, b_i, c_0, d_0)$.

THEOREM 3.1.3

Assume that for each $a \in Y_0$ there holds: $a_i(t,x) = b_i(t,x) = 0$ for a.e. $(t,x) \in \mathbb{R} \times D$, and $c_0(t,x) \leq 0$ for a.e. $(t,x) \in \mathbb{R} \times D$. Then $\Sigma(Y_0) \subset (-\infty, 0]$.

PROOF   Fix $a \in Y_0$ and $u_0 \in L_2(D)$ with $\|u_0\| = 1$, and put $u(t,x) := (U_a(t,0)u_0)(x)$. It follows from Proposition 2.1.4 that

$$\|u(t,\cdot)\|^2 - \|u(0,\cdot)\|^2 = -2\int_0^t B_a(\tau, u(\tau,\cdot), u(\tau,\cdot)) \, d\tau$$

$$\leq -2\int_0^t \int_D \Big(\sum_{i,j=1}^N a_{ij}(\tau,x)\partial_{x_j}u(\tau,x)\partial_{x_j}u(\tau,x) \Big) \, dx \, d\tau \leq 0$$

for any $t > 0$. Consequently, $\|U_a(t,0)u_0\| \leq \|u_0\| = 1$ for all $t > 0$. Therefore $(0,\infty) \subset \rho_+(Y_0)$.   □

In the case of the Dirichlet boundary conditions more can be said.

THEOREM 3.1.4

In the case of the Dirichlet boundary conditions, assume that for each $a \in Y_0$ there holds: $a_i(t,x) = b_i(t,x) = 0$ for a.e. $(t,x) \in \mathbb{R} \times D$, and $c_0(t,x) \leq 0$ for a.e. $(t,x) \in \mathbb{R} \times D$. Then $\lambda_{\max}(Y_0) < 0$.

PROOF   It follows by the Poincaré inequality (see [39, Theorem 3 in Section 5.6]) that there is $\alpha_1 > 0$ such that $\|u\| \leq \alpha_1 \|\nabla u\|$ for any $u \in \mathring{W}_2^1(D)$.

Starting as in the proof of Theorem 3.1.3 we estimate

$$\|u(t,\cdot)\|^2 - \|u(0,\cdot)\|^2 = -2 \int_0^t B_a(\tau, u(\tau,\cdot), u(\tau,\cdot)) \, d\tau$$

$$\leq -2 \int_0^t \int_D \left(\sum_{i,j=1}^N a_{ij}(\tau, x) \partial_{x_j} u(\tau, x) \partial_{x_j} u(\tau, x) \right) dx \, d\tau$$

$$\underset{\leq}{\text{by (A2-1)}} -2\alpha_0 \int_0^t \|\nabla u(\tau,\cdot)\|^2 \, d\tau \leq \frac{-2\alpha_0}{(\alpha_1)^2} \int_0^t \|u(\tau,\cdot)\|^2 \, d\tau.$$

An application of the regular Gronwall inequality gives that

$$\|U_a(t,0)\| \leq e^{-\lambda_0 t}$$

for all $t \geq 0$, where $\lambda_0 := \alpha_0/\alpha_1^2 > 0$. Consequently, $[-\lambda_0, \infty) \subset \rho_+(Y_0)$ and $\lambda_{\max} \leq -\lambda_0$. ☐

Assume that μ is an invariant ergodic Borel probability measure for the topological flow σ on Y_0 (see Lemma 1.2.7 for the existence of invariant ergodic measures for (Y_0, σ)). We have

THEOREM 3.1.5
There exist a Borel set $Y_1 \subset Y_0$ with $\mu(Y_1) = 1$ and a real number $\lambda(\mu)$ such that

$$\lim_{t \to \infty} \frac{\ln \|U_a(t,0)\|}{t} = \lambda(\mu)$$

for all $a \in Y_1$.

PROOF We will prove the theorem by applying Kingman's subadditive ergodic theorem.
We define a sequence of functions $f_n : Y_0 \to \mathbb{R}$, $n = 1, 2, 3, \ldots$, as

$$f_n(a) := \ln \|U_a(n, 0)\|, \qquad n \in \mathbb{N}, \ a \in Y_0.$$

For each $n \in \mathbb{N}$ the function f_n is well-defined and bounded (by Lemma 3.1.2). Further, there holds

$$f_{n+m}(a) \leq f_n(a) + f_m(\sigma_n a), \qquad n, m \in \mathbb{N}, \ a \in Y_0.$$

For each $n \in \mathbb{N}$ the function $[Y_0 \ni a \mapsto \|U_a(n, 0)\| \in (0, \infty)]$ is easily seen to be lower semicontinuous. Therefore the functions f_n are lower semicontinuous, hence $(\mathfrak{B}(Y_0), \mathfrak{B}(\mathbb{R}))$-measurable.
Observe that by Lemma 3.1.4 there are $M > 0$ and $\underline{\lambda} \in \mathbb{R}$ such that $f_n(a) > \ln M + \underline{\lambda} n$ for any $n \in \mathbb{N}$ and any $a \in Y_0$. Consequently,

$$\inf \left\{ \frac{1}{n} \int_{Y_0} f_n \, d\mu : n \in \mathbb{N} \right\} > -\infty.$$

We are now in a position to apply the subadditive ergodic theorem (see, e.g., [67, Theorem 5.3 in Chapter 1]) to conclude that there are a Borel set $\tilde{Y}_0 \subset Y_0$ with $\mu(\tilde{Y}_0) = 1$ and a $(\mathcal{B}(Y_0), \mathcal{B}(\mathbb{R}))$-measurable function $\tilde{\lambda} \colon \tilde{Y}_0 \to \mathbb{R}$ such that

(i) $\lim_{n \to \infty} \frac{f_n(a)}{n} = \tilde{\lambda}(a)$ for any $a \in \tilde{Y}_0$,

(ii) $\tilde{\lambda} \in L_1((Y_0, \mathcal{B}(Y_0), \mu))$, and $\int_{Y_0} \tilde{\lambda} \, d\mu = \inf \left\{ \frac{1}{n} \int_{Y_0} f_n \, d\mu : n \in \mathbb{N} \right\}$,

(iii) $\tilde{\lambda}(a) = \tilde{\lambda}(\sigma_n a)$ for any $a \in \tilde{Y}_0$ and any $n \in \mathbb{N}$.

Put $\tilde{Y}_1 := \bigcap_{k \in \mathbb{Z}} \sigma_k(\tilde{Y}_0)$. $\tilde{Y}_1 \subset \tilde{Y}_0$ is a Borel set with $\mu(\tilde{Y}_1) = 1$. We claim that for any $s \in \mathbb{R}$ and any $a \in \tilde{Y}_1$ there holds

$$\lim_{n \to \infty} \frac{\ln \|U_{a \cdot s}(n, 0)\|}{n} = \tilde{\lambda}(a).$$

By (i), (ii) and the construction of $\tilde{Y}_1$, the above equality holds for any $s \in \mathbb{Z}$ and any $a \in \tilde{Y}_1$. Let $s \in \mathbb{R}$ and let for the moment a be any member of Y_0. We estimate

$$\|U_{a \cdot s}(n, 0)\| = \|U_a(n + s, s)\|$$
$$\leq \|U_a(n + s, n + \lfloor s \rfloor)\| \cdot \|U_a(n + \lfloor s \rfloor, \lfloor s \rfloor + 1)\| \cdot \|U_a(\lfloor s \rfloor + 1, s)\|$$

for $n = 2, 3, \dots$. Since the first and the third term on the right-hand side are bounded above (by Lemma 3.1.2(1)), we have

$$\limsup_{n \to \infty} \frac{\ln \|U_{a \cdot s}(n, 0)\|}{n} \leq \limsup_{n \to \infty} \frac{\ln \|U_{a \cdot (\lfloor s \rfloor + 1)}(n, 0)\|}{n}.$$

Further, we estimate

$$\|U_a(n + \lfloor s \rfloor + 1, \lfloor s \rfloor)\| \leq \|U_a(n + \lfloor s \rfloor + 1, n + s)\| \cdot \|U_a(n + s, s)\| \cdot \|U_a(s, \lfloor s \rfloor)\|$$

for $n = 1, 2, \dots$. Since the first and the third term on the right-hand side are bounded above (by Lemma 3.1.2(1)), we have

$$\liminf_{n \to \infty} \frac{\ln \|U_{a \cdot s}(n, 0)\|}{n} \geq \liminf_{n \to \infty} \frac{\ln \|U_{a \cdot \lfloor s \rfloor}(n, 0)\|}{n}.$$

Let now $a \in \tilde{Y}_1$. There holds

$$\tilde{\lambda}(a) = \lim_{n \to \infty} \frac{\ln \|U_{a \cdot \lfloor s \rfloor}(n, 0)\|}{n} \leq \liminf_{n \to \infty} \frac{\ln \|U_{a \cdot s}(n, 0)\|}{n}$$
$$\leq \limsup_{n \to \infty} \frac{\ln \|U_{a \cdot s}(n, 0)\|}{n} \leq \lim_{n \to \infty} \frac{\ln \|U_{a \cdot (\lfloor s \rfloor + 1)}(n, 0)\|}{n} = \tilde{\lambda}(a),$$

which proves the claim.

Next we show that
$$\lim_{t \to \infty} \frac{\ln \|U_a(t,0)\|}{t} = \tilde{\lambda}(a)$$

for $a \in \tilde{Y}_1$. This follows from the estimates
$$\|U_a(t,0)\| \leq \|U_a(t, \lfloor t \rfloor)\| \cdot \|U_a(\lfloor t \rfloor, 0)\|$$

and
$$\|U_a(\lfloor t \rfloor + 1, 0)\| \leq \|U_a(\lfloor t \rfloor + 1, t)\| \cdot \|U_a(t,0)\|$$

and from Lemma 3.1.2(1).

Finally we prove that there is a Borel set $Y_1 \subset Y_0$ with $\mu(Y_1) = 1$ such that $\tilde{\lambda}(a) = \text{const}$ on Y_1. Let $\lambda^{\pm}(a)$ be defined as follows:

$$\lambda^+(a) := \limsup_{n \to \infty} \frac{\ln \|U_a(n,0)\|}{n} \quad \text{and} \quad \lambda^-(a) := \liminf_{n \to \infty} \frac{\ln \|U_a(n,0)\|}{n}.$$

By the $(\mathcal{B}(Y_0), \mathcal{B}(\mathbb{R}))$-measurability of $[a \mapsto f_n(a) = \|U_a(n,0)\|]$, both λ^+ and λ^- are $(\mathcal{B}(Y_0), \mathcal{B}(\mathbb{R}))$-measurable. From the subadditivity and Lemma 3.1.2(1) it follows that λ^+, consequently λ^-, are bounded above. Lemma 3.1.4 implies that λ^-, consequently λ^+, are bounded below. Therefore λ^+ and λ^- belong to $L_1((Y_0, \mathcal{B}(Y_0), \mu))$. It follows from the Birkhoff Ergodic Theorem (Lemma 1.2.6) that there is a Borel set $Y_2 \subset Y_0$ with $\mu(Y_2) = 1$ such that for each $a \in Y_2$,

$$\lim_{t \to \infty} \frac{1}{t} \int_0^t \lambda^+(a \cdot s) \, ds = \int_{Y_0} \lambda^+ \, d\mu$$

and

$$\lim_{t \to \infty} \frac{1}{t} \int_0^t \lambda^-(a \cdot s) \, ds = \int_{Y_0} \lambda^- \, d\mu.$$

Observe that for any $a \in \tilde{Y}_1$ and $s \in \mathbb{R}$,

$$\lambda^+(a \cdot s) = \lambda^-(a \cdot s) = \tilde{\lambda}(a).$$

This implies that

$$\tilde{\lambda}(a) = \int_{Y_0} \lambda^+ \, d\mu = \int_{Y_0} \lambda^- \, d\mu$$

for any $a \in Y_1 := \tilde{Y}_1 \cap Y_2$. The theorem is thus proved. □

DEFINITION 3.1.3 (Principal Lyapunov exponent) $\lambda(\mu)$ *as defined above is called the* principal Lyapunov exponent *of Π for the ergodic invariant measure μ.*

THEOREM 3.1.6
For any ergodic μ supported on Y_0 the principal Lyapunov exponent $\lambda(\mu)$ belongs to the principal spectrum $[\lambda_{\min}, \lambda_{\max}]$ of Π on Y_0.

PROOF Suppose to the contrary that $\lambda(\mu) < \lambda_{\min}$ for some ergodic μ supported on Y_0. It follows by definition that there is $\epsilon > 0$ such that $\liminf_{t \to \infty} (1/t) \ln \|U_a(t, 0)\| \geq \lambda(\mu) + \epsilon$ for all $a \in Y_0$, whereas Theorem 3.1.5 establishes the existence of $\tilde{a} \in Y_0$ with $\lim_{t \to \infty} (1/t) \ln \|U_{\tilde{a}}(t, 0)\| = \lambda(\mu)$. The case $\lambda(\mu) > \lambda_{\max}$ is excluded in a similar way. □

3.2 Exponential Separation: Definitions and Basic Properties

In the present section we introduce the definitions of exponential separation for (2.0.1)+(2.0.2), or for Π, as well as show basic properties of principal spectrum and principal Lyapunov exponents under the assumption that exponential separation holds.

The standing assumption throughout the present section is that (2.0.1)+(2.0.2) satisfies (A2-1)–(A2-3). Π shall denote the topological linear skew-product semiflow generated on $L_2(D) \times Y$ by (2.0.1)+(2.0.2) (see (2.2.6)).

Let Y_0 be a closed connected invariant subset of Y. By a *one-dimensional (trivial) subbundle* $X^{(1)}$ of $L_2(D) \times Y_0$ we understand a set $\{ (rw(a), a) : r \in \mathbb{R}, a \in Y_0 \}$, where $w \colon Y_0 \to L_2(D)$ is a continuous mapping with the property that $\|w(a)\| = 1$ for all $a \in Y_0$. For $a \in Y_0$ we call the one-dimensional vector subspace $\operatorname{span}\{w(a)\} =: X^{(1)}(a)$ the *fiber* of $X^{(1)}$ over a. A *one-codimensional (trivial) subbundle* $X^{(2)}$ is usually defined as a family of one-codimensional subspaces continuously depending on $a \in Y_0$. For our purposes it suffices to define it as a set $\{ (v, a) \in L_2(D) : \langle v, w^*(a) \rangle = 0, a \in Y_0 \}$, where $w^* \colon Y_0 \to L_2(D)^*$ is a continuous mapping satisfying $\|w^*(a)\| = 1$ for all $a \in Y_0$. For $a \in Y_0$ we call the one-codimensional vector subspace $\{ v \in L_2(D) : \langle v, w^*(a) \rangle = 0 \} =: X^{(2)}(a)$ the *fiber* of $X^{(2)}$ over a. As we will consider only trivial subbundles, we drop the adjective "trivial" from now on.

A one-dimensional subbundle $X^{(1)}$ and a one-codimensional subbundle $X^{(2)}$ are *complementary* if $X^{(1)}(a) \oplus X^{(2)}(a) = L_2(D)$ for each $a \in Y_0$, where $\oplus$ denotes direct sum in the Banach space sense ($X^{(1)}(a)$ and $X^{(2)}(a)$ need not be orthogonal). Notice that $X^{(1)}$ and $X^{(2)}$ are complementary if and only if $\langle w(a), w^*(a) \rangle \neq 0$ for all $a \in Y_0$. If $X^{(1)}$ and $X^{(2)}$ are complementary, we write $X^{(1)} \oplus X^{(2)} = L_2(D) \times Y_0$.

A one-dimensional subbundle $X^{(1)}$ is said to be *invariant* (under Π) if for each $a \in Y_0$ and each $t > 0$ there is a real $r_a(t)$ such that $U_a(t, 0)w(a) = r_a(t)w(a \cdot t)$. This is equivalent to saying that $U_a(t, 0)X^{(1)}(a) \subset X^{(1)}(a \cdot t)$ for each $a \in Y_0$ and each $t > 0$.

A one-codimensional subbundle $X^{(2)}$ is said to be *invariant* (under Π) if for each $a \in Y_0$ and each $t < 0$ there is a real $r_a^*(t)$ such that $U_a^*(t, 0)w^*(a) = r_a^*(t)w^*(a \cdot t)$. This is equivalent to saying that $U_a(t, 0)X^{(2)}(a) \subset X^{(2)}(a \cdot t)$

for each $a \in Y_0$ and each $t > 0$.

For a given one-dimensional subbundle $X^{(1)}$ denote by $(X^{(1)})^*$ the one-codimensional subbundle given by $(X^{(1)})^*(a) := \{ v \in L_2(D) : \langle w(a), v \rangle = 0 \} = X^{(1)}(a)^\perp$, $a \in Y_0$. It is easy to see that $X^{(1)}$ is invariant under Π if and only if $(X^{(1)})^*$ is invariant under Π^*.

Similarly, for a one-codimensional subbundle $X^{(2)}$ denote by $(X^{(2)})^*$ the one-dimensional subbundle given by $(X^{(2)})^*(a) := \mathrm{span}\{w^*(a)\}$. It is easy to see that $X^{(2)}$ is invariant under Π if and only if $(X^{(2)})^*$ is invariant under Π^*.

For more on subbundles, see [97].

DEFINITION 3.2.1 (Exponential separation)   *Let Y_0 be a compact connected invariant subset of Y. We say that Π admits an exponential separation with separating exponent $\gamma_0 > 0$ over Y_0 if there are an invariant one-dimensional subbundle X_1 of $L_2(D) \times Y_0$ with fibers $X_1(a) = \mathrm{span}\{w(a)\}$, and an invariant complementary one-codimensional subbundle X_2 of $L_2(D) \times Y_0$ with fibers $X_2(a) = \{ v \in L_2(D) : \langle v, w^*(a) \rangle = 0 \}$ having the following properties:*

(i) *$w(a) \in L_2(D)^+$ for all $a \in Y_0$,*

(ii) *$X_2(a) \cap L_2(D)^+ = \{0\}$ for all $a \in Y_0$,*

(iii) *There is $M \geq 1$ such that for any $a \in Y_0$ and any $v \in X_2(a)$ with $\|v\| = 1$,*

$$\|U_a(t, 0)v\| \leq Me^{-\gamma_0 t}\|U_a(t, 0)w(a)\| \quad (t > 0).$$

As a consequence of the invariance of X_1 and of Proposition 2.2.9(2) we have the following stronger property.

$$U_a(t, 0)X^{(1)}(a) = X^{(1)}(a \cdot t), \quad a \in Y_0, \ t > 0. \tag{3.2.1}$$

Moreover, for each $a \in Y_0$ and each $t > 0$ there is $r_a(t) > 0$ such that $U_a(t; 0)w(a) = r_a(t)w(a \cdot t)$. Since $r_a(t) = \|U_a(t, 0)w(a)\|$, by Proposition 2.2.13, the function $[Y_0 \times (0, \infty) \ni (a, t) \mapsto r_a(t) \in (0, \infty)]$ is continuous.

We claim that $w^*(a)$ can be taken to belong to $L_2(D)^+$ for each $a \in Y_0$. If $w^*(a) \in -L_2(D)^+$ for each $a \in Y_0$ then we replace $w^*(a)$ with $-w^*(a)$. Suppose to the contrary that there is $a \in Y_0$ such that $w^*(a) \notin L_2(D)^+ \cup -L_2(D)^+$. This means that the Lebesgue measure of the set $D_1 := \{ x \in D : w^*(a)(x) > 0 \}$ is positive, as well as the Lebesgue measure of the set $D_2 := \{ x \in D : w^*(a)(x) < 0 \}$ is positive. Let v be a simple function taking value $(\int_{D_1} w^*(a)(x) \, dx)^{-1}$ on D_1, taking value $-(\int_{D_2} w^*(a)(x) \, dx)^{-1}$ on D_2, and equal to zero elsewhere. Clearly $v \in L_2(D)^+ \setminus \{0\}$ and $\langle v, w^*(a) \rangle = 0$, that is, $v \in X_2(a)$, which contradicts (ii).

As the dual Π^* of the topological linear skew-product semiflow Π is generated by the adjoint equation (2.3.1)+(2.3.2), we have that for each $a \in Y_0$

and each $t < 0$ there is $r_a^*(t) > 0$ such that $U_a^*(t,0)w^*(a) = r_a^*(t)w^*(a \cdot t)$. Further, the function $[\, Y_0 \times (0,\infty) \ni (a,t) \mapsto r_a^*(t) \in (0,\infty)\,]$ is continuous.

LEMMA 3.2.1

Let Π admit an exponential separation over a compact connected invariant subset $Y_0 \subset Y$. For any sequence $(a^{(n)}) \subset Y_0$ and any positive real sequence (t_n), if $\lim_{n \to \infty} a^{(n)} = a$ and $\lim_{n \to \infty} t_n = t$, where $t \geq 0$, then $U_{a^{(n)}}(t_n, 0)w(a^{(n)})$ converge in $L_2(D)$ to $U_a(t,0)w(a)$.

PROOF Observe that

$$U_{a^{(n)}}(t_n, 0)w(a^{(n)}) = \frac{U_{a^{(n)} \cdot (-1)}(t_n + 1, 0)w(a^{(n)} \cdot (-1))}{r_{a^{(n)} \cdot (-1)}(1)},$$

which converges in $L_2(D)$, by Proposition 2.2.12, to

$$\frac{U_{a \cdot (-1)}(t + 1, 0)w(a \cdot (-1))}{r_{a \cdot (-1)}(1)} = U_a(t,0)w(a).$$

$\square$

Denote by $\Pi|_{X_1}$ the restriction of Π to the subbundle X_1. We extend $\Pi|_{X_1}$ to negative times in the following way:

$$(\Pi|_{X_1})_t(w(a), a) := \begin{cases} ((U_{a \cdot t}(-t, 0)|_{X_1(a \cdot t)})^{-1}w(a), a \cdot t), & t < 0 \\ (U_a(t,0)w(a), a \cdot t), & t \geq 0, \end{cases}$$

or, in view of the invariance of X_1:

$$(\Pi|_{X_1})_t(w(a), a) = \begin{cases} \left(\dfrac{w(a \cdot t)}{\|U_{a \cdot t}(-t, 0)w(a \cdot t)\|}, a \cdot t \right), & t < 0 \\ (\|U_a(t,0)w(a)\|w(a \cdot t), a \cdot t), & t \geq 0. \end{cases}$$

One has

$$(\Pi|_{X_1})_0 = \mathrm{Id}_{X_1}$$

and

$$(\Pi|_{X_1})_s \circ (\Pi|_{X_1})_t = (\Pi|_{X_1})_{s+t} \qquad \text{for any } s, t \in \mathbb{R}.$$

Also, from Lemma 3.2.1 it follows that the mapping $[\, \mathbb{R} \times X_1 \ni (t, (v, a)) \mapsto (\Pi|_{X_1})_t(v, a) \in X_1 \,]$ is continuous. Such an object is called a *topological linear skew-product flow* on the bundle X_1 covering the topological flow σ. For a theory of topological linear skew-product flows on (finite-dimensional) vector bundles see [65].

For $a \in Y_0$ fixed, $u \in L_{2,\mathrm{loc}}((-\infty, \infty), V)$ is an *entire positive weak solution* of $(2.0.1)_a + (2.0.2)_a$ if for any $s < t$, $u|_{[s,t]}$ is a weak solution of

$(2.0.1)_a + (2.0.2)_a$ and for any $t \in \mathbb{R}$, $u(t) \in L_2(D)^+ \setminus \{0\}$. Note that the mapping defined as

$$v_a(t) := \begin{cases} (U_{a \cdot t}(-t,0)|_{X_1(a \cdot t)})^{-1} w(a), & t < 0 \\ U_a(t,0) w(a), & t \geq 0 \end{cases}$$

(that is, the projection onto the first axis of $[\,(-\infty, \infty) \ni t \mapsto (\Pi|_{X_1})_t(w(a), a)\,]$) is an entire positive weak solution of the problem $(2.0.1)_a + (2.0.2)_a$.

For $a \in Y_0$, denote by $P_1(a)$ the projection of $L_2(D)$ on $X_1(a)$ along $X_1(a)$, and by $P_2(a)$ the projection of $L_2(D)$ on $X_2(a)$ along $X_1(a)$, $P_2(a) = \mathrm{Id}_{L_2(D)} - P_1(a)$. Notice that

$$P_1(a)u = \frac{\langle u, w^*(a) \rangle}{\langle w(a), w^*(a) \rangle} w(a), \qquad a \in Y_0, \ u \in L_2(D). \tag{3.2.2}$$

The mappings $[\,Y_0 \ni a \mapsto P_1(a) \in \mathcal{L}(L_2(D))\,]$ and $[\,Y_0 \ni a \mapsto P_2(a) \in \mathcal{L}(L_2(D))\,]$ are continuous. Indeed, notice that for any two $a^{(1)}$, $a^{(2)} \in Y_0$ and any $u_0 \in L_2(D)$ the following estimate holds:

$$\|P_1(a^{(1)})u_0 - P_1(a^{(2)})u_0\|$$
$$\leq \left| \frac{1}{\langle w(a^{(1)}), w^*(a^{(1)}) \rangle} - \frac{1}{\langle w(a^{(2)}), w^*(a^{(2)}) \rangle} \right| |\langle u_0, w^*(a^{(1)}) \rangle| \, \|w(a^{(1)})\|$$
$$+ \frac{|\langle u_0, w^*(a^{(1)}) - w^*(a^{(2)}) \rangle|}{\langle w(a^{(2)}), w^*(a^{(2)}) \rangle} \|w(a^{(1)})\|$$
$$+ \frac{|\langle u_0, w^*(a^{(2)}) \rangle|}{\langle w(a^{(2)}), w^*(a^{(2)}) \rangle} \|w(a^{(1)}) - w(a^{(2)})\|,$$

which reduces the issue of the continuity of the former mapping to the continuity of the mappings $w, w^* \colon Y_0 \to L_2(D)$. The continuity of the latter mapping follows by the formula $P_2(a) = \mathrm{Id}_{L_2(D)} - P_1(a)$.

Recall that the dual topological linear skew-product semiflow $\{\Pi^*(t)\}_{t \geq 0}$ (denoted also by Π^*) is defined as

$$\Pi_t^*(v^*, a) := (U_a^*(-t,0)v^*, a \cdot (-t)), \qquad t \geq 0, \ v^* \in L_2(D), \ a \in Y,$$

where

$$U_a^*(-t,0) = (U_{a \cdot (-t)}(t,0))^*, \qquad a \in Y, \ t \geq 0.$$

THEOREM 3.2.1

A topological linear skew-product semiflow Π admits an exponential separation over Y_0, with a one-dimensional subbundle X_1 and a one-codimensional subbundle X_2, if and only if its dual Π^* admits an exponential separation over Y_0, with a one-dimensional subbundle X_2^* and a one-codimensional subbundle X_1^*. Separating exponents can be chosen to be equal.

PROOF   Recall that, for any $a \in Y_0$,

$$v_1^* \in X_1^*(a) \quad \text{if and only if} \quad \langle v_1, v_1^* \rangle = 0 \text{ for each } v_1 \in X_1(a),$$

and

$$X_2^*(a) = \text{span}\{w^*(a)\},$$

where $w^*(a) \in L_2(D)^+$ for any $a \in Y_0$. Also, $X_1^*(a)$ is, for any $a \in Y_0$, the subspace orthogonal to $w(a) \in L_2(D)^+$.

The facts that X_1^* is a one-codimensional subbundle invariant under Π^* and that X_2^* is a one-dimensional subbundle invariant under Π^*, follow from the respective definitions and from the invariance of X_1 and X_2 under Π.

We estimate first the norms of the restrictions of $U_a^*(-t, 0)$ to X_2^*.

$$
\begin{aligned}
\|U_a^*(-t,0)w^*(a)\| &= \sup\{\, |\langle u, U_a^*(-t,0)w^*(a)\rangle| : \|u\| = 1 \,\} \\
&= \sup\{\, |\langle P_1(a)U_{a\cdot(-t)}(t,0)u, w^*(a)\rangle| : \|u\| = 1 \,\} \\
&= \sup\{\, |\langle U_{a\cdot(-t)}(t,0)P_1(a\cdot(-t))u, w^*(a)\rangle| : \|u\| = 1 \,\} \\
&\geq K_1\|U_{a\cdot(-t)}(t,0)w(a\cdot(-t))\|,
\end{aligned}
$$

where $K_1 := \inf\{\, \langle w(\tilde{a}), w^*(\tilde{a})\rangle : \tilde{a} \in Y_0 \,\} > 0$.

Next we estimate the norms of the restrictions of $U_a^*(-t,0)$ to X_1^*. Let $v_1^* \in X_1^*(a)$ with $\|v_1^*\| = 1$.

$$
\begin{aligned}
\|U_a^*(-t,0)v_1^*\| &= \sup\{\, |\langle u, U_a^*(-t,0)v_1^*\rangle| : \|u\| = 1 \,\} \\
&= \sup\{\, |\langle P_2(a)U_{a\cdot(-t)}(t,0)u, v_1^*\rangle| : \|u\| = 1 \,\} \\
&= \sup\{\, |\langle U_{a\cdot(-t)}(t,0)P_2(a\cdot(-t))u, v_1^*\rangle| : \|u\| = 1 \,\} \\
&\leq K_2\sup\{\, \|U_{a\cdot(-t)}(t,0)u\| : u \in X_2(a\cdot(-t)), \ \|u\| = 1 \,\},
\end{aligned}
$$

where $K_2 := \sup\{\, \|P_2(\tilde{a})\| : \tilde{a} \in Y_0 \,\} < \infty$.

Consequently,

$$\frac{\|U_a^*(-t,0)v_1^*\|}{\|U_a^*(-t,0)w^*(a)\|} \leq \frac{K_2}{K_1} \frac{\|U_{a\cdot(-t)}(t,0)|_{X_2(a\cdot(-t))}\|}{\|U_{a\cdot(-t)}(t,0)|_{X_1(a\cdot(-t))}\|}$$

for any $a \in Y$, $t > 0$ and $v_1^* \in X_1^*(a)$ with $v_1^* = 1$. By Definition 3.2.1(iii),

$$\frac{\|U_{a\cdot(-t)}(t,0)|_{X_2(a\cdot(-t))}\|}{\|U_{a\cdot(-t)}(t,0)|_{X_1(a\cdot(-t))}\|} \leq Me^{-\gamma_0 t}$$

for any $a \in Y$ and any $t > 0$. This together with the previous display gives a desired result.

The reverse implication follows by the observation that $(\Pi^*)^* = \Pi$.   ☐

The following easy result will be needed a couple of times, so we formulate it here.

LEMMA 3.2.2
Assume that Π *admits an exponential separation over a compact connected invariant subset* $Y_0 \subset Y$. *Then for each nonzero* $u_0 \in L_2(D)^+$ *there exists* $K > 0$ *such that for each* $a \in Y_0$ *the inequality* $\|P_2(a)u_0\| \leq K\|P_1(a)u_0\|$ *holds.*

PROOF   Observe that for each $a \in Y_0$ and each nonzero $u_0 \in L_2(D)^+$ one has $P_1(a)u_0 \neq 0$, since otherwise u_0 would be in $X_2(a)$, which contradicts the property in Definition 3.2.1(ii). Suppose to the contrary that for each positive integer n there is $a^{(n)} \in Y_0$ such that $\|P_1(a^{(n)})u_0\| < \frac{1}{n}\|P_2(a^{(n)})u_0\|$. We can choose a subsequence of $(a^{(n)})_{n=1}^{\infty}$ converging to some $\tilde{a} \in Y_0$. But by the continuous dependence of the projections P_2 on the base point and the compactness of Y_0, the set $\{\|P_2(a)u_0\| : a \in Y_0\}$ is bounded, consequently $\|P_1(\tilde{a})u_0\| = 0$, which is impossible.   ☐

Sometimes we have an "exponential separation" only for the discrete time. For convenience, we introduce

DEFINITION 3.2.2 (Exponential separation for discrete time)
Let Y_0 *be a compact connected invariant subset of* Y, *and let* $T > 0$. Π *is said to admit an* exponential separation *with separating exponent* $\gamma'_0 > 0$ *for the discrete time* T *over* Y_0 *if there are a one-dimensional subbundle* X_1 *of* $L_2(D) \times Y_0$ *with fibers* $X_1(a) = \text{span}\{w(a)\}$, *and a one-codimensional subbundle* X_2 *of* $L_2(D) \times Y_0$ *with fibers* $X_2(a) = \{v \in L_2(D) : \langle v, w^*(a) \rangle = 0\}$ *having the following properties:*

(a) $U_a(T,0)X_1(a) = X_1(a \cdot T)$ *and* $U_a(T,0)X_2(a) \subset X_2(a \cdot T)$ *for all* $a \in Y_0$,

(b) $w(a) \in L_2(D)^+$ *for all* $a \in Y_0$,

(c) $X_2(a) \cap L_2(D)^+ = \{0\}$ *for all* $a \in Y_0$,

(d) *there are* $M' \geq 1$ *such that for any* $a \in Y_0$ *and any* $v \in X_2(a)$ *with* $\|v\| = 1$,

$$\|U_a(nT,0)v\| \leq M'e^{-\gamma'_0 n}\|U_a(nT,0)w(a)\| \quad (n = 1, 2, 3, \dots).$$

The next result shows that the exponential separation for some discrete time implies exponential separation.

THEOREM 3.2.2
Assume that Π *admits an exponential separation with separating exponent* γ'_0 *for some discrete time* $T > 0$ *over a compact connected invariant subset* $Y_0 \subset Y$. *Then* Π *admits an exponential separation with separating exponent* $\gamma_0 = \gamma'_0$ *over* Y_0.

PROOF   We start by proving that the subbundles X_1 and X_2 having the above properties are invariant. Without loss of generality, assume that $T = 1$. Suppose by way of contradiction that X_1 is not invariant, that is, there are $a \in Y_0$ and $\tau > 0$ such that $U_a(\tau, 0)w(a) \notin X_1(a \cdot \tau)$. Define a continuous function $f \colon \mathbb{R} \to \mathbb{R}$ as

$$
f(t) := \begin{cases} \dfrac{\|P_2(a \cdot t)U_a(t, 0)w(a)\|}{\|P_1(a \cdot t)U_a(t, 0)w(a)\|} & \text{for } t \geq 0 \\[3ex] \dfrac{\|P_2(a \cdot t)U_a(t, \lfloor t \rfloor)w(a \cdot \lfloor t \rfloor)\|}{\|P_1(a \cdot t)U_a(t, \lfloor t \rfloor)w(a \cdot \lfloor t \rfloor)\|} & \text{for } t < 0, \end{cases}
$$

where $P_1(\cdot)$ is as in (3.2.2) and $P_2(\cdot) = \mathrm{Id}_{L_2(D)} - P_1(\cdot)$. As, by Proposition 2.2.9(2) and (b), $(U_{a \cdot s}(t + s, s)w(a \cdot s))(x) > 0$ for all $a \in Y$, $s \in \mathbb{R}$, $t > 0$ and a.e. $x \in D$, the function f is well defined. By part (a) $f(k) = 0$ for any integer k. Moreover, from the continuity and the positivity of the mapping $[\, Y \times [0, 1] \ni (a, t) \mapsto \|P_1(U_a(t, 0)w(a))\| \,]$ and from the continuity of $[\, Y \times [0, 1] \ni (a, t) \mapsto \|P_2(U_a(t, 0)w(a))\| \,]$ it follows that f is bounded from above.

Take a positive integer n_0 so large that $M'e^{-\gamma_0' n_0} < 1$. From (d) it follows that $f(t + n) \leq M'e^{-\gamma_0' n}f(t)$ for all $t \in \mathbb{R}$ and all $n \in \mathbb{N}$. This implies that

$$
f(t) \leq (M'e^{-\gamma_0' n_0})^k f(t - n_0 k)
$$

for all $t \in \mathbb{R}$ and $k \in \mathbb{N}$. It then follows from the nonnegativity and boundedness of f that

$$
0 \leq f(t) \leq \limsup_{k \to \infty} (M'e^{-\gamma_0' n_0})^k f(t - n_0 k) = 0
$$

for all $t \in \mathbb{R}$. Hence $f \equiv 0$. This contradicts the assumption $U_a(a \cdot \tau)w(a) \notin X_1(a \cdot \tau)$. Therefore X_1 is invariant.

The proof of the invariance of X_2 goes along the following lines. We prove first a discrete-time analog of Theorem 3.2.1, that is, that Π^* admits an exponential separation for discrete time $T > 0$ over a compact invariant $Y_0 \subset Y$, with one-dimensional subbundle X_2^* and one-codimensional subbundle X_1^*. The previous paragraph gives that X_2^* is invariant under Π^*, which is equivalent to X_2 being invariant under Π.

The property (iii) in Definition 3.2.1 follows from (d) by applying the $L_2(D)$–$L_2(D)$ estimate in Proposition 2.2.2. $\Box$

THEOREM 3.2.3

Assume that Π admits an exponential separation over Y_0 with a one dimensional bundle X_1 and a one-codimensional bundle X_2, and admits an exponential separation over Y_0 with a one-dimensional bundle $\tilde{X}_1$ and a one-codimensional bundle $\tilde{X}_2$. Then $X_1 = \tilde{X}_1$ and $X_2 = \tilde{X}_2$.

PROOF For $a \in Y_0$ denote by $P_1(a)$ and $P_2(a)$ the projections corresponding to the exponential separation with X_1 and X_2, and by $\tilde{w}(a)$ the unique element of $L_2(D)^+$ such that $\|\tilde{w}(a)\| = 1$ and $\tilde{X}_1 = \text{span}\{\tilde{w}(a)\}$. Further, we define a function $f: Y_0 \to [0, \infty)$ as $f(a) := \|P_2(a)\tilde{w}(a)\|/\|P_1(a)\tilde{w}(a)\|$. Since $\|P_1(a)\tilde{w}(a)\| > 0$ (compare the proof of Lemma 3.2.2), f is well defined. The function f is clearly continuous, so it is bounded. The invariance of the subbundle $\tilde{X}_1$ together with Definition 3.2.1 (for X_1 and X_2) imply that $f(a \cdot t) \leq Me^{-\gamma_0 t} f(a)$ for each $a \in Y$ and $t > 0$, consequently $f(a) \leq Me^{-\gamma_0 t} f(a \cdot (-t))$ for each $a \in Y$ and $t > 0$, which gives $f \equiv 0$. We have obtained thus $X_1 = \tilde{X}_1$.

By Theorem 3.2.1, Π^* admits an exponential separation over Y_0 with a one-dimensional bundle X_2^* and a one-codimensional bundle X_1^*, as well as admits an exponential separation over Y_0 with a one-dimensional bundle $\tilde{X}_2^*$ and a one-codimensional bundle $\tilde{X}_1^*$. The first part of the proof gives that $X_2^* = \tilde{X}_2^*$, which implies $X_2 = \tilde{X}_2$. $\qquad\square$

LEMMA 3.2.3

Assume that Π admits an exponential separation over a nonempty compact connected invariant subset Y_0 of Y.

(1) *For each $a \in Y_0$, each $u_0 \in L_2(D) \setminus X_2(a)$, $\|u_0\| = 1$, and each $\delta_0 > 0$ there is $M_1 = M_1(a, u_0, \delta_0) \in (0, 1)$ such that*

$$\|U_a(t, 0)u_0\| \geq M_1 \|U_a(t, 0)w(a)\|$$

for all $t \geq \delta_0$.

(2) *There is $M_2 \geq 1$ such that*

$$\|U_a(t, 0)u_0\| \leq M_2 \|U_a(t, 0)w(a)\|$$

for all $a \in Y_0$, all $t \geq 0$, and all $u_0 \in L_2(D)$ with $\|u_0\| = 1$. Consequently, for that same $M_2 \geq 1$,

$$\|U_a(t, 0)w(a)\| \leq \|U_a(t, 0)\|^+ = \|U_a(t, 0)\| \leq M_2 \|U_a(t, 0)w(a)\|$$

for all $a \in Y_0$ and all $t \geq 0$.

PROOF (1) Let $P_1(\cdot)$ and $P_2(\cdot)$ denote the projections for the exponential separation. Fix $a \in Y_0$ and $u_0 \in L_2(D) \setminus X_2(a)$ with $\|u_0\| = 1$. Put $K :=$

$\|P_2(a)u_0\|/\|P_1(a)u_0\|$. We estimate

$$\|U_a(t,0)u_0\| \geq \|P_1(a \cdot t)U_a(t,0)u_0\| - \|P_2(a \cdot t)U_a(t,0)u_0\|$$

$$= \left(1 - \frac{\|P_2(a \cdot t)U_a(t,0)u_0\|}{\|P_1(a \cdot t)U_a(t,0)u_0\|}\right)\|P_1(a \cdot t)U_a(t,0)u_0\|$$

$$= \left(1 - \frac{\|U_a(t,0)P_2(a)u_0\|}{\|U_a(t,0)P_1(a)u_0\|}\right)\|U_a(t,0)P_1(a)u_0\|$$

$$\geq (1 - MKe^{-\gamma_0 t})\|U_a(t,0)w(a)\|\,\|P_1(a)u_0\| \quad \text{(by Def. 3.2.1(iii))}$$

$$\geq \frac{1 - MKe^{-\gamma_0 t}}{1+K}\|U_a(t,0)w(a)\| \quad \text{(since } \|u_0\| \leq (1+K)\|P_1(a)u_0\|)$$

for any $a \in Y_0$ and any $t > 0$. Take $T > 0$ to be such that $(1 - MKe^{-\gamma_0 t})/(1+K) > 0$ for all $t \geq T$. If $T \leq \delta_0$ we are done. Assume not. Notice that $U_a(t,0)u_0 \neq 0$ for all $t > 0$. Indeed, $P_1(a)u_0 \neq 0$, consequently $P_1(a \cdot t)U_a(t,0)u_0 = U_a(t,0)P_1(a)u_0 \neq 0$. The equality $U_a(t,0)u_0 = 0$ for some $t > 0$ would imply $P_2(a \cdot t)U_a(t,0)u_0 = -P_1(a \cdot t)U_a(t,0)u_0 \neq 0$. But then $0 \neq P_2(a \cdot t)U_a(t,0)u_0 \in X_1(a \cdot t) \cap X_2(a \cdot t)$, which is impossible. Consequently $\|U_a(t,0)u_0\| > 0$ for all $t > 0$. Since $Y_0 \times [\delta_0, T]$ is compact, Proposition 2.2.12 implies that the set $\{\,\|U_a(t,0)u_0\| : t \in [\delta_0, T]\,\}$ is bounded away from zero. The set $\{\,\|U_a(t,0)w(a)\| : t \in [\delta_0, T]\,\}$ is clearly bounded, hence the conclusion of Part (1) follows.

(2) Put $K_1 := \max\{\,\|P_1(a)\| : a \in Y_0\,\}$, $K_2 := \max\{\,\|P_2(a)\| : a \in Y_0\,\}$. Fix $a \in Y_0$ and $u_0 \in L_2(D)$ with $\|u_0\| = 1$. We estimate

$$\|U_a(t,0)u_0\| \leq \|P_1(a \cdot t)U_a(t,0)u_0\| + \|P_2(a \cdot t)U_a(t,0)u_0\|$$

$$= \|U_a(t,0)P_1(a)u_0\| + \|U_a(t,0)P_2(a)u_0\|$$

$$= \left(\|P_1(a)u_0\| + \frac{\|U_a(t,0)P_2(a)u_0\|}{\|U_a(t,0)w(a)\|}\right)\|U_a(t,0)w(a)\|$$

$$\leq (\|P_1(a)u_0\| + Me^{-\gamma_0 t}\|P_2(a)u_0\|)\|U_a(t,0)w(a)\| \quad \text{(by Def. 3.2.1(iii))}$$

$$\leq (K_1 + K_2 Me^{-\gamma_0 t})\|U_a(t,0)w(a)\|$$

for any $t > 0$. $\qquad\qquad\qquad\qquad\qquad\qquad\qquad\qquad\qquad\qquad\qquad\qquad$ ∎

LEMMA 3.2.4

Assume that Π admits an exponential separation over a nonempty compact connected invariant subset Y_0 of Y. Then for each $u_0 \in L_2(D)^+$, $\|u_0\| = 1$, and each $\delta_0 > 0$ there is $M_1 = M_1(u_0, \delta_0) \in (0,1)$ such that

$$\|U_a(t,0)u_0\| \geq M_1\|U_a(t,0)w(a)\|$$

for all $a \in Y_0$ and all $t \geq \delta_0$.

PROOF   Let $P_1(\cdot)$ and $P_2(\cdot)$ denote the projections for the exponential separation. By Lemma 3.2.2 there is $K = K(u_0) > 0$ such that $\|P_2(a)u_0\| \leq$

$K\|P_1(a)u_0\|$ for all $a \in Y_0$. We estimate

$$
\begin{aligned}
\|U_a(t,0)u_0\| &\geq \|P_1(a \cdot t)U_a(t,0)u_0\| - \|P_2(a \cdot t)U_a(t,0)u_0\| \\
&= \left(1 - \frac{\|P_2(a \cdot t)U_a(t,0)u_0\|}{\|P_1(a \cdot t)U_a(t,0)u_0\|}\right)\|P_1(a \cdot t)U_a(t,0)u_0\| \\
&= \left(1 - \frac{\|U_a(t,0)P_2(a)u_0\|}{\|U_a(t,0)P_1(a)u_0\|}\right)\|U_a(t,0)P_1(a)u_0\| \\
&\geq (1 - MKe^{-\gamma_0 t})\|U_a(t,0)w(a)\|\,\|P_1(a)u_0\| \quad \text{(by Def. 3.2.1(iii))} \\
&\geq \frac{1 - MKe^{-\gamma_0 t}}{1 + K}\|U_a(t,0)w(a)\| \quad \text{(since } \|u_0\| \leq (1 + K)\|P_1(a)u_0\|)
\end{aligned}
$$

for any $a \in Y_0$ and any $t > 0$. Take $T > 0$ to be such that $(1 - MKe^{-\gamma_0 t})/(1 + K) > 0$ for all $t \geq T$. If $T \leq \delta_0$ we are done. Assume not. Then notice that by Proposition 2.2.9(2), $\|U_a(t,0)u_0\| > 0$ for all $a \in Y_0$ and $t > 0$. Since $Y_0 \times [\delta_0, T]$ is compact, Proposition 2.2.12 implies that the set $\{\,\|U_a(t,0)u_0\| : a \in Y_0,\ t \in [\delta_0, T]\,\}$ is bounded away from zero. The set $\{\,\|U_a(t,0)w(a)\| : a \in Y_0,\ t \in [\delta_0, T]\,\}$ is clearly bounded, hence the conclusion follows. $\qquad\square$

From now on until the end of Section 3.2 we assume that Y_0 is a compact connected invariant subset of Y such that Π admits an exponential separation with separating exponent γ_0 over Y_0.

THEOREM 3.2.4

Let μ be an invariant ergodic Borel probability measure for $\sigma|_{Y_0}$. Then there is a Borel set $Y_1 \subset Y_0$, $\mu(Y_1) = 1$, with the following properties.

(1) *For any $a \in Y_1$ and any $u_0 \in L_2(D) \setminus X_2(a)$ one has*

$$
\lim_{t \to \infty} \frac{\ln\|U_a(t,0)u_0\|}{t} = \lambda(\mu). \tag{3.2.3}
$$

(2) *For any $a \in Y_1$ and any $u_0 \in X_2(a) \setminus \{0\}$ one has*

$$
\limsup_{t \to \infty} \frac{\ln\|U_a(t,0)u_0\|}{t} \leq \lambda(\mu) - \gamma_0. \tag{3.2.4}
$$

PROOF By Theorem 3.1.5 there is a Borel set $Y_1 \subset Y_0$ with $\mu(Y_1) = 1$ such that

$$
\lim_{t \to \infty} \frac{\ln\|U_a(t,0)\|}{t} = \lambda(\mu)
$$

for all $a \in Y_1$. Fix $a \in Y_0$ and $u_0 \in L_2(D) \setminus X_2(a)$ with $\|u_0\| = 1$. Lemma 3.2.3 yields the existence of $M_1 = M_1(a, u_0, 1) > 0$ and $M_2 > 0$ such that

$$
\|U_a(t,0)u_0\| \geq M_1\|U_a(t,0)w(a)\| \geq \frac{M_1}{M_2}\|U_a(t,0)\| \geq \frac{M_1}{M_2}\|U_a(t,0)u_0\|
$$

for all $t \geq 1$. Part (1) follows immediately.

Part (2) is a consequence of Part (1) and Definition 3.2.1(iii). ⬜

COROLLARY 3.2.1

Let μ and Y_1 be as in Theorem 3.2.4. Then for any $a \in Y_1$, $u_0 \in X_2(a) \setminus \{0\}$ if and only if

$$\limsup_{t \to \infty} \frac{\ln \|U_a(t, 0)u_0\|}{t} < \lambda(\mu).$$

The following result will be extensively used in Chapter 4.

LEMMA 3.2.5

Let $\lambda \in \mathbb{R}$, $(a^{(n)})_{n=1}^{\infty} \subset Y_0$, and $(s_n)_{n=1}^{\infty} \subset \mathbb{R}$, $(t_n)_{n=1}^{\infty} \subset \mathbb{R}$ with $t_n - s_n \to \infty$. Then the following conditions are equivalent:

(1) $\displaystyle \lim_{n \to \infty} \frac{\ln \|U_{a^{(n)}}(t_n, s_n)w(a^{(n)} \cdot s_n)\|}{t_n - s_n} = \lambda$.

(2) $\displaystyle \lim_{n \to \infty} \frac{\ln \|U_{a^{(n)}}(t_n, s_n)u_0\|}{t_n - s_n} = \lambda$ *for any $u_0 \in L_2(D)^+ \setminus \{0\}$.*

(3) $\displaystyle \lim_{n \to \infty} \frac{\ln \|U_{a^{(n)}}(t_n, s_n)\|}{t_n - s_n} = \lim_{n \to \infty} \frac{\ln \|U_{a^{(n)}}(t_n, s_n)\|^+}{t_n - s_n} = \lambda$.

PROOF Fix $u_0 \in L_2(D)^+$ with $\|u_0\| = 1$. By Lemmas 3.2.4 and 3.2.3(2) there are $M_1 = M_1(u_0, 1) > 0$ and $M_2 > 0$ such that

$$\|U_{a^{(n)}}(t_n, s_n)u_0\| \geq M_1 \|U_{a^{(n)}}(t_n, s_n)w(a^{(n)} \cdot s_n)\| \geq \frac{M_1}{M_2} \|U_{a^{(n)}}(t_n, s_n)u_0\|$$

for any $n \in \mathbb{N}$ such that $t_n - s_n \geq 1$. This implies the equivalence of (1) and (2).

The equivalence of (1) and (3) is a consequence of Lemma 3.2.3(2). ⬜

LEMMA 3.2.6

(1) $\lambda \in \mathbb{R}$ *belongs to the upper principal resolvent of Π over Y_0 if and only if there are $\epsilon > 0$ and $M \geq 1$ such that*

$$\|U_a(t, 0)w(a)\| \leq Me^{(\lambda - \epsilon)t} \quad \text{for } t > 0 \text{ and } a \in Y_0,$$

(2) $\lambda \in \mathbb{R}$ *belongs to the lower principal resolvent of Π over Y_0 if and only if there are $\epsilon > 0$ and $M \in (0, 1)$ such that*

$$\|U_a(t, 0)w(a)\| \geq Me^{(\lambda + \epsilon)t} \quad \text{for } t > 0 \text{ and } a \in Y_0.$$

PROOF   Part (1) is a consequence of Lemma 3.2.3(2). Part (2) is a consequence of Lemma 3.2.3(2) together with Lemma 3.1.3.   ☐

For the topological linear skew-product flow $\Pi|_{X_1}$ on the one-dimensional bundle X_1 its *dynamical spectrum* (or the *Sacker–Sell spectrum*) is defined as the complement of the set of those $\lambda \in \mathbb{R}$ for which either of the conditions in Lemma 3.2.6 holds (see [65]). Therefore the principal spectrum of Π over Y_0 equals the dynamical spectrum of $\Pi|_{X_1}$, which allows us to make use of [65, Theorem 3.3] to prove the following important results.

THEOREM 3.2.5
There exist ergodic invariant measures $\mu_{\min}$ and $\mu_{\max}$ for $\sigma|_{Y_0}$ such that $\lambda_{\min} = \lambda(\mu_{\min})$ and $\lambda_{\max} = \lambda(\mu_{\max})$.

COROLLARY 3.2.2
If $(Y_0, \{\sigma_t\}_{t \in \mathbb{R}})$ is uniquely ergodic then $\lambda_{\min} = \lambda_{\max}$.

REMARK 3.2.1   When $(2.0.1)_a + (2.0.2)_a$ is actually time independent or periodic, then the unique (by Corollary 3.2.2) element of the principal spectrum turns out to be the classical principal eigenvalue.   ☐

THEOREM 3.2.6
Let μ be an invariant ergodic measure for $\sigma|_{Y_0}$. Then there is a Borel set $\tilde{Y}_1 \subset Y_0$, $\mu(\tilde{Y}_1) = 1$, with the property that

$$\lim_{t \to \pm\infty} \frac{\ln \|U_a(t,0)w(a)\|}{t} = \lambda(\mu). \qquad (3.2.5)$$

for any $a \in \tilde{Y}_1$, where, for $t < 0$, $U_a(t,0)w(a)$ denotes the only element $v \in X_1(a \cdot t)$ such that $U_{a \cdot t}(-t,0)v = w(a)$.

PROOF   It follows by an application of [65, Theorem 2.1] to the topological linear skew-product flow $\Pi|_{X_1}$ on the one-dimensional bundle X_1.   ☐

The following result shows that in the case of (Y_0, σ) being topologically transitive, when analyzing the principal spectrum/principal Lyapunov exponent one can restrict oneself to considering solutions of $(2.0.1) + (2.0.2)$ with only one parameter value. It will be extensively used in Chapter 4.

THEOREM 3.2.7
If $Y_0 = \mathrm{cl}\{a^{(0)} \cdot t : t \in \mathbb{R}\}$ for some $a^{(0)} \in Y_0$, where the closure is taken in the weak- topology, then*

(1) (i) *There are sequences* $(s'_n)_{n=1}^{\infty}, (t'_n)_{n=1}^{\infty} \subset \mathbb{R}$, $t'_n - s'_n \to \infty$ *as* $n \to \infty$, *such that*

$$\lambda_{\min} = \lim_{n \to \infty} \frac{\ln \|U_{a^{(0)}}(t'_n, s'_n) w(a^{(0)} \cdot s'_n)\|}{t'_n - s'_n}$$

$$= \lim_{n \to \infty} \frac{\ln \|U_{a^{(0)}}(t'_n, s'_n) u_0\|}{t'_n - s'_n} = \lim_{n \to \infty} \frac{\ln \|U_{a^{(0)}}(t'_n, s'_n)\|}{t'_n - s'_n}$$

for each $u_0 \in L_2(D)^+ \setminus \{0\}$,

(ii) *There are sequences* $(s''_n)_{n=1}^{\infty}, (t''_n)_{n=1}^{\infty} \subset \mathbb{R}$, $t''_n - s''_n \to \infty$ *as* $n \to \infty$, *such that*

$$\lambda_{\max} = \lim_{n \to \infty} \frac{\ln \|U_{a^{(0)}}(t''_n, s''_n) w(a^{(0)} \cdot s''_n)\|}{t''_n - s''_n}$$

$$= \lim_{n \to \infty} \frac{\ln \|U_{a^{(0)}}(t''_n, s''_n) u_0\|}{t''_n - s''_n} = \lim_{n \to \infty} \frac{\ln \|U_{a^{(0)}}(t''_n, s''_n)\|}{t''_n - s''_n}$$

for each $u_0 \in L_2(D)^+ \setminus \{0\}$.

(2) *For any* $u_0 \in L_2(D)^+ \setminus \{0\}$ *there holds*

$$\lambda_{\min} = \liminf_{t-s \to \infty} \frac{\ln \|U_{a^{(0)}}(t, s) w(a^{(0)} \cdot s)\|}{t - s}$$

$$= \liminf_{t-s \to \infty} \frac{\ln \|U_{a^{(0)}}(t, s) u_0\|}{t - s} = \liminf_{t-s \to \infty} \frac{\ln \|U_{a^{(0)}}(t, s)\|}{t - s}$$

$$\leq \limsup_{t-s \to \infty} \frac{\ln \|U_{a^{(0)}}(t, s)\|}{t - s} = \limsup_{t-s \to \infty} \frac{\ln \|U_{a^{(0)}}(t, s) u_0\|}{t - s}$$

$$= \limsup_{t-s \to \infty} \frac{\ln \|U_{a^{(0)}}(t, s) w(a^{(0)} \cdot s)\|}{t - s} = \lambda_{\max}.$$

(3) *For each* $\lambda \in [\lambda_{\min}, \lambda_{\max}]$ *there are sequences* $(k_n)_{n=1}^{\infty}, (l_n)_{n=1}^{\infty} \subset \mathbb{Z}$, $l_n - k_n \to \infty$ *as* $n \to \infty$, *such that*

$$\lambda = \lim_{n \to \infty} \frac{\ln \|U_{a^{(0)}}(l_n, k_n) w(a^{(0)} \cdot k_n)\|}{l_n - k_n}$$

$$= \lim_{n \to \infty} \frac{\ln \|U_{a^{(0)}}(l_n, k_n) u_0\|}{l_n - k_n} = \lim_{n \to \infty} \frac{\ln \|U_{a^{(0)}}(l_n, k_n)\|}{l_n - k_n}$$

for each $u_0 \in L_2(D)^+ \setminus \{0\}$.

PROOF (1) We prove only (i), the other part being similar. By Theorem 3.2.5(2), there is an ergodic invariant measure $\mu_{\min}$ for the topological flow (Y_0, σ) such that $\lambda_{\min}$ equals the principal Lyapunov exponent on Y_0 for $\mu_{\min}$.

Theorem 3.1.5 and Lemma 3.2.5 provide the existence of a Borel set $Y_1 \subset Y_0$ with $\mu_{\min}(Y_1) = 1$ such that

$$\lim_{t \to \infty} \frac{\ln \|U_{\tilde{a}}(t,0)w(\tilde{a})\|}{t} = \lambda_{\min} \quad \text{for any} \quad \tilde{a} \in Y_1.$$

Fix $\tilde{a} \in Y_1$. For each $n = 1, 2, 3, \ldots$, take $\tau_n \geq n$ such that

$$\lambda_{\min} - \frac{1}{2n} < \frac{\ln \|U_{\tilde{a}}(\tau_n, 0)w(\tilde{a})\|}{\tau_n} < \lambda_{\min} + \frac{1}{2n}.$$

As Y_0 equals the closure of $\{ a^{(0)} \cdot t : t \in \mathbb{R} \}$, for any $n = 1, 2, 3, \ldots$ we can find $s'_n \in \mathbb{R}$ with the property that $a^{(0)} \cdot s'_n$ is so close to $\tilde{a}$ that

$$| \ln \|U_{\tilde{a}}(\tau_n, 0)w(\tilde{a})\| - \ln \|U_{a^{(0)} \cdot s'_n}(\tau_n, 0)w(a \cdot s'_n)\| | < \frac{\tau_n}{2n}.$$

Take $t'_n := s'_n + \tau_n$. We have found two sequences $(s'_n)_{n=1}^{\infty}, (t'_n)_{n=1}^{\infty} \subset \mathbb{R}$, $t'_n - s'_n \to \infty$ as $n \to \infty$, such that

$$\lambda_{\min} = \lim_{n \to \infty} \frac{\ln \|U_{a^{(0)}}(t'_n, s'_n)w(a^{(0)} \cdot s'_n)\|}{t'_n - s'_n}.$$

By Lemma 3.2.5 again, the statement follows.

(2) It follows from (1) together with Theorem 3.1.2 and Lemma 3.2.5.

(3) Let $\delta \in [0,1]$ be such that $\lambda = \delta\lambda_{\min} + (1 - \delta)\lambda_{\max}$. Let $(s'_n), (t'_n), (s''_n), (t''_n)$ be sequences as in (1).

For each $n = 1, 2, 3, \ldots$ consider the function

$$\left[\begin{array}{l} [0,1] \ni \delta \\ \\ \mapsto \dfrac{\ln \|U_{a^{(0)}}(\delta t'_n + (1-\delta)t''_n, \delta s'_n + (1-\delta)s''_n)w(a^{(0)} \cdot (\delta s'_n + (1-\delta)s''_n))\|}{\delta(t'_n - s'_n) + (1-\delta)(t''_n - s''_n)} \end{array} \right].$$

The above function is continuous, so there exists $\tilde{\delta} = \tilde{\delta}(n) \in [0,1]$ such that

$$\frac{\ln \|U_{a^{(0)}}(\tilde{\delta} t'_n + (1-\tilde{\delta})t''_n, \tilde{\delta} s'_n + (1-\tilde{\delta})s''_n)w(a^{(0)} \cdot (\tilde{\delta} s'_n + (1-\tilde{\delta})s''_n))\|}{\tilde{\delta}(t'_n - s'_n) + (1-\tilde{\delta})(t''_n - s''_n)}$$

$$= \delta \frac{\ln \|U_{a^{(0)}}(t'_n, s'_n)w(a^{(0)} \cdot s'_n)\|}{t'_n - s'_n} + (1-\delta)\frac{\ln \|U_{a^{(0)}}(t''_n, s''_n)w(a^{(0)} \cdot s''_n)\|}{t''_n - s''_n}.$$

Taking $s_n := \tilde{\delta}(n)s'_n + (1 - \tilde{\delta}(n))s''_n$ and $t_n := \tilde{\delta}(n)t'_n + (1 - \tilde{\delta}(n))t''_n$, we see that we have found sequences $(s_n)_{n=1}^{\infty}, (t_n)_{n=1}^{\infty} \subset \mathbb{R}$, $t_n - s_n \to \infty$ as $n \to \infty$, such that

$$\lim_{n \to \infty} \frac{\ln \|U_{a^{(0)}}(t_n, s_n)w(a^{(0)} \cdot s_n)\|}{t_n - s_n} = \lambda.$$

Put $k_n := \lfloor s_n \rfloor$, $l_n := \lfloor t_n \rfloor$. We have

$$\frac{M_1}{M_2} \|U_{a^{(0)}}(t_n, s_n)w(a^{(0)} \cdot s_n)\| \le \|U_{a^{(0)}}(l_n, k_n)w(a^{(0)} \cdot k_n)\|$$

$$\le \frac{M_2}{M_1} \|U_{a^{(0)}}(t_n, s_n)w(a^{(0)} \cdot s_n)\|$$

for n sufficiently large, where $M_1 := \inf\{\|U_{\tilde{a}}(s,0)w(\tilde{a})\| : s \in [0,1], \tilde{a} \in Y_0\} > 0$, $M_2 := \sup\{\|U_{\tilde{a}}(s,0)w(\tilde{a})\| : s \in [0,1], \tilde{a} \in Y_0\} < \infty$. After simple calculation we have that

$$\lim_{n \to \infty} \frac{\ln \|U_{a^{(0)}}(l_n, k_n)w(a^{(0)} \cdot k_n)\|}{l_n - k_n} = \lambda.$$

An application of Lemma 3.2.5 gives the desired result.   ❑

Recall that for $a \in Y$, $B_a(t, u, v)$ is defined in (2.1.4) in the Dirichlet and Neumann boundary condition cases, and is defined in (2.1.5) in the Robin boundary condition case.

LEMMA 3.2.7

$$\|U_a(t,s)w(a)\| = \exp\left(-\int_s^t B_a(\tau, w(a\cdot\tau), w(a\cdot\tau))\,d\tau\right) \quad \text{for any } a \in Y_0, \ s < t.$$

PROOF   Fix $a \in Y_0$ and $s < t$. Let, for $\tau \in [s,t]$, $\eta(\tau) := \|U_a(\tau,s)w(a)\|$. By Proposition 2.1.4,

$$\frac{1}{2}\left((\eta(\tau))^2 - (\eta(s)))^2\right) = -\int_s^\tau B_a(r, w(a \cdot r), w(a \cdot r))(\eta(r))^2\,dr,$$

hence $(\eta(\cdot))^2$ is absolutely continuous. As it is bounded away from 0, $\ln \eta(\cdot)$ is absolutely continuous on $[s,t]$. We deduce then that

$$\dot{\eta}(\tau) = -B_a(\tau, w(a \cdot \tau), w(a \cdot \tau))\eta(\tau)$$

for a.e. $\tau \in [s,t]$. This implies the statement of the lemma.   ❑

The following results are straightforward corollaries of Lemma 3.2.7 and Theorems 3.2.7 or 3.1.5, respectively.

THEOREM 3.2.8

If $Y_0 = \text{cl}\{a^{(0)} \cdot t : t \in \mathbb{R}\}$ for some $a^{(0)} \in Y_0$, where the closure is taken in *the weak-* topology, then*

$$\lambda_{\min} = \liminf_{t-s\to\infty} \frac{-1}{t-s} \int_s^t B_{a^{(0)}}(\tau, w(a^{(0)} \cdot \tau), w(a^{(0)} \cdot \tau))\, d\tau$$

$$\leq \limsup_{t-s\to\infty} \frac{-1}{t-s} \int_s^t B_{a^{(0)}}(\tau, w(a^{(0)} \cdot \tau), w(a^{(0)} \cdot \tau))\, d\tau = \lambda_{\max}.$$

THEOREM 3.2.9
Let μ be an ergodic invariant measure for $\sigma|_{Y_0}$. Then there exists a Borel set $Y_1 \subset Y_0$ with $\mu(Y_1) = 1$ such that

$$\lambda(\mu) = -\lim_{t\to\infty} \frac{1}{t} \int_0^t B_a(\tau, w(a \cdot \tau), w(a \cdot \tau))\, d\tau$$

for any $a \in Y_1$.

3.3 Existence of Exponential Separation and Entire Positive Solutions

In this section we discuss the existence of exponential separation and the existence and uniqueness of positive solutions.

We show that in general the existence of an entire positive solution and exponential separation follows from certain Harnack inequalities for nonnegative solutions of parabolic problems.

In the present section we assume that assumptions (A2-1) through (A2-3) are satisfied. Y_0 is a compact connected invariant subset of Y.

For $x \in D$ we denote by $d(x)$ the distance of x from the boundary ∂D of D.

We introduce now the assumptions (A3-1) and (A3-2).

(A3-1) (Harnack type inequality for quotients): *For each $\delta_1 > 0$ there is $C_1 = C_1(\delta_1) > 1$ such that*

$$\sup_{x\in D} \frac{(U_a(t,0)u_{01})(x)}{(U_a(t,0)u_{02})(x)} \leq C_1 \inf_{x\in D} \frac{(U_a(t,0)u_{01})(x)}{(U_a(t,0)u_{02})(x)}$$

for any $a \in Y_0$, $t \geq \delta_1$, any $u_{01}, u_{02} \in L_2(D)^+$, where $u_{02} \neq 0$, and any $x \in D$.

(A3-2) (Pointwise Harnack inequality) *There is $\varsigma \geq 0$ such that for each $\delta_2 > 0$ there is $C_2 = C_2(\delta_2) > 0$ with the property that*

$$(U_a(t,0)u_0)(x) \geq C_2(d(x))^\varsigma \|U_a(t,0)u_0\|_\infty \tag{3.3.1}$$

for any $a \in Y_0$, $t \geq \delta_2$, $u_0 \in L_2(D)^+$ and $x \in D$.

Both (2.0.1)+(2.0.2) and (2.3.1)+(2.3.2) satisfy (A3-1) and (A3-2) under quite general conditions. For example, it is proved in [61] that both (A3-1) and (A3-2) are satisfied in the Dirichlet boundary condition case for a Lipschitz domain (see [61, Theorem 2.1 and Lemma 3.9]). It is proved in [59] that (A3-2) is satisfied with $\varsigma = 0$ in certain Neumann and Robin boundary condition cases (see [59, Theorem 2.5]). If (A3-2) is satisfied with $\varsigma = 0$, then (A3-1) is also satisfied. More precisely, we have

LEMMA 3.3.1
(A3-2) *with $\varsigma = 0$ implies* (A3-1).

PROOF   Let u_{01} and u_{02} be as in (A3-1). Put $u_i(t, x) := (U_a(t, 0)u_{0i})(x)$, $i = 1, 2$. By (A3-2) (with $\varsigma = 0$) we have

$$\sup_{x \in D} \frac{u_1(t, x)}{u_2(t, x)} \leq \frac{\sup_{x \in D} u_1(t, x)}{\inf_{x \in D} u_2(t, x)} \leq \frac{1}{C_2^2} \frac{\inf_{x \in D} u_1(t, x)}{\sup_{x \in D} u_2(t, x)} \leq \frac{1}{C_2^2} \inf_{x \in D} \frac{u_1(t, x)}{u_2(t, x)}$$

for $t \geq \delta_2$. Hence (A3-1) holds with $\delta_1 = \delta_2$ and $C_1 = 1/C_2^2$.   ☐

We say that (A3-1) and/or (A3-2) are satisfied by (2.3.1)+(2.3.2) if they are satisfied with $U_a(t, 0)$ being replaced by $U_a^*(-t, 0)$. We have the following theorems about the existence and uniqueness of an entire positive weak solution and the existence of exponential separation.

THEOREM 3.3.1
Consider (2.0.1)+(2.0.2) *and assume* (A3-1)–(A3-2). *Then there exists a continuous function $w: Y_0 \to L_2(D)^+$, $\|w(a)\| = 1$ for each $a \in Y_0$, having the property that for each $a \in Y_0$ the function $[t \mapsto v_a(t) = r_a(t)w(a \cdot t)]$, where*

$$r_a(t) := \begin{cases} \|U_{a \cdot t}(-t, 0)w(a \cdot t)\|^{-1} & t < 0, \\ \|U_a(t, 0)w(a)\| & t \geq 0, \end{cases}$$

is an entire positive weak solution of (2.0.1)$_a$+(2.0.2)$_a$. *Moreover, for any entire positive weak solution v of* (2.0.1)$_a$+(2.0.2)$_a$ *one has $v(t) = \|v(0)\|v_a(t)$, $t \in \mathbb{R}$.*

THEOREM 3.3.2
Consider (2.3.1)+(2.3.2) *and assume that* (A3-1)–(A3-2) *are satisfied by* (2.3.1)+(2.3.2). *Then there exists a continuous function $w^*: Y_0 \to L_2(D)^+$, $\|w^*(a)\| = 1$ for each $a \in Y_0$, having the property that for each $a \in Y_0$ the*

function $[\, t \mapsto v_a^*(t) = r_a^*(t)w^*(a \cdot t)\,]$, *where*

$$r_a^*(t) := \begin{cases} \|U_a^*(t,0)w^*(a)\| & t \leq 0, \\ \|U_{a\cdot t}^*(-t,0)w^*(a\cdot t)\|^{-1} & t > 0, \end{cases}$$

is an entire positive weak solution of $(2.3.1)_a + (2.3.2)_a$. *Moreover, for any entire positive weak solution* v^* *of* $(2.3.1)_a + (2.3.2)_a$ *one has* $v^*(t) = \|v^*(0)\|v_a^*(t)$, $t \in \mathbb{R}$.

THEOREM 3.3.3

Assume that (A3-1) *and* (A3-2) *are satisfied by both* (2.0.1) $+$(2.0.2) *and* (2.3.1) $+$(2.3.2). *Let* w *and* w^* *be as in Theorems* 3.3.1 *and* 3.3.2, *respectively. Then* Π *admits an exponential separation over* Y_0 *with an invariant one-dimensional subbundle given by* $X_1(a) = \mathrm{span}\{w(a)\}$ *and an invariant one-codimensional subbundle given by* $X_2(a) = \{v \in L_2(D) : \langle v, w^*(a) \rangle = 0\}$. *Moreover, for any* $a \in Y_0$ *the fiber* $X_2(a)$ *is characterized as the set of those* $u_0 \in L_2(D)$ *such that the global weak solution* $[\, [0, \infty) \ni t \mapsto U_a(t,0)u_0\,]$ *is neither eventually positive nor eventually negative (plus the trivial solution).*

We remark that Theorems 3.3.1–3.3.3 have been proved for the Dirichlet boundary conditions case in [60] and [61] (see Section 3.6). We shall provide unified proofs of Theorems 3.3.1–3.3.3 (i.e., proofs which apply to all three, Dirichlet, Neumann, Robin, boundary conditions cases). In order to do so, we first introduce three positive constants M_1, M_2, M_3 by the following:

$$\begin{cases} \|U_a(t,0)u_0\| \leq M_1\|u_0\| & \text{if } u_0 \in L_2(D) \\ \|U_a(t,0)u_0\|_\infty \leq M_1\|u_0\|_\infty & \text{if } u_0 \in L_\infty(D) \\ \|U_a(t+1/2,0)u_0\|_\infty \leq M_1\|u_0\| & \text{if } u_0 \in L_2(D) \end{cases} \qquad (3.3.2)$$

for any $a \in Y_0$ and $0 \leq t \leq 1$,

$$\|u\| \leq M_2\|u\|_\infty \qquad (3.3.3)$$

for any $u \in L_\infty(D)$, and

$$M_3 := \|(d(\cdot))^\varsigma\|, \qquad (3.3.4)$$

where ς is a nonnegative constant as in (A3-2).

The existence of M_1 is guaranteed by Proposition 2.2.2.

Observe that for any $u_0 \in L_2(D)$, by Proposition 2.2.2, $U_a(t,0)u_0 \in L_\infty(D)$ for $t > 0$. Notice that if (A3-2) holds, then for any positive weak solution u on $[0, \infty) \times D$

$$\|u(t)\| \geq M_4\|u(t)\|_\infty \qquad (3.3.5)$$

for $t \geq \delta_2$, where $M_4 = C_2 M_3$.

We now show several lemmas, among which, some lemmas follow from the arguments in [60] and [61]. For convenience, we provide proofs here. They will be formulated for the problem (2.0.1)+(2.0.2) only. Their analogs for the adjoint problem (2.3.1)+(2.3.2) are straightforward.

First of all, for convenience, we restate the interior Harnack inequality (see Proposition 2.2.8)

LEMMA 3.3.2
For any $t_1 > 0$ there is $0 < \delta_3 < 1$ such that for any $0 < \delta < \delta_3$ there is $C_3 > 0$ with the property that

$$(U_a(t, 0)u_0)(y) \leq C_3\,(U_a(t + \tau, 0)u_0)(x)$$

for any $a \in Y_0$, $t \geq \delta^2$, $\delta^2 \leq \tau \leq t_1$, $u_0 \in L_2(D)^+$, and any $x, y \in D^\delta := \{\xi \in D : d(\xi) > \delta\}$.

LEMMA 3.3.3
Assume (A3-2). Then there are $0 < \delta_4 \leq 1$ and $C_4 > 0$ such that

$$\|U_a(t + \tau, 0)u_0\|_\infty \geq C_4 \|U_a(t, 0)u_0\|_\infty \tag{3.3.6}$$

for any $a \in Y_0$, $t \geq \delta_4$, $\tau \in [0, 2]$, and $u_0 \in L_2(D)^+$.

PROOF   In Lemma 3.3.2 take $t_1 = 2$, fix $0 < \delta < \delta_3$ and fix an $x_0 \in D^\delta$. We have the existence of $C_3 > 0$ such that

$$(U_a(t + \tau, 0)u_0)(x) \geq \frac{1}{C_3}(U_a(t, 0)u_0)(x_0)$$

for any $a \in Y_0$, $t \geq \delta^2$, $\delta^2 \leq \tau \leq 2$, $u_0 \in L_2(D)^+$, and $x \in D^\delta$. Further, fix $0 < \delta_2 \leq \delta^2$. By (A3-2) there is $C_2 > 0$ such that

$$(U_a(t, 0)u_0)(x_0) \geq C_2(d(x_0))^\varsigma \|U_a(t, 0)u_0\|_\infty$$

for any $a \in Y_0$, any $u_0 \in L_2(D)$, and any $t \geq \delta_2$. Consequently,

$$(U_a(t + \tau, 0)u_0)(x) \geq \frac{C_2}{C_3}(d(x_0))^\varsigma \|U_a(t, 0)u_0\|_\infty$$

for any $a \in Y_0$, $t \geq \delta^2$, $\delta^2 \leq \tau \leq 2$, $u_0 \in L_2(D)^+$, and $x \in D^\delta$, hence

$$\|U_a(t + \tau, 0)u_0\|_\infty \geq \frac{C_2}{C_3}(d(x_0))^\varsigma \|U_a(t, 0)u_0\|_\infty \tag{3.3.7}$$

for any $a \in Y_0$, $t \geq \delta^2$, $\delta^2 \leq \tau \leq 2$, and $u_0 \in L_2(D)^+$.
On the other hand, by (3.3.2) there holds

$$\|U_a(t + \delta^2, 0)u_0\|_\infty \leq M_1 \|U_a(t + \tau, 0)u_0\|_\infty \tag{3.3.8}$$

for any $a \in Y_0$, $t > 0$, $0 \le \tau \le \delta^2$, and $u_0 \in L_2(D)^+$. It then follows that (3.3.6) holds with $\delta_4 = \delta^2$ and $C_4 = C_2(d(x_0))^\varsigma / C_3$ (notice that $M_1 \ge 1$). □

LEMMA 3.3.4
Assume (A3-2). Then there are $0 < \delta_4 \le 1$ and $\tilde{C}_4 > 0$ such that

$$\|U_a(t+1,0)u_0\| \ge \tilde{C}_4 \|U_a(t,0)u_0\|$$

for any $a \in Y_0$, $t \ge \delta_4$, and $u_0 \in L_2(D)^+$.

PROOF By (3.3.6) there is $\delta_4 > 0$ such that

$$\|U_a(t+2,0)u_0\|_\infty \ge C_4 \|U_a(t,0)u_0\|_\infty$$

for $t \ge \delta_4$. (3.3.2) gives

$$\|U_a(t+2,0)u_0\|_\infty \le M_1 \|U_a(t+1,0)u_0\|$$

for $t \ge 0$. By (3.3.3),

$$\|U_a(t,0)u_0\| \le M_2 \|U_a(t,0)u_0\|_\infty$$

for $t \ge \delta_4$. It then follows that

$$\|U_a(t+1,0)u_0\| \ge \tilde{C}_4 \|U_a(t,0)u_0\|$$

for $t \ge \delta_4$, where $\tilde{C}_4 = \frac{C_4}{M_1 M_2}$. □

LEMMA 3.3.5
Assume (A3-1). Then for each $\delta_1 > 0$ there is $C_1 > 1$ such that for any $a \in Y_0$, if u_1, u_2 are global weak solutions of $(2.0.1)_a + (2.0.2)_a$ on $[0, \infty) \times D$, with u_2 being positive $(u_1(0), u_2(0) \in L_2(D))$, then the functions $\varrho_{\min}, \varrho_{\max} : (0, \infty) \to \mathbb{R}$ defined as

$$\varrho_{\min}(t) := \inf_{x \in D} \frac{u_1(t)(x)}{u_2(t)(x)}, \qquad \varrho_{\max}(t) := \sup_{x \in D} \frac{u_1(t)(x)}{u_2(t)(x)}$$

enjoy the following properties:

(1) $\varrho_{\min}(\cdot)$ *is nondecreasing and* $\varrho_{\max}(\cdot)$ *is nonincreasing on* $(0, \infty)$.

(2) $\varrho_{\max}(t) - \varrho_{\min}(t) \le \left(1 - \frac{1}{C_1}\right)(\varrho_{\max}(\tau) - \varrho_{\min}(\tau))$ *for* $0 < \tau < \tau + \delta_1 \le t$.

PROOF (1) Fix a $t_0 > 0$ and put $w(t) := u_1(t) - \varrho_{\min}(t_0)u_2(t)$. Then $w(t_0) \ge 0$, hence by Proposition 2.2.7, $w(t) \ge 0$ for $t > t_0$. It then follows that $\varrho_{\min}(t) \geq \varrho_{\min}(t_0)$ for $t > t_0$, consequently, $\varrho_{\min}(\cdot)$ is nondecreasing. In a similar way we prove that $\varrho_{\max}(\cdot)$ is nonincreasing.

(2) Put $\varrho(t) := \varrho_{\max}(t) - \varrho_{\min}(t)$, $t > 0$. Fix a $\tau > 0$. Put $\hat{u}_1(t) := u_1(t) - \varrho_{\min}(\tau)u_2(t)$, and define

$$\hat{\varrho}_{\min}(t) := \inf_{x \in D} \frac{\hat{u}_1(t)(x)}{u_2(t)(x)}, \quad \hat{\varrho}_{\max}(t) := \sup_{x \in D} \frac{\hat{u}_1(t)(x)}{u_2(t)(x)}, \quad \hat{\varrho}(t) := \hat{\varrho}_{\max}(t) - \hat{\varrho}_{\min}(t),$$

for $t \geq \tau$. We have $\hat{u}_1(\tau) \in L_2(D)^+$. Assume first that $\hat{u}_1(\tau) \neq 0$. Then $\hat{u}_1$ and u_2 are positive weak solutions on $[\tau, \infty) \times D$. Notice that $\hat{\varrho}_{\min}(t) = \varrho_{\min}(t) - \varrho_{\min}(\tau)$ and $\hat{\varrho}_{\max}(t) = \varrho_{\max}(t) - \varrho_{\min}(\tau)$ for $t \geq \tau$, consequently $\hat{\varrho}(t) = \varrho(t)$ for any $t \geq \tau$. With the help of (A3-1) we obtain

$$\varrho(t) = \hat{\varrho}(t) \leq \left(1 - \frac{1}{C_1}\right)\hat{\varrho}_{\max}(t) = \left(1 - \frac{1}{C_1}\right)(\varrho_{\max}(t) - \varrho_{\min}(\tau)) \leq \left(1 - \frac{1}{C_1}\right)\varrho(\tau)$$

for any $t \geq \tau + \delta_1$. The case $\hat{u}_1(\tau) = 0$ means that the solutions u_1 and u_2 are proportional on $[\tau, \infty)$, which implies that $\varrho(t) = 0$ for all $t \in [\tau, \infty)$. ⬛

LEMMA 3.3.6

Assume (A3-2). There are $\tilde{C}_1, \tilde{C}_2 > 0$ such that if u_1 is a weak solution on $[0, \infty) \times D$ with $u_1(0) \in L_2(D)$ and u_2 is a positive weak solution on $[-1, \infty) \times D$ with $u_2(-1) \in L_2(D)^+$ then

(1) $\dfrac{\|u_1(t)\|_\infty}{\|u_2(t)\|_\infty} \leq \tilde{C}_1 \dfrac{\|u_1(1/2)\|}{\|u_2(1/2)\|}$ *for $t \in [1, 2]$,*

(2) $\dfrac{\|u_1(t)\|}{\|u_2(t)\|} \leq \tilde{C}_2 \dfrac{\|u_1(0)\|}{\|u_2(0)\|}$ *for $t \in [0, 1]$.*

PROOF   (1) First of all, by (3.3.2) $\|u_1(t)\|_\infty \leq M_1\|u_1(1/2)\|$ for $t \in [1, 2]$. By (3.3.6) and (3.3.3),

$$\|u_2(t)\|_\infty \geq C_4\|u_2(1/2)\|_\infty \geq \frac{C_4}{M_2}\|u_2(1/2)\|$$

for $t \in [1, 2]$. (1) therefore follows with $\tilde{C}_1 = \frac{M_1 M_2}{C_4}$.

(2) By (3.3.2), $\|u_1(t)\| \leq M_1\|u_1(0)\|$ for $t \in [0, 1]$. By (3.3.5) with $\delta_2 = 1$, (3.3.6) and (3.3.3),

$$\|u_2(t)\| \geq M_4\|u_2(t)\|_\infty \geq C_4 M_4\|u_2(0)\|_\infty \geq \frac{C_4 M_4}{M_2}\|u_2(0)\|$$

for $t \in [0, 1]$. (2) then follows with $\tilde{C}_2 = \frac{M_1 M_2}{C_3 M_4}$.   ⬛

LEMMA 3.3.7

Assume (A3-1). There is $\tilde{C}_0 > 0$ such that if u_1 is a weak solution on $[0, \infty) \times D$ ($u_1(0) \in L_2(D)$) which is neither eventually positive nor eventually

*negative, u_2 is a positive weak solution on $[-1, \infty) \times D$ $(u_2(-1) \in L_2(D)^+)$,
and*

$$\|u_1(k)\|_\infty \leq \eta_0 \|u_1(k+1)\|_\infty$$

for some $\eta_0 > 0$ and some $k \geq 1$, then

$$\sup_{x \in D} \frac{|u_1(k+1)(x)|}{u_2(k+1)(x)} \leq \tilde{C}_0 \eta_0 \frac{\|u_1(k+1)\|_\infty}{\|u_2(k+1)\|_\infty}.$$

PROOF It can be proved by arguments similar to those in [61, Lemma 6.1] (for parabolic equations with Dirichlet boundary conditions). For completeness, we provide a proof here.

Put $u(t) := U_a(t, k)(u_1)_+(k)$, $v(t) := U_a(t, k)(u_1)_-(k)$, $t \geq k$. Then by (A3-1) with $\delta_1 = 1$

$$\sup_{x \in D} \frac{v(k+1)(x)}{u(k+1)(x)} \leq C_1 \inf_{x \in D} \frac{v(k+1)(x)}{u(k+1)(x)}.$$

Without loss of generality we may assume that

$$\inf_{x \in D} \frac{v(k+1)(x)}{u(k+1)(x)} \leq 1$$

(otherwise, we just exchange the roles of u and v).

By (3.3.2) and the assumption of the lemma,

$$\|u(k+1)\|_\infty \leq M_1 \|u(k)\|_\infty = M_1 \|(u_1)_+(k)\|_\infty$$
$$\leq M_1 \|u_1(k)\|_\infty \leq M_1 \eta_0 \|u_1(k+1)\|_\infty.$$

Note that

$$\inf_{x \in D} \frac{u(k+1)(x)}{u_2(k+1)(x)} \leq \frac{\|u(k+1)\|_\infty}{\|u_2(k+1)\|_\infty} \leq \sup_{x \in D} \frac{u(k+1)(x)}{u_2(k+1)(x)}.$$

Consequently

$$\sup_{x \in D} \frac{|u_1(k+1)(x)|}{u_2(k+1)(x)} = \sup_{x \in D} \left(\frac{u(k+1)(x)}{u_2(k+1)(x)} \frac{|u_1(k+1)(x)|}{u(k+1)(x)} \right)$$
$$\leq \sup_{x \in D} \frac{u(k+1)(x)}{u_2(k+1)(x)} \cdot \sup_{x \in D} \frac{|u_1(k+1)(x)|}{u(k+1)(x)}$$
$$\leq C_1 \inf_{x \in D} \frac{u(k+1)(x)}{u_2(k+1)(x)} \cdot \sup_{x \in D} \frac{u(k+1)(x) + v(k+1)(x)}{u(k+1)(x)}$$
$$\leq C_1 (1 + C_1) \frac{\|u(k+1)\|_\infty}{\|u_2(k+1)\|_\infty}$$
$$\leq M_1 C_1 (1 + C_1) \eta_0 \frac{\|u_1(k+1)\|_\infty}{\|u_2(k+1)\|_\infty}.$$

The lemma then follows with $\tilde{C}_0 = M_1 C_1 (1 + C_1)$. $\square$

LEMMA 3.3.8
*Assume (A3-1)–(A3-2). There are $\tilde{M} > 0$ and $\gamma_0 > 0$ such that if u_1 is a weak
solution on $[0, \infty) \times D$ ($u_1(0) \in L_2(D)$) which is neither eventually positive
nor eventually negative and u_2 is a positive weak solution on $[-1, \infty) \times D$
($u_2(-1) \in L_2(D)^+$) then*

$$\frac{\|u_1(t)\|_\infty}{\|u_2(t)\|_\infty} \le \tilde{M} e^{-\gamma_0 t} \frac{\|u_1(1)\|_\infty}{\|u_2(1)\|_\infty} \quad \text{for} \quad t > 1.$$

PROOF   It can be proved by arguments similar to those in [61, Theorem 2.2] (for parabolic equations with Dirichlet boundary conditions). For completeness, we provide a proof here.

First of all, by (3.3.2),

$$\|u_1(t)\|_\infty \le M_1 \|u_1([t])\|_\infty$$

for any $t \ge 1$, which together with (3.3.6) implies that

$$\frac{\|u_1(t)\|_\infty}{\|u_2(t)\|_\infty} \le \frac{M_1}{C_4} \frac{\|u_1(\lfloor t \rfloor)\|_\infty}{\|u_2(\lfloor t \rfloor)\|_\infty}, \quad t \ge 1. \tag{3.3.9}$$

Put $\varrho_0 := 1 - \frac{1}{C_1}$, where C_1 is as in (A3-1) with $\delta_1 = 1$. Clearly we have either
(i) For all $k \in \mathbb{N}$, $\frac{\|u_1(k+1)\|_\infty}{\|u_1(k)\|_\infty} < C_4 \varrho_0$,
or
(ii) For some $k \in \mathbb{N}$, $\frac{\|u_1(k+1)\|_\infty}{\|u_1(k)\|_\infty} \ge C_4 \varrho_0$.

Assume that (i) holds. Note that by (3.3.6),

$$\frac{\|u_2(k+1)\|_\infty}{\|u_2(k)\|_\infty} \ge C_4$$

for $k \in \mathbb{N}$. Hence

$$\frac{\|u_1(k+1)\|_\infty}{\|u_2(k+1)\|_\infty} \le \varrho_0 \frac{\|u_1(k)\|_\infty}{\|u_2(k)\|_\infty}$$

for $k \in \mathbb{N}$. Therefore

$$\frac{\|u_1(t)\|_\infty}{\|u_2(t)\|_\infty} \le \frac{M_1}{C_4} \frac{\|u_1(\lfloor t \rfloor)\|_\infty}{\|u_2(\lfloor t \rfloor)\|_\infty} \le \frac{M_1}{C_4} \varrho_0^{\lfloor t \rfloor - 1} \frac{\|u_1(1)\|_\infty}{\|u_2(1)\|_\infty}$$

$$\le \frac{M_1}{C_4} \varrho_0^{-2} e^{-(-\ln \varrho_0) t} \frac{\|u_1(1)\|_\infty}{\|u_2(1)\|_\infty}$$

for any $t \ge 1$.

Assume that (ii) holds. Let

$$k_0 := \inf\{\, k \in \mathbb{N} : \|u_1(k+1)\|_\infty \ge C_4 \varrho_0 \|u_1(k)\|_\infty \,\}.$$

Arguments as in (i) give

$$\frac{\|u_1(t)\|_\infty}{\|u_2(t)\|_\infty} \le \frac{M_1}{C_4}\varrho_0^{-2}e^{-(-\ln\varrho_0)t}\frac{\|u_1(1)\|_\infty}{\|u_2(1)\|_\infty}$$

for $1 \le t \le k_0 + 1$. By (3.3.9) and Lemma 3.3.5 with $\delta_1 = 1$, for $t \ge k_0 + 1$ we have

$$\begin{aligned}
\frac{\|u_1(t)\|_\infty}{\|u_2(t)\|_\infty} &\le \frac{M_1}{C_4}\frac{\|u_1(\lfloor t \rfloor)\|_\infty}{\|u_2(\lfloor t \rfloor)\|_\infty} \le \frac{M_1}{C_4}\sup_{x\in D}\frac{|u_1(\lfloor t \rfloor)(x)|}{u_2(\lfloor t \rfloor)(x)} \\
&\le \frac{M_1}{C_4}\left(\sup_{x\in D}\frac{u_1(\lfloor t \rfloor)(x)}{u_2(\lfloor t \rfloor)(x)} - \inf_{x\in D}\frac{u_1(\lfloor t \rfloor)(x)}{u_2(\lfloor t \rfloor)(x)}\right) \\
&\le \frac{M_1}{C_4}\varrho_0^{\lfloor t \rfloor - k_0 - 1}\left(\sup_{x\in D}\frac{u_1(k_0 + 1)(x)}{u_2(k_0 + 1)(x)} - \inf_{x\in D}\frac{u_1(k_0 + 1)(x)}{u_2(k_0 + 1)(x)}\right) \\
&\le 2\frac{M_1}{C_4}\varrho_0^{\lfloor t \rfloor - k_0 - 1}\sup_{x\in D}\frac{|u_1(k_0 + 1)(x)|}{u_2(k_0 + 1)(x)}.
\end{aligned}$$

This together with Lemma 3.3.7 implies that

$$\frac{\|u_1(t)\|_\infty}{\|u_2(t)\|_\infty} \le 2M_1\tilde{C}_0\varrho_0^{\lfloor t \rfloor - k_0}\frac{\|u_1(k_0 + 1)\|_\infty}{\|u_2(k_0 + 1)\|_\infty}.$$

But we have already proved that

$$\frac{\|u_1(k_0 + 1)\|_\infty}{\|u_2(k_0 + 1)\|_\infty} \le \frac{M_1}{C_4}\varrho_0^{k_0 - 1}\frac{\|u_1(1)\|_\infty}{\|u_2(1)\|_\infty},$$

which gives

$$\begin{aligned}
\frac{\|u_1(t)\|_\infty}{\|u_2(t)\|_\infty} &\le \frac{2M_1^2\tilde{C}_0}{C_4}\varrho_0^{\lfloor t \rfloor - 1}\frac{\|u_1(1)\|_\infty}{\|u_2(1)\|_\infty} \\
&\le \frac{2M_1^2\tilde{C}_0}{C_4}\varrho_0^{-2}e^{-(-\ln\varrho_0)t}\frac{\|u_1(1)\|_\infty}{\|u_2(1)\|_\infty}
\end{aligned}$$

for $t \ge k_0 + 1$. The lemma then follows with $\tilde{M} = \frac{M_1}{\varrho_0^2 C_4}\max\{1, 2M_1\tilde{C}_0\}$ and $\gamma_0 = -\ln\varrho_0$. □

PROOF (Proof of Theorem 3.3.1)　We prove first the existence of an entire positive solution. Fix $a \in Y_0$ and $u_0 \in L_2(D)^+$ with $\|u_0\| = 1$. Define a sequence $(u_n)_{n=1}^\infty \subset L_2(D)^+ \setminus \{0\}$ by

$$u_n := \frac{U_a(0, -n)u_0}{\|U_a(-1, -n)u_0\|} = U_a(0, -1)\frac{U_a(-1, -n)u_0}{\|U_a(-1, -n)u_0\|}.$$

By Proposition 2.2.2, the set $\{\|u_n\| : n = 1, 2, \dots\}$ is bounded above. Moreover, from the local regularity (Proposition 2.2.4) we deduce that there is a sequence $(n_k)_{k=1}^\infty$ such that $\lim_{k\to\infty} n_k = \infty$ and $\lim_{k\to\infty} u_{n_k} = \tilde{u}_0^{(0)}$ in $L_2(D)$. By Lemma 3.3.4 the set $\{\|u_n\| : n = 1, 2, \dots\}$ is bounded below by $\tilde{C}_4 > 0$, consequently $\tilde{u}_0^{(0)} \in L_2(D)^+ \setminus \{0\}$.

We claim that there is an entire positive weak solution $\hat{u}$ of $(2.0.1)_a + (2.0.2)_a$ such that $\hat{u}(0) = \tilde{u}_0^{(0)}$. We start by finding a sequence $(\tilde{u}_0^{(-l)})_{l=1}^\infty \subset L_2(D)^+ \setminus \{0\}$ and a sequence $(r_l)_{l=1}^\infty \subset (0, \infty)$ such that $U_a(-l+1, -l)\tilde{u}_0^{(-l)} = r_l\tilde{u}_0^{(-l+1)}$ for $l = 1, 2, \dots$. Such sequences are constructed by induction on l. We show only the first step (that is, finding $\tilde{u}_0^{(-1)}$ and r_1), the remaining being similar. Put

$$v_k := \frac{U_a(-1, -n_k)u_0}{\|U_a(-2, -n_k)u_0\|} = U_a(-1, -2)\frac{U_a(-2, -n_k)u_0}{\|U_a(-2, -n_k)u_0\|}.$$

By the same reason as above, there are a subsequence $(k_m)_{m=1}^\infty$ and $\tilde{v}_0 \in L_2(D)^+$ with $\|\tilde{v}_0\| \geq \tilde{C}_4$ such that $\lim_{m\to\infty} k_m = \infty$ and $\lim_{m\to\infty} v_{k_m} = \tilde{v}$ in $L_2(D)$. Note that

$$u_{n_k} = r^{(k)} \cdot U_a(0, -1)v_k$$

for each $k \in \mathbb{N}$, where

$$r^{(k)} = \frac{\|U_a(-2, -n_k)u_0\|}{\|U_a(-1, -n_k)u_0\|}.$$

From $(3.3.2)$ it follows that the set $\{r^{(k)}; k = 1, 2, \dots\}$ is bounded below by $(M_1)^{-1} > 0$, and from Lemma 3.3.4 it follows that the set $\{r^{(k)} : k = 1, 2, \dots\}$ is bounded above by $(\tilde{C}_4)^{-1}$. Therefore we may assume without loss of generality that $\lim_{m\to\infty} r^{(k_m)} =: r > 0$. Then we have

$$\tilde{u}_0^{(0)} = r \cdot U_a(0, -1)\tilde{v}_0$$

It suffices to take $\tilde{u}_0^{(-1)} := \tilde{v}_0$ and $r_1 := r$. By a diagonal process we obtain sequences $(\tilde{u}_0^{(-l)})$ and a sequence (r_l) sought for.

The function $\hat{u}(\cdot)$ defined as

$$\hat{u}(\cdot) := \begin{cases} \dfrac{1}{r_{\lfloor t \rfloor} r_{\lfloor t \rfloor - 1} \dots r_1} U_a(t, \lfloor t \rfloor)\tilde{u}_0^{(\lfloor t \rfloor)} & t < 0, \\ U_a(t, 0)\tilde{u}_0^{(0)} & t \geq 0 \end{cases}$$

is an entire positive weak solution of $(2.0.1)_a + (2.0.2)_a$.

To prove uniqueness, fix $a \in Y_0$ and suppose that $\hat{u}_1$ and $\hat{u}_2$ are two entire positive weak solutions of $(2.0.1)_a + (2.0.2)_a$. Without loss of generality, assume that $\|\hat{u}_1(0)\| = \|\hat{u}_2(0)\| = 1$. Let $\hat{u}(t, x) := \hat{u}_1(t)(x) - \hat{u}_2(t)(x)$. We first claim that for any $t < 0$, there is $x(t) \in D$ such that $\hat{u}(t, x(t)) = 0$. For otherwise, if there is $t < 0$ such that $\hat{u}(t, x) > 0$ for all $x \in D$, then $\hat{u}(0, x) > 0$ or $\hat{u}_1(0)(x) > \hat{u}_2(0)(x)$ for all $x \in D$. This implies $\|\hat{u}_1(0)\| > \|\hat{u}_2(0)\|$. This is a contradiction. Similarly, if $\hat{u}(t, x) < 0$ for some $t < 0$ and all $x \in D$, then $\hat{u}_1(0)(x) < \hat{u}_2(0)(x)$ for all $x \in D$. This implies that $\|\hat{u}_1(0)\| < \|\hat{u}_2(0)\|$, a contradiction again. Therefore for any $t < 0$, there is $x(t) \in D$ such that $\hat{u}(t, x(t)) = 0$. This implies that

$$\varrho_{\min}(t) := \inf_{x \in D} \frac{\hat{u}_1(t)(x)}{\hat{u}_2(t)(x)} \leq 1$$

for all $t < 0$. Then by (A3-1),

$$\varrho_{\max}(t) := \sup_{x \in D} \frac{\hat{u}_1(t)(x)}{\hat{u}_2(t)(x)} \leq C_1$$

for all $t < 0$. Hence $\varrho_{\max}(t) - \varrho_{\min}(t)$ is bounded for $t < 0$. Define $\varrho(t) := \varrho_{\max}(t) - \varrho_{\min}(t)$, $(t > 0)$. By Lemma 3.3.5(2) with $\delta_1 = 1$

$$\varrho(t) \leq \left(1 - \frac{1}{C_1}\right)\varrho(s)$$

for $t \geq s + 1$, $t, s \in \mathbb{R}$. This implies that $\varrho_{\max}(t) = \varrho_{\min}(t)$ for $t \in \mathbb{R}$, hence $\hat{u}_1(t)(x) = \hat{u}_2(t)(x)$ for all $t \in \mathbb{R}$ and all $x \in D$.

Now, for each $a \in Y_0$ denote by $w(a)$ the value at time 0 of the unique positive entire weak solution of $(2.0.1)_a + (2.0.2)_a$, normalized so that $\|w(a)\| = 1$. We want to show that $w \colon Y_0 \to L_2(D)$ is continuous. Fix a sequence $(a^{(n)}) \subset Y_0$ converging, as $n \to \infty$, to $\tilde{a} \in Y_0$. By the uniqueness of entire positive solutions,

$$w(a^{(n)}) = \frac{U_{a_n}(0, -1)w(a^{(n)} \cdot (-1))}{\|U_{a^{(n)}}(0, -1)w(a^{(n)} \cdot (-1))\|}$$

and

$$U_{a^{(n)}}(0, -1)w(a^{(n)} \cdot (-1)) = \frac{U_{a^{(n)}}(0, -2)w(a^{(n)} \cdot (-2))}{\|U_{a^{(n)}}(-1, -2)w(a^{(n)} \cdot (-2))\|}.$$

From the local regularity (Proposition 2.2.4) we obtain that there is a sequence $(n_k)_{k=1}^\infty$ such that $\lim_{k \to \infty} n_k = \infty$ and $\lim_{k \to \infty} U_{a^{(n_k)}}(0, -1)w(a^{(n_k)} \cdot (-1)) = \tilde{u}_0$ in $L_2(D)$. Lemma 3.3.3 implies

$$\|U_{a^{(n)}}(0, -1)w(a^{(n)} \cdot (-1))\| \geq \tilde{C}_4$$

for all $n \in \mathbb{N}$. Consequently, by extracting again a subsequence, if necessary, we can assume that $w(a^{(n_k)}) \to w_0$ in $L_2(D)$, where $\|w_0\| = 1$.

By a diagonal process as in the proof of the existence we obtain that there are a subsequence $(n_{k_m})_{m=1}^\infty$ with limit ∞, a positive entire weak solution $\tilde{u}$ of $(2.0.1)_a + (2.0.2)_a$ and a sequence $(r_l)_{l=0}^\infty$ such that

$$\tilde{u}(-l) = r_l \lim_{m \to \infty} w(a^{(n_{k_m})} \cdot (-l))$$

for each $l = 0, 1, 2, \dots$. In particular, $\tilde{u}(0) = w_0$. By uniqueness, $w_0 = w(\tilde{a})$.

□

PROOF (Proof of Theorem 3.3.2)   It can be proved by arguments similar to those in Theorem 3.3.1.   ☐

PROOF (Proof of Theorem 3.3.3)   First, we prove that Π admits an exponential separation over Y_0 with an invariant one-dimensional subbundle given by $X_1(a) = \mathrm{span}\{w(a)\}$ and an invariant one-codimensional subbundle given by $X_2(a) = \{\, v \in L_2(D) : \langle v, w^*(a)\rangle = 0 \,\}$.

First of all, $\langle w(a), w^*(a)\rangle > 0$ for any $a \in Y_0$. Therefore $L_2(D) = X_1(a) \oplus X_2(a)$ for any $a \in Y_0$.

Clearly, X_1 and X_2 are invariant.

Now, for any $u_0 \in X_2(a)$, by Lemma 3.3.6(2),

$$\frac{\|U_a(t,0)u_0\|}{\|U_a(t,0)w(a)\|} \leq \tilde{C}_2 \frac{\|u_0\|}{\|w(a)\|}$$

for $0 < t \leq 1$. By Lemma 3.3.8,

$$\frac{\|U_a(t,0)u_0\|_\infty}{\|U_a(t,0)w(a)\|_\infty} \leq \tilde{M} e^{-\gamma_0 t} \frac{\|U_a(1,0)u_0\|_\infty}{\|U_a(1,0)w(a)\|_\infty}$$

for $t \geq 1$. Applying Lemma 3.3.6 again we get

$$\frac{\|U_a(1,0)u_0\|_\infty}{\|U_a(1,0)w(a)\|_\infty} \leq \tilde{C}_1 \frac{\|U_a(1/2,0)u_0\|}{\|U_a(1/2,0)w(a)\|} \leq \tilde{C}_1 \tilde{C}_2 \frac{\|u_0\|}{\|w(a)\|}.$$

Note that by (3.3.3),

$$\|U_a(t,0)u_0\| \leq M_2 \|U_a(t,0)u_0\|_\infty$$

and by (3.3.5),

$$M_4 \|U_a(t,0)w(a)\|_\infty \leq \|U_a(t,0)w(a)\|$$

for $t \geq 1$. Therefore we have

$$\frac{\|U_a(t,0)u_0\|}{\|U_a(t,0)w(a)\|} \leq \frac{M_2 \tilde{C}_1 \tilde{C}_2 \tilde{M}}{M_4} e^{-\gamma_0 t} \frac{\|u_0\|}{\|w(a)\|}$$

for $t > 1$.

It then follows that

$$\frac{\|U_a(t,0)u_0\|}{\|U_a(t,0)w(a)\|} \leq M e^{-\gamma_0 t} \frac{\|u_0\|}{\|w(a)\|}$$

for $t > 0$, where $M = \max\{\tilde{C}_2 e^{\gamma_0}, \frac{M_2 \tilde{C}_1 \tilde{C}_2 \tilde{M}}{M_4}\}$. Therefore Π admits an exponential separation over Y_0 with an invariant one-dimensional subbundle $X_1(a) = \mathrm{span}\{w(a)\}$ and an invariant one-codimensional subbundle $X_2(a) = \{\, v \in L_2(D) : \langle v, w^*(a)\rangle = 0 \,\}$.

Next, we prove that $X_2(a)$ is characterized as the set of those $u_0 \in L_2(D)$ such that the global weak solution $[\,[0,\infty) \ni t \mapsto U_a(t,0)u_0\,]$ is neither eventually positive nor eventually negative (plus the trivial solution). First, for a given $a \in Y_0$, by the invariance of X_2, if a nonzero $u_0 \in X_2(a)$ then $U_a(t,0)u_0 \notin L_2(D)^+ \cup (-L_2(D)^+)$ for any $t \geq 0$. Conversely, suppose that $a \in Y_0$ and $u_0 \in L_2(D)$ are such that $U_a(t,0)u_0 \notin L_2(D)^+ \cup (-L_2(D)^+)$ for any $t \geq 0$. If $u_0 \notin X_2(a)$ then there is a nonzero $c \in \mathbb{R}$ such that $u_0 - cw(a) \in X_2(a)$. It then follows that

$$\frac{\|U_a(t,0)(u_0 - cw(a))\|}{\|U_a(t,0)w(a)\|} \leq Me^{-\gamma_0 t}\frac{\|u_0 - cw(a)\|}{\|w(a)\|},$$

which gives

$$\|U_a(t,0)u_0\| \geq \|U_a(t,0)(cw(a))\| - \|U_a(t,0)(u_0 - cw(a))\|$$
$$\geq \left(|c| - e^{-\gamma_0 t}\frac{\|u_0 - cw(a)\|}{\|w(a)\|}\right)\|U_a(t,0)w(a)\|,$$

for $t > 0$. This implies that

$$\liminf_{t\to\infty}\frac{1}{t}\ln\|U_a(t,0)u_0\| \geq \liminf_{t\to\infty}\frac{1}{t}\ln\|U_a(t,0)w(a)\|.$$

On the other hand, by Lemma 3.3.8, we have

$$\liminf_{t\to\infty}\frac{1}{t}\ln\|U_a(t,0)u_0\|_\infty < \liminf_{t\to\infty}\frac{1}{t}\ln\|U_a(t,0)w(a)\|_\infty.$$

Note that, by the inequality $\|U_a(t,0)w(a)\| \leq |D|^{1/2}\|U_a(t,0)w(a)\|_\infty$ and Eq. (3.3.5) with $\delta_2 = 1$, there holds

$$\liminf_{t\to\infty}\frac{1}{t}\ln\|U_a(t,0)w(a)\| = \liminf_{t\to\infty}\frac{1}{t}\ln\|U_a(t,0)w(a)\|_\infty.$$

Further, by the inequality $\|U_a(t,0)u_0\| \leq |D|^{1/2}\|U_a(t,0)u_0\|_\infty$ we have

$$\liminf_{t\to\infty}\frac{1}{t}\ln\|U_a(t,0)u_0\| \leq \liminf_{t\to\infty}\frac{1}{t}\ln\|U_a(t,0)u_0\|_\infty.$$

Consequently,

$$\liminf_{t\to\infty}\frac{1}{t}\ln\|U_a(t,0)u_0\| \leq \liminf_{t\to\infty}\frac{1}{t}\ln\|U_a(t,0)u_0\|_\infty$$
$$< \liminf_{t\to\infty}\frac{1}{t}\ln\|U_a(t,0)w(a)\|_\infty = \liminf_{t\to\infty}\frac{1}{t}\ln\|U_a(t,0)w(a)\|$$
$$\leq \liminf_{t\to\infty}\frac{1}{t}\ln\|U_a(t,0)u_0\|,$$

which is a contradiction. Therefore $u_0 \in X_2(a)$. $\qquad\square$

THEOREM 3.3.4

Assume that (A3-1) *and* (A3-2) *are satisfied by both* (2.0.1)+(2.0.2) *and* (2.3.1)+(2.3.2). *Then for each* $\delta_0 > 0$ *there exists* $K = K(\delta_0) > 0$ *such that*

$$\frac{\|P_2(a \cdot t)U_a(t,0)u_0\|}{\|P_1(a \cdot t)U_a(t,0)u_0\|} \leq K$$

for all $t \geq \delta_0$, *all* $a \in Y_0$, *and all* $u_0 \in L_2(D)^+ \setminus \{0\}$.

PROOF   Eq. (3.2.2) gives

$$P_1(a)u = \frac{\langle u, w^*(a) \rangle}{\langle w(a), w^*(a) \rangle} w(a)$$

for any $a \in Y_0$ and any $u \in L_2(D)$. By the pointwise Harnack inequality (A3-2), there are $C_2 > 0$ and $\varsigma \geq 0$, $\varsigma' \geq 0$ such that

$$(U_a(t,0)u_0)(x) \geq C_2(d(x))^\varsigma \|U_a(t,0)u_0\|_\infty$$

for any $t \geq \delta_0/2$, $a \in Y_0$, $x \in D$ and nonzero $u_0 \in X^+$, and

$$w^*(a)(x) \geq C_2(d(x))^{\varsigma'} \|w^*(a)\|_\infty$$

for any $a \in Y_0$ and $x \in D$. Further, (3.3.3) yields $\|w(a)\|_\infty \geq (M_2)^{-1}$ and $\|w^*(a)\|_\infty \geq (M_2)^{-1}$ for all $a \in Y_0$. It follows that

$$\langle U_a(t,0)u_0, w^*(a \cdot t) \rangle = \int_D w^*(a \cdot t)(x) (U_a(t,0)u_0)(x) \, dx \geq \tilde{M}_1 \|U_a(t,0)u_0\|_\infty$$

for any $a \in Y_0$, $t \geq \delta_0/2$ and any nonzero $u_0 \in L_2(D)^+$, where

$$\tilde{M}_1 := \frac{(C_2)^2}{M_2} \int_D (d(x))^{\varsigma+\varsigma'} \, dx > 0.$$

We have thus obtained that

$$(P_1(a \cdot t)U_a(t,0)u_0)(x) \geq \tilde{M}_2 \|U_a(t,0)u_0\|_\infty w(a)(x)$$

for all $u_0 \in L_2(D)^+ \setminus \{0\}$, $a \in Y_0$, $t \geq \delta_0/2$ and a.e. $x \in D$, where

$$\tilde{M}_2 := \frac{\tilde{M}_1}{\sup\{\langle w(a), w^*(a) \rangle : a \in Y_0\}} > 0.$$

Consequently,

$$\|P_1(a \cdot (\delta_0/2))U_a(\delta_0/2, 0)u_0\| \geq \tilde{M}_2 \|U_a(\delta_0/2, 0)u_0\|_\infty$$

$$\geq \frac{\tilde{M}_2}{M_2} \|U_a(\delta_0/2, 0)u_0\| \qquad \text{by (3.3.3)}$$

for all $u_0 \in L_2(D)^+ \setminus \{0\}$ and $a \in Y_0$. Let $\tilde{M}_3 := \sup\{\|P_2(a)\| : a \in Y_0\}$ $(< \infty)$. There holds

$$\frac{\|P_2(a \cdot (\delta_0/2))U_a(\delta_0/2, 0)u_0\|}{\|P_1(a \cdot (\delta_0/2))U_a(\delta_0/2, 0)u_0\|} \leq \frac{M_2 \tilde{M}_3}{\tilde{M}_2}$$

for all nonzero $u_0 \in L_2(D)^+$ and all $a \in Y_0$.

An exponential separation (Definition 3.2.1(iii)) gives

$$\frac{\|P_2(a \cdot t)U_a(t, 0)u_0\|}{\|P_1(a \cdot t)U_a(t, 0)u_0\|} \leq \frac{MM_2 \tilde{M}_3}{\tilde{M}_2} e^{-\gamma_0 \delta_0/2}$$

for all nonzero $u_0 \in L_2(D)^+$, all $t \geq \delta_0$ and all $a \in Y_0$. $\qquad\qquad \square$

3.4   Multiplicative Ergodic Theorems

In this section we will give applications of multiplicative ergodic theorems (Oseledets-type theorems) to the linear topological skew-product semiflow $\Pi = \{\Pi_t\}_{t \geq 0}$ generated on $L_2(D) \times Y$ by (2.0.1)+(2.0.2),

$$\Pi_t(u_0, a) = (U_a(t, 0)u_0, a \cdot t)$$

for $u_0 \in L_2(D)$ and $a \in Y$ (see (2.2.6)).

Throughout the present section we assume that (2.0.1)+(2.0.2) satisfies (A2-1)–(A2-3). Assume that Y_0 is a compact connected invariant subset of Y, and μ is an ergodic invariant measure on Y_0. Moreover, we assume that Π admits an exponential separation over Y_0 with an invariant one-dimensional subbundle X_1 and an invariant one-codimensional subbundle X_2. We also assume that for each $a \in Y$ and $t > 0$ the linear operator $U_a(t, 0)$ is injective.

For $t < 0$, $a \in Y_0$ and $u_0 \in L_2(D)$ the symbol $U_a(t, 0)u_0$ stands for $v_0 \in L_2(D)$ such that $U_{a \cdot t}(-t, 0)v_0 = u_0$. By injectivity, such a v_0, if it exists, is unique.

Let $\lambda(\mu)$ be the principal Lyapunov exponent of Π for the ergodic invariant measure μ. For $a \in Y_0$ let $w(a) \in L_2(D)^+$ be as in Definition 3.2.1.

THEOREM 3.4.1

There exists a Borel set $\tilde{Y}_0 \subset Y_0$, $\mu(\tilde{Y}_0) = 1$, with the property that one of the following (mutually exclusive) cases holds:

(1) *There are:*

     (a) *$k \ (\geq 1)$ real numbers $\lambda_1(\mu) > \cdots > \lambda_k(\mu)$, and*

(b) *k measurable families* $\{E_1(\mu; a)\}_{a \in \tilde{Y}_0}, \ldots, \{E_k(\mu; a)\}_{a \in \tilde{Y}_0}$, *of vector subspaces of constant finite dimensions, and a measurable family* $\{F_\infty(\mu; a)\}_{a \in \tilde{Y}_0}$ *of infinite dimensional vector subspaces such that*

- $U_a(t, 0)E_i(\mu; a) = E_i(\mu; a \cdot t)$ $(i = 1, 2, \ldots)$ *and* $U_a(t, 0)F_\infty(\mu; a)$ $\subset F_\infty(\mu; a \cdot t)$, *for any* $a \in \tilde{Y}_0$ *and* $t \geq 0$,
- $E_1(\mu; a) \oplus \cdots \oplus E_k(\mu; a) \oplus F_\infty(\mu; a) = L_2(D)$ *for any* $a \in \tilde{Y}_0$,
- $\lim_{t \to \infty} (1/t) \ln \|U_a(t, 0)u_0\| = \lambda_i(\mu)$ *for any* $a \in \tilde{Y}_0$ *and any nonzero* $u_0 \in E_i(\mu; a)$ $(i = 1, \ldots, k)$, *and*
- $\lim_{t \to \infty} (1/t) \ln \|U_a(t, 0)u_0\| = -\infty$ *for any* $a \in \tilde{Y}_0$ *and any nonzero* $u_0 \in F_\infty(\mu; a)$.

Further, for each $i = 1, \ldots, k$ *and each* $a \in \tilde{Y}_0$, $E_i(\mu; a) \setminus \{0\}$ *is characterized as the set of those nonzero* $u_0 \in L_2(D)$ *for which* $U_a(t, 0)u_0$ *exists for all* $t \in \mathbb{R}$ *and*

$$\lim_{t \to \pm\infty} \frac{\ln \|U_a(t, 0)u_0\|}{t} = \lambda_i(\mu)$$

holds.

(2) *There are:*

(a) *a sequence of real numbers* $\lambda_1(\mu) > \cdots > \lambda_i(\mu) > \lambda_{i+1}(\mu) > \ldots$ *having limit* $-\infty$, *and*

(b) *countably many measurable families* $\{E_1(\mu; a)\}_{a \in \tilde{Y}_0}, \{E_2(\mu; a)\}_{a \in \tilde{Y}_0}$, $\ldots$, *of vector subspaces of constant finite dimensions, and countably many measurable families* $\{F_1(\mu; a)\}_{a \in \tilde{Y}_0}, \{F_2(\mu; a)\}_{a \in \tilde{Y}_0}, \ldots$, *of vector subspaces of constant finite codimensions such that there holds:*

- $U_a(t, 0)E_i(\mu; a) = E_i(\mu; a \cdot t)$ *and* $U_a(t, 0)F_i(\mu; a) \subset F_i(\mu; a \cdot t)$ $(i = 1, 2, \ldots)$, *for any* $a \in \tilde{Y}_0$ *and* $t \geq 0$,
- $E_1(\mu; a) \oplus \ldots E_i(\mu; a) \oplus F_i(\mu; a) = L_2(D)$ *and* $F_i(\mu; a) = E_{i+1}(\mu; a) \oplus F_{i+1}(\mu; a)$ *for any* $a \in \tilde{Y}_0$ $(i = 1, 2, \ldots)$,
- $\lim_{t \to \infty} (1/t) \ln \|U_a(t, 0)u_0\| = \lambda_i(\mu)$ *for any* $a \in \tilde{Y}_0$ *and any nonzero* $u_0 \in E_i(\mu; a)$ $(i = 1, 2, \ldots)$, *and*
- $\lim_{t \to \infty} (1/t) \ln \|U_a(t, 0)|_{F_i(\mu; a)}\| = \lambda_{i+1}(\mu)$ *for any* $a \in \tilde{Y}_0$ $(i = 1, 2, \ldots)$.

Further, for each $i \in \mathbb{N}$ *and each* $a \in \tilde{Y}_0$, $E_i(\mu; a) \setminus \{0\}$ *is characterized as the set of those nonzero* $u_0 \in L_2(D)$ *for which* $U_a(t, 0)u_0$ *exists for all* $t \in \mathbb{R}$ *and*

$$\lim_{t \to \pm\infty} \frac{\ln \|U_a(t, 0)u_0\|}{t} = \lambda_i(\mu)$$

holds.

In both cases
$$\lambda_1(\mu) = \lambda(\mu).$$

For the meaning of measurability, see [72]; compare [75].

The decompositions of $L_2(D)$ as in Theorem 3.4.1 are called the *Oseledets splittings* for μ. The real numbers $\lambda_1(\mu) > \lambda_2(\mu) > \ldots$ are referred to as the *Lyapunov exponents* for μ.

PROOF (Proof of Theorem 3.4.1)   The theorem follows from the Multiplicative Ergodic Theorem for Continuous Time Linear Random Dynamical Systems in [72, Theorem 3.3].

We check the applicability of that theorem, namely, the measurability of $[Y_0 \ni a \mapsto U_a(1,0)u_0 \in L_2(D)]$ for $u_0 \in L_2(D)$, the injectivity of $U_a(1,0)$, and the integrability of $f_1(\cdot)$ and $f_2(\cdot)$, where

$$f_1(a) := \sup_{0 \le s \le 1} \ln^+ \|U_a(s,0)\|, \qquad a \in Y_0,$$

and

$$f_2(a) := \sup_{0 \le s \le 1} \ln^+ \|U_{a \cdot s}(1-s,0)\|, \qquad a \in Y_0,$$

where $\ln^+(\cdot) := \max\{\ln(\cdot), 0\}$.

First, for each $u_0 \in L_2(D)$ the mapping

$$[Y_0 \ni a \mapsto U_a(1,0)u_0 \in L_2(D)]$$

is continuous (by Proposition 2.2.12), hence $(\mathfrak{B}(Y_0), \mathfrak{B}(L_2(D)))$-measurable. The injectivity of $U_a(1,0)$ is just an assumption.

We proceed now to show that the mappings $f_1(\cdot)$ and $f_2(\cdot)$ belong to $L_1((Y_0, \mathfrak{B}(Y_0), \mu))$.

We show that both f_1 and f_2 are lower semicontinuous, consequently $(\mathfrak{B}(Y_0), \mathfrak{B}(\mathbb{R}))$-measurable. Assume $a^{(n)} \to a$ in Y_0. For any $\epsilon > 0$ there are $s_\epsilon \in [0,1]$ and $u_\epsilon \in L_2(D)$ with $\|u_\epsilon\| = 1$ such that

$$\ln^+ \|U_a(s_\epsilon,0)u_\epsilon\| + \epsilon \ge f_1(a).$$

Since $\|U_{a^{(n)}}(s_\epsilon,0)u_\epsilon\| \to \|U_a(s_\epsilon,0)u_\epsilon\|$ as $n \to \infty$ (this is obvious if $s_\epsilon = 0$, and follows from Proposition 2.2.13 otherwise), there is $n_1 > 0$ such that for $n \ge n_1$,

$$\ln^+ \|U_{a^{(n)}}(s_\epsilon,0)u_\epsilon\| + 2\epsilon \ge f_1(a).$$

Therefore

$$f_1(a^{(n)}) + 2\epsilon \ge f_1(a)$$

for $n \ge n_1$, which implies that f_1 is lower semicontinuous. Note that

$$f_2(a) = \ln^+ \|U_a(1,s)\|.$$

By similar arguments we prove that f_2 is lower semicontinuous, too.

By the L_2–L_2 estimates (see Proposition 2.2.2), we have that f_1 and f_2 are bounded. Therefore they belong to $L_1((Y_0, \mathfrak{B}(Y_0), \mu))$.

The final part of the proof is to exclude the case that

$$\lim_{t \to \infty} \frac{\ln \|U_a(t, 0)\|}{t} = -\infty$$

for μ-a.e. $a \in Y_0$, which is done by applying Lemma 3.1.4. □

In the light of the properties of Lyapunov exponents contained in Theorem 3.4.1 the following corollary is a consequence of Theorem 3.2.4.

COROLLARY 3.4.1

There exists a Borel set $\hat{Y}_1 \subset Y_0$, $\mu(\hat{Y}_1) = 1$, such that either

- $X_1(a) = E_1(\mu; a)$ *and* $X_2(a) = F_\infty(\mu; a)$ *for all $a \in \hat{Y}_1$ (if (1) in Theorem 3.4.1 holds with $k = 1$), or*

- $X_1(a) = E_1(\mu; a)$ *and* $X_2(a) = E_2(\mu; a) \oplus \cdots \oplus E_k(\mu; a) \oplus F_\infty(\mu; a)$ *for all $a \in \hat{Y}_1$ (if (1) in Theorem 3.4.1 holds with $k > 1$), or else*

- $X_1(a) = E_1(\mu; a)$ *and* $X_2(a) = F_1(\mu; a)$ *for all $a \in \hat{Y}_1$ (if (2) in Theorem 3.4.1 holds).*

PROOF   Put $\hat{Y}_1 := Y_1 \cap \tilde{Y}_1 \cap \tilde{Y}_0$, where Y_1 is as in Theorem 3.2.4, $\tilde{Y}_1$ is as in Theorem 3.2.6 and $\tilde{Y}_0$ is as in Theorem 3.4.1. The characterization of $E_1(\mu; a)$ given in Theorem 3.4.1 together with Theorem 3.2.6 implies that $E_1(\mu; a) = X_1(a)$ for all $a \in \hat{Y}_1$.

For each $a \in \hat{Y}_1$ define

$$X_2'(a) := \begin{cases} F_\infty(\mu; a) \\ E_2(\mu; a) \oplus \cdots \oplus E_k(\mu; a) \oplus F_\infty(\mu; a) \\ F_1(\mu; a), \end{cases}$$

depending on which property in Theorem 3.4.1 holds. In each case $X_2'(a)$ is a subspace of $L_2(D)$ of codimension one.

Fix $a \in \hat{Y}_0$. Take any nonzero $u_0 \in X_2'(a)$. If in Theorem 3.4.1 (2) holds then $\limsup_{t \to \infty} (1/t) \ln \|U_a(t, 0)u_0\| \le \lambda_2(\mu) < \lambda(\mu)$, hence $u_0 \in X_2(a)$ by Corollary 3.2.1. If in Theorem 3.4.1 (1) holds with $k = 1$ then we have $\lim_{t \to \infty} (1/t) \ln \|U_a(t, 0)u_0\| = -\infty$, hence $u_0 \in X_2(a)$ by Corollary 3.2.1.

Assume now that (1) in Theorem 3.4.1 holds with $k > 1$. Write u_0 in the $E_2(\mu; a) \oplus \cdots \oplus E_k(\mu; a) \oplus F_\infty(\mu; a)$ decomposition as $u_{j(1)} + \cdots + u_{j(m)}$, where $1 < j(1) < \cdots < j(m) \le \infty$ and all $u_{j(1)}, \ldots, u_{j(m)}$ are nonzero. Take some $\lambda^* \in (\lambda_2(\mu), \lambda_1(\mu))$. There exists $T \ge 0$ such that $\|U_a(t, 0)u_{j(l)}\| \le e^{\lambda^* t}$ for all

$t \geq T$ and all $l = 1, \ldots, m$. Consequently, $\|U_a(t,0)u_0\| \leq me^{\lambda^* t}$ for all $t \geq T$, hence $\limsup_{t \to \infty} (1/t) \ln \|U_a(t,0)u_0\| \leq \lambda^* < \lambda(\mu)$. Therefore $u_0 \in X_2(a)$ by Corollary 3.2.1.

We have proved that $X_2'(a)$ is a one-codimensional subspace contained in the one-codimensional subspace $X_2(a)$. As a consequence, $X_2'(a) = X_2(a)$. □

3.5   The Smooth Case

Consider (2.0.1)+(2.0.2) and assume (A2-5). Then (A2-1)–(A2-3) as well as (A3-1) and (A3-2) are satisfied for both (2.0.1)+(2.0.2) and (2.3.1)+ (2.3.2) (see [59] and [61]).

Let Y_0 be a compact connected invariant subset of Y.

By Theorem 3.3.3, Π admits an exponential separation over Y_0 with invariant complementary subbundles X_1 and X_2, where $X_1(a) = \operatorname{span}\{w(a)\}$ and $X_2(a) = \{v \in L_2(D) : \langle v, w^*(a) \rangle = 0\}$, with $\|w(a)\| = \|w^*(a)\| = 1$ for all $a \in Y_0$.

From now on, until the end of the section, X stands for a Banach space such that

$$W_p^2(D) \hookrightarrow X \hookrightarrow L_2(D) \tag{3.5.1}$$

(recall that $\hookrightarrow$ denotes a compact embedding).

Examples of spaces X are interpolation spaces V_p^β and $\tilde{V}_p^\beta$ (when $p \geq 2$ and $0 < \beta < 1$), $C(\bar{D})$ (when $p > N/2$), and $C^1(\bar{D})$ (when $p > N$), see Lemma 2.5.5.

For a space X satisfying (3.5.1) we have the following stronger exponential separation theorem.

THEOREM 3.5.1
Assume that X is a Banach space satisfying (3.5.1). Then for any $\delta_0 > 0$ there is $\check{M} = \check{M}(\delta_0, X) > 0$ such that

$$\frac{\|U_a(t,0)u_0\|_X}{\|U_a(t,0)w(a)\|_X} \leq \check{M} e^{-\gamma_0 t} \frac{\|u_0\|}{\|w(a)\|}$$

for any $a \in Y_0$, $u_0 \in X_2(a)$, and $t \geq \delta_0$, where $\gamma_0 > 0$ is as in Theorem 3.3.3.

THEOREM 3.5.2
Assume that $X = V_p^\beta$ or $X = \tilde{V}_p^\beta$ with $p \geq 2$ and $0 < \beta < 1$, and denote, for each $a \in Y$, $X(a) := V_p^\beta(a)$ or $X(a) := \tilde{V}_p^\beta(a)$, respectively. Then there is $\widehat{M} = \widehat{M}(X) > 0$ such that

$$\frac{\|U_a(t,0)u_0\|_X}{\|U_a(t,0)w(a)\|_X} \leq \widehat{M} e^{-\gamma_0 t} \frac{\|u_0\|_X}{\|w(a)\|_X}$$

for any $a \in Y_0$, $u_0 \in X_2(a) \cap X(a)$, and $t \geq 0$, where $\gamma_0 > 0$ is as in Theorem 3.3.3.

REMARK 3.5.1 We remark that the exponential separation in the smooth case does not follow from the abstract result in [94] directly due to the lack of the proper continuity of $U_a(t,0)$ at $t = 0$, which results from the time dependence of the boundary conditions. But under some additional assumption on the continuous dependence of the evolution operator $U_a(t,0)u_0$ on $a \in Y_0$, $t \geq 0$, and $u_0 \in L_2(D)$, the existence of exponential separation can be proved by the abstract results in [94] together with Theorem 3.2.2 (this approach is utilized in Chapter 6 and in the paper [84]). □

To prove Theorems 3.5.1 and 3.5.2, we first show the following two lemmas.

LEMMA 3.5.1
The functions w, $w^* \colon Y_0 \to X$ are continuous.

PROOF Recall that $r_{a \cdot (-1)}(1) = \|U_{a \cdot (-1)}(1,0)w(a \cdot (-1))\|$. The function $[Y_0 \ni a \mapsto r_{a \cdot (-1)}(1)]$ is bounded above and bounded away from 0. Now

$$w(a) = U_{a \cdot (-1)}(1,0)w(a \cdot (-1))/r_{a \cdot (-1)}(1).$$

It then follows from Proposition 2.5.2 that the set $\{ w(a) : a \in Y_0 \}$ is bounded in $W_p^2(D)$, consequently is relatively compact in X.

Assume that $a^{(n)} \to a$. Then $w(a^{(n)}) \to w(a)$ in $L_2(D)$. By the above arguments, there are a subsequence $(n_k)_{k=1}^{\infty}$ and $u^* \in X$ such that $w(a^{(n_k)}) \to u^*$ in X. Therefore we must have $u^* = w(a)$ and $w(a^{(n)}) \to w(a)$ in X.

In a similar way we prove that $w^* \colon Y_0 \to X$ is continuous. □

Observe that by the above lemma and the compactness of Y_0 there is $\widetilde{M_1} = \widetilde{M_1}(X) \geq 1$ such that

$$\frac{1}{\widetilde{M_1}}\|w(a)\| \leq \|w(a)\|_X \leq \widetilde{M_1}\|w(a)\| \quad \text{for any} \quad x \in Y_0. \tag{3.5.2}$$

LEMMA 3.5.2
For each $\delta_0 > 0$ there is $C^* = C^*(\delta_0) > 0$ such that

$$\|U_a(\delta_0,0)\|_{L_2(D),W_p^2(D)} \leq C^*(\delta_0) \qquad \text{for any } a \in Y.$$

PROOF It is a consequence of the L_2–L_p estimates (Proposition 2.2.2) and Proposition 2.5.2. □

PROOF (Proof of Theorem 3.5.1) By Lemma 3.5.2

$$\|U_a(t,0)u_0\|_{W_p^2(D)} \le C^*(\delta_0)\|U_a(t-\delta_0,0)u_0\|$$

for $t \ge \delta_0$. Further, we have

$$\|U_a(t,0)u_0\|_X \le \check{M}_1\|U_a(t,0)u_0\|_{W_p^2(D)}$$

for $t \ge \delta_0$, where $\check{M}_1$ denotes the norm of the embedding $W_p^2(D) \hookrightarrow X$. Since

$$\|U_a(t,0)w(a)\|_X \ge \frac{1}{\widetilde{M}_1}\|U_a(t,0)w(a)\| \ge \frac{\check{M}_2}{\widetilde{M}_1}\|U_a(t-\delta_0,0)w(a)\|$$

for all $t \in \mathbb{R}$, where $\check{M}_2 := \inf\{\, \|U_a(\delta_0,0)w(a)\| : a \in Y_0 \,\} > 0$, we have

$$\frac{\|U_a(t,0)u_0\|_X}{\|U_a(t,0)w(a)\|_X} \le \frac{C^*(\delta_0)\check{M}_1\widetilde{M}_1}{\check{M}_2}\frac{\|U_a(t-\delta_0,0)u_0\|}{\|U_a(t-\delta_0,0)w(a)\|}$$

for $t \ge \delta_0$. Theorem 3.3.3 provides the existence of $M > 0$ and $\gamma_0 > 0$ such that

$$\frac{\|U_a(t-\delta_0,0)u_0\|}{\|U_a(t-\delta_0,0)w(a)\|} \le Me^{-\gamma_0(t-\delta_0)}\frac{\|u_0\|}{\|w(a)\|}$$

for $t \ge \delta_0$. Consequently,

$$\frac{\|U_a(t,0)u_0\|_X}{\|U_a(t,0)w(a)\|_X} \le \check{M}e^{-\gamma_0 t}\frac{\|u_0\|}{\|w(a)\|}$$

for $t \ge \delta_0$, where $\check{M} = MC^*(\delta_0)\check{M}_1\widetilde{M}_1e^{\gamma_0}/\check{M}_2$. □

PROOF (Proof of Theorem 3.5.2) By Proposition 2.5.3, there is $C > 0$ such that

$$\|U_a(t,0)u_0\|_X \le C\|u_0\|_X$$

for any $a \in Y$, $0 \le t \le 2$ and $u_0 \in X(a)$. Consequently

$$\|U_a(t,0)u_0\|_X \le Ce^{2\gamma_0}e^{-\gamma_0 t}\|u_0\|_X$$

for $a \in Y$, $0 \le t \le 2$ and $u_0 \in X(a)$. On the other hand,

$$\|U_a(t,0)w(a)\|_X \ge \frac{\check{M}_3}{\widetilde{M}_1^2}\|w(a)\|_X$$

for $0 \le t \le 2$, where $\check{M}_3 := \inf\{\, \|U_a(t,0)w(a)\| : a \in Y_0,\ t \in [0,2] \,\} > 0$. Hence

$$\frac{\|U_a(t,0)u_0\|_X}{\|U_a(t,0)w(a)\|_X} \le \frac{C\widetilde{M}_1^2e^{2\gamma_0}}{\check{M}_3}e^{-\gamma_0 t}\frac{\|u_0\|_X}{\|w(a)\|_X}$$

for $0 \leq t \leq 2$, provided that $u_0 \in X(a)$.

For $t > 2$ we have, with the help of Lemma 3.5.2,

$$\frac{\|U_a(t,0)u_0\|_X}{\|U_a(t,0)w(a)\|_X} \leq \check{M}_4 \frac{\|U_a(t,0)u_0\|_{W_p^2(D)}}{\|U_a(t,0)w(a)\|_X} \leq \frac{\check{M}_4\widetilde{M}_1 C^*(1)}{\check{M}_3} \frac{\|U_a(t-1,0)u_0\|}{\|U_a(t-1,0)w(a)\|},$$

where $\check{M}_4 > 0$ stands for the norm of the embedding $W_p^2(D) \hookrightarrow X$. By Theorem 3.3.3,

$$\frac{\|U_a(t-1,0)u_0\|}{\|U_a(t-1,0)w(a)\|} \leq Me^{-\gamma_0(t-2)} \frac{\|U_a(1,0)u_0\|}{\|U_a(1,0)w(a)\|}.$$

Further,

$$\frac{\|U_a(1,0)u_0\|}{\|U_a(1,0)w(a)\|} \leq \frac{\check{M}_5\widetilde{M}_1}{\check{M}_3} \frac{\|u_0\|_X}{\|w(a)\|_X},$$

where $\check{M}_5 := \sup\{\|U_a(1,0)\|_{X,L_2(D)} : a \in Y_0\} < \infty$ (by $X \hookrightarrow L_2(D)$ and the L_2–L_2 estimates in Proposition 2.2.2).

As a consequence,

$$\frac{\|U_a(t,0)u_0\|_X}{\|U_a(t,0)w(a)\|_X} \leq \widehat{M}e^{-\gamma_0 t} \frac{\|u_0\|_X}{\|w(a)\|_X}$$

for $t \geq 0$ and $u_0 \in X(a)$, where $\widehat{M} = \frac{\widetilde{M}_1^2 e^{2\gamma_0}}{\check{M}_3} \max\{C, \frac{C^*(1)\check{M}_4\check{M}_5}{\check{M}_3}\}$. $\quad\square$

For each $a \in Y_0$, let $\kappa(a)$ be defined by

$$\kappa(a) := -B_a(0, w(a), w(a)). \tag{3.5.3}$$

By (A2-5) and Lemma 3.5.1(2), the function $\kappa \colon Y_0 \to \mathbb{R}$ is well defined and continuous. Furthermore, we have the following two lemmas.

LEMMA 3.5.3
For $a \in Y$ and $t \in \mathbb{R}$, put $\eta(t;a) := \|U_a(t,0)w(a)\|$. Then

$$\dot{\eta}(t;a) = \kappa(a \cdot t)\eta(t;a)$$

for any $a \in Y$ and $t \in \mathbb{R}$, where dot denotes the derivative in t.

PROOF Let $u(t) := U_a(t,0)w(a)$ (recall that for $t < 0$, $U_a(t,0)w(a) = \frac{w(a \cdot t)}{\|U_{a \cdot t}(-t,0)w(a \cdot t)\|}$). Then by Proposition 2.1.4, we have

$$(\eta(t;a))^2 - (\eta(s;a))^2 = -2\int_s^t B_a(\tau, u(\tau), u(\tau))\, d\tau$$

$$= -2\int_s^t B_{a \cdot \tau}(0, \eta(\tau;a)w(a \cdot \tau), \eta(\tau;a)w(a \cdot \tau))\, d\tau$$

$$= 2\int_s^t \kappa(a \cdot \tau)(\eta(\tau;a))^2\, d\tau$$

for any $a \in Y$ and $s \leq t$. It then follows that

$$\dot{\eta}(t; a) = \kappa(a \cdot t)\eta(t; a)$$

for any $a \in Y$ and $t \in \mathbb{R}$. $\square$

LEMMA 3.5.4
For $a \in Y$, $t \in \mathbb{R}$ and $x \in \bar{D}$, put $v(t, x; a) := w(a \cdot t)(x)$. Then $v(\cdot, \cdot)$ is an entire classical solution of the parabolic equation

$$
\begin{cases}
\dfrac{\partial v}{\partial t} = \displaystyle\sum_{i=1}^{N} \dfrac{\partial}{\partial x_i}\left(\sum_{j=1}^{N} a_{ij}(t, x)\dfrac{\partial v}{\partial x_j} + a_i(t, x)v\right) \\[4mm]
\qquad + \displaystyle\sum_{i=1}^{N} b_i(t, x)\dfrac{\partial v}{\partial x_i} + c_0(t, x)v - \kappa(a \cdot t)v, \quad t \in \mathbb{R}, \ x \in D \\[4mm]
\mathcal{B}_a(t)v = 0, \qquad\qquad\qquad\qquad\qquad\qquad\quad t \in \mathbb{R}, \ x \in \partial D.
\end{cases}
$$

PROOF   Observe that $U_a(t, 0)w(a) = \|U_a(t, 0)w(a)\|w(a \cdot t) = \eta(t; a)w(a \cdot t)$. The lemma then follows from Lemma 3.5.3. $\square$

Similarly to Theorems 3.1.5 and 3.2.4 we have

THEOREM 3.5.3
Let μ be an ergodic invariant measure for $\sigma|_{Y_0}$. There exists a Borel set $\tilde{Y} \subset Y_0$ with $\mu(\tilde{Y}) = 1$ such that

$$
\lambda(\mu) = \lim_{t \to \infty} \frac{\ln\|U_a(t, 0)w(a)\|}{t} = \lim_{t \to \infty} \frac{\ln\|U_a(t, 0)w(a)\|_X}{t}
$$

$$
= \lim_{t \to \infty} \frac{1}{t}\int_0^t \kappa(a \cdot \tau)\, d\tau = \int_{Y_0} \kappa\, d\mu
$$

for all $a \in \tilde{Y}$.

COROLLARY 3.5.1
Assume that μ is an ergodic invariant measure for $\sigma|_{Y_0}$. Then for μ-a.e. $a \in Y_0$ and each nonzero $u_0 \in L_2(D)^+$ one has

$$
\lambda(\mu) = \lim_{t \to \infty} \frac{\ln\|U_a(t, 0)u_0\|}{t} = \lim_{t \to \infty} \frac{\ln\|U_a(t, 0)u_0\|_X}{t}.
$$

PROOF (Proof of Theorem 3.5.3)   The existence of a Borel set $Y_1 \subset Y_0$ with $\mu(Y_1) = 1$ such that

$$
\lambda(\mu) = \lim_{t \to \infty} \frac{\ln\|U_a(t, 0)w(a)\|}{t} = \lim_{t \to \infty} \frac{1}{t}\int_0^t \kappa(a \cdot \tau)\, d\tau
$$

for all $a \in Y_1$ follows by Theorem 3.1.5 and Proposition 3.2.9. The fact that

$$\lim_{t \to \infty} \frac{\ln \|U_a(t,0)w(a)\|}{t} = \lim_{t \to \infty} \frac{\ln \|U_a(t,0)w(a)\|_X}{t}$$

is a consequence of (3.5.2).

The use of Birkhoff's Ergodic Theorem (Lemma 1.2.6) establishes the existence of a Borel set $Y_2 \subset Y_0$ with $\mu(Y_2) = 1$ such that

$$\lim_{t \to \infty} \frac{1}{t} \int_0^t \kappa(a \cdot \tau) \, d\tau = \int_{Y_0} \kappa \, d\mu$$

for all $a \in Y_2$. It suffices now to put $\tilde{Y} := Y_1 \cap Y_2$.   ☐

PROOF (Proof of Corollary 3.5.1)   The proof goes along the lines of the proof of Theorem 3.2.4, with the $L_2(D)$-norm replaced by the X-norm, and Definition 3.2.1 replaced by Theorem 3.5.1.   ☐

The following result is an analog of Theorem 3.3.4.

THEOREM 3.5.4
For any $\delta_0 > 0$ there exists $\tilde{K} = \tilde{K}(\delta_0, X) > 0$ such that

$$\frac{\|P_2(a \cdot t)U_a(t,0)u_0\|_X}{\|P_1(a \cdot t)U_a(t,0)u_0\|_X} \leq \tilde{K}$$

for all $t \geq \delta_0$, all $a \in Y_0$, and all nonzero $u_0 \in L_2(D)^+$.

PROOF   By Theorem 3.3.4, there exists $K > 0$ such that

$$\frac{\|P_2(a \cdot \frac{\delta_0}{2})U_a(\frac{\delta_0}{2},0)u_0\|}{\|P_1(a \cdot \frac{\delta_0}{2})U_a(\frac{\delta_0}{2},0)u_0\|} \leq K$$

for any $a \in Y_0$ and any nonzero $u_0 \in L_2(D)^+$. Theorem 3.5.1 implies the existence of $\widehat{M} > 0$ such that

$$\frac{\|P_2(a \cdot t)U_a(t,0)u_0\|_X}{\|P_1(a \cdot t)U_a(t,0)u_0\|_X} \leq \widehat{M} e^{-\gamma_0(t-\frac{\delta_0}{2})} \frac{\|P_2(a \cdot \frac{\delta_0}{2})U_a(\frac{\delta_0}{2},0)u_0\|}{\|P_1(a \cdot \frac{\delta_0}{2})U_a(\frac{\delta_0}{2},0)u_0\|}$$

for any $a \in Y_0$, any nonzero $u_0 \in L_2(D)^+$, and any $t \geq \delta_0$. It suffices to put $\tilde{K} := K\widehat{M}e^{\gamma_0\delta_0/2}$.   ☐

3.6 Remarks on the General Nondivergence Case

Consider

$$\frac{\partial u}{\partial t} = \sum_{i,j=1}^{N} a_{ij}(t,x)\frac{\partial^2 u}{\partial x_i \partial x_j} + \sum_{i=1}^{N} b_i(t,x)\frac{\partial u}{\partial x_i} + c_0(t,x)u, \quad t > s, \ x \in D,$$

(3.6.1)

complemented with the boundary conditions

$$\mathcal{B}(t)u = 0, \qquad t > s, \ x \in \partial D,$$ (3.6.2)

where $D \subset \mathbb{R}^N$ is a bounded domain, $s \in \mathbb{R}$ is an initial time, and $\mathcal{B}$ is a boundary operator of either the Dirichlet or Neumann or Robin type, that is,

$$\mathcal{B}(t)u = \begin{cases} u & \text{(Dirichlet)} \\ \sum_{i=1}^{N} \partial_{x_i} u \bar{\nu}_i(t,x) & \text{(Neumann)} \\ \sum_{i=1}^{N} \partial_{x_i} u \bar{\nu}_i(t,x) + d_0(t,x)u, & \text{(Robin)} \end{cases}$$

where (in the Neumann or Robin cases) $(\bar{\nu}_1, \ldots, \bar{\nu}_N)$ is (in general time dependent) a vector field on ∂D pointing out of D.

First of all, if both the domain D and the coefficients are sufficiently smooth and, in the Neumann or Robin cases, $\bar{\nu}_i(t,x) = \sum_{j=1}^{N} a_{ji}(t,x)\nu_j(x)$, $1 \le i \le N$, (that is, the derivative is conormal), then (3.6.1)+(3.6.2) can be written in the divergence form and then the results in the previous sections hold for (3.6.1)+(3.6.2).

Historically, when the domain D and the coefficients are sufficiently smooth and the boundary conditions are independent of time, the existence of exponential separation has been proved in [94] (see also [76], [102]; compare [95] for a finite-dimensional counterpart). The existence and uniqueness of entire positive solutions has been proved in [77], [78], [92]. Recently, in [84] the authors proved the exponential separation as well as the existence and uniqueness of entire positive solutions in a general nondivergence case and with time dependent boundary conditions but assuming that the domain and the coefficients are smooth enough.

In [60] Húska and Poláčik proved the existence of exponential separation and existence and uniqueness of entire positive solutions in a general divergence case with the Dirichlet boundary condition, with weak assumptions on the regularity of the coefficients.

As regards a general nondivergence case, recently Húska, Poláčik, and Safonov in [61] proved the existence of exponential separation and existence and uniqueness of entire positive solutions of (3.6.1)+(3.6.2) when the boundary condition is Dirichlet. In [59], Húska studied (3.6.1) with a general oblique boundary conditions and showed the exponential separation between a posi- tive solution and sign-changing solutions and the uniqueness of entire positive solutions, both under the condition that an entire positive solution exists. However, the question of existence of entire positive solutions was not ad- dressed in [59].

3.7 Appendix: The Case of One-Dimensional Spatial Domain

In the present appendix we consider the case when the domain D is one- dimensional.

Let $Y_0 = Y$ be defined as

$$Y := \{\, a = (1, 0, c, 0) : \|c\|_{L_\infty(\mathbb{R} \times [0, \pi])} \le R \,\}$$

for some $R > 0$. The set Y is (norm-)bounded in $L_\infty(\mathbb{R} \times [0, \pi], \mathbb{R}^4)$. Y is considered with the weak-* topology. It is a compact connected metrizable space.

For $a \in Y$ and $t \in \mathbb{R}$ there holds $a \cdot t \in Y$.

Consider the family of partial differential equations

$$\begin{cases} \dfrac{\partial u}{\partial t} = \dfrac{\partial^2 u}{\partial x^2} + c(t, x)u, & t > 0, \ x \in (0, \pi), \\ u(t, 0) = u(t, \pi) = 0, & t > 0, \end{cases} \tag{3.7.1}$$

parameterized by $a = (1, 0, c, 0) \in Y$. For $a \in Y$ and $u_0 \in L_2((0, \pi))$ denote by $[\, [0, \infty) \ni t \mapsto U_a(t, 0)u_0 \in L_2(0, \pi)) \,]$ the unique (weak) solution of (3.7.1) satisfying the initial condition $u(0, \cdot) = u_0$. For a fixed $a \in Y$ we will write $(3.7.1)_a$.

Denote by A_2 the realization in $L_2((0, \pi))$ of the operator $[\, u \mapsto \frac{\partial^2 u}{\partial x^2} \,]$ with the Dirichlet boundary conditions. Denote by $\{e^{A_2 t}\}_{t \ge 0}$ the analytic semi- group generated on $L_2((0, \pi))$ by A_2. A continuous function $u \colon [0, \infty) \to L_2((0, \pi))$ is a *mild solution* of $(3.7.1)_a$ satisfying the initial condition $u(0, \cdot) = u_0$ if for any $t > 0$ there holds

$$u(t) = e^{A_2 t} u_0 + \int_0^t e^{A_2(t - \tau)} (C(\tau)u(\tau)) \, d\tau,$$

where, for $\tau \in (0, \tau]$, $(C(\tau)u(\tau))(x) := c(t, x)u(t)(x)$ for a.e. $x \in (0, \pi)$. We claim that for each $a \in Y$ and $u_0 \in L_2((0, \pi))$ the weak and mild solutions are the same. Indeed, if c is sufficiently smooth then both are classical solutions, and the claim follows from the uniqueness of classical solutions. For a general c we use approximation along the lines of the proof of [33, Proposition 4.2]. Consequently, we can use results in [22], where solutions of (3.7.1) were defined as mild solutions.

Recall that $\mathring{C}^1([0,\pi])$ denotes the (closed) vector subspace consisting of those $\phi \in C^1([0,\pi])$ for which $\phi(0) = \phi(\pi) = 0$.

The following result was proved as [22, Theorem 3.3].

PROPOSITION 3.7.1

(1) *For each $T > 0$ the mapping*

$$[Y \times L_2((0,\pi)) \ni (a, u_0) \mapsto U_a(\cdot, 0)u_0|_{[0,T]} \in C([0,T], L_2((0,\pi)))]$$

is continuous.

(2) *For any $a \in Y$, $t > 0$, and $u_0 \in L_2((0,\pi))$ there holds $U_a(t,0)u_0 \in \mathring{C}^1([0,\pi])$. Moreover, for any $0 < t_1 \le t_2$ the mapping*

$$[Y \times L_2((0,\pi)) \ni (a, u_0) \mapsto U_a(\cdot, 0)u_0|_{[t_1,t_2]} \in C([t_1, t_2], \mathring{C}^1([0,\pi]))]$$

is continuous, and there is $M = M(t_1, t_2) > 0$ such that

$$\|U_a(t,0)\|_{L_2((0,\pi)), \mathring{C}^1([0,\pi])} \le M$$

for each $t \in [t_1, t_2]$.

For a given nonzero $\phi \in C([0,\pi])$ such that $\phi(0) = \phi(\pi) = 0$ define the *lap* or *Matano number* of ϕ to be

$$z(\phi) = \sup\{\, l \ge 1 : \text{ there exist } 0 < x_1 < x_2 < \cdots < x_l < \pi$$
$$\text{such that } \phi(x_k)\phi(x_{k+1}) < 0 \quad \text{for} \quad 1 \le k \le l-1 \}$$

with $z(\phi) = 1$ if either $\phi(x) \ge 0$ or $\phi(x) \le 0$ for all $x \in [0, \pi]$.

We have the following result (see [22, p. 257] and the references contained therein).

LEMMA 3.7.1

(i) *$z(U_a(t,0)u_0) < \infty$ for any $a \in Y$, any $t > 0$, and any nonzero $u_0 \in L_2((0,\pi))$.*

(ii) *For any $a \in Y$ and any nonzero $u_0 \in L_2((0,\pi))$ the function $[(0,\infty) \ni t \mapsto z(U_a(t,0)u_0) \in \mathbb{N}]$ is nonincreasing.*

As a byproduct of Lemma 3.7.1(i) we have that for each $a \in Y$ and each $t > 0$ the operator $U_a(t, 0)$ is injective.

Define an *exponentially bounded solution* of (3.7.1) to be an entire solution $u \colon \mathbb{R} \to L_2((0, \pi))$ such that there are constants K_1, K_2 with

$$\|u(t)\| \leq K_1 e^{K_2 |t|} \quad \text{for } t \in \mathbb{R}.$$

For $a \in Y$ and $i = 1, 2, \ldots$ let $X_i'(a) \subset L_2((0, \pi))$ be defined by

$$X_i'(a) := \{\, \phi \in L_2((0, \pi)) : \phi = u(0) \text{ for some exponentially bounded solution}$$
$$u \colon \mathbb{R} \to L_2((0, \pi)) \text{ of } (3.7.1)_a \text{ satisfying } z(u(t)) = i \text{ for all } t \in \mathbb{R} \,\} \cup \{0\}.$$

By [22, Theorem 5.1 and Proposition 5.2], $X_i'(a)$ is a one-dimensional vector subspace, for each $i = 1, 2, \ldots$ and each $a \in Y$. This allows us to define, for $a \in Y$ and $i = 1, 2, \ldots$,

$$X_i''(a) := \mathrm{cl}\left(\bigoplus_{j=i}^{\infty} X_j'(a) \right),$$

where the closure is considered in the $L_2((0, \pi))$-norm.

PROPOSITION 3.7.2

For each $i = 1, 2, \ldots$ the following holds.

(i) $U_a(t, 0) X_i'(a) = X_i'(a \cdot t)$ *and* $U_a(t, 0) X_{i+1}''(a) \subset X_{i+1}''(a \cdot t)$ *for each $a \in Y$ and each $t \geq 0$.*

(ii) *There exists a continuous function $w_i \colon Y \to L_2((0, \pi))$ such that for each $a \in Y$, $\|w_i(a)\| = 1$, $\frac{\partial w_i(a)}{\partial x}(0) > 0$, and $X_i'(a) = \mathrm{span}\{w_i(a)\}$; moreover, such a function is unique.*

(iii) $z(\phi) \geq i + 1$ *for any $a \in Y$ and any nonzero $\phi \in X_{i+1}''(a)$.*

(iv) $L_2(D) = X_1'(a) \oplus X_2'(a) \oplus \cdots \oplus X_i'(a) \oplus X_{i+1}''(a)$ *for any $a \in Y$.*

(v) *There are constants $K_i > 0$ and $\gamma_i > 0$ such that*

$$\frac{\|U_a(t, 0) u_0\|}{\|U_a(t, 0) w_i(a)\|} \leq K_i e^{-\gamma_i t}$$

for any $a \in Y$, any $t \geq 0$, and any $u_0 \in X_{i+1}''(a)$ with $\|u_0\| = 1$.

From (i) and (ii) it follows that $X_i' := \bigcup_{a \in Y} (\{a\} \times X_i'(a))$ is a trivial one-dimensional subbundle of $L_2(D) \times Y$, invariant under Π.

PROOF See [22, Proposition 4.6, Theorem 5.1, Proposition 6.2 and Theorem 7.1] ⧠

Notice that from the above proposition it follows that Π admits an exponential separation. Indeed, we take $X_1(a) = X_1'(a)$ and $X_2(a) = X_2''(a)$ (in fact, $X_2''(a)$ being equal to $\{ v \in L_2((0,\pi)) : \langle v, w^*(a) \rangle = 0 \}$ for some $w^*(a) \in L_2((0,\pi))^+$ is not explicitly mentioned in Proposition 3.7.2, but it follows from the method of proving relevant theorems in [22].)

THEOREM 3.7.1

Assume that μ is an ergodic invariant measure for $\sigma|_{Y_0}$. Then condition (2) in Theorem 3.4.1 holds. More precisely, there exist:

- *a Borel set $\tilde{Y}_0 \subset Y_0$, $\mu(\tilde{Y}_0) = 1$, and*

- *a sequence of real numbers $\lambda_1(\mu) > \cdots > \lambda_i(\mu) > \lambda_{i+1}(\mu) > \dots$ having limit $-\infty$,*

such that

$$\lim_{t \to \pm \infty} \frac{\ln \| U_a(t,0) w_i(a) \|}{t} = \lambda_i(\mu)$$

for each $a \in \tilde{Y}_0$ and each $i = 1, 2, \dots$.
Further, for each $a \in \tilde{Y}_0$ and each $i = 1, 2, \dots$ there holds

$$E_i(\mu; a) = X_i'(a) \qquad and \qquad F_i(\mu; a) = X_{i+1}''(a).$$

PROOF For each $i = 1, 2, \dots$ an application of [65, Theorem 2.1] to the topological linear skew-product flow $\Pi|_{X_i'}$ on the one-dimensional bundle X_i' gives the existence of Borel set $\hat{Y}_{0,i} \subset Y_0$, $\mu(\hat{Y}_{0,i}) = 1$, and a real number λ_i' such that

$$\lim_{t \to \pm \infty} \frac{\ln \| U_a(t,0) w_i(a) \|}{t} = \lambda_i'$$

for each $a \in \hat{Y}_{0,i}$. Put $\hat{Y}_0 := \bigcap_{i=1}^{\infty} \hat{Y}_{0,i}$. Obviously $\hat{Y}_0$ is a Borel set, with $\mu(\hat{Y}_0) = 1$.

It is a consequence of Corollary 3.4.1 and the remarks below Proposition 3.7.2 that $\lambda_1' = \lambda_1(\mu)$, $E_1(\mu; a) = X_1'(a)$ and $F_1(\mu; a) = X_2''(a)$, for μ-a.e. $a \in \hat{Y}_0$.

Assume that we already have that, for some $i = 1, 2, \dots$, there holds $\lambda_i' = \lambda_i(\mu)$, $E_i(\mu; a) = X_i'(a)$ and $F_i(\mu; a) = X_{i+1}''(a)$, for μ-a.e. $a \in \hat{Y}_0$. By Theorem 3.4.1,

$$\lambda_{i+1}(\mu) = \lim_{t \to \infty} (1/t) \ln \| U_a(t,0)|_{F_i(\mu;a)} \|$$

for μ-a.e. $a \in \hat{Y}_0$. From Proposition 3.7.2 we deduce that

$$\lambda_{i+1}' = \lim_{t \to \infty} (1/t) \ln \| U_a(t,0)|_{X_{i+1}''(\mu;a)} \|$$

for all $a \in \hat{Y}_0$. So our induction assumption gives that $\lambda'_{i+1} = \lambda_{i+1}(\mu)$. Further, from Theorem 3.4.1 it follows that $E_{i+1}(\mu; a) = X'_{i+1}(a)$ for μ-a.e. $a \in \hat{Y}_0$.

Suppose to the contrary that for some $a \in \hat{Y}_0$ there is $v \in F_{i+1}(\mu; a) \setminus X''_{i+2}(a)$. Decompose $v = v_{i+1} + v_{i+2}$, where $v_{i+1} \in X'_{i+1}(a) \setminus \{0\}$ and $v_{i+2} \in X''_{i+2}(a)$. It follows from Proposition 3.7.2 that $\lim_{t\to\infty}(1/t)\ln\|U_a(t,0)v\| = \lim_{t\to\infty}(1/t)\ln\|U_a(t,0)v_{i+1}\| = \lambda_{i+1}$, consequently

$$\lim_{t\to\infty}(1/t)\ln\|U_a(t,0)|_{F_{i+1}(\mu;a)}\| \geq \lambda_{i+1} > \lambda_{i+2},$$

which is impossible. Therefore, $F_{i+1}(\mu; a) \subset X''_{i+2}(a)$. But both are subspaces of $X''_{i+1}(a)$, of relative codimension one, so they must be equal. □

Chapter 4

Spectral Theory in Nonautonomous and Random Cases

In this chapter, we consider principal spectrum and principal Lyapunov exponents of nonautonomous and random parabolic equations. First in Section 4.1, we introduce basic assumptions for a given random (nonautonomous) parabolic equation and associate a proper family of parabolic equations with the given equation. Then based on the notions introduced in Chapters 2 and 3 in the general setting, we introduce the concepts of principal spectrum and principal Lyapunov exponents for a given random (nonautonomous) parabolic equation in terms of the associated family of parabolic equations, which naturally extends the classical concept of principal eigenvalue for the elliptic and periodic parabolic problems. Also applying the general theories developed in Chapters 2 and 3, we present basic properties of principal spectrum and principal Lyapunov exponents of random (nonautonomous) parabolic equations. In addition, we provide some examples which satisfy the basic assumptions in this section. In Section 4.2, we investigate the monotonicity of principal spectrum and principal Lyapunov exponents of random (nonautonomous) parabolic equations with respect to zero order terms. We also study the relation among the principal spectrum and principal Lyapunov exponents for random (nonautonomous) parabolic equations with different types of boundary conditions. Sections 4.3 and 4.4 concern the continuous dependence of principal spectrum and principal Lyapunov exponents of random (nonautonomous) parabolic equations with respect to the coefficients. Because of the speciality of the zero order coefficients, the continuous dependence with respect to these coefficients are considered in Section 4.3 first. In Section 4.4, the general continuous dependence is then discussed. Throughout Section 4.1 to Section 4.4, many results and arguments for random and nonautonomous equations are similar. However, considering that different readers may be interested in different types of equations, for convenience, in each section from Section 4.1 to Section 4.4, we treat these two types of equations in different subsections and provide proofs for most similar results. This chapter ends up with some historical remarks in Section 4.5.

4.1   Principal Spectrum and Principal Lyapunov Exponents in Random and Nonautonomous Cases

This section is to introduce basic assumptions, concepts, and properties. We first introduce basic assumptions for a given random (nonautonomous) parabolic equation and associate a proper family of parabolic equations with the given equation. Next, based on the notions introduced in Chapters 2 and 3 in the general setting, we introduce the concepts of principal spectrum and principal Lyapunov exponents for a given random (nonautonomous) parabolic equation in terms of the associated family of parabolic equations, which naturally extends the classical concept of principal eigenvalue for the elliptic and periodic parabolic problems. We provide some examples which satisfy the basic assumptions. Then applying the general theories developed in Chapters 2 and 3, we present basic properties of principal spectrum and principal Lyapunov exponents of random (nonautonomous) parabolic equations. For the reader's convenience, random and nonautonomous cases are treated separately.

4.1.1   The Random Case

Assume that $((\Omega, \mathfrak{F}, \mathbb{P}), \{\theta_t\}_{t\in\mathbb{R}})$ is an ergodic metric dynamical system. Consider the following random linear parabolic equation:

$$\frac{\partial u}{\partial t} = \sum_{i=1}^{N} \frac{\partial}{\partial x_i} \left(\sum_{j=1}^{N} a_{ij}(\theta_t\omega, x) \frac{\partial u}{\partial x_j} + a_i(\theta_t\omega, x)u \right)$$

$$+ \sum_{i=1}^{N} b_i(\theta_t\omega, x)\frac{\partial u}{\partial x_i} + c_0(\theta_t\omega, x)u, \quad x \in D, \qquad (4.1.1)$$

endowed with the boundary condition

$$\mathcal{B}_\omega(t)u = 0, \quad x \in \partial D, \qquad (4.1.2)$$

where $\mathcal{B}_\omega(t) = \mathcal{B}_{a^\omega}(t)$, $\mathcal{B}_{a^\omega}(t)$ is as in (2.0.3) with a being replaced by a^ω, $a^\omega(t, x) = (a_{ij}(\theta_t\omega, x), a_i(\theta_t\omega, x), b_i(\theta_t\omega, x), c_0(\theta_t\omega, x), d_0(\theta_t\omega, x))$, and $d_0(\omega, x) \geq 0$ for all $\omega \in \Omega$ and a.e. $x \in \partial D$. To emphasize the coefficients in (4.1.1)+(4.1.2), we will write $(4.1.1)_a+(4.1.2)_a$.

Our first assumption in the present subsection concerns measurability of the coefficients of the equation and of the boundary conditions (recall that for a metric space S the symbol $\mathfrak{B}(S)$ stands for the countably additive algebra of Borel sets):

(A4-R1) (Measurability) *The functions* $a_{ij}(= a_{ji})$ $(i, j = 1, \ldots, N)$, a_i $(i = 1, \ldots, N)$, b_i $(i = 1, \ldots, N)$, *and* c_0 *are* $(\mathfrak{F} \times \mathfrak{B}(D), \mathfrak{B}(\mathbb{R}))$-*measurable, and the function* d_0 *is* $(\mathfrak{F} \times \mathfrak{B}(\partial D), \mathfrak{B}(\mathbb{R}))$-*measurable.*

Among others, (4.1.1)+(4.1.2) arise from linearization of random nonlinear parabolic equations at a certain entire solution (i.e., a solution which exists for all $t \in \mathbb{R}$) as well as from linearization of autonomous nonlinear equations at some invariant set of solutions.

For each $\omega \in \Omega$, let $a_{ij}^{\omega}(t,x) := a_{ij}(\theta_t\omega, x)$, $a_i^{\omega}(t,x) := a_i(\theta_t\omega, x)$, $b_i^{\omega}(t,x) :=$ $b_i(\theta_t\omega, x)$, $c_0^{\omega}(t,x) := c_0(\theta_t\omega, x)$, $d_0^{\omega}(t,x) := d_0(\theta_t\omega, x)$.

The functions

$$[\, \Omega \times \mathbb{R} \times D \ni (\omega, t, x) \mapsto a_{ij}^{\omega}(t,x) \in \mathbb{R} \,]$$

are $(\mathfrak{F} \times \mathfrak{B}(\mathbb{R}) \times \mathfrak{B}(D), \mathfrak{B}(\mathbb{R}))$-measurable (as composites of Borel measurable functions). Similarly, the functions $a_i^{\omega}(t,x)$, $b_i^{\omega}(t,x)$, and c_0^{ω} are $(\mathfrak{F} \times \mathfrak{B}(\mathbb{R}) \times \mathfrak{B}(D), \mathfrak{B}(\mathbb{R}))$-measurable and the function $d_0^{\omega}(t,x)$ is $(\mathfrak{F} \times \mathfrak{B}(\mathbb{R}) \times \mathfrak{B}(\partial D), \mathfrak{B}(\mathbb{R}))$-measurable.

As sections of Borel measurable functions, the functions a_{ij}^{ω}, a_i^{ω}, b_i^{ω}, and c_0^{ω} are $(\mathfrak{B}(\mathbb{R}) \times \mathfrak{B}(D), \mathfrak{B}(\mathbb{R}))$-measurable, for any fixed $\omega \in \Omega$. Similarly, the function d_0^{ω} is $(\mathfrak{B}(\mathbb{R}) \times \mathfrak{B}(\partial D), \mathfrak{B}(\mathbb{R}))$-measurable, for any fixed $\omega \in \Omega$.

We write $a^{\omega} := ((a_{ij}^{\omega})_{i,j=1}^N, (a_i^{\omega})_{i=1}^N, (b_i^{\omega})_{i=1}^N, c_0^{\omega}, d_0^{\omega})$. Sometimes we write the random problem (4.1.1)+(4.1.2) as (4.1.1)$_a$+(4.1.2)$_a$.

Our second assumption regards uniform boundedness of the coefficients of the equations (and of the boundary conditions):

(A4-R2) (Boundedness)
For each $\omega \in \Omega$, a^{ω} belongs to $L_{\infty}(\mathbb{R} \times D, \mathbb{R}^{N^2+2N+1}) \times L_{\infty}(\mathbb{R} \times \partial D, \mathbb{R})$. Moreover, the set $\{\, a^{\omega} : \omega \in \Omega \,\}$ is bounded in the $L_{\infty}(\mathbb{R} \times D, \mathbb{R}^{N^2+2N+1}) \times L_{\infty}(\mathbb{R} \times \partial D, \mathbb{R})$-norm by $M \geq 0$.

Define the mapping $E_a : \Omega \to L_{\infty}(\mathbb{R} \times D, \mathbb{R}^{N^2+2N+1}) \times L_{\infty}(\mathbb{R} \times \partial D, \mathbb{R})$ as

$$E_a(\omega) := a^{\omega}.$$

Put
$$\tilde{Y}(a) := \text{cl}\{\, E_a(\omega) : \omega \in \Omega \,\} \tag{4.1.3}$$

with the weak-* topology, where the closure is taken in the weak-* topology. The set $\tilde{Y}(a)$ is a compact metrizable space and $(\tilde{Y}(a), \{\sigma_t\}_{t\in\mathbb{R}})$ is a compact flow, where $\sigma_t \tilde{a}(\cdot, \cdot) = \tilde{a}(\cdot + t, \cdot)$.

LEMMA 4.1.1
The mapping E_a is $(\mathfrak{F}, \mathfrak{B}(\tilde{Y}(a)))$-measurable.

PROOF   Recall that $\{g_1, g_2, \dots\}$ is a countable dense subset of the unit ball in $L_1(\mathbb{R} \times D, \mathbb{R}^{N^2+2N+1}) \times L_1(\mathbb{R} \times \partial D, \mathbb{R})$ (see (1.3.1)). It is clear that for each $\tilde{a} \in \tilde{Y}(a)$ and $k \in \mathbb{N}$ the function

$$[\, \Omega \ni \omega \mapsto \langle g_k, (\tilde{a} - a^{\omega}) \rangle_{L_1, L_{\infty}} \in \mathbb{R} \,]$$

is $(\mathfrak{F}, \mathfrak{B}(\mathbb{R}))$-measurable. This implies that E_a is $(\mathfrak{F}, \mathfrak{B}(\tilde{Y}(a)))$-measurable. $\square$

An important property of the mapping E_a is the following

$$\sigma_t \circ E_a = E_a \circ \theta_t \qquad \text{for each } t \in \mathbb{R}. \tag{4.1.4}$$

It follows that $E_a(\Omega)$ is $\{\sigma_t\}$-invariant. Consequently, $\tilde{Y}(a)$ is $\{\sigma_t\}$-invariant, too.

The mapping E_a is a homomorphism of the measurable flow $((\Omega, \mathfrak{F}), \{\theta_t\}_{t\in\mathbb{R}})$ into the measurable flow $((\tilde{Y}(a), \mathfrak{B}(\tilde{Y}(a))), \{\sigma_t\}_{t\in\mathbb{R}})$. Denote by $\tilde{\mathbb{P}}$ the image of the measure $\mathbb{P}$ under E_a: for any Borel set $A \in \mathfrak{B}(\tilde{Y}(a))$, $\tilde{\mathbb{P}}(A) := \mathbb{P}(E_a^{-1}(A))$. $\tilde{\mathbb{P}}$ is a $\{\sigma_t\}$-invariant ergodic Borel measure on $\tilde{Y}(a)$. So, E_a is a homomorphism of the metric flow $((\Omega, \mathfrak{F}, \mathbb{P}), \{\theta_t\}_{t\in\mathbb{R}})$ into the metric flow $((\tilde{Y}(a), \mathfrak{B}(\tilde{Y}(a)), \tilde{\mathbb{P}}), \{\sigma_t\}_{t\in\mathbb{R}})$.

We will consider $(\tilde{Y}(a), \{\sigma_t\}_{t\in\mathbb{R}})$ a *topological* flow, with an ergodic invariant measure $\tilde{\mathbb{P}}$. Put

$$\tilde{Y}_0(a) := \operatorname{supp} \tilde{\mathbb{P}} \tag{4.1.5}$$

($\tilde{a} \in \tilde{Y}_0(a)$ if and only if for any neighborhood U of $\tilde{a}$ in $\tilde{Y}(a)$ one has $\tilde{\mathbb{P}}(U) > 0$). $\tilde{Y}_0(a)$ is a closed (hence compact) and $\{\sigma_t\}$-invariant subset of $\tilde{Y}(a)$, with $\tilde{\mathbb{P}}(\tilde{Y}_0(a)) = 1$. Also, $\tilde{Y}_0(a)$ is connected, since otherwise there would exist two open sets $U_1, U_2 \subset \tilde{Y}(a)$ such that $\tilde{Y}_0(a) \cap U_1$ and $\tilde{Y}_0(a) \cap U_2$ are nonempty, compact and disjoint, and their union equals $\tilde{Y}_0(a)$. The sets $\tilde{Y}_0(a) \cap U_1$ and $\tilde{Y}_0(a) \cap U_2$ are invariant, and, by the definition of support, each of them has $\tilde{\mathbb{P}}$-measure positive, which contradicts the ergodicity of $\tilde{\mathbb{P}}$.

LEMMA 4.1.2

There exists $\Omega_0 \subset \Omega$ with $\mathbb{P}(\Omega_0) = 1$ such that $\tilde{Y}_0(a) = \operatorname{cl}\{E(\theta_t\omega) : t \in \mathbb{R}\}$ for any $\omega \in \Omega_0$, where the closure is taken in the weak-* topology on Y.

PROOF   By [89, Theorem 9.27], there exists a Borel set $Y' \subset \tilde{Y}_0(a)$ with $\tilde{\mathbb{P}}(Y') = 1$ with the property that for each $\tilde{a} \in Y'$ there holds

$$\lim_{t\to\infty} \frac{1}{t} \int_0^t h(\sigma_s \tilde{a}) \, ds = \int_{\tilde{Y}_0(a)} h(\cdot) \, d\tilde{\mathbb{P}}(\cdot),$$

for any $h \in C(\tilde{Y}_0(a))$. We claim that $\operatorname{cl}\{\sigma_t\tilde{a} : t \geq 0\} = \tilde{Y}_0(a)$ for any $\tilde{a} \in Y'$. Suppose not. Then there are $\tilde{a} \in Y'$ and $\bar{a} \in \tilde{Y}_0(a)$ such that $\bar{a} \notin \operatorname{cl}\{\sigma_t\tilde{a} : t \geq 0\} =: Y''$. By the Urysohn lemma, there is a nonnegative $h \in C(\tilde{Y}_0(a))$ such that $h(\hat{a}) = 0$ for any $\hat{a} \in Y''$ and $h(\bar{a}) > 0$. From the former property it follows that $\lim_{t\to\infty} \frac{1}{t} \int_0^t h(\sigma_s \tilde{a}) \, ds = 0$. By continuity, there is a relative neighborhood V of $\bar{a}$ in $\tilde{Y}_0(a)$ such that $h(\check{a}) > 0$ for any $\check{a} \in V$. Since $\bar{a}$ belongs to the support of μ, we have $\int_{\tilde{Y}_0(a)} h(\cdot) \, d\tilde{\mathbb{P}}(\cdot) \geq \int_V h(\cdot) \, d\tilde{\mathbb{P}}(\cdot) > 0$, a contradiction.

It suffices to put $\Omega_0 := E_a^{-1}(Y')$. $\square$

If Assumptions (A2-1)–(A2-3) are satisfied for Y replaced with $\tilde{Y}(a)$, we will denote by $\Pi(a) = \{\Pi(a)_t\}_{t \geq 0}$ the *topological linear skew-product semiflow generated by* (4.1.1)+(4.1.2) on the product Banach bundle $L_2(D) \times \tilde{Y}(a)$:

$$\Pi(a)(t; u_0, \tilde{a}) = \Pi(a)_t(u_0, \tilde{a}) := (U_{\tilde{a}}(t, 0)u_0, \sigma_t \tilde{a})$$

for $t \geq 0$, $\tilde{a} \in \tilde{Y}(a)$, and $u_0 \in L_2(D)$, where $U_{\tilde{a}}(t, 0)u_0$ stands for the weak solution of $(2.0.1)_{\tilde{a}} + (2.0.2)_{\tilde{a}}$ with the initial condition $u(0, x) = u_0(x)$, $x \in D$. (Here, $\tilde{a} = (\tilde{a}_{ij}, \tilde{a}_i, \tilde{b}_i, \tilde{c}_0, \tilde{d}_0)$.)

Moreover, define

$$\tilde{\Pi}(t; u_0, \omega) := (U_{E_a(\omega)}(t, 0)u_0, \theta_t \omega), \quad t \geq 0, \ \omega \in \Omega, \ u_0 \in L_2(D).$$

We have

LEMMA 4.1.3
If (A2-1)–(A2-3) *are satisfied, then $\tilde{\Pi}$ is a random linear skew-product semiflow on the measurable Banach bundle $L_2(D) \times \Omega$, covering the metric flow $((\Omega, \mathfrak{F}, \mathbb{P}), \{\theta_t\}_{t \in \mathbb{R}})$.*

PROOF   It follows from (4.1.4) and the definitions of Π and $\tilde{\Pi}$ that for each $t \geq 0$ the diagram

$$
\begin{array}{ccc}
L_2(D) \times \Omega & \xrightarrow{\ \tilde{\Pi}_t\ } & L_2(D) \times \Omega \\
{\scriptstyle (\mathrm{Id}_{L_2(D)}, E)} \Big\downarrow & & \Big\downarrow {\scriptstyle (\mathrm{Id}_{L_2(D)}, E)} \\
L_2(D) \times Y & \xrightarrow{\ \Pi_t\ } & L_2(D) \times Y
\end{array}
$$

commutes. Consequently, the properties (RSP1) and (RSP2) of the random skew-product semiflow are satisfied. Obviously, for any $t \geq 0$ and $\omega \in \Omega$ the mapping $U_{E_a(\omega)}(t, 0)$ belongs to $\mathcal{L}(L_2(D))$.

It remains to prove that the mapping

$$[(t, u_0, \omega) \mapsto U_{E_a(\omega)}(t, 0)u_0]$$

is $(\mathfrak{B}([0, \infty)) \times \mathfrak{B}(L_2(D)) \times \mathfrak{F}, \mathfrak{B}(L_2(D)))$-measurable. Indeed, for $n \in \mathbb{N}$ denote

$$\tilde{\Pi}^{[n]}(t; u_0, \omega) := (\tilde{U}^{[n]}(t; u_0, \omega), \theta_t \omega) \quad \text{for} \quad t \geq 0, \ u_0 \in L_2(D), \ \omega \in \Omega,$$

where

$$\tilde{U}^{[n]}(t; u_0, \omega) := \begin{cases} U_{E_a(\omega)}(1/n, 0)u_0 & \text{for } t \in [0, 1/n], \ u_0 \in L_2(D), \ \omega \in \Omega, \\ U_{E_a(\omega)}(t, 0)u_0 & \text{for } t \in [1/n, \infty), \ u_0 \in L_2(D), \ \omega \in \Omega, \end{cases}$$

and

$$\Pi^{[n]}(t; u_0, \tilde{a}) := (U^{[n]}(t; u_0, \tilde{a}), \sigma_t \tilde{a}) \quad \text{for} \quad t \geq 0, \ u_0 \in L_2(D), \ \tilde{a} \in \tilde{Y}(a),$$

where

$$U^{[n]}(t; u_0, \tilde{a}) := \begin{cases} U_{\tilde{a}}(1/n, 0)u_0 & \text{for } t \in [0, 1/n], \ u_0 \in L_2(D), \ \tilde{a} \in \tilde{Y}(a), \\ U_{\tilde{a}}(t, 0)u_0 & \text{for } t \in [1/n, \infty), \ u_0 \in L_2(D), \ \tilde{a} \in \tilde{Y}(a). \end{cases}$$

One has

$$\tilde{U}^{[n]} = U^{[n]} \circ (\mathrm{Id}_{\mathbb{R}_+}, \mathrm{Id}_{L_2(D)}, E).$$

Since Assumption (A2-3) is satisfied, the mapping $U^{[n]}$ is continuous. Further, $(\mathrm{Id}_{[0,\infty)}, \mathrm{Id}_{L_2(D)}, E)$ is $(\mathfrak{B}([0, \infty)) \times \mathfrak{B}(L_2(D)) \times \mathfrak{F}, \mathfrak{B}([0, \infty)) \times \mathfrak{B}(L_2(D)) \times \mathfrak{B}(\tilde{Y}(a)))$-measurable. Consequently, $\tilde{\Pi}^{[n]}$ is $(\mathfrak{B}([0, \infty)) \times \mathfrak{B}(L_2(D)) \times \mathfrak{F}, \mathfrak{B}(L_2(D)) \times \mathfrak{F})$-measurable. As $\tilde{\Pi}^{[n]}$ converge pointwise to $\tilde{\Pi}$, the latter is $(\mathfrak{B}([0, \infty)) \times \mathfrak{B}(L_2(D)) \times \mathfrak{F}, \mathfrak{B}(L_2(D)) \times \mathfrak{F})$-measurable, too. □

A next assumption regards the satisfaction of (A2-1)–(A2-3) by $(4.1.1)_a +$ $(4.1.2)_a$ or $\tilde{Y}(a)$:

(A4-R) (Satisfaction of (A2-1)–(A2-3) and (A4-R1), (A4-R2)) *The assumptions (A4-R1)–(A4-R2) are fulfilled and assumptions (A2-1)–(A2-3) are satisfied for Y replaced with $\tilde{Y}(a)$ defined by (4.1.3).*

Sometimes we say simply that a or $(4.1.1)_a + (4.1.2)_a$ or $\tilde{\Pi}$ or Π *satisfies property* (A4-R).

DEFINITION 4.1.1 *The* principal spectrum *of the random problem* $(4.1.1)_a + (4.1.2)_a$ *satisfying property* (A4-R) *equals the principal spectrum of the topological linear skew-product semiflow* $\Pi(a)$ *over* $\tilde{Y}_0(a)$. *We will denote the principal spectrum by* $\Sigma(a) = [\lambda_{\min}(a), \lambda_{\max}(a)]$.

DEFINITION 4.1.2 *The* principal Lyapunov exponent *of the random problem* $(4.1.1)_a + (4.1.2)_a$ *satisfying property* (A4-R) *equals the principal Lyapunov exponent of the topological linear skew-product semiflow* $\Pi(a)$ *over* $\tilde{Y}_0(a)$ *for the ergodic invariant measure* $\tilde{\mathbb{P}}$. *We will denote the principal Lyapunov exponent by* $\lambda(a)$.

LEMMA 4.1.4
There exists $\Omega_1 \subset \Omega_0$ *with* $\mathbb{P}(\Omega_1) = 1$, *where* Ω_0 *is as in Lemma 4.1.2, such that*

$$\lim_{t \to \infty} \frac{\ln \|U_{E(\omega)}(t, 0)\|}{t} = \lambda(a) \quad \text{for any} \quad \omega \in \Omega_1.$$

PROOF   By Theorem 3.1.5, there is a Borel set $Y_1 \subset \tilde{Y}_0$ with $\tilde{\mathbb{P}}(Y_1) = 1$ such that

$$\lim_{t \to \infty} \frac{\ln \|U_{\tilde{a}}(t, 0)\|}{t} = \lambda(a)$$

for any $\tilde{a} \in Y_1$. It suffices to take $\Omega_1 := E^{-1}(Y_1) \cap \Omega_0$. $\square$

The following assumption is about exponential separation:

(A4-R-ES) (Exponential separation) $(4.1.1)_a + (4.1.2)_a$ *has property* (A4-R) *and, moreover, the topological linear skew-product semiflow* $\Pi(a)$ *generated by* $(4.1.1)_a + (4.1.2)_a$ *on* $L_2(D) \times \tilde{Y}(a)$ *admits an exponential separation over* $\tilde{Y}_0(a)$ *defined by* (4.1.5).

Sometimes we say simply that *a satisfies property* (A4-R-ES).
We now give two examples of a satisfying the property (A4-R-ES).

EXAMPLE 4.1.1 (Only zero-order terms depend on t)
Consider the following random linear parabolic equation:

$$\frac{\partial u}{\partial t} = \sum_{i=1}^{N} \frac{\partial}{\partial x_i} \left(\sum_{j=1}^{N} a_{ij}(x) \frac{\partial u}{\partial x_j} + a_i(x)u \right)$$

$$+ \sum_{i=1}^{N} b_i(x) \frac{\partial u}{\partial x_i} + c_0(\theta_t \omega, x)u, \quad x \in D, \qquad (4.1.6)$$

endowed with the boundary condition

$$\mathcal{B}_\omega(t)u = 0, \quad x \in \partial D, \qquad (4.1.7)$$

where $\mathcal{B}_\omega(t) = \mathcal{B}_{a^\omega}(t)$ and $a^\omega(t, x) = (a_{ij}(x), a_i(x), b_i(x), c_0(\theta_t \omega, x), d_0(x))$.

We make the following assumptions:
In the Dirichlet case:

- $D \subset \mathbb{R}^N$ is a bounded domain with Lipschitz boundary.

- $a_{ij}(= a_{ji}), a_i, b_i \colon D \to \mathbb{R}$ are $(\mathfrak{B}(D), \mathfrak{B}(\mathbb{R}))$-measurable and $a_{ij}(= a_{ji})$, $a_i, b_i \in L_\infty(D)$. Moreover, (A2-1) is satisfied.

- $c_0 \colon \Omega \times D \to \mathbb{R}$ is $(\mathfrak{F} \times \mathfrak{B}(D), \mathfrak{B}(\mathbb{R}))$-measurable. Moreover, for any $\omega \in \Omega$, the function $c_0^\omega(\cdot)$ belongs to $L_\infty(\mathbb{R} \times D)$, with the $L_\infty(\mathbb{R} \times D)$-norm bounded uniformly in $\omega \in \Omega$.

In the Neumann or Robin cases:

- $D \subset \mathbb{R}^N$ is a bounded domain, where its boundary is an $(N-1)$-dimensional manifold of class C^2.

- $a_{ij}(= a_{ji}), a_i, b_i \in C^1(\bar{D})$, $d_0 \in C^1(\partial D)$. Moreover, (A2-1) is satisfied, and $d_0(x) \geq 0$ for all $x \in \partial D$.

- $c_0 : \Omega \times D \to \mathbb{R}$ is $(\mathfrak{F} \times \mathfrak{B}(D), \mathfrak{B}(\mathbb{R}))$-measurable. Moreover, for any $\omega \in \Omega$, the function $c_0^\omega(\cdot)$ belongs to $L_\infty(\mathbb{R} \times D)$, with the $L_\infty(\mathbb{R} \times D)$-norm bounded uniformly in $\omega \in \Omega$.

We claim that a random problem (4.1.6)+(4.1.7) satisfying the above requirements has the property (A4-R-ES). Indeed, assumptions (A4-R1)–(A4-R2) and (A2-1)–(A2-2) are formulated explicitly. As a_{ij}, a_i, b_i, and d_0 are independent of time, the condition (A2-4) is satisfied, so, by Theorem 2.4.1, the assumption (A2-3) holds.

In the Dirichlet case, the inequalities (A3-1) and (A3-2) hold for any $\tilde{a} \in \tilde{Y}_0(a)$, for both $(2.0.1)_{\tilde{a}}+(2.0.2)_{\tilde{a}}$ and its adjoint, by [61, Theorem 2.1 and Lemma 3.9], so the topological linear skew-product flow $\Pi(a)$ admits an exponential separation over $\tilde{Y}_0(a)$.

In the Neumann and Robin cases, we first show that any weak solution of $(2.0.1)_{\tilde{a}}+(2.0.2)_{\tilde{a}}$, as well as of its adjoint equation, is in fact a strong solution on any interval away from the initial time.

In order to do so, fix $\tilde{a} \in \tilde{Y}_0$. We approximate $\tilde{a} = (a_{ij}, a_i, b_i, \tilde{c}_0, d_0)$ by a sequence $(a^{(n)})_{n=1}^\infty \subset L_\infty(\mathbb{R} \times D, \mathbb{R}^{N^2+2N+1}) \times L_\infty(\mathbb{R} \times \partial D, \mathbb{R})$ such that (where we write $a^{(n)} = (a_{ij}^{(n)}, a_i^{(n)}, b_i^{(n)}, c_0^{(n)}, d_0^{(n)})$):

(i) $a_{ij}^{(n)}(= a_{ji}^{(n)})$, $a_i^{(n)}$, $b_i^{(n)} \in C^2(\bar{D})$ $(i, j = 1, 2, \ldots, N)$, $c_0^{(n)} \in C^2(\mathbb{R} \times \bar{D})$, $d_0^{(n)} \in C^2(\partial D)$; moreover, $d_0^{(n)}(x) \geq 0$ for all $n \in \mathbb{N}$ and all $x \in \partial D$,

(ii) $\sup\{\|a^{(n)}\|_\infty : n \in \mathbb{N}\} < \infty$, where $\|\cdot\|_\infty$ denotes the $L_\infty(\mathbb{R} \times D, \mathbb{R}^{N^2+2N+1}) \times L_\infty(\mathbb{R} \times \partial D, \mathbb{R})$-norm,

(iii) $a_{ij}^{(n)}(x)$, $a_i^{(n)}(x)$, $b_i^{(n)}(x)$ $(i, j = 1, 2, \ldots, N)$ converge respectively to $a_{ij}(x)$, $a_i(x)$, $b_i(x)$, for a.e. $x \in D$,

(iv) $c_0^{(n)}$ converge to $\tilde{c}_0$ in the $L_{2,\mathrm{loc}}(\mathbb{R} \times \bar{D})$ sense; moreover, the sections $c_0^{(n)}(t, \cdot)$ converge in the $L_2(D)$-norm to the section $\tilde{c}_0(t, \cdot)$ for a.e. $t > 0$,

(v) $d_0^{(n)}(x)$ converge to $d_0(x)$ for a.e. $x \in \partial D$.

Let $U_{a^{(n)}}(t, s)$ stand for the solution operator of (2.0.1)+(2.0.2) with a replaced by $a^{(n)}$. Further, let $U_a^0(t, s)$ denote the solution operator of (2.0.1)+(2.0.2) with a replaced by $(a_{ij}, a_i, b_i, 0, d_0)$, and let $U_{a^{(n)}}^0(t, s)$ denote the solution operator of (2.0.1)+(2.0.2) with a replaced by $(a_{ij}^{(n)}, a_i^{(n)}, b_i^{(n)}, 0, d_0^{(n)})$.

For a given $u_0 \in L_2(D)$, let $(u_n)_{n=1}^\infty \subset \mathring{C}(\bar{D})$ be such that $u_n \to u_0$ in $L_2(D)$. $U_{a^{(n)}}(\cdot, 0)u_n$ is, for any $n \in \mathbb{N}$, a classical solution, consequently it satisfies the following integral equation

$$U_{a^{(n)}}(t, 0)u_n = U_{a^{(n)}}^0(t, 0)u_n + \int_0^t U_{a^{(n)}}^0(t, \tau)(c_0^{(n)}(\tau, \cdot)U_{a^{(n)}}(\tau, 0)u_n) \, d\tau$$

for all $t > 0$ (here and in the sequel, we identify the operator of multiplying a function from $L_2(D)$ by the section $c_0^{(n)}(\tau, \cdot)$ with that section).

Applying the ideas used in the proofs of Theorem 2.4.1 and Proposition 2.2.13 we see that

$$\|U_{a^{(n)}}(t,0)u_n - U_{\tilde{a}}(t,0)u_0\| \to 0$$

and

$$\|U_{a^{(n)}}^0(t,0)u_n - U_a^0(t,0)u_0\| \to 0$$

for any $t > 0$.

Put $u_n(t) := U_{a^{(n)}}(t,0)u_n$ and $u(t) := U_{\tilde{a}}(t,0)u_0$, for any $t \geq 0$. Fix $t > 0$. The sequence $(a^{(n)})$ is so chosen that we get, via the L_2–L_2 estimates (Proposition 2.2.2) the following:

$$\|U_{a^{(n)}}^0(t,\tau)(c_0^{(n)}(\tau,\cdot)u_n(\tau)) - U_a^0(t,\tau)(\tilde{c}_0(\tau,\cdot)u(\tau))\|$$
$$\leq \|U_{a^{(n)}}^0(t,\tau)((c_0^{(n)}(\tau,\cdot) - \tilde{c}_0(\tau,\cdot))u_n(\tau))\|$$
$$\quad + \|U_{a^{(n)}}^0(t,\tau)(\tilde{c}_0(\tau,\cdot)u_n(\tau)) - U_a^0(t,\tau)(\tilde{c}_0(\tau,\cdot)u(\tau))\|$$
$$\leq Me^{\gamma(t-\tau)}\|(c_0^{(n)}(\tau,\cdot) - \tilde{c}_0(\tau,\cdot))(u_n(\tau))\|$$
$$\quad + \|U_{a^{(n)}}^0(t,\tau)(\tilde{c}_0(\tau,\cdot)u_n(\tau)) - U_a^0(t,\tau)(\tilde{c}_0(\tau,\cdot)u(\tau))\|$$
$$\leq Me^{\gamma(t-\tau)}\|c_0^{(n)}(\tau,\cdot) - \tilde{c}_0(\tau,\cdot)\|\,\|u_n(\tau)\|$$
$$\quad + \|U_{a^{(n)}}^0(t,\tau)(\tilde{c}_0(\tau,\cdot)u_n(\tau)) - U_a^0(t,\tau)(\tilde{c}_0(\tau,\cdot)u(\tau))\|$$
$$\to 0$$

as $n \to \infty$ for a.e. $\tau \in (0,t)$. By the L_2–L_2 estimates (Proposition 2.2.2), the set

$$\{\,\|U_{a^{(n)}}^0(t,\tau)(c_0^{(n)}(\tau,0)u_n(\tau))\| : \tau \in [0,t],\ n \in \mathbb{N}\,\}$$

is bounded. It then follows that

$$U_{\tilde{a}}(t,0)u_0 = U_a^0(t,0)u_0 + \int_0^t U_a^0(t,\tau)(\tilde{c}_0(\tau,\cdot)U_{\tilde{a}}(\tau,0)u_0)\,d\tau$$

for $t > 0$ (in other words, $[t \mapsto U_{\tilde{a}}(t,0)u_0]$ is a mild solution). Therefore by the arguments in [92, Section 2] and the Sobolev embedding theorems we have that $[[0,T] \ni t \mapsto U_{\tilde{a}}(t,0)u_0] \in W_p^{1,2}((0,T) \times D)$ for any $T > 0$ and $p > 1$, and $U_{\tilde{a}}(t,0)u_0$ is a strong solution on (t_0, T) for any $0 < t_0 < T$.

Now we use [59, Theorem 2.5] to conclude that the inequality (A3-2) holds with $\varsigma = 0$, which by Lemma 3.3.1 implies the assumption (A3-1). Similarly, (A3-1) and (A3-2) hold for the adjoint problem of $(2.0.1)_{\tilde{a}} + (2.0.2)_{\tilde{a}}$. Consequently, the topological linear skew-product flow $\Pi(a)$ admits an exponential separation over $\tilde{Y}_0(a)$.

EXAMPLE 4.1.2 (The classical case)

Consider the linear random parabolic equation (4.1.1) endowed with the boundary conditions (4.1.2), where we make the following assumptions:

- $D \subset \mathbb{R}^N$ is a bounded domain, where its boundary is an $(N-1)$-dimensional manifold of class $C^{3+\alpha}$ for some $\alpha > 0$.

- The functions $a_{ij} (= a_{ji})$, a_i, b_i, and c_0 are $(\mathfrak{F} \times \mathfrak{B}(D), \mathfrak{B}(\mathbb{R}))$-measurable, and the function d_0 is $(\mathfrak{F} \times \mathfrak{B}(\partial D), \mathfrak{B}(\mathbb{R}))$-measurable.

- There is $\alpha \in (0,1)$ such that for any $\omega \in \Omega$, the functions a_{ij}^ω and a_i^ω $(i, j = 1, 2, \ldots, N)$ belong to $C^{2+\alpha, 2+\alpha}(\mathbb{R} \times \bar{D})$, the functions b_i^ω $(i = 1, 2, \ldots, N)$ and c_0^ω belong to $C^{2+\alpha, 1+\alpha}(\mathbb{R} \times \bar{D})$ and the function d_0^ω belongs to $C^{2+\alpha, 2+\alpha}(\mathbb{R} \times \partial D)$. Moreover, there is $M > 0$ such that for any $\omega \in \Omega$, the $C^{2+\alpha, 2+\alpha}(\mathbb{R} \times \bar{D})$-norms of a_{ij}^ω and a_i^ω $(i, j = 1, 2, \ldots, N)$, the $C^{2+\alpha, 1+\alpha}(\mathbb{R} \times \bar{D})$-norms of b_i^ω $(i = 1, 2, \ldots, N)$ and c_0^ω, and the $C^{2+\alpha, 2+\alpha}(\mathbb{R} \times \partial D)$-norms of d_0^ω, are bounded by M.

- There exists $\alpha_0 > 0$ such that $\sum_{i,j=1}^N a_{ij}(\omega, x) \xi_i \xi_j \geq \alpha_0 \sum_{i=1}^N \xi_i^2$ for all $\omega \in \Omega$, $x \in \bar{D}$, and $\xi \in \mathbb{R}^N$, and $d_0(\omega, x) \geq 0$ for all $\omega \in \Omega$ and $x \in \partial D$.

The problem (4.1.1)+(4.1.2) satisfies (A4-R-ES). Indeed, (A2-1)–(A2-2) and (A4-R1)–(A4-R2) are explicitly stated. By a standard reasoning making use of the Ascoli–Arzelà theorem we see that all the estimates on the $C^{2+\alpha, 2+\alpha}(\mathbb{R} \times \bar{D})$-norms, etc., carry over to the elements of $\tilde{Y}(a)$. Consequently, (A2-5) is satisfied (see Section 2.5), so (A2-3) is fulfilled.

As in Example 4.1.1, in the Dirichlet case, the inequalities (A3-1) and (A3-2) hold for any $\tilde{a} \in \hat{Y}_0(a)$, for both $(2.0.1)_{\tilde{a}}+(2.0.2)_{\tilde{a}}$ and its adjoint, by [61, Theorem 2.1 and Lemma 3.9], so the topological linear skew-product flow $\Pi(a)$ admits an exponential separation over $\hat{Y}_0(a)$.

In the Neumann and Robin cases, by Proposition 2.5.1, $U_{\tilde{a}}(t, 0)u_0$ is a classical solution on $[t_0, T]$ for any $0 < t_0 < T$. Then we use [59, Theorem 2.5] again to conclude that the inequality (A3-2) holds with $\varsigma = 0$, which by Lemma 3.3.1 implies the assumption (A3-1). Similarly, (A3-1) and (A3-2) hold for the adjoint problem of $(2.0.1)_{\tilde{a}}+(2.0.2)_{\tilde{a}}$. Consequently, the topological linear skew-product flow $\Pi(a)$ admits an exponential separation over $\hat{Y}_0(a)$.

From now until the end of the present subsection we assume that a is such that property (A4-R-ES) holds.

As a is fixed, we will suppress its symbol: We write E, $\tilde{Y}$, $\tilde{Y}_0$, Π for E_a, $\tilde{Y}(a)$, $\tilde{Y}_0(a)$, $\Pi(a)$, respectively. Also, instead of $U_{E(\omega)}(t, s)$ we will write $U_\omega(t, s)$.

The following results are simple consequences of the results in Chapter 3.

PROPOSITION 4.1.1

Let Ω_1 be as in Lemma 4.1.4. Then for any $\omega \in \Omega_1$ and any $u_0 \in L_2(D)^+ \setminus \{0\}$ one has

$$\lim_{t \to \infty} \frac{\ln \|U_\omega(t, 0)u_0\|}{t} = \lambda(a). \tag{4.1.8}$$

PROOF   See Lemmas 4.1.4 and 3.2.5.   ▯

PROPOSITION 4.1.2

For any sequence $(\omega^{(n)})_{n=1}^\infty \subset \Omega_0$, where Ω_0 is as in Lemma 4.1.2, any $u_0 \in L_2(D)^+ \setminus \{0\}$, and any real sequences $(s_n)_{n=1}^\infty$, $(t_n)_{n=1}^\infty$ such that $t_n - s_n \to \infty$ one has

$$\lambda_{\min}(a)$$

$$\leq \liminf_{n \to \infty} \frac{\ln \|U_{\omega^{(n)}}(t_n, s_n)w(E(\omega^{(n)}) \cdot s_n)\|}{t_n - s_n} = \liminf_{n \to \infty} \frac{\ln \|U_{\omega^{(n)}}(t_n, s_n)u_0\|}{t_n - s_n}$$

$$= \liminf_{n \to \infty} \frac{\ln \|U_{\omega^{(n)}}(t_n, s_n)\|}{t_n - s_n} \leq \limsup_{n \to \infty} \frac{\ln \|U_{\omega^{(n)}}(t_n, s_n)\|}{t_n - s_n}$$

$$= \limsup_{n \to \infty} \frac{\ln \|U_{\omega^{(n)}}(t_n, s_n)u_0\|}{t_n - s_n} = \limsup_{n \to \infty} \frac{\ln \|U_{\omega^{(n)}}(t_n, s_n)w(E(\omega^{(n)}) \cdot s_n)\|}{t_n - s_n}$$

$$\leq \lambda_{\max}(a).$$

PROOF   See Theorem 3.1.2(1) and Lemma 3.2.5.   ▯

PROPOSITION 4.1.3

For each $\omega \in \Omega_0$, where Ω_0 is as in Lemma 4.1.2,

(i) *there are sequences $(s_n')_{n=1}^\infty, (t_n')_{n=1}^\infty \subset \mathbb{R}$, $t_n' - s_n' \to \infty$ as $n \to \infty$, such that*

$$\lambda_{\min}(a) = \lim_{n \to \infty} \frac{\ln \|U_\omega(t_n', s_n')w(E(\omega) \cdot s_n')\|}{t_n' - s_n'}$$

$$= \lim_{n \to \infty} \frac{\ln \|U_\omega(t_n', s_n')u_0\|}{t_n' - s_n'} = \lim_{n \to \infty} \frac{\ln \|U_\omega(t_n', s_n')\|}{t_n' - s_n'}$$

for each $u_0 \in L_2(D)^+ \setminus \{0\}$,

(ii) *there are sequences $(s_n'')_{n=1}^\infty, (t_n'')_{n=1}^\infty \subset \mathbb{R}$, $t_n'' - s_n'' \to \infty$ as $n \to \infty$, such that*

$$\lambda_{\max}(a) = \lim_{n \to \infty} \frac{\ln \|U_\omega(t_n'', s_n'')w(E(\omega) \cdot s_n'')\|}{t_n'' - s_n''}$$

$$= \lim_{n \to \infty} \frac{\ln \|U_\omega(t_n'', s_n'')u_0\|}{t_n'' - s_n''} = \lim_{n \to \infty} \frac{\ln \|U_\omega(t_n'', s_n'')\|}{t_n'' - s_n''}$$

for each $u_0 \in L_2(D)^+ \setminus \{0\}$.

PROOF   Note that for each $\omega \in \Omega_0$, $\tilde{Y}_0 = \text{cl}\{\, E(\omega) \cdot t : t \in \mathbb{R} \,\}$. The proposition then follows from Theorem 3.2.7(1).   ∎

A consequence of Propositions 4.1.2 and 4.1.3 is the following.

PROPOSITION 4.1.4

For any $\omega \in \Omega_0$, where Ω_0 is as in Lemma 4.1.2, and any $u_0 \in L_2(D)^+ \setminus \{0\}$ there holds

$$
\begin{aligned}
\lambda_{\min}(a) &= \liminf_{t-s \to \infty} \frac{\ln \|U_\omega(t,s)w(E(\omega) \cdot s)\|}{t - s} \\
&= \liminf_{t-s \to \infty} \frac{\ln \|U_\omega(t,s)u_0\|}{t - s} = \liminf_{t-s \to \infty} \frac{\ln \|U_\omega(t,s)\|}{t - s} \\
&\leq \limsup_{t-s \to \infty} \frac{\ln \|U_\omega(t,s)\|}{t - s} = \limsup_{t-s \to \infty} \frac{\ln \|U_\omega(t,s)u_0\|}{t - s} \\
&= \limsup_{t-s \to \infty} \frac{\ln \|U_\omega(t,s)w(E(\omega) \cdot s)\|}{t - s} = \lambda_{\max}(a).
\end{aligned}
$$

PROPOSITION 4.1.5

For each $\omega \in \Omega_0$, where Ω_0 is as in Lemma 4.1.2, and each $\lambda \in [\lambda_{\min}(a), \lambda_{\max}(a)]$ there are sequences $(k_n)_{n=1}^{\infty}, (l_n)_{n=1}^{\infty} \subset \mathbb{Z}$, $l_n - k_n \to \infty$ as $n \to \infty$, such that

$$
\begin{aligned}
\lambda &= \lim_{n \to \infty} \frac{\ln \|U_\omega(l_n, k_n)w(E(\omega) \cdot k_n)\|}{l_n - k_n} \\
&= \lim_{n \to \infty} \frac{\ln \|U_\omega(l_n, k_n)u_0\|}{l_n - k_n} = \lim_{n \to \infty} \frac{\ln \|U_\omega(l_n, k_n)\|}{l_n - k_n}
\end{aligned}
$$

for each $u_0 \in L_2(D)^+ \setminus \{0\}$.

PROOF   Note that for $\omega \in \Omega_0$, $\tilde{Y}_0 = \text{cl}\{\, E(\omega) \cdot t : t \in \mathbb{R} \,\}$. The proposition then follows from Theorem 3.2.7(3).   ∎

In the light of Proposition 4.1.5, Proposition 4.1.4 has the following strengthening.

PROPOSITION 4.1.6

For any $\omega \in \Omega_0$, where Ω_0 is as in Lemma 4.1.2, and any $u_0 \in L_2(D)^+ \setminus \{0\}$

there holds

$$\lambda_{\min}(a)$$

$$= \liminf_{t-s \to \infty} \frac{\ln \|U_\omega(t,s)w(E(\omega) \cdot s)\|}{t-s} = \liminf_{\substack{l-k \to \infty \\ k,l \in \mathbb{Z}}} \frac{\ln \|U_\omega(l,k)w(E(\omega) \cdot k)\|}{l-k}$$

$$= \liminf_{t-s \to \infty} \frac{\ln \|U_\omega(t,s)u_0\|}{t-s} = \liminf_{\substack{l-k \to \infty \\ k,l \in \mathbb{Z}}} \frac{\ln \|U_\omega(l,k)u_0\|}{l-k}$$

$$\leq \limsup_{\substack{l-k \to \infty \\ k,l \in \mathbb{Z}}} \frac{\ln \|U_\omega(l,k)u_0\|}{l-k} = \limsup_{t-s \to \infty} \frac{\ln \|U_\omega(t,s)u_0\|}{t-s}$$

$$= \limsup_{\substack{l-k \to \infty \\ k,l \in \mathbb{Z}}} \frac{\ln \|U_\omega(l,k)w(E(\omega) \cdot k)\|}{l-k} = \limsup_{t-s \to \infty} \frac{\ln \|U_\omega(t,s)w(E(\omega) \cdot s)\|}{t-s}$$

$$= \lambda_{\max}(a).$$

4.1.2   The Nonautonomous Case

Consider the following nonautonomous linear parabolic equation:

$$\frac{\partial u}{\partial t} = \sum_{i=1}^{N} \frac{\partial}{\partial x_i}\left(\sum_{j=1}^{N} a_{ij}(t,x)\frac{\partial u}{\partial x_j} + a_i(t,x)u\right)$$

$$+ \sum_{i=1}^{N} b_i(t,x)\frac{\partial u}{\partial x_i} + c_0(t,x)u, \quad x \in D, \tag{4.1.9}$$

endowed with the boundary condition

$$\mathcal{B}_a(t)u = 0, \quad x \in \partial D, \tag{4.1.10}$$

where $\mathcal{B}_a(t)$ is a boundary operator of either the Dirichlet or Neumann or Robin type as in (2.0.3), $a = (a_{ij}, a_i, b_i, c_0, d_0)$ and $d_0(t,x) \geq 0$ for a.e. $(t,x) \in \mathbb{R} \times \partial D$. Sometimes we write the nonautonomous problem (4.1.9)+(4.1.10) as (4.1.9)$_a$ +(4.1.10)$_a$.

Our first assumption regards boundedness of the coefficients of the equations (and of the boundary conditions):

(A4-N1) (Boundedness) *a belongs to* $L_\infty(\mathbb{R} \times D, \mathbb{R}^{N^2+2N+1}) \times L_\infty(\mathbb{R} \times \partial D, \mathbb{R})$.

Among others, (4.1.9)+(4.1.10) arise from linearization of nonautonomous nonlinear parabolic equations at certain entire solution as well as from linearization of autonomous nonlinear parabolic equations at some entire time dependent solution.

In the rest of the present subsection it is assumed that a satisfies (A4-N1).

Put

$$\tilde{Y}(a) := \mathrm{cl}\,\{\,a \cdot t : t \in \mathbb{R}\,\} \tag{4.1.11}$$

with the weak-* topology, where the closure is taken in the weak-* topology. The set $\tilde{Y}(a)$ is a compact connected metrizable space.

If Assumptions (A2-1)–(A2-3) are satisfied for Y replaced with $\tilde{Y}(a)$, we will denote by $\Pi(a) = \{\Pi(a)_t\}_{t\geq 0}$ the *topological linear skew-product semiflow generated by* (4.1.9)+(4.1.10) on the product Banach bundle $L_2(D) \times \tilde{Y}(a)$:

$$\Pi(a)(t; u_0, \tilde{a}) = \Pi(a)_t(u_0, \tilde{a}) := (U_{\tilde{a}}(t, 0)u_0, \sigma_t \tilde{a})$$

for $t \geq 0$, $\tilde{a} \in \tilde{Y}(a)$, $u_0 \in L_2(D)$, where $U_{\tilde{a}}(t, 0)u_0$ stands for the weak solution of $(2.0.1)_{\tilde{a}}+(2.0.2)_{\tilde{a}}$ with initial condition $u(0, x) = u_0(x)$.

The next assumption is about the satisfaction of (A2-1)–(A2-3).

(A4-N) (Satisfaction of (A2-1)–(A2-3) and (A4-N1)) *The assumption* (A4-N1) *is fulfilled, and assumptions* (A2-1)–(A2-3) *are satisfied for Y replaced with $\tilde{Y}(a)$ defined by* (4.1.11).

Sometimes we say simply that a or $(4.1.9)_a+(4.1.10)_a$ or $\Pi(a)$ *satisfies property* (A4-N).

DEFINITION 4.1.3 *The* principal spectrum *of the nonautonomous problem* $(4.1.9)_a+(4.1.10)_a$ *satisfying property* (A4-N) *equals the principal spectrum of the topological linear skew-product semiflow $\Pi(a)$ over $\tilde{Y}(a)$. We will denote the principal spectrum by $\Sigma(a) = [\lambda_{\min}(a), \lambda_{\max}(a)]$.*

The following assumption is about the exponential separation.

(A4-N-ES) (Exponential separation) $(4.1.9)_a+(4.1.10)_a$ *has property* (A4-N) *and, moreover, the topological linear skew-product semiflow $\Pi(a)$ generated by* $(4.1.9)_a+(4.1.10)_a$ *on $L_2(D) \times \tilde{Y}(a)$ admits an exponential separation over $\tilde{Y}(a)$.*

Sometimes we say simply that a *satisfies property* (A4-N-ES).
We now give two examples of a satisfying the property (A4-N-ES).

EXAMPLE 4.1.3 (Only zero-order terms depend on t)
Consider the following nonautonomous linear parabolic equation:

$$\frac{\partial u}{\partial t} = \sum_{i=1}^{N} \frac{\partial}{\partial x_i}\left(\sum_{j=1}^{N} a_{ij}(x)\frac{\partial u}{\partial x_j} + a_i(x)u\right)$$

$$+ \sum_{i=1}^{N} b_i(x)\frac{\partial u}{\partial x_i} + c_0(t, x)u, \quad x \in D, \tag{4.1.12}$$

endowed with the boundary condition

$$\mathcal{B}_a(t)u = 0, \quad x \in \partial D, \tag{4.1.13}$$

where $\mathcal{B}_a$ is as in (2.0.3) with $a(t, x) = (a_{ij}(x), a_i(x), b_i(x), c_0(t, x), d_0(x))$.

We make the following assumptions:

In the Dirichlet case:

- $D \subset \mathbb{R}^N$ is a bounded domain with Lipschitz boundary,

- $a_{ij}(= a_{ji}), a_i, b_i \in L_\infty(D)$. Moreover, (A2-1) is satisfied,

- $c_0 \in L_\infty(\mathbb{R} \times D)$.

In the Neumann or Robin cases:

- $D \subset \mathbb{R}^N$ is a bounded domain, where its boundary is an $(N-1)$-dimensional manifold of class C^2,

- $a_{ij}(= a_{ji}), a_i, b_i \in C^1(\bar{D}), d_0 \in C^1(\partial D)$. Moreover, (A2-1) is satisfied, and $d_0(x) \geq 0$ for all $x \in \partial D$,

- $c_0 \in L_\infty(\mathbb{R} \times D)$.

By arguments similar to those in Example 4.1.1, the nonautonomous problem (4.1.12)+(4.1.13) satisfying the above requirements has the property (A4-N-ES).

EXAMPLE 4.1.4 (The classical case)

Consider the linear nonautonomous parabolic equation (4.1.9) endowed with the boundary conditions (4.1.10), where we make the following assumptions:

- $D \subset \mathbb{R}^N$ is a bounded domain, where its boundary is an $(N-1)$-dimensional manifold of class $C^{3+\alpha}$ for some $\alpha > 0$.

- There is $\alpha > 0$ such that the functions $a_{ij}(= a_{ji})$ and a_i $(i, j = 1, 2, \ldots, N)$ belong to $C^{2+\alpha, 2+\alpha}(\mathbb{R} \times \bar{D})$, the functions b_i $(i = 2, \cdots, N)$ and c_0 belong to $C^{2+\alpha, 1+\alpha}(\mathbb{R} \times \bar{D})$, and the function d_0 belongs to $C^{2+\alpha, 2+\alpha}(\mathbb{R} \times \partial D)$.

- There exists $\alpha_0 > 0$ such that $\sum_{i,j=1}^N a_{ij}(t, x) \xi_i \xi_j \geq \alpha_0 \sum_{i=1}^N \xi_i^2$ for all $t \in \mathbb{R}$, $x \in \bar{D}$, and $\xi \in \mathbb{R}^N$, and $d_0(t, x) \geq 0$ for all $t \in \mathbb{R}$ and $x \in \partial D$.

By similar arguments to those in Example 4.1.2, the nonautonomous problem (4.1.12)+(4.1.13) satisfying the above requirements has the property (A4-N-ES).

From now until the end of the present subsection we assume that a is such that property (A4-N-ES) holds.

As a is fixed, we will suppress its symbol: We write $\tilde{Y}$, Π for $\tilde{Y}(a)$, $\Pi(a)$, respectively. Also, instead of $U_a(t,s)$ we will write $U(t,s)$.

The following results are simple consequences of the results in Chapter 3.

PROPOSITION 4.1.7

For any $u_0 \in L_2(D)^+ \setminus \{0\}$ and any real sequences $(s_n)_{n=1}^\infty$, $(t_n)_{n=1}^\infty$ such that $t_n - s_n \to \infty$ one has

$$\lambda_{\min}(a) \le \liminf_{n\to\infty} \frac{\ln\|U(t_n, s_n)w(a \cdot s_n)\|}{t_n - s_n} = \liminf_{n\to\infty} \frac{\ln\|U(t_n, s_n)u_0\|}{t_n - s_n}$$

$$= \liminf_{n\to\infty} \frac{\ln\|U(t_n, s_n)\|}{t_n - s_n} \le \limsup_{n\to\infty} \frac{\ln\|U(t_n, s_n)\|}{t_n - s_n}$$

$$= \limsup_{n\to\infty} \frac{\ln\|U(t_n, s_n)u_0\|}{t_n - s_n} = \limsup_{n\to\infty} \frac{\ln\|U(t_n, s_n)w(a \cdot s_n)\|}{t_n - s_n} \le \lambda_{\max}(a).$$

PROOF   See Theorem 3.1.2(1) and Lemma 3.2.5.   ◻

PROPOSITION 4.1.8

(i) *There are sequences $(s_n')_{n=1}^\infty, (t_n')_{n=1}^\infty \subset \mathbb{R}$, $t_n' - s_n' \to \infty$ as $n \to \infty$, such that*

$$\lambda_{\min}(a) = \lim_{n\to\infty} \frac{\ln\|U(t_n', s_n')w(a \cdot s_n')\|}{t_n' - s_n'}$$

$$= \lim_{n\to\infty} \frac{\ln\|U(t_n', s_n')u_0\|}{t_n' - s_n'} = \lim_{n\to\infty} \frac{\ln\|U(t_n', s_n')\|}{t_n' - s_n'}$$

for each $u_0 \in L_2(D)^+ \setminus \{0\}$.

(ii) *There are sequences $(s_n'')_{n=1}^\infty, (t_n'')_{n=1}^\infty \subset \mathbb{R}$, $t_n'' - s_n'' \to \infty$ as $n \to \infty$, such that*

$$\lambda_{\max}(a) = \lim_{n\to\infty} \frac{\ln\|U(t_n'', s_n'')w(a \cdot s_n'')\|}{t_n'' - s_n''}$$

$$= \lim_{n\to\infty} \frac{\ln\|U(t_n'', s_n'')u_0\|}{t_n'' - s_n''} = \lim_{n\to\infty} \frac{\ln\|U(t_n'', s_n'')\|}{t_n'' - s_n''}$$

for each $u_0 \in L_2(D)^+ \setminus \{0\}$.

PROOF   Note that $\tilde{Y} = \mathrm{cl}\{a \cdot t : t \in \mathbb{R}\}$. The proposition then follows from Theorem 3.2.7(1).   ◻

A consequence of Propositions 4.1.7 and 4.1.8 is the following.

PROPOSITION 4.1.9

For any $u_0 \in L_2(D)^+ \setminus \{0\}$ there holds

$$
\lambda_{\min}(a) = \liminf_{t-s \to \infty} \frac{\ln \|U(t,s)w(a \cdot s)\|}{t-s} = \liminf_{t-s \to \infty} \frac{\ln \|U(t,s)u_0\|}{t-s}
$$

$$
= \liminf_{t-s \to \infty} \frac{\ln \|U(t,s)\|}{t-s} \leq \limsup_{t-s \to \infty} \frac{\ln \|U(t,s)\|}{t-s} = \limsup_{t-s \to \infty} \frac{\ln \|U(t,s)u_0\|}{t-s}
$$

$$
= \limsup_{t-s \to \infty} \frac{\ln \|U(t,s)w(a \cdot s)\|}{t-s} = \lambda_{\max}(a).
$$

PROPOSITION 4.1.10

For each $\lambda \in [\lambda_{\min}(a), \lambda_{\max}(a)]$ there are sequences $(k_n)_{n=1}^{\infty}, (l_n)_{n=1}^{\infty} \subset \mathbb{Z}$, $l_n - k_n \to \infty$ as $n \to \infty$, such that

$$
\lambda = \lim_{n \to \infty} \frac{\ln \|U(l_n, k_n)w(a \cdot k_n)\|}{l_n - k_n}
$$

$$
= \lim_{n \to \infty} \frac{\ln \|U(l_n, k_n)u_0\|}{l_n - k_n} = \lim_{n \to \infty} \frac{\ln \|U(l_n, k_n)\|}{l_n - k_n}
$$

for each $u_0 \in L_2(D)^+ \setminus \{0\}$.

PROOF   Note that $\tilde{Y} = \mathrm{cl}\{a \cdot t : t \in \mathbb{R}\}$. The proposition then follows from Theorem 3.2.7(3). □

In the light of Proposition 4.1.10, Proposition 4.1.9 has the following strengthening.

PROPOSITION 4.1.11

For any $u_0 \in L_2(D)^+ \setminus \{0\}$ there holds

$$
\lambda_{\min}(a) = \liminf_{t-s \to \infty} \frac{\ln \|U(t,s)w(a \cdot s)\|}{t-s} = \liminf_{\substack{l-k \to \infty \\ k,l \in \mathbb{Z}}} \frac{\ln \|U(l,k)w(a \cdot k)\|}{l-k}
$$

$$
= \liminf_{t-s \to \infty} \frac{\ln \|U(t,s)u_0\|}{t-s} = \liminf_{\substack{l-k \to \infty \\ k,l \in \mathbb{Z}}} \frac{\ln \|U(l,k)u_0\|}{l-k}
$$

$$
\leq \limsup_{\substack{l-k \to \infty \\ k,l \in \mathbb{Z}}} \frac{\ln \|U(l,k)u_0\|}{l-k} = \limsup_{t-s \to \infty} \frac{\ln \|U(t,s)u_0\|}{t-s}
$$

$$
= \limsup_{\substack{l-k \to \infty \\ k,l \in \mathbb{Z}}} \frac{\ln \|U(l,k)w(a \cdot k)\|}{l-k} = \limsup_{t-s \to \infty} \frac{\ln \|U(t,s)w(a \cdot s)\|}{t-s} = \lambda_{\max}(a).
$$

4.2 Monotonicity with Respect to the Zero Order Terms

In this section, we investigate the monotonicity of principal spectrum and principal Lyapunov exponents of random (nonautonomous) parabolic equations with respect to zero order terms. We also study the relation among the principal spectrum and principal Lyapunov exponents for random (nonautonomous) parabolic equations with different types of boundary conditions. Similarly, we treat random equations and nonautonomous equations separately.

4.2.1 The Random Case

Assume that $((\Omega, \mathfrak{F}, \mathbb{P}), \{\theta_t\}_{t\in\mathbb{R}})$ is an ergodic metric dynamical system. Let $a^{(1)}$, $a^{(2)}$ satisfy property (A4-R).

Throughout the present subsection we assume moreover that there is $\tilde{\Omega} \subset \Omega$ with $\mathbb{P}(\tilde{\Omega}) = 1$ such that for each $\omega \in \tilde{\Omega}$:

- $a_{ij}^{(1),\omega}(\cdot,\cdot) = a_{ij}^{(2),\omega}(\cdot,\cdot)$, $a_i^{(1),\omega}(\cdot,\cdot) = a_i^{(2),\omega}(\cdot,\cdot)$, $b_i^{(1),\omega}(\cdot,\cdot) = b_i^{(2),\omega}(\cdot,\cdot)$, for a.e. $(t,x) \in \mathbb{R} \times D$, and

- one of the following conditions, (M-Ra), (M-Rb), (M-Rc), (M-Rd), or (M-Re), holds:

(M-Ra) both $a^{(1)}$ and $a^{(2)}$ are endowed with the Dirichlet boundary conditions, and

　　$*$ $c_0^{(1),\omega}(\cdot,\cdot) \leq c_0^{(2),\omega}(\cdot,\cdot)$　for a.e. $(t,x) \in \mathbb{R} \times D$,

(M-Rb) both $a^{(1)}$ and $a^{(2)}$ are endowed with the Robin boundary conditions, and

　　$*$ $c_0^{(1),\omega}(\cdot,\cdot) \leq c_0^{(2),\omega}(\cdot,\cdot)$　for a.e. $(t,x) \in \mathbb{R} \times D$,

　　$*$ $d_0^{(1),\omega}(\cdot,\cdot) \geq d_0^{(2),\omega}(\cdot,\cdot)$　for a.e. $(t,x) \in \mathbb{R} \times \partial D$,

(M-Rc) both $a^{(1)}$ and $a^{(2)}$ are endowed with the Neumann boundary conditions, and

　　$*$ $c_0^{(1),\omega}(\cdot,\cdot) \leq c_0^{(2),\omega}(\cdot,\cdot)$　for a.e. $(t,x) \in \mathbb{R} \times D$,

(M-Rd) $a^{(1)}$ is endowed with the Dirichlet boundary conditions and $a^{(2)}$ is endowed with Robin boundary conditions, and

　　$*$ $c_0^{(1),\omega}(\cdot,\cdot) = c_0^{(2),\omega}(\cdot,\cdot)$　for a.e. $(t,x) \in \mathbb{R} \times D$,

(M-Re) $a^{(1)}$ is endowed with the Robin boundary conditions and $a^{(2)}$ is endowed with the Neumann conditions, and

　　$*$ $c_0^{(1),\omega}(\cdot,\cdot) = c_0^{(2),\omega}(\cdot,\cdot)$　for a.e. $(t,x) \in \mathbb{R} \times D$.

For $\tilde{a} \in \tilde{Y}(a^{(k)})$, $s < t$, and $u_0 \in L_2(D)$, denote by $U_{\tilde{a}}^{(k)}(t,s)u_0$, $k = 1,2$, the weak solution of $(2.0.1)_{\tilde{a}} + (2.0.2)_{\tilde{a}}$ with initial condition $u(0,x) = u_0(x)$. For $\omega \in \Omega$, instead of $U_{E_{a^{(k)}}(\omega)}^{(k)}(t,s)u_0$ we write $U_{\omega}^{(k)}(t,s)u_0$.

THEOREM 4.2.1
$\lambda(a^{(1)}) \leq \lambda(a^{(2)})$.

PROOF   Let $\Omega_1^{(k)} \subset \Omega$, $k = 1,2$, be sets such that $\mathbb{P}(\Omega_1^{(k)}) = 1$ and

$$\lambda(a^{(k)}) = \lim_{t \to \infty} \frac{\ln \|U_{\omega}^{(k)}(t,0)u_0\|}{t}$$

for any $\omega \in \Omega_1^{(k)}$ and any $u_0 \in L_2(D)^+ \setminus \{0\}$ (see Lemmas 4.1.4 and 3.1.1). Fix $\omega \in \Omega_1^{(1)} \cap \Omega_1^{(2)} \cap \tilde{\Omega}$. As a consequence of Proposition 2.2.10, $0 \leq (U_{\omega}^{(1)}(t,0)u_0)(x) \leq (U_{\omega}^{(2)}(t,0)u_0)(x)$ for each nonzero $u_0 \in L_2(D)^+$, each $t > 0$, and each $x \in D$. The monotonicity of the $L_2(D)$-norm gives the desired result.   ☐

THEOREM 4.2.2
$\lambda_{\min}(a^{(1)}) \leq \lambda_{\min}(a^{(2)})$ *and* $\lambda_{\max}(a^{(1)}) \leq \lambda_{\max}(a^{(2)})$.

PROOF   We prove only the first inequality, the proof of the other being similar. Let $\Omega_0^{(k)} \subset \Omega$, $k = 1,2$, be sets such that $\mathbb{P}(\Omega_0^{(k)}) = 1$ and $\tilde{Y}_0(a^{(k)}) = \text{cl}\{ E_{a^{(1)}}(\theta_t \omega) : t \in \mathbb{R} \}$ for $\omega \in \Omega_0^{(k)}$ (see Lemma 4.1.2). Fix $\omega \in \Omega_0^{(1)} \cap \Omega_0^{(2)} \cap \tilde{\Omega}$.

By Theorem 3.1.2(2A), there are sequences $(\tilde{a}^{(n)})_{n=1}^{\infty} \subset \tilde{Y}_0(a^{(2)})$, $(t_n)_{n=1}^{\infty} \subset \mathbb{R}$, $(s_n)_{n=1}^{\infty} \subset \mathbb{R}$, with $t_n - s_n \to \infty$ as $n \to \infty$, such that

$$\lim_{n \to \infty} \frac{\ln \|U_{\tilde{a}^{(n)}}^{(2)}(t_n,s_n)\|^+}{t_n - s_n} = \lambda_{\min}(a^{(2)}).$$

For each $\tilde{a}^{(n)}$ there is a real sequence $(\tau_l)_{l=1}^{\infty}$ (depending on n) such that $E_{a^{(2)}}(\omega) \cdot \tau_l$ converge in $\tilde{Y}(a^{(2)})$ to $\tilde{a}^{(n)}$. From (τ_l) we can extract a subsequence (denoted again by (τ_l)) such that $E_{a^{(1)}}(\omega) \cdot \tau_l$ converge in $\tilde{Y}(a^{(1)})$ to some $\hat{a}^{(n)}$. Proposition 2.2.10 implies that for each $u_0 \in L_2(D)^+$ there holds

$$\|U_{E_{a^{(1)}}(\omega) \cdot \tau_l}^{(1)}(t_n,s_n)u_0\| \leq \|U_{E_{a^{(2)}}(\omega) \cdot \tau_l}^{(2)}(t_n,s_n)u_0\|.$$

From Proposition 2.2.13 we deduce that $\|U_{\hat{a}^{(n)}}^{(1)}(t_n,s_n)u_0\| \leq \|U_{\tilde{a}^{(n)}}^{(2)}(t_n,s_n)u_0\|$ for each $u_0 \in L_2(D)^+$, which implies $\|U_{\hat{a}^{(n)}}^{(1)}(t_n,s_n)\|^+ \leq \|U_{\tilde{a}^{(n)}}^{(2)}(t_n,s_n)\|^+$. By

Theorem 3.1.2 and Lemma 3.1.1,

$$\lambda_{\min}(a^{(1)}) \leq \liminf_{n\to\infty} \frac{\ln \|U^{(1)}_{\tilde{a}^{(n)}}(t_n, s_n)\|^+}{t_n - s_n}$$

$$\leq \lim_{n\to\infty} \frac{\ln \|U^{(2)}_{\tilde{a}^{(n)}}(t_n, s_n)\|^+}{t_n - s_n} = \lambda_{\min}(a^{(2)}).$$

$\square$

4.2.2 The Nonautonomous Case

Let $a^{(1)}$, $a^{(2)}$ satisfy property (A4-N).
Throughout the present subsection we assume moreover that:

- $a^{(1)}_{ij}(\cdot, \cdot) = a^{(2)}_{ij}(\cdot, \cdot)$, $a^{(1)}_i(\cdot, \cdot) = a^{(2)}_i(\cdot, \cdot)$, $b^{(1)}_i(\cdot, \cdot) = b^{(2)}_i(\cdot, \cdot)$, for a.e. $(t, x) \in \mathbb{R} \times D$, and

- one of the following conditions, (M-Na), (M-Nb), (M-Nc), (M-Nd), or (M-Ne) holds:

(M-Na) both $a^{(1)}$ and $a^{(2)}$ are endowed with the Dirichlet boundary conditions, and

    ∗ $c^{(1)}_0(\cdot, \cdot) \leq c^{(2)}_0(\cdot, \cdot)$ for a.e. $(t, x) \in \mathbb{R} \times D$,

(M-Nb) both $a^{(1)}$ and $a^{(2)}$ are endowed with the Robin boundary conditions, and

    ∗ $c^{(1)}_0(\cdot, \cdot) \leq c^{(2)}_0(\cdot, \cdot)$ for a.e. $(t, x) \in \mathbb{R} \times D$,
    ∗ $d^{(1)}_0(\cdot, \cdot) \geq d^{(2)}_0(\cdot, \cdot)$ for a.e. $(t, x) \in \mathbb{R} \times \partial D$,

(M-Nc) both $a^{(1)}$ and $a^{(2)}$ are endowed with the Neumann boundary conditions, and

    ∗ $c^{(1)}_0(\cdot, \cdot) \leq c^{(2)}_0(\cdot, \cdot)$ for a.e. $(t, x) \in \mathbb{R} \times D$,

(M-Nd) $a^{(1)}$ is endowed with the Dirichlet boundary conditions and $a^{(2)}$ is endowed with the Robin boundary conditions, and

    ∗ $c^{(1)}_0(\cdot, \cdot) = c^{(2)}_0(\cdot, \cdot)$ for a.e. $(t, x) \in \mathbb{R} \times D$,

(M-Ne) $a^{(1)}$ is endowed with the Robin boundary conditions and $a^{(2)}$ is endowed with the Neumann boundary conditions, and

    ∗ $c^{(1)}_0(\cdot, \cdot) = c^{(2)}_0(\cdot, \cdot)$ for a.e. $(t, x) \in \mathbb{R} \times D$.

For $\tilde{a} \in \tilde{Y}(a^{(k)})$, $s < t$, and $u_0 \in L_2(D)$, denote by $U_{\tilde{a}}^{(k)}(t,s)u_0$, $k = 1, 2$, the weak solution of $(2.0.1)_{\tilde{a}} + (2.0.2)_{\tilde{a}}$ with initial condition $u(0,x) = u_0(x)$.

THEOREM 4.2.3
$\lambda_{\min}(a^{(1)}) \le \lambda_{\min}(a^{(2)})$ and $\lambda_{\max}(a^{(1)}) \le \lambda_{\max}(a^{(2)})$.

PROOF   We prove only the second inequality, the proof of the other being similar.

By Theorem 3.1.2(2A), there are sequences $(\tilde{a}^{(n)})_{n=1}^{\infty} \subset \tilde{Y}(a^{(1)})$, $(t_n)_{n=1}^{\infty} \subset \mathbb{R}$, $(s_n)_{n=1}^{\infty} \subset \mathbb{R}$, with $t_n - s_n \to \infty$ as $n \to \infty$, such that

$$\lim_{n \to \infty} \frac{\ln \|U_{\tilde{a}^{(n)}}^{(1)}(t_n, s_n)\|^+}{t_n - s_n} = \lambda_{\max}(a^{(1)}).$$

For each $\tilde{a}^{(n)}$ there is a real sequence $(\tau_l)_{l=1}^{\infty}$ (depending on n) such that $a^{(1)} \cdot \tau_l$ converge in $\tilde{Y}(a^{(1)})$ to $\tilde{a}^{(n)}$. From (τ_l) we can extract a subsequence (denoted again by (τ_l)) such that $a^{(2)} \cdot \tau_l$ converge in $\tilde{Y}(a^{(2)})$ to some $\hat{a}^{(n)}$. Proposition 2.2.10 implies that for each $u_0 \in L_2(D)^+$ there holds

$$\|U_{a^{(1)} \cdot \tau_l}^{(1)}(t_n, s_n)u_0\| \le \|U_{a^{(2)} \cdot \tau_l}^{(2)}(t_n, s_n)u_0\|.$$

From Proposition 2.2.13 we deduce that

$$\|U_{\tilde{a}^{(n)}}^{(1)}(t_n, s_n)u_0\| \le \|U_{\hat{a}^{(n)}}^{(2)}(t_n, s_n)u_0\|$$

for each $u_0 \in L_2(D)^+$, which implies $\|U_{\tilde{a}^{(n)}}^{(1)}(t_n, s_n)\|^+ \le \|U_{\hat{a}^{(n)}}^{(2)}(t_n, s_n)\|^+$. By Theorem 3.1.2 and Lemma 3.1.1,

$$\lambda_{\max}(a^{(1)}) = \lim_{n \to \infty} \frac{\ln \|U_{\tilde{a}^{(n)}}^{(1)}(t_n, s_n)\|^+}{t_n - s_n}$$
$$\le \limsup_{n \to \infty} \frac{\ln \|U_{\hat{a}^{(n)}}^{(2)}(t_n, s_n)\|^+}{t_n - s_n} \le \lambda_{\max}(a^{(2)}).$$

$\square$

4.3   Continuity with Respect to the Zero Order Coefficients

In the present section we investigate the continuous dependence of the principal spectrum/principal Lyapunov exponent on the zero order terms.

For any $\tilde{a} = (\tilde{a}_{ij}, \tilde{a}_i, \tilde{b}_i, \tilde{c}_0, \tilde{d}_0) \in L_\infty(\mathbb{R} \times D, \mathbb{R}^{N^2+2N+1}) \times L_\infty(\mathbb{R} \times \partial D, \mathbb{R})$ and any $R \in \mathbb{R}$, put

$$\tilde{a} + R := (\tilde{a}_{ij}, \tilde{a}_i, \tilde{b}_i, \tilde{c}_0 + R, \tilde{d}_0).$$

It is straightforward that if $\tilde{a}^{(n)}$ converges to $\tilde{a}$ in the weak-* topology, then $\tilde{a}^{(n)} + R$ converges to $\tilde{a} + R$ in the weak-* topology. For $R \in \mathbb{R}$ we write sometimes $T_R \tilde{a}$ instead of $\tilde{a} + R$.

We collect now some facts that will be useful in the sequel.

LEMMA 4.3.1

Assume that Y is such that (A2-1)–(A2-3) are satisfied for Y. Let $R \in \mathbb{R}$. Then

(i) *The assumptions (A2-1)–(A2-3) are satisfied for $T_R Y$.*

(ii) *For any $\tilde{a} \in Y$ and any $s \leq t$,*

$$U_{\tilde{a}+R}(t, s) = e^{R(t-s)} U_{\tilde{a}}(t, s). \tag{4.3.1}$$

PROOF   The fulfillment of (A2-1)–(A2-2) for $T_R Y$ is obvious. The bilinear forms $B_{\tilde{a}}(\cdot, \cdot, \cdot)$ and $B_{\tilde{a}+R}(\cdot, \cdot, \cdot)$ are related, for a.e. $t \in \mathbb{R}$, in the following way:

$$B_{\tilde{a}+R}(t, u, v) = B_{\tilde{a}}(t, u, v) - R\langle u, v \rangle, \qquad u, v \in V$$

(see Section 2.1). Fix $s \in \mathbb{R}$ and $u_0 \in L_2(D)$, and denote $u(t) := U_{\tilde{a}}(t, s)u_0$, $t \geq s$. One has

$$-\int_s^t \langle u(\tau), v \rangle \, \dot{\psi}(\tau) \, d\tau + \int_s^t B_{\tilde{a}}(\tau, u(\tau), v)\psi(\tau) \, d\tau - \langle u_0, v \rangle \, \psi(s) = 0 \quad (4.3.2)$$

for all $v \in V$ and $\psi \in \mathcal{D}([s, t))$ (see Definition 2.1.6). We have to show that

$$-\int_s^t \langle e^{R(\tau-s)} u(\tau), v \rangle \, \dot{\psi}(\tau) \, d\tau + \int_s^t B_{\tilde{a}+R}(\tau, e^{R(\tau-s)} u(\tau), v)\psi(\tau) \, d\tau$$
$$- \langle u_0, v \rangle \, \psi(s) = 0,$$

that is,

$$-\int_s^t \langle e^{R(\tau-s)} u(\tau), v \rangle \, \dot{\psi}(\tau) \, d\tau + \int_s^t B_{\tilde{a}}(\tau, e^{R(\tau-s)} u(\tau), v)\psi(\tau) \, d\tau$$
$$- \int_s^t \langle R e^{R(\tau-s)} u(\tau), v \rangle \, \psi(\tau) \, d\tau - \langle u_0, v \rangle \, \psi(s) = 0$$

for all $v \in V$ and $\psi \in \mathcal{D}([s, t))$. This follows from (4.3.2) by replacing the test function ψ with $e^{R(t-s)}\psi(t)$.

To prove that (A2-3) is satisfied for $T_R Y$, notice that $\tilde{a}^{(n)} \to \tilde{a}$ in Y if and only if $\tilde{a}^{(n)} + R \to \tilde{a} + R$ in $T_R Y$, and apply (4.3.1).   ∎

4.3.1   The Random Case

Assume that $((\Omega, \mathfrak{F}, \mathbb{P}), \{\theta_t\}_{t\in\mathbb{R}})$ is an ergodic metric dynamical system. Let $a^{(1)}$, $a^{(2)}$ satisfy property (A4-R).

Throughout the present subsection we assume moreover that there is $\tilde{\Omega} \subset \Omega$ with $\mathbb{P}(\tilde{\Omega}) = 1$ such that for each $\omega \in \tilde{\Omega}$ the following holds:

- $a_{ij}^{(1),\omega}(\cdot,\cdot) = a_{ij}^{(2),\omega}(\cdot,\cdot),\ a_i^{(1),\omega}(\cdot,\cdot) = a_i^{(2),\omega}(\cdot,\cdot),\ b_i^{(1),\omega}(\cdot,\cdot) = b_i^{(2),\omega}(\cdot,\cdot)$
  a.e. on $\mathbb{R} \times D$,

- $d_0^{(2),\omega}(\cdot,\cdot) = d_0^{(1),\omega}(\cdot,\cdot)$ a.e. on $\mathbb{R} \times \partial D$.

Denote

$$r := \operatorname{ess\,sup}\{\,|c_0^{(2),\omega}(t,x) - c_0^{(1),\omega}(t,x)| : \omega \in \tilde{\Omega},\ t \in \mathbb{R},\ x \in D\,\}.$$

THEOREM 4.3.1
$|\lambda(a^{(1)}) - \lambda(a^{(2)})| \le r.$

THEOREM 4.3.2
$|\lambda_{\min}(a^{(1)}) - \lambda_{\min}(a^{(2)})| \le r$ and $|\lambda_{\max}(a^{(1)}) - \lambda_{\max}(a^{(2)})| \le r.$

PROOF (Proofs of Theorems 4.3.1 and 4.3.2)   The assumptions (A4-R1) and (A4-R2) are satisfied for $a^{(1)} - r$, $a^{(1)} + r$.

As the mappings $T_{\pm r}$ are continuous in the weak-* topology and $T_{\pm r} \circ \sigma_t = \sigma_t \circ T_{\pm r}$ for all $t \in \mathbb{R}$, we have that $\tilde{Y}(a^{(1)} \pm r) = T_{\pm r}\tilde{Y}(a^{(1)})$. Further, it follows that $T_{\pm r}$ sends, in a one-to-one way, invariant measures on $\tilde{Y}(a^{(1)})$ onto invariant measures on $T_{\pm r}\tilde{Y}(a^{(1)})$, consequently $\tilde{Y}_0(a^{(1)} \pm r) = T_{\pm r}\tilde{Y}_0(a^{(1)})$.

By Lemma 4.3.1, $a^{(1)} \pm r$ satisfy (A4-R), and

$$\lambda(a^{(1)} \pm r) = \lambda(a^{(1)}) \pm r,$$
$$\lambda_{\min}(a^{(1)} \pm r) = \lambda_{\min}(a^{(1)}) \pm r,$$
$$\lambda_{\min}(a^{(1)} \pm r) = \lambda_{\min}(a^{(1)}) \pm r.$$

It follows from Theorems 4.2.1 and 4.2.2 that

$$\lambda(a^{(1)} - r) \le \lambda(a^{(2)}) \le \lambda(a^{(1)} + r)$$

as well as

$$\lambda_{\min}(a^{(1)} - r) \le \lambda_{\min}(a^{(2)}) \le \lambda_{\min}(a^{(1)} + r)$$

and

$$\lambda_{\max}(a^{(1)} - r) \le \lambda_{\max}(a^{(2)}) \le \lambda_{\max}(a^{(1)} + r),$$

which gives the desired result.   ☐

4.3.2　The Nonautonomous Case

Let $a^{(1)}$, $a^{(2)}$ satisfy property (A4-N).

Throughout the present subsection we assume moreover that the following holds:

- $a_{ij}^{(1)}(\cdot,\cdot) = a_{ij}^{(2)}(\cdot,\cdot)$, $a_i^{(1)}(\cdot,\cdot) = a_i^{(2)}(\cdot,\cdot)$, $b_i^{(1)}(\cdot,\cdot) = b_i^{(2)}(\cdot,\cdot)$, $\quad$ a.e. on $\mathbb{R} \times D$,

- $d_0^{(2)}(\cdot,\cdot) = d_0^{(1)}(\cdot,\cdot)$ $\quad$ a.e. on $\mathbb{R} \times \partial D$.

Denote

$$r := \operatorname{ess\,sup}\{\,|c_0^{(2)}(t,x) - c_0^{(1)}(t,x)| : t \in \mathbb{R},\ x \in D\,\}.$$

THEOREM 4.3.3
$|\lambda_{\min}(a^{(1)}) - \lambda_{\min}(a^{(2)})| \le r$ *and* $|\lambda_{\max}(a^{(1)}) - \lambda_{\max}(a^{(2)})| \le r$.

PROOF　The assumption (A4-N1) is satisfied for $a^{(1)} - r$, $a^{(1)} + r$.

As the mappings $T_{\pm r}$ are continuous in the weak-* topology and $T_{\pm r} \circ \sigma_t = \sigma_t \circ T_{\pm r}$ for all $t \in \mathbb{R}$, we have that $\check{Y}(a^{(1)} \pm r) = T_{\pm r}\check{Y}(a^{(1)})$.

By Lemma 4.3.1, $a^{(1)} \pm r$ satisfy (A4-N), and

$$\lambda_{\min}(a^{(1)} \pm r) = \lambda_{\min}(a^{(1)}) \pm r,$$
$$\lambda_{\min}(a^{(1)} \pm r) = \lambda_{\min}(a^{(1)}) \pm r.$$

It follows from Theorem 4.2.3 that

$$\lambda_{\min}(a^{(1)} - r) \le \lambda_{\min}(a^{(2)}) \le \lambda_{\min}(a^{(1)} + r)$$

and

$$\lambda_{\max}(a^{(1)} - r) \le \lambda_{\max}(a^{(2)}) \le \lambda_{\max}(a^{(1)} + r),$$

which gives the desired result.　　　　　　　　　　　　　　　　□

4.4　General Continuity with Respect to the Coefficients

In the present section, L_∞ denotes the Banach space $L_\infty(\mathbb{R} \times D, \mathbb{R}^{N^2+2N+1}) \times L_\infty(\mathbb{R} \times \partial D, \mathbb{R})$, and $\|\cdot\|_\infty$ stands for the norm in L_∞.

We start with the following simple result.

LEMMA 4.4.1
For any $a^{(1)}, a^{(2)} \in L_\infty$ *one has* $d(a^{(1)} \cdot t, a^{(2)} \cdot t) \le \|a^{(1)} - a^{(2)}\|_\infty$ *for each* $t \in \mathbb{R}$, *where* d *stands for the metric given by* (1.3.1).

PROOF

$$d(a^{(1)} \cdot t, a^{(2)} \cdot t) = \sum_{k=1}^{\infty} \frac{1}{2^k} |\langle g_k, a^{(1)} \cdot t - a^{(2)} \cdot t \rangle_{L_1, L_\infty}| \leq \|a^{(1)} - a^{(2)}\|_\infty.$$

$\square$

In the rest of the present subsection let Y be such that (A2-1)–(A2-3) are satisfied and that the linear skew-product flow Π admits an exponential separation over Y.

Recall that, for $\tilde{a} \in Y$ and $u_0 \in L_2(D)$, $[[0, \infty) \ni t \mapsto U_{\tilde{a}}(t, 0)u_0 \in L_2(D)]$ denotes the weak solution of $(2.0.1)_{\tilde{a}} + (2.0.2)_{\tilde{a}}$ with the initial condition $u(0, x) = u_0(x)$ $(x \in D)$. For $s < t$, $U_{\tilde{a}}(t, s)$ stands for $U_{\tilde{a} \cdot s}(t - s, 0)$.

LEMMA 4.4.2
For each $\epsilon > 0$ there is $\delta > 0$ with the following property. Let $\hat{a}, \check{a} \in Y$ be such that $d(\hat{a} \cdot t, \check{a} \cdot t) < \delta$ for all $t \in \mathbb{R}$. Then, for any integer sequences $(k_n)_{n=1}^{\infty}$, $(l_n)_{n=1}^{\infty}$, such that $l_n - k_n \to \infty$ as $n \to \infty$ and

$$\lim_{n \to \infty} \frac{\ln \|U_{\hat{a}}(l_n, k_n) w(\hat{a} \cdot k_n)\|}{l_n - k_n} = \lambda,$$

one has

$$\lambda - \epsilon \leq \liminf_{n \to \infty} \frac{\ln \|U_{\check{a}}(l_n, k_n) w(\check{a} \cdot k_n)\|}{l_n - k_n}$$

$$\leq \limsup_{n \to \infty} \frac{\ln \|U_{\check{a}}(l_n, k_n) w(\check{a} \cdot k_n)\|}{l_n - k_n} \leq \lambda + \epsilon.$$

PROOF As the mapping $[Y \ni \tilde{a} \mapsto \|U_{\tilde{a}}(1, 0) w(\tilde{a})\| \in (0, \infty)]$ is continuous on a compact set, we have $M_1 := \inf \{ \|U_{\tilde{a}}(1, 0) w(\tilde{a})\| : \tilde{a} \in Y \} > 0$. Moreover, that mapping is uniformly continuous. Fix $\epsilon > 0$, and take $\delta > 0$ such that for any $\tilde{a}^{(1)}, \tilde{a}^{(2)} \in Y$, if $d(\tilde{a}^{(1)}, \tilde{a}^{(2)}) < \delta$ then $| \|U_{\tilde{a}^{(1)}}(1, 0) w(\tilde{a}^{(1)})\| - \|U_{\tilde{a}^{(2)}}(1, 0) w(\tilde{a}^{(2)})\| | < M_1(1 - e^{-\epsilon})$.

We have

$$e^{-\epsilon} \|U_{\check{a}}(k+1, k) w(\check{a} \cdot k)\| \leq \|U_{\check{a}}(k+1, k) w(\check{a} \cdot k)\| - M_1(1 - e^{-\epsilon})$$

$$< \|U_{\hat{a}}(k+1, k) w(\hat{a} \cdot k)\| < \|U_{\check{a}}(k+1, k) w(\check{a} \cdot k)\| + M_1(1 - e^{-\epsilon})$$

$$\leq (2 - e^{-\epsilon}) \|U_{\check{a}}(k+1, k) w(\check{a} \cdot k)\| < e^{\epsilon} \|U_{\check{a}}(k+1, k) w(\check{a} \cdot k)\|$$

for all $k \in \mathbb{Z}$. Since $U_{\check{a}}(l, k) w(\check{a} \cdot k) = U_{\check{a}}(l, l-1) U(l-1, l-2) \ldots U(k+1, k) w(\check{a} \cdot k))$ and $U_{\hat{a}}(l, k) w(\hat{a} \cdot k) = U_{\hat{a}}(l, l-1) U(l-1, l-2) \ldots U(k+1, k) w(\hat{a} \cdot k))$, for any $k < l$, $k, l \in \mathbb{Z}$, there holds

$$e^{-\epsilon(l-k)} \|U_{\check{a}}(l, k) w(\check{a} \cdot k)\| < \|U_{\hat{a}}(l, k) w(\hat{a} \cdot k)\| < e^{\epsilon(l-k)} \|U_{\check{a}}(l, k) w(\check{a} \cdot k)\|$$

for any $k, l \in \mathbb{Z}$, $k < l$. The lemma follows easily.

$\square$

4.4.1　The Random Case

Let $((\Omega, \mathfrak{F}, \mathbb{P}), \{\theta_t\}_{t \in \mathbb{R}})$ be an ergodic metric dynamical system.

We say that the random problem $(4.1.1)_a + (4.1.2)_a$ (or, simply, a) is Y-admissible if it satisfies property (A4-R-ES) and $\tilde{Y}_0(a) \subset Y$.

In the present subsection we will investigate the continuous dependence of the principal Lyapunov exponent and the principal spectrum on the coefficients in the norm topology in the random case.

For the rest of the subsection we fix a Y-admissible $a^{(0)}$.

For $\omega \in \Omega$, we write $U_\omega^{(0)}(t, s)$ instead of $U_{E_{a^{(0)}}(\omega)}(t, s)$.

THEOREM 4.4.1

For each $\epsilon > 0$ there is $\delta > 0$ such that for any Y-admissible a, if $\|E_a(\omega) - E_{a^{(0)}}(\omega)\|_\infty < \delta$ for $\mathbb{P}$-a.e. $\omega \in \Omega$ then

$$|\lambda(a) - \lambda(a^{(0)})| < \epsilon.$$

THEOREM 4.4.2

For each $\epsilon > 0$ there is $\delta > 0$ such that for any Y-admissible a, if $\|E_a(\omega) - E_{a^{(0)}}(\omega)\|_\infty < \delta$ for $\mathbb{P}$-a.e. $\omega \in \Omega$ then

$$|\lambda_{\min}(a) - \lambda_{\min}(a^{(0)})| < \epsilon \quad and \quad |\lambda_{\max}(a) - \lambda_{\max}(a^{(0)})| < \epsilon.$$

PROOF (Proof of Theorems 4.4.1 and 4.4.2)　　Fix $\epsilon > 0$, and fix a Y-admissible a such that $\|E_a(\omega) - E_{a^{(0)}}(\omega)\|_\infty < \delta$ for $\mathbb{P}$-a.e. $\omega \in \Omega$, where $\delta > 0$ is as in Lemma 4.4.2. For $\omega \in \Omega$, we write $U_\omega(t, s)$ instead of $U_{E_a(\omega)}(t, s)$.

We start with the proof of Theorem 4.4.2. Let $\Omega_0^{(0)} \subset \Omega$, $\mathbb{P}(\Omega_0^{(0)}) = 1$, be such that for any $\omega \in \Omega_0^{(0)}$ and any $\lambda \in [\lambda_{\min}(a^{(0)}), \lambda_{\max}(a^{(0)})]$ there are sequences $(k_n)_{n=1}^\infty, (l_n)_{n=1}^\infty \subset \mathbb{Z}$, $l_n - k_n \to \infty$ as $n \to \infty$, such that

$$\lambda = \lim_{n \to \infty} \frac{\ln \|U_\omega^{(0)}(l_n, k_n) w(E_{a^{(0)}}(\omega) \cdot k_n)\|}{l_n - k_n}$$

(see Proposition 4.1.5). Similarly, let $\Omega_0 \subset \Omega$, $\mathbb{P}(\Omega_0) = 1$, be such that

$$\lambda_{\min}(a) = \liminf_{t-s \to \infty} \frac{\ln \|U_\omega(t, s) w(E_a(\omega) \cdot s)\|}{t - s}$$

$$\leq \limsup_{t-s \to \infty} \frac{\ln \|U_\omega(t, s) w(E_a(\omega) \cdot s)\|}{t - s} = \lambda_{\max}(a)$$

for any $\omega \in \Omega_0$ (see Proposition 4.1.4).

Fix $\omega \in \Omega_0^{(0)} \cap \Omega_0$ such that $\|E_a(\omega) - E_{a^{(0)}}(\omega)\|_\infty < \delta$. It is a consequence of Lemma 4.4.1 that $d(E_{a^{(0)}}(\omega) \cdot t, E_a(\omega) \cdot t) < \delta$ for all $t \in \mathbb{R}$. It follows from Lemma 4.4.2 that for each $\lambda \in [\lambda_{\min}(a^{(0)}), \lambda_{\max}(a^{(0)})]$ there is $\tilde{\lambda} \in [\lambda_{\min}(a), \lambda_{\max}(a)]$ with $|\tilde{\lambda} - \lambda| < \epsilon$.

By interchanging the roles of $a^{(0)}$ and a we obtain that for each $\tilde{\lambda} \in [\lambda_{\min}(a),$ $\lambda_{\max}(a)]$ there is $\lambda \in [\lambda_{\min}(a^{(0)}), \lambda_{\max}(a^{(0)})]$ with $|\tilde{\lambda} - \lambda| < \epsilon$. Hence the Hausdorff distance between $[\lambda_{\min}(a), \lambda_{\max}(a)]$ and $[\lambda_{\min}(a^{(0)}), \lambda_{\max}(a^{(0)})]$ is less than ϵ, which is equivalent to the statement of Theorem 4.4.2.

We proceed now to the proof of Theorem 4.4.1. Let $\Omega_1^{(0)}$ be such that $\mathbb{P}(\Omega_1^{(0)}) = 1$ and for any $\omega \in \Omega_1^{(0)}$ there holds

$$\lim_{t \to \infty} \frac{\ln \|U_\omega^{(0)}(t,0)w(E_{a^{(0)}}(\omega))\|}{t} = \lambda(a^{(0)}).$$

Similarly, let Ω_1 be such that $\mathbb{P}(\Omega_1) = 1$ and for any $\omega \in \Omega_1$ there holds

$$\lim_{t \to \infty} \frac{\ln \|U_\omega(t,0)w(E_a(\omega))\|}{t} = \lambda(a)$$

(see Lemma 4.1.4).

Fix $\omega \in \Omega_1^{(0)} \cap \Omega_1$ such that $\|E_a(\omega) - E_{a^{(0)}}(\omega)\|_\infty < \delta$. It is a consequence of Lemma 4.4.1 that $d(E_{a^{(0)}}(\omega) \cdot t, E_a(\omega) \cdot t) < \delta$ for all $t \in \mathbb{R}$. The statement of Theorem 4.4.1 follows now from Lemma 4.4.2. $\qquad\square$

4.4.2 The Nonautonomous Case

We say that the nonautonomous problem $(4.1.9)_a + (4.1.10)_a$ (or, simply, a) is Y-*admissible* if it satisfies property (A4-N-ES) and $\tilde{Y}(a) \subset Y$.

In the present subsection we will investigate the continuous dependence of principal spectrum on the coefficients in the norm topology in the nonautonomous case.

For the rest of the subsection we fix a Y-admissible $a^{(0)}$.

We write $U^{(0)}(t,s)$ instead of $U_{a^{(0)}}(t,s)$.

THEOREM 4.4.3

For each $\epsilon > 0$ there is $\delta > 0$ such that for any Y-admissible a, if $\|a - a^{(0)}\|_\infty < \delta$ then

$$|\lambda_{\min}(a) - \lambda_{\min}(a^{(0)})| < \epsilon \quad and \quad |\lambda_{\max}(a) - \lambda_{\max}(a^{(0)})| < \epsilon.$$

PROOF Fix $\epsilon > 0$, and fix a Y-admissible a such that $\|a - a^{(0)}\|_\infty < \delta$, where $\delta > 0$ is as in Lemma 4.4.2. We write $U(t,s)$ instead of $U_a(t,s)$.

For any $\lambda \in [\lambda_{\min}(a^{(0)}), \lambda_{\max}(a^{(0)})]$ there are sequences $(k_n)_{n=1}^\infty, (l_n)_{n=1}^\infty \subset \mathbb{Z}$, $l_n - k_n \to \infty$ as $n \to \infty$, such that

$$\lambda = \lim_{n \to \infty} \frac{\ln \|U^{(0)}(l_n, k_n)w(a^{(0)} \cdot k_n)\|}{l_n - k_n}$$

(see Proposition 4.1.10). Further, we have

$$\lambda_{\min}(a) = \liminf_{t-s\to\infty} \frac{\ln \|U(t,s)w(a\cdot s)\|}{t-s}$$

$$\leq \limsup_{t-s\to\infty} \frac{\ln \|U(t,s)w(a\cdot s)\|}{t-s} = \lambda_{\max}(a)$$

(see Proposition 4.1.9).

It is a consequence of Lemma 4.4.1 that $d(a^{(0)}\cdot t, a\cdot t) < \delta$ for all $t \in \mathbb{R}$. It follows from Lemma 4.4.2 that for each $\lambda \in [\lambda_{\min}(a^{(0)}), \lambda_{\max}(a^{(0)})]$ there is $\tilde{\lambda} \in [\lambda_{\min}(a), \lambda_{\max}(a)]$ with $|\tilde{\lambda} - \lambda| < \epsilon$.

By interchanging the roles of $a^{(0)}$ and a we obtain that for each $\tilde{\lambda} \in [\lambda_{\min}(a), \lambda_{\max}(a)]$ there is $\lambda \in [\lambda_{\min}(a^{(0)}), \lambda_{\max}(a^{(0)})]$ with $|\tilde{\lambda} - \lambda| < \epsilon$. Hence the Hausdorff distance between $[\lambda_{\min}(a), \lambda_{\max}(a)]$ and $[\lambda_{\min}(a^{(0)}), \lambda_{\max}(a^{(0)})]$ is less than ϵ, which is equivalent to the statement of Theorem 4.4.3. $\Box$

4.5 Historical Remarks

In this section, we present some historical works on the principal spectrum of time independent, periodic, as well as general time dependent parabolic problems.

Consider the following nonautonomous linear parabolic equation:

$$\frac{\partial u}{\partial t} = \sum_{i=1}^{N} \frac{\partial}{\partial x_i} \left(\sum_{j=1}^{N} a_{ij}(t,x) \frac{\partial u}{\partial x_j} + a_i(t,x)u \right)$$

$$+ \sum_{i=1}^{N} b_i(t,x) \frac{\partial u}{\partial x_i} + c_0(t,x)u, \quad x \in D, \tag{4.5.1}$$

endowed with the boundary condition

$$\mathcal{B}_a(t)u = 0, \quad x \in \partial D, \tag{4.5.2}$$

where $\mathcal{B}_a(t)$ is a boundary operator of either the Dirichlet or Neumann or Robin type as in (2.0.3) with $a(t,x) = (a_{ij}(t,x), a_i(t,x), b_i(t,x), c_0(t,x), d_0(t,x))$, $d_0(t,x) \geq 0$ for a.e. $(t,x) \in \mathbb{R} \times \partial D$.

We first outline the principal eigenvalue theory for (4.5.1)+(4.5.2) in the time independent and periodic cases, and then give a review of the principal spectrum theory in the general time dependent case.

4.5.1 The Time Independent and Periodic Case

In this subsection, we assume that all the coefficients in (4.5.1)+(4.5.2) are time independent or periodic with period T. Recall that D is a domain, hence is connected.

The eigenvalue problem associated to (4.5.1)+(4.5.2) with time T-periodic coefficients reads as follows:

$$
\begin{cases}
-\dfrac{\partial u}{\partial t} + \displaystyle\sum_{i=1}^{N} \dfrac{\partial}{\partial x_i} \left(\sum_{j=1}^{N} a_{ij}(t,x) \dfrac{\partial u}{\partial x_j} + a_i(t,x)u \right) \\[2mm]
\qquad + \displaystyle\sum_{i=1}^{N} b_i(t,x) \dfrac{\partial u}{\partial x_i} + c_0(t,x)u = \lambda u, \quad x \in D, \\[4mm]
\mathcal{B}_a(t)u = 0, \quad x \in \partial D, \\[2mm]
u(0,\cdot) = u(T,\cdot).
\end{cases} \qquad (4.5.3)
$$

When the domain D and the coefficients of (4.5.1)+(4.5.2) are sufficiently smooth, based on the Kreĭn–Rutman theorem, it can be proved that there is a unique $\lambda_{\mathrm{princ}} \in \mathbb{R}$ such that (4.5.3) with $\lambda = \lambda_{\mathrm{princ}}$ has a positive solution u (λ_{princ} is called the *principal eigenvalue* of (4.5.1)+(4.5.2) and u is a *principal eigenfunction*) (see [50, Proposition 14.4]). It is not difficult to prove that $[\lambda_{\min}(a), \lambda_{\max}(a)] = \{\lambda_{\mathrm{princ}}\}$. Observe that for any other eigenvalue λ of (4.5.3), $\operatorname{Re}\lambda < \lambda_{\mathrm{princ}}$.

When D is a general bounded domain and $a = (a_{ij}, a_i, b_i, c_0, d_0)$ is a time independent or periodic function in $L_\infty(\mathbb{R}\times D, \mathbb{R}^{N^2+2N+1}) \times L_\infty(\mathbb{R}\times\partial D, \mathbb{R})$, the associated eigenvalue problem, in particular, the associated principal eigenvalue problem of (4.5.1)+(4.5.2) has also been extensively studied (see [11], [14], [29], [31], [32], etc.). Under quite general conditions, it is shown that the principal eigenvalue λ_{princ} of (4.5.1)+(4.5.2) exists (λ_{princ} is called a principal eigenvalue of (4.5.3) if (4.5.3) with $\lambda = \lambda_{\mathrm{princ}}$ has a nontrivial nonnegative solution (in weak sense), such an eigenvalue if exists is unique). Write λ_{princ} as $\lambda_{\mathrm{princ}}(a, D)$. The continuous dependence of $\lambda_{\mathrm{princ}}(a, D)$ with respect to a and D has also been widely studied (see [11], [14], [29], [31], [32] and references therein). For use in Chapter 5 we state some results from [29].

Assume that $a = (a_{ij}, a_i, b_i, c_0, 0)$ and $a^{(n)} = (a_{ij}^{(n)}, a_i^{(n)}, b_i^{(n)}, c_0^{(n)}, 0)$ satisfy (A4-N). Moreover, we assume that the $L_\infty(\mathbb{R}\times D, \mathbb{R}^{N^2+2N+1}) \times L_\infty(\mathbb{R}\times \partial D, \mathbb{R})$-norms of $a^{(n)}$ are bounded in $n \in \mathbb{N}$.

Let $D_n \subset \mathbb{R}^N$ be bounded domains. We say that D_n *converges to* D, denoted by $D_n \to D$, if

- $\lim_{n\to\infty} |D_n \setminus \bar{D}| = 0$, where $|\cdot|$ denotes the N-dimensional Lebesgue measure.

- There exists a compact set $K \subset D$ of capacity zero such that for each compact set $K' \subset D \setminus K$ there exists $n_0 \in \mathbb{N}$ such that $K' \subset D_n$ for $n \geq n_0$.

Recall that the *capacity* of a compact subset $K \subset \mathbb{R}^N$ with respect to a set B_0, denoted by $\mathrm{cap}_{B_0}(K)$, is defined by

$$\mathrm{cap}_{B_0}(K) := \inf\{\,\|\phi\|^2_{W^1_2(B_0)} : \phi \in \mathcal{D}(B_0) \quad \text{and} \quad \phi \geq 1 \quad \text{on} \quad K\,\}$$

(see [28, (2.15)]). If for one open set B_0, $\mathrm{cap}_{B_0}(K) = 0$, then for any open set B, $\mathrm{cap}_B(K) = 0$. In this case, K is said to have *zero capacity*.

Observe that if D is a Lipschitz domain, $D_n \subset D$, and for each compact set $K \subset D$ there exists $n_0 \in \mathbb{N}$ such that $K \subset D_n$ for $n \geq n_0$, then $D_n \to D$.

LEMMA 4.5.1

Consider a Dirichlet boundary condition problem. Assume that D is a Lipschitz domain, $D_n \subset D$ converge to D, and $a_{ij}^{(n)}$, $a_i^{(n)}$, $b_i^{(n)}$ $(i,j = 1, 2, \ldots, N)$, $c_0^{(n)}$ converge respectively to a_{ij}, a_i, b_i, c_0 in $L_{2,\mathrm{loc}}((0,T) \times \mathbb{R}^N)$. Then $\lambda_{\mathrm{princ}}(a^{(n)}, D_n) \to \lambda_{\mathrm{princ}}(a, D)$.

In the above, the coordinates of $a^{(n)}$ are understood to be equal to 0 outside D_n.

PROOF   See [29, Theorem 2.10].   ☐

LEMMA 4.5.2

Let D be a Lipschitz domain in $\mathbb{R}^N$. Then there is a sequence $(D_n)_{n=1}^{\infty}$ of C^∞ domains satisfying $D_n \subset D$ and $D_n \to D$ as $n \to \infty$.

PROOF   See the proof of [60, Lemma 4.1].   ☐

4.5.2   The General Time Dependent Case

In this subsection, we assume that a in (4.5.1)+(4.5.2) is a general time dependent function.

When D and a are sufficiently smooth, based on the abstract work [94] the principal spectrum and exponential separation theory for (4.5.1)+(4.5.2) has been well established (see [62], [79], [81], [82], [84], [92], etc.), which extends the principal eigenvalue and principal eigenfunction theory for elliptic and time periodic parabolic problems to general nonautonomous problems.

Recently, principal spectrum and exponential separation for general time dependent parabolic problems on general domain have been investigated in several papers (see [59], [60], [61], etc.), mostly for the Dirichlet boundary condition case.

It should be remarked that in some of the papers mentioned above the equations in the nondivergence form are allowed, also the derivatives in the Neumann or Robin boundary conditions need not be conormal.

Chapter 5

Influence of Spatial-Temporal Variations and the Shape of Domain

Consider the following random linear parabolic equation:

$$
\begin{cases}
\dfrac{\partial u}{\partial t} = \displaystyle\sum_{i=1}^{N} \dfrac{\partial}{\partial x_i} \left(\sum_{j=1}^{N} a_{ij}(\theta_t \omega, x)\dfrac{\partial u}{\partial x_j} + a_i(\theta_t \omega, x)u \right) \\
\qquad + \displaystyle\sum_{i=1}^{N} b_i(\theta_t \omega, x)\dfrac{\partial u}{\partial x_i} + c_0(\theta_t \omega, x)u, \qquad\qquad x \in D, \\
\mathcal{B}_\omega(t)u = 0, \qquad\qquad\qquad\qquad\qquad\qquad\qquad\quad\; x \in \partial D,
\end{cases}
\tag{5.0.1}
$$

where $\mathcal{B}_\omega(t) = \mathcal{B}_{a^\omega}(t)$, $\mathcal{B}_{a^\omega}$ is the boundary operator in (2.0.3) with a being replaced by $a^\omega(t, x) = (a_{ij}(\theta_t \omega, x), a_i(\theta_t \omega, x), b_i(\theta_t \omega, x), c_0(\theta_t, x), d_0(\theta_t \omega, x))$, $d_0(\omega, x) \geq 0$ for all $\omega \in \Omega$ and a.e. $x \in \partial D$, $((\Omega, \mathfrak{F}, \mathbb{P}), \{\theta_t\}_{t \in \mathbb{R}})$ is an ergodic metric dynamical system, and the functions a_{ij} $(i, j = 1, \ldots, N)$, a_i $(i = 1, \ldots, N)$, b_i $(i = 1, \ldots, N)$ and c_0 are $(\mathfrak{F} \times \mathfrak{B}(D), \mathfrak{B}(\mathbb{R}))$-measurable, and the function d_0 is $(\mathfrak{F} \times \mathfrak{B}(\partial D), \mathfrak{B}(\mathbb{R}))$-measurable; and consider the following nonautonomous linear parabolic equation:

$$
\begin{cases}
\dfrac{\partial u}{\partial t} = \displaystyle\sum_{i=1}^{N} \dfrac{\partial}{\partial x_i} \left(\sum_{j=1}^{N} a_{ij}(t, x)\dfrac{\partial u}{\partial x_j} + a_i(t, x)u \right) \\
\qquad + \displaystyle\sum_{i=1}^{N} b_i(t, x)\dfrac{\partial u}{\partial x_i} + c_0(t, x)u, \qquad\qquad\quad x \in D, \\
\mathcal{B}_a(t)u = 0, \qquad\qquad\qquad\qquad\qquad\qquad\qquad\; x \in \partial D,
\end{cases}
\tag{5.0.2}
$$

where $\mathcal{B}_a$ is the boundary operator in (2.0.3) with $a(t, x) = (a_{ij}(t, x), a_i(t, x), b_i(t, x), c_0(t, x), d_0(t, x))$, $d_0(t, x) \geq 0$ for a.e. $(t, x) \in \mathbb{R} \times \partial D$.

In the present chapter, we investigate the influence of spatial and temporal variations of the zero order terms of (5.0.1) and (5.0.2) on their principal spectrum and principal Lyapunov exponents. We show that spatial and temporal variations cannot reduce the principal spectrum and principal Lyapunov exponents. Indeed, if the coefficients and the domain are sufficiently regular, spatial and temporal variations increase the principal spectrum and principal Lyapunov exponents except in the degenerate cases. In the biological context these results mean that invasion by a new species (see [16], p. 220) is always easier in the space and time dependent case.

We also investigate the influence of the shape of the domain of (5.0.1) and (5.0.2) on their principal spectrum and principal Lyapunov exponents and extend the so called Faber–Krahn inequalities for elliptic and periodic parabolic problems to general time dependent and random ones.

This chapter is organized as follows. In Section 5.1, we introduce notions and basic assumptions and present some lemmas for the use in later sections. We study the influence of temporal variations of the zero order terms of (5.0.1) and (5.0.2) on their principal spectrum and principal Lyapunov exponents in Section 5.2. Section 5.3 is devoted to the investigation of the influence of spatial variations of the zero order terms of (5.0.1) and (5.0.2) on their principal spectrum and principal Lyapunov exponents. In Section 5.4 the influence of the shape of the domain of (5.0.1) and (5.0.2) on their principal spectrum and principal Lyapunov exponents is explored.

5.1   Preliminaries

In this section, we introduce notions and basic assumptions and establish lemmas which will be used in later sections.

5.1.1   Notions and Basic Assumptions

First, consider (5.0.1). Write $a^\omega := ((a_{ij}^\omega)_{i,j=1}^N, (a_i^\omega)_{i=1}^N, (b_i^\omega)_{i=1}^N, c_0^\omega, d_0^\omega)$, where $a_{ij}^\omega(t,x) = a_{ij}(\theta_t \omega, x)$, etc. We assume that for each $\omega \in \Omega$, a^ω belongs to $L_\infty(\mathbb{R} \times D, \mathbb{R}^{N^2+2N+1}) \times L_\infty(\mathbb{R} \times \partial D, \mathbb{R})$. Moreover, the set $\{\, a^\omega : \omega \in \Omega \,\}$ is bounded in the $L_\infty(\mathbb{R} \times D, \mathbb{R}^{N^2+2N+1}) \times L_\infty(\mathbb{R} \times \partial D, \mathbb{R})$-norm by $M \geq 0$.

Define the mapping $E_a : \Omega \to L_\infty(\mathbb{R} \times D, \mathbb{R}^{N^2+2N+1}) \times L_\infty(\mathbb{R} \times \partial D, \mathbb{R})$ as

$$E_a(\omega) := a^\omega.$$

Put

$$\tilde{Y}(a) := \mathrm{cl}\,\{\, E_a(\omega) : \omega \in \Omega \,\} \tag{5.1.1}$$

with the weak-* topology, where the closure is taken in the weak-* topology. The set $\tilde{Y}(a)$ is a compact metrizable space.

Note that E_a is $(\mathfrak{F}, \mathfrak{B}(\tilde{Y}(a)))$-measurable (see Lemma 4.1.1).

Denote by $\tilde{\mathbb{P}}$ the image of the measure $\mathbb{P}$ under E_a: for any Borel set $A \in \mathfrak{B}(\tilde{Y}(a))$, $\tilde{\mathbb{P}}(A) := \mathbb{P}(E_a^{-1}(A))$. $\tilde{\mathbb{P}}$ is a $\{\sigma_t\}$-invariant ergodic Borel measure on $\tilde{Y}(a)$. Put

$$\tilde{Y}_0(a) := \mathrm{supp}\,\tilde{\mathbb{P}}. \tag{5.1.2}$$

Then $\tilde{Y}_0(a)$ is a closed (hence compact) and $\{\sigma_t\}$-invariant subset of $\tilde{Y}(a)$, with $\tilde{\mathbb{P}}(\tilde{Y}_0(a)) = 1$. Also, $\tilde{Y}_0(a)$ is connected (see Subsection 4.1.1 for detail).

If Assumptions (A2-1)–(A2-3) are satisfied for Y replaced with $\tilde{Y}(a)$, we will denote by $\Pi(a) = \{\Pi(a)_t\}_{t \geq 0}$ the *topological linear skew-product semiflow generated by* (5.0.1) *on the product Banach bundle* $L_2(D) \times \tilde{Y}(a)$:

$$\Pi(a)(t; u_0, \tilde{a}) = \Pi(a)_t(u_0, \tilde{a}) := (U_{\tilde{a}}(t, 0)u_0, \sigma_t \tilde{a})$$

for $t \geq 0$, $\tilde{a} \in \tilde{Y}(a)$, $u_0 \in L_2(D)$.

Throughout this chapter, when speaking about the random equation (5.1.1) we assume that

(A5-R1) *The assumptions* (A2-1)–(A2-3) *are satisfied for* Y *replaced with* $\tilde{Y}(a)$ *defined by* (5.1.1), *and the topological linear skew-product semiflow* $\Pi(a)$ *generated by* (5.0.1) *on* $L_2(D) \times \tilde{Y}(a)$ *satisfies* (A3-1) *and* (A3-2) *and admits an exponential separation over* $\tilde{Y}_0(a)$ *defined by* (5.1.2). *In the case of the Neumann or Robin boundary conditions, the exponent* ς *in* (A3-2) *equals zero.*

It should be remarked that from (A5-R1) the assumption (A4-R-ES) follows.

We denote by $\lambda(a)$ the principal Lyapunov exponent of (5.0.1) (see Section 4.1 for definitions). Sometimes we may denote $\lambda(a)$ by $\lambda(a, D)$ to indicate the dependence of $\lambda(a)$ on the domain D. We denote by $w \colon \tilde{Y}_0(a) \to L_2(D)^+$ with $\|w(\tilde{a})\| = 1$ for any $\tilde{a} \in \tilde{Y}_0(a)$ the unique function such that for any $\tilde{a} \in \tilde{Y}_0(a)$ the fiber $X_1(\tilde{a})$ of the one-dimensional bundle X_1 in the exponential separation equals $\mathrm{span}\{w(\tilde{a})\}$.

We observe that, as (A3-1) and (A3-2) hold, it follows by Theorem 3.3.1 that if, for some $\omega \in \Omega$ with $E_a(\omega) \in \tilde{Y}_0(a)$, a function $u = u(t, x)$ is an entire positive solution of (5.0.1) then $w(E_a(\omega) \cdot t) = u(t, \cdot)/\|u(t, \cdot)\|$ for each $t \in \mathbb{R}$.

At some places we also assume the following:

(A5-R2) *In the case of Dirichlet boundary conditions, D is a Lipschitz domain. In the case of Neumann and Robin boundary conditions, for each* $\tilde{a} \in \tilde{Y}_0(a)$, $w(\cdot, \cdot; \tilde{a}) \in W^{1,0}((S, T) \times D)$ *and* $w(\cdot, \cdot; \tilde{a})|_{\partial D} \in L_2((S, T) \times \partial D)$ *for any* $S < T$, *where* $w(t, x; \tilde{a}) := w(\tilde{a} \cdot t)(x)$.

It will be pointed out explicitly where (A5-R2) is assumed. Recall that by Proposition 2.2.11, for any $\tilde{a} \in \tilde{Y}_0(a)$, $w(\cdot, \cdot; \tilde{a}) \in W^{0,1}((S, T) \times D)$.

Consider (5.0.2). We write $a = (a_{ij}, a_i, b_i, c_0, d_0)$ and assume that a belongs to $L_\infty(\mathbb{R} \times D, \mathbb{R}^{N^2 + 2N + 1}) \times L_\infty(\mathbb{R} \times \partial D, \mathbb{R})$.

Put

$$\tilde{Y}(a) := \mathrm{cl}\{a \cdot t : t \in \mathbb{R}\} \qquad (5.1.3)$$

with the weak-* topology, where the closure is taken in the weak-* topology. The set $\tilde{Y}(a)$ is a compact connected metrizable space.

If Assumptions (A2-1)–(A2-3) are satisfied for Y replaced with $\tilde{Y}(a)$, we will denote by $\Pi(a) = \{\Pi(a)_t\}_{t \geq 0}$ the *topological linear skew-product semiflow generated by* (5.0.2) *on the product Banach bundle* $L_2(D) \times \tilde{Y}(a)$:

$$\Pi(a)(t; u_0, \tilde{a}) = \Pi(a)_t(u_0, \tilde{a}) := (U_{\tilde{a}}(t, 0)u_0, \sigma_t \tilde{a})$$

for $t \geq 0$, $\tilde{a} \in \tilde{Y}(a)$, $u_0 \in L_2(D)$.

Throughout this chapter, when speaking about the nonautonomous equation (5.1.3) we assume that

(A5-N1) *The assumptions* (A2-1)–(A2-3) *are satisfied for* Y *replaced with* $\tilde{Y}(a)$ *defined by* (5.1.3) *and the topological linear skew-product semiflow* $\Pi(a)$ *generated by* (5.0.2) *on* $L_2(D) \times \tilde{Y}(a)$ *satisfies* (A3-1) *and* (A3-2) *and admits an exponential separation over* $\tilde{Y}(a)$. *In the case of the Neumann or Robin boundary conditions, the exponent* ς *in* (A3-2) *equals zero.*

It should be remarked that from (A5-N1) the assumption (A4-N-ES) follows.

We denote by $[\lambda_{\min}(a), \lambda_{\max}(a)]$ the principal spectrum interval of (5.0.2) (see Section 4.1 for definitions). Sometimes we may denote $\lambda_{\min}(a)$ and $\lambda_{\max}(a)$ by $\lambda_{\min}(a, D)$ and $\lambda_{\max}(a, D)$ to indicate the dependence of $\lambda_{\min}(a)$ and $\lambda_{\max}(a)$ on the domain D. We denote by $w \colon \tilde{Y}(a) \to L_2(D)^+$ with $\|w(\tilde{a})\| = 1$ for any $\tilde{a} \in \tilde{Y}(a)$ the unique function such that for any $\tilde{a} \in \tilde{Y}(a)$ the fiber $X_1(a)$ of the one-dimensional bundle X_1 in the exponential separation of (5.0.2) equals span$\{w(\tilde{a})\}$.

We observe that, as (A3-1) and (A3-2) hold, it follows by Theorem 3.3.1 that if a function $u = u(t, x)$ is an entire positive solution of (5.0.2) then $w(a \cdot t) = u(t, \cdot)/\|u(t, \cdot)\|$ for each $t \in \mathbb{R}$.

At some places we also assume the following:

(A5-N2) *In the case of Dirichlet boundary conditions, D is a Lipschitz domain. In the case of Neumann and Robin boundary conditions, for each* $\tilde{a} \in \tilde{Y}(a)$, $w(\cdot, \cdot; \tilde{a}) \in W^{1,0}((S, T) \times D)$ *and* $w(\cdot, \cdot; \tilde{a})|_{\partial D} \in L_2((S, T) \times \partial D)$ *for any* $S < T$, *where* $w(t, x; \tilde{a}) := w(\tilde{a} \cdot t)(x)$.

Again, it will be pointed out explicitly where (A5-N2) is assumed. Recall also that by Proposition 2.2.11, for any $\tilde{a} \in \tilde{Y}(a)$, $w(\cdot, \cdot; \tilde{a}) \in W^{0,1}((S, T) \times D)$.

DEFINITION 5.1.1

(1) *Let a be as in* (5.0.2). *We say that a is* uniquely ergodic *if the compact flow* $(\tilde{Y}(a), \sigma)$ *is uniquely ergodic.*

(2) *Let a be as in* (5.0.2). *We say that a is* minimal *or* recurrent *if the compact flow* $(\tilde{Y}(a), \sigma)$ *is minimal.*

(3) *Let $g \in L_\infty(\mathbb{R}, \mathbb{R})$ and $H(g) := \mathrm{cl}\{g(t + \cdot) : t \in \mathbb{R}\}$ with the weak-* topology, where the closure is taken under the weak-* topology. We say g is* minimal *or* recurrent *if the compact flow* $(H(g), \tilde{\sigma})$ *is minimal, where* $\tilde{\sigma}_t \tilde{g}(\cdot) := \tilde{g}(t + \cdot)$ *for any* $\tilde{g} \in H(g)$ *and* $t \in \mathbb{R}$.

DEFINITION 5.1.2   Let $g \in L_\infty(\mathbb{R}, \mathbb{R})$ and a be as in (5.0.2). We say that g is recurrent with at least the same recurrence *as a if both g and a*

are recurrent and for any $(t_n)_{n=1}^{\infty} \subset \mathbb{R}$, if $\lim_{n \to \infty} a \cdot t_n$ exists, then so does $\lim_{n \to \infty} g \cdot t_n$ (in the weak-* topology of $L_{\infty}(\mathbb{R}, \mathbb{R})$).

5.1.2 Auxiliary Lemmas

LEMMA 5.1.1

(1) *Let $h_i \colon [0, T] \times D \to \mathbb{R}$ $(i = 1, 2, \ldots, N)$ be square-integrable in $t \in [0, T]$ and $a_{ij} = a_{ji} \colon D \to \mathbb{R}$ $(i, j = 1, 2, \ldots, N)$ satisfy*

$$\sum_{i,j=1}^{N} a_{ij}(x) \xi_i \xi_j \geq \alpha_0 \sum_{i=1}^{N} \xi_i^2$$

for some $\alpha_0 > 0$ and a.e. $x \in D$, $\xi = (\xi_1, \xi_2, \ldots, \xi_N)^{\top} \in \mathbb{R}^N$. Then for a.e. $x \in D$,

$$\sum_{i,j=1}^{N} a_{ij}(x) \frac{1}{T} \int_0^T h_i(t, x) \, dt \, \frac{1}{T} \int_0^T h_j(t, x) \, dt$$

$$\leq \sum_{i,j=1}^{N} a_{ij}(x) \frac{1}{T} \int_0^T h_i(t, x) h_j(t, x) \, dt.$$

Moreover, the equality holds at some $x_0 \in D$ if and only if $h_i(t, x_0) = \tilde{h}_i(x_0)$ for some $\tilde{h}_i(x_0)$ $(i = 1, 2, \ldots, N)$ and a.e. $t \in [0, T]$.

(2) *Let $h_i \colon \Omega \times D \to \mathbb{R}$ $(i = 1, 2, \cdots, N)$ be square-integrable in $\omega \in \Omega$ and $a_{ij} = a_{ji} \colon D \to \mathbb{R}$ $(i, j = 1, 2, \ldots, N)$ satisfy*

$$\sum_{i,j=1}^{N} a_{ij}(x) \xi_i \xi_j \geq \alpha_0 \sum_{i=1}^{N} \xi_i^2$$

for some $\alpha_0 > 0$ and a.e. $x \in D$, $\xi = (\xi_1, \xi_2, \cdots, \xi_N)^{\top} \in \mathbb{R}^N$. Then for any a.e. $x \in D$,

$$\sum_{i,j=1}^{N} a_{ij}(x) \int_{\Omega} h_i(\omega, x) \, d\mathbb{P}(\omega) \int_{\Omega} h_j(\omega, x) \, d\mathbb{P}(\omega)$$

$$\leq \sum_{i,j=1}^{N} a_{ij}(x) \int_{\Omega} h_i(\omega, x) h_j(\omega, x) \, d\mathbb{P}(\omega).$$

Moreover, the equality holds at some $x_0 \in D$ if and only if $h_i(\omega, x_0) = \tilde{h}_i(x_0)$ for some $\tilde{h}_i(x_0)$ $(i = 1, 2, \ldots, N)$ and $\mathbb{P}$-a.e. $\omega \in \Omega$.

PROOF (1) is proved in [62, Lemma 2.2] and (2) is proved in [81, Lemma 3.5]. For completeness, we provide a proof of (2) in the following. (1) can be proved by similar arguments.

(2) First, note that for any fixed $x \in D$, there is an orthogonal matrix L such that $A = (a_{ij}(x))_{N \times N} = L^\top \operatorname{diag}(d_i) L$, where $d_i > 0$. Let

$$(y_1(\omega, x), y_2(\omega, x), \ldots, y_N(\omega, x))^\top := L(h_1(\omega, x), h_2(\omega, x), \ldots, h_N(\omega, x))^\top.$$

Then we have

$$\sum_{i,j=1}^{N} a_{ij}(x) \left(\int_\Omega h_i(\omega, x) \, d\mathbb{P}(\omega) \int_\Omega h_j(\omega, x) \, d\mathbb{P}(\omega) - \int_\Omega h_i(\omega, x) h_j(\omega, x) \, d\mathbb{P}(\omega) \right)$$

$$= \sum_{i=1}^{N} d_i \left(\left(\int_\Omega y_i(\omega, x) \, d\mathbb{P}(\omega) \right)^2 - \int_\Omega y_i^2(\omega, x) \, d\mathbb{P}(\omega) \right).$$

By the Schwarz inequality,

$$\left(\int_\Omega y_i(\omega, x) \, d\mathbb{P}(\omega) \right)^2 \leq \int_\Omega y_i^2(\omega, x) \, d\mathbb{P}(\omega)$$

and the equality holds for some $x_0 \in D$ if and only if $y_i(\omega, x_0) = \tilde{y}_i(x_0)$ for some $\tilde{y}_i(x_0)$ $(i = 1, 2, \ldots, N)$ and $\mathbb{P}$-a.e. $\omega \in \Omega$. Hence

$$\sum_{i=1}^{N} d_i \left(\left(\int_\Omega y_i(\omega, x) \, d\mathbb{P}(\omega) \right)^2 - \int_\Omega y_i^2(\omega, x) \, d\mathbb{P}(\omega) \right) \leq 0$$

and then

$$\sum_{i,j=1}^{N} a_{ij}(x) \int_\Omega h_i(\omega, x) \, d\mathbb{P}(\omega) \int_\Omega h_j(\omega, x) \, d\mathbb{P}(\omega)$$

$$\leq \sum_{i,j=1}^{N} a_{ij}(x) \int_\Omega h_i(\omega, x) h_j(\omega, x) \, d\mathbb{P}(\omega)$$

and the equality holds at some $x_0 \in D$ if and only if $h_i(\omega, x_0) = \tilde{h}_i(x_0)$ for some $\tilde{h}_i(x_0)$ $(i = 1, 2, \ldots, N)$ and $\mathbb{P}$-a.e. $\omega \in \Omega$. ∎

LEMMA 5.1.2
Let $(\Omega, \mathfrak{F}, \mathbb{P})$ be a probability space, and let $E \subset \mathbb{R}^N$. Assume that $h: \Omega \times E \to \mathbb{R}$ (resp. $h: \Omega \times \bar{E} \to \mathbb{R}$) has the following properties:

(i) *$h(\cdot, x)$ belongs to $L^1((\Omega, \mathfrak{F}, \mathbb{P}))$, for each $x \in E$,*

(ii) *For each $x \in E$ (resp. $x \in \bar{E}$) and each $\epsilon > 0$ there is $\delta > 0$ such that if $y \in E$ (resp. $y \in \bar{E}$), $\omega \in \Omega$ and $\|x - y\| < \delta$ then $|h(\omega, x) - h(\omega, y)| < \epsilon$, where $\|\cdot\|$ stands for the norm in $\mathbb{R}^N$.*

Denote, for each $x \in E$ (resp. $x \in \bar{E}$), $\hat{h}(x) := \int_\Omega h(\omega, x) \, d\mathbb{P}(\omega)$. Then

(a) *for any $x \in E$ (resp. $x \in \bar{E}$) and any $\epsilon > 0$ there is $\delta > 0$ (the same as in (ii)) such that if $y \in E$ (resp. $y \in \bar{E}$), $\omega \in \Omega$ and $\|x - y\| < \delta$ then $|\hat{h}(x) - \hat{h}(y)| < \epsilon$,*

(b) *there is a measurable $\Omega' \subset \Omega$ with $\mathbb{P}(\Omega') = 1$ such that*

$$\lim_{T \to \infty} \frac{1}{T} \int_0^T h(\theta_t \omega, x) \, dt = \hat{h}(x)$$

for all $\omega \in \Omega'$ and all $x \in E$ (resp. $x \in \bar{E}$). Moreover the convergence is uniform in $x \in E_0$, for any compact $E_0 \subset E$ (resp. uniform in $x \in \bar{E}$).

PROOF This is, in fact, [84, Lemma 2.3]. For completeness we give a proof here.

Part (a) follows easily by the fact that the continuity is uniform in $\omega \in \Omega$. To prove (b), take a countable dense set $\{x_l\}_{l=1}^\infty$ in E. By Birkhoff's Ergodic Theorem (Lemma 1.2.6), for each $l \in \mathbb{N}$ there is $\Omega_l \subset \Omega$ with $\mathbb{P}(\Omega_l) = 1$ such that

$$\lim_{T \to \infty} \frac{1}{T} \int_0^T h(\theta_t \omega, x_l) \, dt = \hat{h}(x_l)$$

for each $\omega \in \Omega_l$. Take $\Omega' := \bigcap_{l=1}^\infty \Omega_l$.

Fix $x \in E$ (resp. $x \in \bar{E}$). For $\epsilon > 0$ take $\delta > 0$ such that if $\|x - y\| < \delta$ then $|h(\omega, x) - h(\omega, y)| < \epsilon/3$ and $|\hat{h}(x) - \hat{h}(y)| < \epsilon/3$. Let x_l be such that $\|x - x_l\| < \delta$, and let $T_0 > 0$ be such that

$$\left| \frac{1}{T} \int_0^T h(\theta_t \omega, x_l) \, dt - \hat{h}(x_l) \right| < \frac{\epsilon}{3}$$

for all $T > T_0$. Then

$$\left| \frac{1}{T} \int_0^T h(\theta_t \omega, x) \, dt - \hat{h}(x) \right| < \epsilon$$

for all $T > T_0$. (b) then follows. $\qquad \square$

For a given bounded Lipschitz domain $D \subset \mathbb{R}^N$, the so-called *Schwarz symmetrized domain* $D_{\text{sym}} \subset \mathbb{R}^N$ of D is the open ball in $\mathbb{R}^N$ with center 0 and the same volume as D. For a Lebesgue measurable function $u \colon D \to \mathbb{R}^+$, u_{sym} defined by

$$u_{\text{sym}}(x) := \sup \{ c \in \mathbb{R} : x \in (D_c)_{\text{sym}} \} \quad \text{for} \quad x \in D_{\text{sym}}$$

is called the *Schwarz symmetrization* of u, where $D_c := \{ x \in D : u(x) \geq c \}$ (see [66]).

LEMMA 5.1.3 (Symmetrization)

(1) *For every* $u \in L_2(D)^+$, $\int_D (u(x))^2 \, dx = \int_{D_{\mathrm{sym}}} (u_{\mathrm{sym}}(x))^2 \, dx$.

(2) *For every* $u \in \mathring{W}_2^1(D)^+$, $\int_D |\nabla u(x)|^2 \, dx \geq \int_{D_{\mathrm{sym}}} |\nabla u_{\mathrm{sym}}(x)|^2 \, dx$. *Moreover, if u is analytic in D and ∂D is analytic, then the equality holds if and only if $D = D_{\mathrm{sym}}$ up to translation and $u = u_{\mathrm{sym}}$ up to phase shift.*

PROOF (1) See [66, Properties (C) and (P1)].
(2) See [66, Properties (G1) and (G2g)]. ☐

5.2 Influence of Temporal Variation on Principal Lyapunov Exponents and Principal Spectrum

In this section, we study the influence of temporal variations of the zeroth order terms in (5.0.1) and (5.0.2) on their principal Lyapunov exponents and principal spectrum. We assume $a_{ij}(\omega, x) = a_{ij}(x)$, $a_i(\omega, x) = a_i(x)$, $b_i(\omega, x) = b_i(x)$ in (5.0.1) and $a_{ij}(t, x) = a_{ij}(x)$, $a_i(t, x) = a_i(x)$, $b_i(t, x) = b_i(x)$ in (5.0.2). That is, in this section, we consider the random equation of the form

$$
\begin{cases}
\dfrac{\partial u}{\partial t} = \displaystyle\sum_{i=1}^{N} \dfrac{\partial}{\partial x_i} \left(\sum_{j=1}^{N} a_{ij}(x) \dfrac{\partial u}{\partial x_j} + a_i(x) u \right) \\
\qquad + \displaystyle\sum_{i=1}^{N} b_i(x) \dfrac{\partial u}{\partial x_i} + c_0(\theta_t \omega, x) u, & x \in D, \qquad (5.2.1) \\
\mathcal{B}_\omega(t) u = 0, & x \in \partial D,
\end{cases}
$$

where $\mathcal{B}_\omega(t) u$ is as in (5.0.1) with $a_{ij}(\omega, x) = a_{ij}(x)$, $a_i(\omega, x) = a_i(x)$, $b_i(\omega, x) = b_i(x)$, and $((\Omega, \mathfrak{F}, \mathbb{P}), \theta_t)$ is an ergodic metric dynamical system, and consider the nonautonomous equation of the form

$$
\begin{cases}
\dfrac{\partial u}{\partial t} = \displaystyle\sum_{i=1}^{N} \dfrac{\partial}{\partial x_i} \left(\sum_{j=1}^{N} a_{ij}(x) \dfrac{\partial u}{\partial x_j} + a_i(x) u \right) \\
\qquad + \displaystyle\sum_{i=1}^{N} b_i(x) \dfrac{\partial u}{\partial x_i} + c_0(t, x) u, & x \in D, \qquad (5.2.2) \\
\mathcal{B}_a(t) u = 0, & x \in \partial D,
\end{cases}
$$

where $\mathcal{B}_a(t)u$ is as in (5.0.2) with $a_{ij}(t,x) = a_{ij}(x)$, $a_i(t,x) = a_i(x)$, $b_i(t,x) = b_i(x)$.

Consider (5.2.1). Write $a = (a_{ij}(\cdot), a_i(\cdot), b_i(\cdot), c_0(\cdot, \cdot), d_0(\cdot, \cdot))$. We assume that a satisfies (A5-R1) and

(A5-R3) *There exists* $\tilde{\Omega} \subset \Omega$ *with* $\mathbb{P}(\tilde{\Omega}) = 1$ *such that for each* $\omega \in \tilde{\Omega}$ *the limits* $\lim_{T\to\infty} \frac{1}{T} \int_0^T c_0(\theta_t\omega, x)\, dt$ *and* $\lim_{T\to\infty} \frac{1}{T} \int_0^T d_0(\theta_t\omega, x)\, dt$ *exist for a.e.* $x \in D$ *and* $x \in \partial D$, *respectively; moreover,*

$$\lim_{T\to\infty} \frac{1}{T} \int_0^T c_0(\theta_t\omega, x)\, dt = \int_\Omega c_0(\cdot, x)\, d\mathbb{P}(\cdot) \quad \text{for a.e. } x \in D,$$

$$\lim_{T\to\infty} \frac{1}{T} \int_0^T d_0(\theta_t\omega, x)\, dt = \int_\Omega d_0(\cdot, x)\, d\mathbb{P}(\cdot) \quad \text{for a.e. } x \in \partial D.$$

We call $\hat{a} = (a_{ij}(\cdot), a_i(\cdot), b_i(\cdot), \hat{c}_0(\cdot), \hat{d}_0(\cdot)) \in L_\infty(D, \mathbb{R}^{N^2+2N+1}) \times L_\infty(\partial D, \mathbb{R})$ the *time averaged function* of a if

$$\hat{c}_0(x) = \int_\Omega c_0(\cdot, x)\, d\mathbb{P}(\cdot) \quad \text{for a.e. } x \in D,$$

$$\hat{d}_0(x) = \int_\Omega d_0(\cdot, x)\, d\mathbb{P}(\cdot) \quad \text{for a.e. } x \in \partial D.$$

The time independent equation

$$\begin{cases} \dfrac{\partial u}{\partial t} = \displaystyle\sum_{i=1}^N \dfrac{\partial}{\partial x_i}\left(\sum_{j=1}^N a_{ij}(x)\dfrac{\partial u}{\partial x_j} + a_i(x)u\right) \\ \qquad + \displaystyle\sum_{i=1}^N b_i(x)\dfrac{\partial u}{\partial x_i} + \hat{c}_0(x)u, \qquad\qquad x \in D, \\ \mathcal{B}_{\hat{a}} u = 0, \qquad\qquad\qquad\qquad\qquad\qquad x \in \partial D, \end{cases} \tag{5.2.3}$$

where $\mathcal{B}_{\hat{a}} \equiv \mathcal{B}_{\hat{a}}(t)$ is as in (2.0.3) with a being replaced by $\hat{a} = (a_{ij}(\cdot), a_i(\cdot), b_i(\cdot), \hat{c}_0(\cdot), \hat{d}_0(\cdot))$ is called the *time averaged equation* of (5.2.1) if $\hat{a}$ is the time averaged function of a. Note that under assumption (A5-R3), the averaged equation of (5.2.1) exists.

Consider (5.2.2). Let $a = (a_{ij}(\cdot), a_i(\cdot), b_i(\cdot), c_0(\cdot, \cdot), d_0(\cdot, \cdot))$. We assume that a satisfies (A5-N1) and

(A5-N3) *The weak-* convergence of*

$$\lim_{n\to\infty} \frac{1}{T_n - S_n} \int_{S_n}^{T_n} c_0(t, x)\, dt \quad \text{and} \quad \lim_{n\to\infty} \frac{1}{T_n - S_n} \int_{S_n}^{T_n} d_0(t, x)\, dt$$

in $L_\infty(D, \mathbb{R})$ *and* $L_\infty(\partial D, \mathbb{R})$ *imply pointwise convergence for a.e.* $x \in D$ *and* $x \in \partial D$, *respectively, for any* $T_n - S_n \to \infty$.

We call $\hat{a} = (a_{ij}(\cdot), a_i(\cdot), b_i(\cdot), \hat{c}_0(\cdot), \hat{d}_0(\cdot)) \in L_\infty(D, \mathbb{R}^{N^2+2N+1}) \times L_\infty(\partial D, \mathbb{R})$ a *time averaged function* of a if

$$\hat{c}_0(x) = \lim_{n\to\infty} \frac{1}{T_n - S_n} \int_{S_n}^{T_n} c_0(t, x)\, dt \quad \text{for a.e. } x \in D,$$

$$\hat{d}_0(x) = \lim_{n\to\infty} \frac{1}{T_n - S_n} \int_{S_n}^{T_n} d_0(t, x)\, dt \quad \text{for a.e. } x \in \partial D$$

for some $T_n - S_n \to \infty$. Equation (5.2.3) is called a *time averaged equation* of (5.2.2) if $\hat{a} = (a_{ij}(\cdot), a_i(\cdot), b_i(\cdot), \hat{c}_0(\cdot), \hat{d}_0(\cdot))$ is a time averaged function of a. Note that under assumption (A5-N3), the averaged equations of (5.2.2) exist.

Observe that the eigenvalue problem associated to (5.2.3) reads as follows:

$$\begin{cases} \displaystyle\sum_{i=1}^{N} \frac{\partial}{\partial x_i} \left(\sum_{j=1}^{N} a_{ij}(x) \frac{\partial u}{\partial x_j} + a_i(x)u \right) \\ \quad + \displaystyle\sum_{i=1}^{N} b_i(x) \frac{\partial u}{\partial x_i} + \hat{c}_0(x)u = \lambda u, \quad x \in D, \\ \mathcal{B}_{\hat{a}} u = 0, \quad\quad\quad\quad\quad\quad\quad\quad\quad x \in \partial D. \end{cases} \quad (5.2.4)$$

It is well known that (5.2.4) has a unique eigenvalue, denoted by $\lambda_{\text{princ}}(\hat{a})$, which satisfies that it is real, simple, has an eigenfunction $\varphi_{\text{princ}}(\hat{a}) \in L_2(D)^+$ associated to it, and for any other eigenvalue λ of (5.2.4), Re $\lambda < \lambda_{\text{princ}}(\hat{a})$ (see [29], [31]). We call $\lambda_{\text{princ}}(\hat{a})$ the *principal eigenvalue* of (5.2.3) and $\varphi_{\text{princ}}(\hat{a})$ a *principal eigenfunction* (in the literature, sometimes, $-\lambda_{\text{princ}}(a)$ is called the principal eigenvalue of (5.2.3)).

Our objective in this section is to compare the principal Lyapunov exponents and principal spectrum of (5.2.1) and (5.2.2) with the principal eigenvalue of their averaged equations. We will consider the smooth case (both the domain and the coefficients are sufficiently smooth) and the nonsmooth case separately.

5.2.1 The Smooth Case

In this subsection we assume that (5.2.1) satisfies (A5-R1) and (A2-5) (i.e., $\tilde{Y}(a)$ satisfies (A2-5), which implies that both (A5-R2) and (A5-R3) are satisfied), and that (5.2.2) satisfies (A5-N1) and (A2-5) (i.e., $\tilde{Y}(a)$ satisfies (A2-5), which also implies that both (A5-N2) and (A5-N3) are satisfied). We show that spatial and temporal variations cannot reduce the principal spectrum and principal Lyapunov exponents, and indeed, spatial and temporal variations increase the principal spectrum and principal Lyapunov exponents except in the degenerate cases.

Consider (5.2.1). Let λ be the principal Lyapunov exponent. An application of Lemma 5.1.2 to appropriate derivatives of c_0 and d_0, together with (A2-5), gives that $\hat{c}_0 \in C^{1+\alpha}(\bar{D})$ and $\hat{d}_0 \in C^{2+\alpha}(\partial D)$. Denote $\hat{\lambda} := \lambda_{\text{princ}}(\hat{a})$, where $\lambda_{\text{princ}}(\hat{a})$ is the principal eigenvalue of (5.2.3).

Let

$$\kappa(\tilde{a}) = -B_{\tilde{a}}(0, w(\tilde{a}), w(\tilde{a})) \qquad (5.2.5)$$

for $\tilde{a} = (a_{ij}, a_i, b_i, \tilde{c}_0, \tilde{d}_0) \in \tilde{Y}_0(a)$, where $B_a(\cdot, u, v)$ is as in (2.1.4) in the Dirichlet and Neumann boundary condition cases, and is as in (2.1.5) in the Robin boundary condition case. Note that $\kappa(\tilde{a})$ is well defined under the smoothness assumption (A2-5). Moreover, by the fact that $w(\tilde{a}) = U_{\tilde{a}\cdot(-1)}(1, 0)w(\tilde{a} \cdot (-1))/\|U_{\tilde{a}\cdot(-1)}(1, 0)w(\tilde{a} \cdot (-1))\|$ and Proposition 2.5.4, the function $[\tilde{Y}_0(a) \ni \tilde{a} \mapsto \kappa(\tilde{a}) \in (0, \infty)]$ is continuous.

We have

THEOREM 5.2.1
Consider (5.2.1).

(1) $\lambda \geq \hat{\lambda}$.

(2) $\lambda = \hat{\lambda}$ *if and only if* $c_0(\theta_t\omega, x) = c_{01}(x) + c_{02}(\theta_t\omega)$ *for* $\mathbb{P}$-*a.e.* $\omega \in \Omega$ *and* $d_0(\theta_t\omega, x) = d_0(x)$ *for* $\mathbb{P}$-*a.e.* $\omega \in \Omega$.

Consider (5.2.2). Let $a = (a_{ij}(\cdot), a_i(\cdot), b_i(\cdot), c_0(\cdot, \cdot), d_0(\cdot, \cdot))$, with $d_0 \equiv 0$ in the Dirichlet and Neumann boundary condition cases. Let

$$\hat{Y}(a) := \{\hat{a} : \exists S_n < T_n \quad \text{with} \quad T_n - S_n \to \infty \quad \text{such that}$$

$$\hat{c}_0(x) = \lim_{n \to \infty} \frac{1}{T_n - S_n} \int_{S_n}^{T_n} c_0(t, x)\, dt \quad \text{for} \quad x \in D,$$

$$\hat{d}_0(x) = \lim_{n \to \infty} \frac{1}{T_n - S_n} \int_{S_n}^{T_n} d_0(t, x)\, dt \quad \text{for} \quad x \in \partial D\}.$$

It follows from (A2-5), via the Ascoli–Arzelà theorem, that $\hat{Y}(a)$ is nonempty; further, the $C^{1+\alpha}(\bar{D})$-norms of $\hat{c}_0$ and the $C^{2+\alpha}(\partial D)$-norms of $\hat{d}_0$ are bounded uniformly in $\hat{Y}(a)$.

Denote by $\Sigma(a) = [\lambda_{\min}(a), \lambda_{\max}(a)]$ the principal spectrum interval of (5.2.2). We have

THEOREM 5.2.2
Consider (5.2.2).

(1) *There is* $\hat{a} \in \hat{Y}(a)$ *such that* $\lambda_{\min}(a) \geq \lambda_{\text{princ}}(\hat{a})$.

(2) $\lambda_{\max}(a) \geq \lambda_{\text{princ}}(\hat{a})$ *for any* $\hat{a} \in \hat{Y}(a)$.

(3) *If* a *is uniquely ergodic and minimal, then* $\lambda_{\min}(a) = \lambda_{\max}(a)$ *and* $\lambda_{\min}(a) = \lambda_{\text{princ}}(\hat{a})$ *for some* $\hat{a} \in \hat{Y}(a)$ *($\hat{Y}(a)$ is necessarily a singleton in this case) if and only if* $c_0(t, x) = c_{01}(x) + c_{02}(t)$ *and* $d_0(t, x) = d_0(x)$.

The above theorems are proved in [84] for general smooth case (see also [81] for the case that the boundary condition is Dirichlet type or Neumann type or Robin type with d_0 being independent of t). For the completeness, we will provide proofs of the theorems. To do so, we first prove the following lemma.

LEMMA 5.2.1
Let $Y_0 = \tilde{Y}(a)$ in the nonautonomous case and $Y_0 = \tilde{Y}_0(a)$ in the random case. For any $\tilde{a} \in Y_0$ and $S < T$, let $\tilde{v}(t, x) = \tilde{v}(t, x; \tilde{a}) := w(\tilde{a} \cdot t)(x)$ and

$$\hat{w}(x; S, T) = \hat{w}(x; S, T, \tilde{a}) := \exp\Big(\frac{1}{T-S}\int_S^T \ln w(\tilde{a} \cdot t)(x)\, dt\Big).$$

Then $\hat{w}(x; S, T)$ satisfies

$$\sum_{i=1}^N \frac{\partial}{\partial x_i}\Big(\sum_{j=1}^N a_{ij}(x)\frac{\partial \hat{w}}{\partial x_j} + a_i(x)\hat{w}\Big) + \sum_{i=1}^N b_i(x)\frac{\partial \hat{w}}{\partial x_i}$$

$$\le \Big(\frac{1}{T-S}\int_S^T \frac{1}{\tilde{v}}\frac{\partial \tilde{v}}{\partial t}(t, x)\, dt\Big)\hat{w}$$

$$+ \Big(\frac{1}{T-S}\int_S^T \kappa(\tilde{a}\cdot t)\, dt - \frac{1}{T-S}\int_S^T \tilde{c}_0(t, x)\, dt\Big)\hat{w} \qquad (5.2.6)$$

for $x \in D$ and

$$\hat{\mathcal{B}}_{\tilde{a}}(S, T)\hat{w} = 0$$

for $x \in \partial D$, where

$$\hat{\mathcal{B}}_{\tilde{a}}(S, T)\hat{w} := \begin{cases} \hat{w} & (Dirichlet) \\[2mm] \sum_{i=1}^N\Big(\sum_{j=1}^N a_{ij}(x)\partial_{x_j}\hat{w} + a_i(x)\hat{w}\Big)\nu_i & (Neumann) \\[2mm] \sum_{i=1}^N\Big(\sum_{j=1}^N a_{ij}(x)\partial_{x_j}\hat{w} + a_i(x)\hat{w}\Big)\nu_i & \\[2mm] \quad + \Big(\frac{1}{T-S}\int_S^T \tilde{d}_0(t, x)\, dt\Big)\hat{w} & (Robin). \end{cases} \qquad (5.2.7)$$

PROOF   First, fix $\tilde{a} \in Y_0$. For given $S < T$, let $\eta(t; S) := \|U_{\tilde{a}\cdot S}(t - S, 0)w(\tilde{a}\cdot S)\|$. Let $\bar{v}(t, x; S) = \tilde{v}(t+S, x)$. Then $\eta(t; S)$ satisfies

$$\eta_t(t; S) = \kappa(\tilde{a}\cdot(t+S))\eta(t; S) \qquad (5.2.8)$$

where $\kappa(\cdot)$ is defined in (5.2.5) (see Lemma 3.5.3) and $\bar{v}(t, x; S)$ satisfies

$$
\begin{cases}
\dfrac{\partial \bar{v}}{\partial t} = \displaystyle\sum_{i=1}^{N} \dfrac{\partial}{\partial x_i}\left(\sum_{j=1}^{N} a_{ij}(x)\dfrac{\partial \bar{v}}{\partial x_j} + a_i(x)\bar{v}\right) \\
\quad + \displaystyle\sum_{i=1}^{N} b_i(x)\dfrac{\partial \bar{v}}{\partial x_i} + \tilde{c}_0(t+S, x)\bar{v} - \kappa(\tilde{a}\cdot(t+S))\bar{v}, \quad x \in D, \\
\mathcal{B}_{\tilde{a}}(t+S)\bar{v} = 0, \qquad\qquad\qquad\qquad\qquad\qquad\qquad x \in \partial D,
\end{cases}
\tag{5.2.9}
$$

(see Lemma 3.5.4).

By Proposition 2.5.1, we can differentiate $\hat{w}$ twice and have

$$
\dfrac{\partial \hat{w}}{\partial x_i}(x; S, T) = \hat{w}(x; S, T)\dfrac{1}{T-S}\int_S^T \left(\dfrac{1}{w(\tilde{a}\cdot t)(x)}\dfrac{\partial w(\tilde{a}\cdot t)(x)}{\partial x_i}\right) dt, \tag{5.2.10}
$$

$$
\dfrac{\partial^2 \hat{w}}{\partial x_i \partial x_j}(x; S, T) = \hat{w}(x; S, T)\left(\dfrac{1}{(T-S)^2}\int_S^T \left(\dfrac{1}{w(\tilde{a}\cdot t)(x)}\dfrac{\partial w(\tilde{a}\cdot t)}{\partial x_i}(x)\right) dt\right)
$$
$$
\cdot\left(\int_S^T \left(\dfrac{1}{w(\tilde{a}\cdot t)(x)}\dfrac{\partial w(\tilde{a}\cdot t)}{\partial x_j}(x)\right) dt\right)
$$
$$
+ \hat{w}(x; S, T)\dfrac{1}{T-S}\int_S^T \left(\dfrac{1}{w(\tilde{a}\cdot t)(x)}\dfrac{\partial^2 w(\tilde{a}\cdot t)}{\partial x_i \partial x_j}(x)\right.
$$
$$
\left. - \dfrac{1}{w^2(\tilde{a}\cdot t)(x)}\dfrac{\partial w(\tilde{a}\cdot t)}{\partial x_i}(x)\dfrac{\partial w(\tilde{a}\cdot t)}{\partial x_j}(x)\right) dt \tag{5.2.11}
$$

for $x \in D$. Hence $\hat{w} = \hat{w}(x; S, T)$ satisfies

$$
\sum_{i=1}^{N} \dfrac{\partial}{\partial x_i}\left(\sum_{j=1}^{N} a_{ij}(x)\dfrac{\partial \hat{w}}{\partial x_j} + a_i(x)\hat{w}\right) + \sum_{i=1}^{N} b_i(x)\dfrac{\partial \hat{w}}{\partial x_i}
$$
$$
= \dfrac{1}{T-S}\left(\int_0^{T-S} \dfrac{1}{\bar{v}}\dfrac{\partial \bar{v}}{\partial t}(t, x; S)\, dt + \int_S^T \kappa(\tilde{a}\cdot t)\, dt - \int_s^T \tilde{c}_0(t, x)\, dt\right)\hat{w}
$$
$$
+ \hat{w}\sum_{i,j=1}^{N} a_{ij}(x)\dfrac{1}{(T-S)^2}\left[\int_S^T \left(\dfrac{1}{w(\tilde{a}\cdot t)}\dfrac{\partial w(\tilde{a}\cdot t)}{\partial x_i}\right) dt\right.
$$
$$
\left.\cdot\int_S^T \left(\dfrac{1}{w(\tilde{a}\cdot t)}\dfrac{\partial w(\tilde{a}\cdot t)}{\partial x_j}\right) dt\right]
$$
$$
- \hat{w}\sum_{i,j=1}^{N} a_{ij}(x)\dfrac{1}{T-S}\int_S^T \left(\dfrac{1}{w^2(\tilde{a}\cdot t)}\dfrac{\partial w(\tilde{a}\cdot t)}{\partial x_i}\dfrac{\partial w(\tilde{a}\cdot t)}{\partial x_j}\right) dt \tag{5.2.12}
$$

for $x \in D$ and

$$
\hat{\mathcal{B}}_{\tilde{a}}(S, T)\hat{w} = 0
$$

on ∂D, where $\hat{\mathcal{B}}_{\tilde{a}}(S, T)$ is as defined in (5.2.7).

By Lemma 5.1.1 we have

$$\sum_{i=1}^{N} \frac{\partial}{\partial x_i} \Big(\sum_{j=1}^{N} a_{ij}(x) \frac{\partial \hat{w}}{\partial x_j} + a_i(x)\hat{w} \Big) + \sum_{i=1}^{N} b_i(x) \frac{\partial \hat{w}}{\partial x_i}$$

$$\leq \frac{1}{T-S} \Big(\int_0^{T-S} \frac{1}{\bar{v}} \frac{\partial \bar{v}}{\partial t}(t, x; S)\, dt + \int_S^T \kappa(\tilde{a} \cdot t)\, dt - \int_s^T \tilde{c}_0(t, x)\, dt \Big) \hat{w}$$

for $x \in D$. The lemma thus follows. $\qquad\square$

PROOF (Proof of Theorem 5.2.1)   In the following proof, we write $U_\omega(t, 0)$ for $U_{E_a(\omega)}(t, 0)$, write $w(\omega)$ for $w(E_a(\omega))$, and write $\kappa(\omega)$ for $\kappa(E_a(\omega))$ for $\omega \in \Omega$ with $E_a(\omega) \in \tilde{Y}_0(a)$.

(1) First of all, since $c_0(\omega, x)$ and $d_0(\omega, x)$ are continuous in x uniformly with respect to ω, an application of Lemma 5.1.2 to c_0 and d_0 gives the existence of $\Omega_1 \subset \Omega$ with $\mathbb{P}(\Omega_1) = 1$ such that for each $\omega \in \Omega_1$,

$$\hat{c}_0(x) = \lim_{T \to \infty} \frac{1}{T} \int_0^T c_0(\theta_t\omega, x)\, dt \quad \text{and} \quad \hat{d}_0(x) = \lim_{T \to \infty} \frac{1}{T} \int_0^T d_0(\theta_t\omega, x)\, dt$$

uniformly for $x \in \bar{D}$ and $x \in \partial D$, respectively. Let $\eta(t; \omega) := \|U_\omega(t, 0)w(\omega)\|$. Then $U_\omega(t, 0)w(\omega) = \eta(t; \omega)w(\omega)(\cdot)$ and by (5.2.8) $\eta(t; \omega)$ satisfies

$$\eta_t(t; \omega) = \kappa(\theta_t\omega)\eta(t; \omega). \tag{5.2.13}$$

It follows from Theorem 3.5.3 that there is $\Omega_2 \subset \Omega$ with $\mathbb{P}(\Omega_2) = 1$ such that

$$\lambda = \lim_{T \to \infty} \frac{1}{T} \int_0^T \kappa(\theta_t\omega)\, dt = \int_\Omega \kappa(\cdot)\, d\mathbb{P}(\cdot) \quad \text{for} \quad \omega \in \Omega_2. \tag{5.2.14}$$

Let $\hat{a} := (a_{ij}, a_i, b_i, \hat{c}_0, \hat{d}_0)$. We claim that $\lambda(a) \geq \lambda(\hat{a})$. In fact, take $\omega \in \Omega_1 \cap \Omega_2$ and $T_n \to \infty$. Let $\hat{c}_0^{(n)}(x) := \frac{1}{T_n} \int_0^{T_n} c_0(\theta_t\omega, x)\, dt$, $\hat{d}_0^{(n)}(x) := \frac{1}{T_n} \int_0^{T_n} d_0(\theta_t\omega t, x)\, dt$, and $\hat{a}^{(n)} := (a_{ij}, a_i, b_i, \hat{c}_0^{(n)}, \hat{d}_0^{(n)})$. Then $\hat{a}^{(n)} \to \hat{a}$ in the open-compact topology as $n \to \infty$ and $\frac{1}{T_n} \int_0^{T_n} \kappa(\theta_t\omega)\, dt \to \lambda(a)$ and $n \to \infty$.

Recall that $\tilde{v}(t, x) = w(\theta_t\omega)(x)$. It follows from Proposition 2.5.1 that for each $x \in D$ the set $\{ w(\theta_t\omega)(x) : t \in \mathbb{R} \}$ is bounded away from zero. Consequently,

$$\lim_{T \to \infty} \frac{1}{T} \int_0^T \frac{1}{\tilde{v}} \frac{\partial \tilde{v}}{\partial t}(t, x)\, dt = \lim_{T \to \infty} \frac{1}{T} (\ln \tilde{v}(T, x) - \ln \tilde{v}(0, x)) = 0 \tag{5.2.15}$$

for any $x \in D$. Again by an application of Proposition 2.5.1 we may assume without loss of generality that there is $w^*(x)$ $(w^*(x) = 0$ for $x \in \partial D$ in the Dirichlet boundary condition case) such that

$$\lim_{n \to \infty} \hat{w}(x; T_n) = w^*(x) \tag{5.2.16}$$

$$\lim_{n\to\infty} \frac{\partial \hat{w}}{\partial x_i}(x;T_n) = \frac{\partial w^*}{\partial x_i}(x) \tag{5.2.17}$$

$$\lim_{n\to\infty} \frac{\partial^2 \hat{w}}{\partial x_i \partial x_j}(x;T_n) = \frac{\partial^2 w^*}{\partial x_i \partial x_j}(x) \tag{5.2.18}$$

for $i,j = 1,2,\ldots,N$ and $x \in D$, where $\hat{w}(x;T_n) = \hat{w}(x;0,T_n,a^\omega)$. Moreover, the limit in (5.2.16) is uniform for x in $\bar{D}$, and the limits in (5.2.17), (5.2.18) are uniform for x in any compact subset D_0 of D. In the Neumann and Robin cases, the limit in (5.2.17) is also uniform for $x \in \bar{D}$.

Then by Lemma 5.2.1 and (5.2.15)–(5.2.18) we have

$$\begin{cases} \displaystyle\sum_{i=1}^{N} \frac{\partial}{\partial x_i}\Big(\sum_{j=1}^{N} a_{ij}(x)\frac{\partial w^*}{\partial x_j} + a_i(x)w^*\Big) + \sum_{i=1}^{N} b_i(x)\frac{\partial w^*}{\partial x_i} \\ \qquad + (\hat{c}_0(x) - \lambda)w^* \le 0, & x \in D, \\[2mm] \mathcal{B}_{\hat{a}}w^* = 0, & x \in \partial D, \end{cases} \tag{5.2.19}$$

where $\mathcal{B}_{\hat{a}}$ is as in (5.2.3).

This implies that $w(t,x) := w^*(x)$ is a supersolution of

$$\begin{cases} \displaystyle\frac{\partial u}{\partial t} = \sum_{i=1}^{N} \frac{\partial}{\partial x_i}\Big(\sum_{j=1}^{N} a_{ij}(x)\frac{\partial u}{\partial x_j} + a_i(x)u\Big) + \sum_{i=1}^{N} b_i(x)\frac{\partial u}{\partial x_i} \\ \qquad + (\hat{c}_0(x) - \lambda)u, & x \in D, \\[2mm] \mathcal{B}_{\hat{a}}u = 0, & x \in \partial D. \end{cases} \tag{5.2.20}$$

Let $u(t,x;w^*)$ be the solution of (5.2.20) satisfying the initial condition $u(0,\cdot;w^*) = w^*$. Then

$$u(t,x;w^*) \le w^*(x) \quad \text{for } t > 0, \ x \in \bar{D}. \tag{5.2.21}$$

Note that $\hat{\lambda} - \lambda$ is the principal eigenvalue of (5.2.20). Then by (5.2.21), we must have $\hat{\lambda} - \lambda \le 0$, hence $\hat{\lambda} \le \lambda$.

(2) First, suppose that $c_0(\theta_t\omega,x) = c_{01}(x) + c_{02}(\theta_t\omega)$ for any $t \in \mathbb{R}$, $x \in D$ and $\omega \in \Omega^*$, where $\mathbb{P}(\Omega^*) = 1$. Without loss of generality we can assume $\int_\Omega c_{02}(\cdot)\,d\mathbb{P}(\cdot) = 0$ (for otherwise we replace $c_{01}(x)$ with $c_{01}(x) + \int_\Omega c_{02}(\cdot)\,d\mathbb{P}(\cdot)$ and replace $c_{02}(\omega)$ with $c_{02}(\omega) - \int_\Omega c_{02}(\cdot)\,d\mathbb{P}(\cdot)$). Suppose also that $d_0(\theta_t\omega,x) = d_0(x)$. One has $\hat{c}_0(x) = c_{01}(x)$ for $x \in \bar{D}$, and $\hat{d}_0(x) = d_0(x)$ for $x \in \partial D$. Let $u = \varphi_{\text{princ}}(\hat{a})(x)$ be the positive principal eigenfunction of (5.2.3) normalized so that $\|\varphi_{\text{princ}}(\hat{a})\| = 1$. Define $v(t,x;\omega) := \varphi_{\text{princ}}(\hat{a})(x)\exp\big(\hat{\lambda}t + \int_0^t c_{02}(\theta_s\omega)\,ds\big)$ for $t \in \mathbb{R}$, $x \in \bar{D}$ and $\omega \in \Omega^*$. It is easy to see that for each $\omega \in \Omega^*$ there holds

$$v(t,x;\omega) = (U_\omega(t,0)u)(x), \qquad t \ge 0, \ x \in \bar{D}.$$

By Proposition 4.1.1, $\lambda = \lim_{t\to\infty}(1/t)\ln\|v(t,\cdot;\omega)\|$ for $\mathbb{P}$-a.e. $\omega \in \Omega$. An application of Birkhoff's Ergodic Theorem (Lemma 1.2.6) to c_{02} (bounded, hence $\in L_1((\Omega,\mathfrak{F},\mathbb{P}))$) yields that $\lim_{t\to\infty}(1/t)\ln\|v(t,\cdot;\omega)\| = \hat{\lambda}$ for $\mathbb{P}$-a.e. $\omega \in \Omega$.

Conversely, suppose that $\lambda = \hat{\lambda}$. Let Ω_1 and Ω_2 be as (1). Let $\bar{v}(s,x;\omega) := w(\theta_s\omega)(x)$ and

$$\phi(x;\omega) := \limsup_{t\to\infty}\exp\left(\frac{1}{t}\int_0^t \ln w(\theta_s\omega)(x)\,ds\right) \quad \text{for } x \in \bar{D}$$

in the case of Neumann or Robin boundary condition,

$$\phi(x;\omega) := \begin{cases} \limsup_{t\to\infty}\exp\left(\dfrac{1}{t}\displaystyle\int_0^t \ln w(\theta_s\omega)(x)\,ds\right) & \text{for } x \in D \\ 0 & \text{for } x \in \partial D \end{cases}$$

in the case of Dirichlet boundary condition. Since $w(\omega)(x)$ are continuous in x uniformly in ω, an application of Lemma 5.1.2 provides the existence of $\Omega_3 \subset \Omega$ with $\mathbb{P}(\Omega_3) = 1$ such that

$$\phi(x;\omega) = \lim_{t\to\infty}\exp\left(\frac{1}{t}\int_0^t \ln w(\theta_s\omega)(x)\,ds\right)$$
$$= \exp\left(\int_\Omega \ln w(\cdot)(x)\,d\mathbb{P}(\cdot)\right) \tag{5.2.22}$$

for any $\omega \in \Omega_3$ and $x \in D$. Clearly, $\phi(x;\omega) > 0$ for $x \in D$ and is independent of $\omega \in \Omega_3$.

Observe that $\dfrac{\partial w(\omega)(x)}{\partial x_i}$ $(i = 1,2,\ldots,N)$ $\left(\dfrac{\partial^2 w(\omega)(x)}{\partial x_i \partial x_j}, \; i,j = 1,2,\ldots,N\right)$ are locally Hölder continuous in $x \in \bar{D}$ ($x \in D$) uniformly in $\omega \in \Omega$ and are integrable in $\omega \in \Omega$. Hence, again by Lemma 5.1.2, there is $\Omega_4 \subset \Omega$ with $\mathbb{P}(\Omega_4) = 1$ such that

$$\frac{\partial \phi}{\partial x_i}(x;\omega) = \phi(x;\omega)\lim_{t\to\infty}\frac{1}{t}\int_0^t\left(\frac{1}{w(\theta_s\omega)(x)}\frac{\partial w(\theta_s\omega)(x)}{\partial x_i}\right)ds$$
$$= \phi(x;\omega)\int_\Omega\left(\frac{1}{w(\cdot)(x)}\frac{\partial w(\cdot)(x)}{\partial x_i}\right)d\mathbb{P}(\cdot), \tag{5.2.23}$$

$$\frac{\partial^2 \phi}{\partial x_i \partial x_j}(x;\omega) = \phi(x;\omega) \lim_{t \to \infty} \left(\frac{1}{t^2} \int_0^t \left(\frac{1}{w(\theta_s \omega)(x)} \frac{\partial w(\theta_s \omega)}{\partial x_i}(x) \right) ds \right) \cdot$$

$$\left(\int_0^t \left(\frac{1}{w(\theta_s \omega)(x)} \frac{\partial w(\theta_s \omega)}{\partial x_j}(x) \right) ds \right)$$

$$+ \phi(x;\omega) \lim_{t \to \infty} \frac{1}{t} \int_0^t \left(\frac{1}{w(\theta_s \omega)(x)} \frac{\partial^2 w(\theta_s \omega)}{\partial x_i \partial x_j}(x) \right.$$

$$\left. - \frac{1}{w^2(\theta_s \omega)(x)} \frac{\partial w(\theta_s \omega)}{\partial x_i}(x) \frac{\partial w(\theta_s \omega)}{\partial x_j}(x) \right) ds$$

$$= \phi(x;\omega) \left[\int_\Omega \left(\frac{1}{w(\cdot)(x)} \frac{\partial w(\cdot)}{\partial x_i}(x) \right) d\mathbb{P}(\cdot) \right.$$

$$\left. \cdot \int_\Omega \left(\frac{1}{w(\cdot)(x)} \frac{\partial w(\cdot)}{\partial x_j}(x) \right) d\mathbb{P}(\cdot) \right]$$

$$+ \phi(x;\omega) \int_\Omega \left(\frac{1}{w(\cdot)(x)} \frac{\partial^2 w(\cdot)}{\partial x_i \partial x_j}(x) \right.$$

$$\left. - \frac{1}{w^2(\cdot)(x)} \frac{\partial w(\cdot)}{\partial x_i}(x) \frac{\partial w(\cdot)}{\partial x_j}(x) \right) d\mathbb{P}(\cdot) \tag{5.2.24}$$

for $\omega \in \Omega_4$, $x \in D$, and

$$\mathcal{B}_{\hat{a}} \phi = 0 \quad \text{for } x \in \partial D, \quad \omega \in \Omega_4,$$

where $\mathcal{B}_{\hat{a}}$ is as in (5.2.3).

Let $\Omega_0 := \Omega_1 \cap \Omega_2 \cap \Omega_3 \cap \Omega_4$. By Lemma 3.5.4, $\bar{v}(t, x; \omega) := w(\theta_t \omega)(x)$ satisfies

$$\begin{cases} \dfrac{\partial \bar{v}}{\partial t} = \displaystyle\sum_{i=1}^N \frac{\partial}{\partial x_i} \left(\sum_{j=1}^N a_{ij}(x) \frac{\partial \bar{v}}{\partial x_j} + a_i(x) \bar{v} \right) + \sum_{i=1}^N b_i(x) \frac{\partial \bar{v}}{\partial x_i} \\ \qquad + c_0(\theta_t \omega, x) \bar{v} - \kappa(\theta_t \omega) \bar{v}, \qquad\qquad\qquad x \in D \\ \mathcal{B}_\omega(t) \bar{v} = 0, \qquad\qquad\qquad\qquad\qquad\qquad\qquad x \in \partial D \end{cases} \tag{5.2.25}$$

for all $\omega \in \Omega_0$.

(5.2.22)–(5.2.25) yield that

$$\sum_{i=1}^N \frac{\partial}{\partial x_i} \left(\sum_{j=1}^N a_{ij}(x) \frac{\partial \phi}{\partial x_j} + a_i(x) \phi \right) + \sum_{i=1}^N b_i(x) \frac{\partial \phi}{\partial x_i}$$

$$= (\lambda - \hat{c}_0(x)) \phi$$

$$+ \phi \sum_{i,j=1}^N a_{ij}(x) \int_\Omega \left(\frac{1}{w(\cdot)} \frac{\partial w(\cdot)}{\partial x_i} \right) d\mathbb{P}(\cdot) \int_\Omega \left(\frac{1}{w(\cdot)} \frac{\partial w(\cdot)}{\partial x_j} \right) d\mathbb{P}(\cdot)$$

$$- \phi \sum_{i,j=1}^N a_{ij}(x) \int_\Omega \left(\frac{1}{w^2(\cdot)} \frac{\partial w(\cdot)}{\partial x_i} \frac{\partial w(\cdot)}{\partial x_j} \right) d\mathbb{P}(\cdot) \tag{5.2.26}$$

for $w \in \Omega_0$ and $x \in D$, and

$$\mathcal{B}_{\hat{a}}\phi = 0 \quad \text{for } w \in \Omega_0, \ x \in \partial D.$$

Consider

$$\begin{cases} \dfrac{\partial u}{\partial t} = \displaystyle\sum_{i=1}^{N} \dfrac{\partial}{\partial x_i}\Big(\sum_{j=1}^{N} a_{ij}(x)\dfrac{\partial u}{\partial x_j} + a_i(x)u\Big) \\ \qquad + \displaystyle\sum_{i=1}^{N} b_i(x)\dfrac{\partial u}{\partial x_i} + (\hat{c}_0(x) - \lambda)u, \quad x \in D, \\ \mathcal{B}_{\hat{a}}u = 0, \qquad\qquad\qquad\qquad\qquad\qquad\quad x \in \partial D. \end{cases} \tag{5.2.27}$$

By $\lambda = \hat{\lambda}$, we have that 0 is the principal eigenvalue of (5.2.27). Let $\hat{\varphi}_{\mathrm{princ}}$ be a positive principal eigenfunction of (5.2.27), and let $u(t, x; \phi)$ be the solution of (5.2.27) with initial condition $u(0, x; \phi) = \phi(x)$. By Lemma 5.1.1(2),

$$\sum_{i,j=1}^{N} a_{ij}(x) \int_\Omega \Big(\frac{1}{w(\cdot)}\frac{\partial w(\cdot)}{\partial x_i}\Big)\, d\mathbb{P}(\cdot) \int_\Omega \Big(\frac{1}{w(\cdot)}\frac{\partial w(\cdot)}{\partial x_j}\Big)\, d\mathbb{P}(\cdot)$$

$$- \sum_{i,j=1}^{N} a_{ij}(x) \int_\Omega \Big(\frac{1}{w^2(\cdot)}\frac{\partial w(\cdot)}{\partial x_i}\frac{\partial w(\cdot)}{\partial x_j}\Big)\, d\mathbb{P}(\cdot) \le 0$$

for all $x \in D$. This together with (5.2.26) implies that $\phi(x)$ is a supersolution of (5.2.27) and hence

$$u(t, x; \phi) \le \phi(x) \quad \text{for} \quad x \in D, \ t \ge 0. \tag{5.2.28}$$

Let $w^*(x)$ be a positive principal eigenfunction of the adjoint problem of (5.2.27). We then have that $\langle \phi, w^* \rangle > 0$ and $\langle \hat{\varphi}_{\mathrm{princ}}, w^* \rangle > 0$. By taking $\alpha := \langle \phi, w^* \rangle / \langle \hat{\varphi}_{\mathrm{princ}}, w^* \rangle \ (> 0)$ we see that

$$\phi = \alpha\hat{\varphi}_{\mathrm{princ}} + \hat{v},$$

where $\hat{v} \in L_2(D)$ is such that $\langle \hat{v}, w^* \rangle = 0$. Note that $u(t, x; \phi) = \alpha\hat{\varphi}_{\mathrm{princ}}(x) + u(t, x; \hat{v})$, where $u(t, x; \hat{v})$ is the solution of (5.2.27) with $u(0, x; \hat{v}) = \hat{v}(x)$. Due to the exponential separation, $\|u(t, \cdot; \hat{v})\| \to 0$ as $t \to \infty$. It then follows from (5.2.28) that $\alpha\hat{\varphi}_{\mathrm{princ}}(x) \le \phi(x)$ for $x \in D$ and then $\hat{v}(x) \ge 0$ for $x \in D$. This implies that $\hat{v}(x) = 0$ for $x \in D$, hence $\alpha\hat{\varphi}_{\mathrm{princ}}(x) = \phi(x)$ for $x \in D$. Therefore we must have

$$\sum_{i,j=1}^{N} a_{ij}(x) \int_\Omega \Big(\frac{1}{w(\omega)}\frac{\partial w(\omega)}{\partial x_i}\Big)\, d\mathbb{P}(\omega) \int_\Omega \Big(\frac{1}{w(\omega)}\frac{\partial w(\omega)}{\partial x_j}\Big)\, d\mathbb{P}(\omega)$$

$$= \sum_{i,j=1}^{N} a_{ij}(x) \int_\Omega \Big(\frac{1}{w^2(\omega)}\frac{\partial w(\omega)}{\partial x_i}\frac{\partial w(\omega)}{\partial x_j}\Big)\, d\mathbb{P}(\omega)$$

for all $x \in D$. Then by Lemma 5.1.1 and continuity of $\dfrac{1}{w(\omega)(x)} \dfrac{\partial w(\omega)(x)}{\partial x_i}$ in $x \in D$, there are $\Omega_5 \subset \Omega_0$ with $\mathbb{P}(\Omega_5) = 1$ and $F_i = F_i(x)$ such that

$$\frac{1}{w(\omega)(x)} \frac{\partial w(\omega)(x)}{\partial x_i} = F_i(x)$$

for $i = 1, 2, \ldots, N$, $x \in D$ and $\omega \in \Omega_5$. Hence

$$\nabla \ln w(\omega)(x) = (F_1(x), F_2(x), \ldots, F_N(x))^\top$$

for $x \in D$ and $\omega \in \Omega_5$. This implies that there are a continuous $F \colon \bar{D} \to \mathbb{R}$ and a measurable $G \colon \Omega_5 \to \mathbb{R}$ such that $F(x) > 0$, $G(\omega) > 0$ and $w(\omega)(x) = F(x)G(\omega)$ for any $x \in D$ and any $\omega \in \Omega_5$.

Let $\Omega_6 := \bigcap_{r \in \mathbb{Q}} \theta_r \Omega_5$, where $\mathbb{Q}$ is the set of rational numbers. Clearly, $\mathbb{P}(\Omega_6) = 1$ and $w(\theta_t \omega)(x) = F(x)G(\theta_t \omega)$ for $t \in \mathbb{Q}$, $x \in D$, and $\omega \in \Omega_6$. The continuity of $w(\theta_t \omega)(x)$ in $t \in \mathbb{R}$ then implies that $w(\theta_t \omega)(x) = F(x)G(\theta_t \omega)$ for any $t \in \mathbb{R}$, $x \in D$, and $\omega \in \Omega_6$. Therefore, by the first equation in (5.2.25),

$$F(x)\frac{dG(\theta_t \omega)}{dt} = \left(\sum_{i,j=1}^{N} \frac{\partial}{\partial x_i} \left(a_{ij}(x) \frac{\partial F}{\partial x_j}(x) + a_i(x)F(x) \right) \right.$$
$$\left. + \sum_{i=1}^{N} b_i(x) \frac{\partial F}{\partial x_i}(x) + c_0(\theta_t \omega, x)F(x) - \kappa(\theta_t \omega)F(x) \right) G(\theta_t \omega)$$

for $t \in \mathbb{R}$, $x \in D$ and $\omega \in \Omega_6$.

After simple calculation we obtain that there are functions $c_{01} \colon \bar{D} \to \mathbb{R}$ and $c_{02} \colon \Omega_6 \to \mathbb{R}$ such that $c_0(\omega, x) = c_{01}(x) + c_{02}(\theta_t \omega)$ for all $t \in \mathbb{R}$, $\omega \in \Omega_6$ and $x \in \bar{D}$. Both functions c_{01} and c_{02} are bounded, consequently c_{02} is $\mathbb{P}$-integrable. Applying a similar reasoning to the boundary condition $B_\omega(t)F = 0$, we see that $d_0(\theta_t \omega, x) = d_0(x)$ for $x \in \partial D$, $t \in \mathbb{R}$ and $\mathbb{P}$-a.e. $\omega \in \Omega$. ☐

In order not to interrupt the presentation, before giving the proof of Theorem 5.2.2 we formulate and prove the following result.

LEMMA 5.2.2
Consider (5.2.2). Assume that a is uniquely ergodic. Then the limits

$$\lim_{T-S \to \infty} \frac{1}{T-S} \int_S^T c_0(t, x)\, dt =: \hat{c}_0(x)$$

and

$$\lim_{T-S \to \infty} \frac{1}{T-S} \int_S^T d_0(t, x)\, dt =: \hat{d}_0(x)$$

exist for each $x \in \bar{D}$ and each $x \in \partial D$, respectively. In particular, it follows that $\hat{Y}(a) = \{(a_{ij}, a_i, b_i, \hat{c}_0, \hat{d}_0)\}$.

PROOF Let $\mathbb{P}$ be the unique ergodic measure on $(\tilde{Y}(a), \sigma)$. Write $c_0(\tilde{a}, x)$ for $\tilde{c}_0(0, x)$ and write $d_0(\tilde{a}, x)$ for $\tilde{d}_0(0, x)$, where $\tilde{a} = (a_{ij}, a_i, b_i, \tilde{c}_0, \tilde{d}_0)$. Put $\hat{c}_0(x) := \int_{\tilde{Y}(a)} c_0(\tilde{a}, x) \, d\mathbb{P}(\tilde{a})$ $(x \in \bar{D})$ and $\hat{d}_0(x) := \int_{\tilde{Y}(a)} d_0(\tilde{a}, x) \, d\mathbb{P}(\tilde{a})$ $(x \in \bar{D})$. As $(\tilde{Y}(a), \sigma)$ is uniquely ergodic, for any $g \in C(\tilde{Y}(a))$ and any $\epsilon > 0$ there is $T_0 = T_0(g, \epsilon) > 0$ such that

$$\left| \frac{1}{t} \int_0^t g(\tilde{a} \cdot s) \, ds - \int_{\tilde{Y}(a)} g(\cdot) \, d\mathbb{P}(\cdot) \right| < \epsilon$$

for each $t > T_0$ and each $\tilde{a} \in \tilde{Y}(a)$ (see [90]). In particular, for any $g \in C(\tilde{Y}(a))$ there holds

$$\lim_{T-S \to \infty} \frac{1}{T-S} \int_S^T g(a \cdot t) \, dt = \int_{\tilde{Y}(a)} g(\tilde{a}) \, d\mathbb{P}(\tilde{a}).$$

By taking, for a fixed $x \in \bar{D}$, a function $g \in C(\tilde{Y}(a))$ given by $g(\tilde{a}) := c_0(\tilde{a}, x)$ we obtain

$$\lim_{T-S \to \infty} \frac{1}{T-S} \int_S^T c_0(t, x) \, dt = \hat{c}_0(x).$$

Similarly, by taking, for a fixed $x \in \partial D$, a function $g \in C(\tilde{Y}(a))$ given by $g(\tilde{a}) := d_0(\tilde{a}, x)$ we obtain

$$\lim_{T-S \to \infty} \frac{1}{T-S} \int_S^T \tilde{d}_0(t, x) \, dt = \hat{d}_0(x).$$

□

PROOF (Proof of Theorem 5.2.2) (1) First, for given $S < T$, let

$$\eta(t; S) := \|U_a(t, S)w(a \cdot S)\|$$

and

$$\hat{w}(x; S, T) := \exp\left(\frac{1}{T-S} \int_S^T \ln w(a \cdot t)(x) \, dt \right).$$

There are $S_n < T_n$ with $T_n - S_n \to \infty$ such that $\frac{\ln \eta(T_n; S_n)}{T_n - S_n} \to \lambda_{\min}(a)$. It follows from (A2-5) with the help of the Ascoli–Arzelà theorem that without loss of generality we may assume that $\lim_{n \to \infty} \frac{1}{T_n - S_n} \int_{S_n}^{T_n} c_0(t, x) \, dt$ and $\lim_{n \to \infty} \frac{1}{T_n - S_n} \int_{S_n}^{T_n} d_0(t, x) \, dt$ exist, and the limits are uniform in $x \in \bar{D}$ and in $x \in \partial D$, respectively.

Denote

$$\hat{c}_0(x) := \lim_{n \to \infty} \frac{1}{T_n - S_n} \int_{S_n}^{T_n} c_0(t, x) \, dt$$

and

$$\hat{d}_0(x) := \lim_{n \to \infty} \frac{1}{T_n - S_n} \int_{S_n}^{T_n} d_0(t, x) \, dt.$$

Let $\hat{a} := (a_{ij}, a_i, b_i, \hat{c}_0, \hat{d}_0)$.

We claim that $\lambda_{\min}(a) \geq \lambda_{\mathrm{princ}}(\hat{a})$. Denote $\bar{v}(t, x) := w(a \cdot t)(x)$. It follows from Proposition 2.5.1 that for each $x \in D$ the set $\{ w(a \cdot t)(x) : t \in \mathbb{R} \}$ is bounded away from zero. Consequently,

$$\lim_{T - S \to \infty} \frac{1}{T - S} \int_S^T \frac{1}{\bar{v}} \frac{\partial \bar{v}}{\partial t}(t, x) \, dt = \lim_{T - S \to \infty} \frac{1}{T - S} (\ln \bar{v}(T, x) - \ln \bar{v}(S, x))$$
$$= 0 \tag{5.2.29}$$

for any $x \in D$. Again by an application of Proposition 2.5.1 we may assume without loss of generality that there is $w^*(x)$ ($w^*(x) = 0$ for $x \in \partial D$ in the Dirichlet boundary condition case) such that

$$\lim_{n \to \infty} \hat{w}(x; S_n, T_n) = w^*(x) \tag{5.2.30}$$

$$\lim_{n \to \infty} \frac{\partial \hat{w}}{\partial x_i}(x; S_n, T_n) = \frac{\partial w^*}{\partial x_i}(x) \tag{5.2.31}$$

$$\lim_{n \to \infty} \frac{\partial^2 \hat{w}}{\partial x_i \partial x_j}(x; S_n, T_n) = \frac{\partial^2 w^*}{\partial x_i \partial x_j}(x) \tag{5.2.32}$$

for $i, j = 1, 2, \ldots, N$ and $x \in D$. Moreover, the limit in (5.2.30) is uniform for x in $\bar{D}$, and the limits in (5.2.31), (5.2.32) are uniform for x in any compact subset D_0 of D. In the Neumann and Robin cases, the limit in (5.2.31) is also uniform for x in $\bar{D}$.

Then by Lemma 5.2.1 and (5.2.29)–(5.2.32) we have

$$\begin{cases} \displaystyle\sum_{i=1}^N \frac{\partial}{\partial x_i} \left(\sum_{j=1}^N a_{ij}(x) \frac{\partial w^*}{\partial x_j} + a_i(x) w^* \right) + \sum_{i=1}^N b_i(x) \frac{\partial w^*}{\partial x_i} \\ \qquad + (\hat{c}_0(x) - \lambda_{\min}(a)) w^* \leq 0, \qquad\qquad x \in D, \qquad (5.2.33) \\ \\ \mathcal{B}_{\hat{a}} w^* = 0, \qquad\qquad\qquad\qquad\qquad\qquad\qquad x \in \partial D. \end{cases}$$

It then follows from arguments similar to those in the proof of Theorem 5.2.1(1) that $\lambda_{\min}(a) \geq \lambda_{\mathrm{princ}}(\hat{a})$.

(2) For any $\hat{a} = (a_{ij}, a_i, b_i, \hat{c}_0, \hat{d}_0) \in \hat{Y}(a)$ there are $S_n < T_n$ with $T_n - S_n \to \infty$ such that

$$\frac{1}{T_n - S_n} \int_{S_n}^{T_n} c_0(t, x) \, dt \to \hat{c}_0(x) \quad \text{and} \quad \frac{1}{T_n - S_n} \int_{S_n}^{T_n} d_0(t, x) \, dt \to \hat{d}_0(x)$$

uniformly in $x \in \bar{D}$ and in $x \in \partial D$, respectively. Without loss of generality, assume that there is λ_0 such that

$$\lim_{n \to \infty} \frac{1}{T_n - S_n} \int_{S_n}^{T_n} \kappa(a \cdot t) \, dt = \lambda_0.$$

By arguments similar to those in the proof of (1), $\lambda_0 \geq \lambda_{\mathrm{princ}}(\hat{a})$. It follows from Theorem 3.1.2 and Lemmas 3.2.5 and 3.2.7 that $\lambda_{\max}(a) \geq \lambda_0$. Then we have $\lambda_{\max}(a) \geq \lambda_{\mathrm{princ}}(\hat{a})$.

(3) Let $\mathbb{P}$ be the unique ergodic measure on $\tilde{Y}(a)$. By Lemma 5.2.2, $\hat{Y}(a) = \{\hat{a}\}$, where $\hat{a} = (a_{ij}, a_i, b_i, \hat{c}_0, \hat{d}_0)$.

Assume that the equality $\lambda_{\min}(a) = \lambda_{\mathrm{princ}}(\hat{a})$ holds. By Theorem 5.2.1, there are $Y_0(a) \subset \tilde{Y}(a)$ with $\mathbb{P}(Y_0(a)) = 1$, a continuous $c_{01} \colon \bar{D} \to \mathbb{R}$, a $\mathbb{P}$-integrable $\bar{c}_{02} \colon Y_0(a) \to \mathbb{R}$ with $\int_{Y_0(a)} \bar{c}_{02}(\tilde{a}) \, d\mathbb{P}(\tilde{a}) = 0$, and a continuous $d_0 \colon \partial D \to \mathbb{R}$ such that for any $\tilde{a} = (a_{ij}, a_i, b_i, \tilde{c}_0, \tilde{d}_0) \in Y_0(a)$ there holds $\tilde{c}_0(t, x) = c_{01}(x) + \bar{c}_{02}(\tilde{a} \cdot t)$ for any $t \in \mathbb{R}$ and $x \in D$, and $\tilde{d}_0(t, x) = d_0(x)$ for any $t \in \mathbb{R}$ and $x \in \partial D$.

Fix some $\tilde{a} \in Y_0(a)$. Since the compact flow $(\tilde{Y}(a), \sigma)$ is minimal, the orbit $\{\tilde{a} \cdot t : t \in \mathbb{R}\}$ is dense in $\tilde{Y}(a)$, consequently there is a real sequence $(s_n)_{n=1}^{\infty}$ such that $\tilde{a} \cdot s_n$ converges in the topology of $\tilde{Y}(a)$ to a, as $n \to \infty$. In particular, $\tilde{c}_0(t + s_n, x)$ converges to $c_0(t, x)$, for any $t \in \mathbb{R}$ and any $x \in \bar{D}$. As a consequence, $\bar{c}_{02}(\tilde{a} \cdot (t + s_n))$ converges, for each $t \in \mathbb{R}$, to some $c_{02}(t)$. Therefore $c_0(t, x) = c_{01}(x) + c_{02}(t)$ for all $t \in \mathbb{R}$ and all $x \in \bar{D}$. The fact that $d_0(t, x) = d_0(x)$ for all $t \in \mathbb{R}$ and all $x \in \partial D$ follows in much the same way.

Let $c_0(t, x) = c_{01}(x) + c_{02}(t)$ and $d_0(t, x) = d_0(x)$. Lemma 5.2.2 implies that $\lim_{t \to \infty} \frac{1}{t} \int_0^t c_{02}(s) \, ds$ exists. Without loss of generality we can assume that this limit equals 0. One has $\hat{c}_0(x) = c_{01}(x)$ for $x \in \bar{D}$, and $\hat{d}_0(x) = d_0(x)$ for $x \in \partial D$. Let $u = u(x)$ be the positive principal eigenfunction of (5.2.3) normalized so that $\|u\| = 1$. Define $v(t, x) := u(x) \exp\left(\lambda_{\mathrm{princ}}(\hat{a})t + \int_0^t c_{02}(s) \, ds\right)$ for $t \in \mathbb{R}$ and $x \in \bar{D}$. A straightforward computation shows that

$$\lim_{t \to \infty} \frac{1}{t} \ln \|v(t, \cdot)\| = \lambda_{\mathrm{princ}}(\hat{a}).$$

It is easy to see that there holds

$$v(t, x) = (U_a(t, 0)u)(x), \qquad t \geq 0, \; x \in \bar{D}.$$

Then by Proposition 4.1.7, $\lambda_{\min}(a) = \lambda_{\max}(a) = \lim_{t \to \infty} \frac{1}{t} \ln \|v(t, \cdot)\|$. $\quad \Box$

5.2.2 The Nonsmooth Case

In this subsection, we study the extension of Theorems 5.2.1 and 5.2.2 about influence of temporal variations on principal spectrum and principal Lyapunov exponents of (5.2.1) and (5.2.2) to the nonsmooth case.

First, consider (5.2.1). Under the assumptions (A5-R1) and (A5-R3) the functions

$$\hat{c}_0(x) = \int_\Omega c_0(\omega, x)\, d\mathbb{P}(\omega) \quad \text{and} \quad \hat{d}_0(x) = \int_\Omega d_0(\omega, x)\, d\mathbb{P}(\omega)$$

are defined for a.e. $x \in D$ and a.e. $x \in \partial D$, respectively. Recall that $\hat{a} = (a_{ij}, a_i, b_i, \hat{c}_0, \hat{d}_0)$ and that $\lambda_{\mathrm{princ}}(\hat{a})$ stands for the principal eigenvalue of the time averaged equation (5.2.3). As in the smooth case we write λ for $\lambda(a)$, and $\hat{\lambda}$ for $\lambda_{\mathrm{princ}}(\hat{a})$.

We have

THEOREM 5.2.3
Consider (5.2.1) and assume (A5-R1)–(A5-R3). There holds $\lambda \geq \hat{\lambda}$.

Next, consider (5.2.2). Note that for any $s_n < t_n$ with $t_n - s_n \to \infty$, there are subsequences s_{n_k} and t_{n_k} such that $\frac{1}{t_{n_k} - s_{n_k}} \int_{s_{n_k}}^{t_{n_k}} c_0(t, x)\, dt$ and $\frac{1}{t_{n_k} - s_{n_k}} \int_{s_{n_k}}^{t_{n_k}} d_0(t, x)\, dt$ converge in the weak-* topology. Under (A5-N3), the weak-* convergence of $\frac{1}{t_{n_k} - s_{n_k}} \int_{s_{n_k}}^{t_{n_k}} c_0(t, x)\, dt$ and $\frac{1}{t_{n_k} - s_{n_k}} \int_{s_{n_k}}^{t_{n_k}} d_0(t, x)\, dt$ implies the pointwise convergence almost everywhere. Let

$$\hat{Y}(a) := \{\, \hat{a} = (a_{ij}(\cdot), a_i(\cdot), b_i(\cdot), \hat{c}_0(\cdot), \hat{d}_0(\cdot)) :$$
$$\exists s_n < t_n \text{ with } t_n - s_n \to \infty \text{ such that}$$
$$\hat{c}_0(x) = \lim_{n \to \infty} \frac{1}{t_n - s_n} \int_{s_n}^{t_n} c_0(t, x)\, dt \text{ for a.e. } x \in D$$
$$\hat{d}_0(x) = \lim_{n \to \infty} \frac{1}{t_n - s_n} \int_{s_n}^{t_n} d_0(t, x)\, dt \text{ for a.e. } x \in \partial D \,\}.$$

We have

THEOREM 5.2.4
Consider (5.2.2) and assume (A5-N1)–(A5-N3).

(1) *There is $\hat{a} \in \hat{Y}(a)$ such that $\lambda_{\min}(a) \geq \lambda_{\mathrm{princ}}(\hat{a})$.*

(2) *$\lambda_{\max}(a) \geq \lambda_{\mathrm{princ}}(\hat{a})$ for any $\hat{a} \in \hat{Y}(a)$.*

To prove the theorems, we first show three lemmas. In the following, let $Y_0 = \tilde{Y}_0(a)$ in the random case, where $\tilde{Y}_0(a)$ is defined in (5.1.2), and $Y_0 = \tilde{Y}(a)$ in the nonautonomous case, where $\hat{Y}(a)$ is defined in (5.1.3).

LEMMA 5.2.3

(1) *In the Dirichlet boundary condition case, there is $M > 0$ such that for any $\tilde{a} \in Y_0$, $w(\tilde{a})(x) \leq M$ for a.e. $x \in D$.*

(2) *In the Neumann or Robin boundary condition case, there are $M, m > 0$ such that for any $\tilde{a} \in Y_0$, $m \leq w(\tilde{a})(x) \leq M$ for a.e. $x \in D$.*

PROOF   First by the L_2–L_∞ estimates (Proposition 2.2.2), there is $C_1 > 0$ such that for any $\tilde{a} \in Y_0$, $\|U_{\tilde{a}}(1,0)w(\tilde{a})\|_\infty \leq C_1$. Further, by the compactness of Y_0 and the continuity of w, there is $C_2 > 0$ such that for any $\tilde{a} \in Y_0$, $\|U_{\tilde{a}}(1,0)w(\tilde{a})\| \geq C_2$. Note that

$$w(\tilde{a}) = U_{\tilde{a}\cdot(-1)}(1,0)w(\tilde{a}\cdot(-1))/\|U_{\tilde{a}\cdot(-1)}(1,0)w(\tilde{a}\cdot(-1))\|.$$

Hence for any $\tilde{a} \in Y_0$, $w(\tilde{a})(x) \leq M$ for a.e. $x \in D$, where $M = C_1/C_2$.

It remains to prove the first inequality in (2). Since $\|U_{\tilde{a}}(1,0)w(\tilde{a})\| \geq C_2$ for all $\tilde{a} \in Y_0$, there is $C_3 > 0$ such that $\|U_{\tilde{a}}(1,0)w(\tilde{a})\|_\infty \geq C_3$ for any $\tilde{a} \in Y_0$. By (A3-2) with $\varsigma = 0$, there is $C_4 > 0$ such that for any $\tilde{a} \in Y_0$, $(U_{\tilde{a}}(1,0)w(\tilde{a}))(x) \geq C_4$ for a.e. $x \in D$. By the L_2–L_2 estimates (Proposition 2.2.2), there is $C_5 > 0$ such that for any $\tilde{a} \in Y_0$, $\|U_{\tilde{a}}(1,0)w(\tilde{a})\| \leq C_5$. Hence for any $\tilde{a} \in Y_0$, $w(\tilde{a})(x) \geq m$ for a.e. $x \in D$, where $m = C_4/C_5$. ▯

For any $\tilde{a} \in Y_0$ and any $T > S$, let

$$\hat{w}(x; S, T, \tilde{a}) := \exp\left(\frac{1}{T-S}\int_S^T \ln w(\tilde{a}\cdot\tau)(x)\,d\tau\right)$$

for $x \in D$. We claim that the function is well defined. By Proposition 2.2.9, $w(\tilde{a})(x) > 0$ for all $x \in D$. Further, Proposition 2.2.4 implies that we can integrate in the definition.

It is a simple consequence of Lemma 5.2.3 that

$$0 < \hat{w}(x; S, T, \tilde{a}) \leq M \tag{5.2.34}$$

for a.e. $x \in D$, in the Dirichlet case, and

$$m \leq \hat{w}(x; S, T, \tilde{a}) \leq M \tag{5.2.35}$$

for a.e. $x \in D$, in the Neumann and Robin cases. In particular, $\hat{w}(\cdot; S, T, \tilde{a}) \in L_\infty(D)^+$ for any $S < T$ and any $\tilde{a} \in Y_0$.

Moreover, as a consequence of Lemma 5.2.3(2) we have

LEMMA 5.2.4
Assume the Neumann or Robin boundary conditions and (A5-R1)–(A5-R3) or (A5-N1)–(A5-N3). For given $\tilde{a} \in Y_0$ and $S < T$ we have that the derivatives

$\frac{\partial \hat{w}}{\partial x_i}(x; T, S, \tilde{a}) =: \partial_{x_i} \hat{w}(x; T, S, \tilde{a})$ $(i = 1, \ldots, N)$ *are well defined and satisfy*

$$\frac{\partial \hat{w}}{\partial x_i}(x; T, S, \tilde{a}) = \frac{\hat{w}(x; T, S, \tilde{a})}{T - S} \int_S^T \frac{1}{w(\tilde{a} \cdot \tau)(x)} \frac{\partial w(\tilde{a} \cdot \tau)}{\partial x_i}(x) \, d\tau$$

for any $x \in D$. Further, $\partial_{x_i} \hat{w}(\cdot; S, T, \tilde{a}) \in L_2(D)$.

LEMMA 5.2.5
Assume the Neumann or Robin boundary conditions and (A5-R1)–(A5-R3) or (A5-N1)–(A5-N3). For given $\tilde{a} \in Y_0$, $S < T$, and $v(\cdot) \in V \cap L_\infty(D)$ (see (2.1.2) for the definition of V), there holds $\frac{1}{u(\cdot, \cdot)}$, $\frac{v(\cdot)\hat{w}(\cdot)}{u(\cdot, \cdot)} \in W(S, T; V, V^*)$ (see (2.1.3) for the definition of $W(S, T; V, V^*)$), where $\hat{w}(x) = \hat{w}(x; T, S, \tilde{a})$ and $u(t, x) = (U_{\tilde{a}}(t, S)w(\tilde{a} \cdot S))(x)$.

PROOF   We only prove that $\frac{v(\cdot)\hat{w}(\cdot)}{u(\cdot, \cdot)} \in W(S, T; V, V^*)$.

Observe that

$$\frac{\partial}{\partial x_i}\left(\frac{v(x)\hat{w}(x)}{u(t, x)}\right) = \frac{\frac{\partial v(x)}{\partial x_i}\hat{w}(x)u(t, x) + v(x)\frac{\partial \hat{w}(x)}{\partial x_i}u(t, x) - v(x)\hat{w}(x)\frac{\partial u(t, x)}{\partial x_i}}{u^2(t, x)}$$

and

$$\frac{\partial}{\partial t}\left(\frac{v(x)\hat{w}(x)}{u(t, x)}\right) = -\frac{v(x)\hat{w}(x)\frac{\partial u(t, x)}{\partial t}}{u^2(t, x)}.$$

It then follows from (A5-R2) or (A5-N2), Lemmas 5.2.3 and 5.2.4, and the boundedness of $v(\cdot)$, that $\frac{v(x)\hat{w}(x)}{u(t, x)} \in L_2((S, T), V)$ and $\frac{\partial}{\partial t}\left(\frac{v(x)\hat{w}(x)}{u(t, x)}\right) \in L((S, T), V^*)$. Therefore $\frac{v(x)\hat{w}(x)}{u(t, x)} \in W(S, T; V, V^*)$. □

In the following, we first prove Theorem 5.2.4. We consider the Dirichlet boundary condition and the Neumann and Robin boundary conditions separately.

PROOF (Proof of Theorem 5.2.4 in the Dirichlet boundary condition case)

(1) By Lemma 3.2.3, there is $M_2 \geq 1$ such that

$$\|U_{\tilde{a}}(T, S)u\| \leq M_2\|U_{\tilde{a}}(T, S)w(\tilde{a} \cdot S)\|$$

for all $\tilde{a} \in \tilde{Y}(a)$, all $S < T$, and all $u \in L_2(D)$ with $\|u\| = 1$. Therefore, for any $\epsilon > 0$, there is $K_\epsilon > 0$ such that for any $S < T$ with $T - S > K_\epsilon$ there holds

$$\frac{\ln\|U_{\tilde{a}}(T, S)u\|}{T - S} \leq \frac{\ln\|U_{\tilde{a}}(T, S)w(\tilde{a} \cdot S)\|}{T - S} + \epsilon \tag{5.2.36}$$

for all $\tilde{a} \in \tilde{Y}(a)$ and $u \in L_2(D)$ with $\|u\| = 1$.

Next, it follows from Proposition 4.1.8 that for a given $\epsilon > 0$ there are $S_\epsilon < T_\epsilon$ with $T_\epsilon - S_\epsilon > K_\epsilon$ such that

$$\lambda_{\min}(a) \geq \frac{\ln \|U_a(T_\epsilon, S_\epsilon)w(a \cdot S_\epsilon)\|}{T_\epsilon - S_\epsilon} - \epsilon. \tag{5.2.37}$$

Let $a_\epsilon \in L_\infty(\mathbb{R} \times D, \mathbb{R}^{N^2+2N+1}) \times L_\infty(\mathbb{R} \times \partial D, \mathbb{R})$, $a_\epsilon = a_\epsilon(t, x)$, be the function periodic in t with period $T_\epsilon - S_\epsilon$ such that $a_\epsilon(t, x) = a(t, x)$ for $S_\epsilon \leq t < T_\epsilon$. We then have

$$U_{a_\epsilon}(T_\epsilon, S_\epsilon) = U_a(T_\epsilon, S_\epsilon),$$

where the symbol $U_{a_\epsilon}(\cdot, \cdot)$ has the obvious meaning. Let $\lambda_\epsilon := \lambda_{\mathrm{princ}}(a_\epsilon)$ and let u_ϵ be the nonnegative principal eigenfunction associated to λ_ϵ, normalized so that $\|u_\epsilon\| = 1$. Then

$$U_{a_\epsilon}(T_\epsilon, S_\epsilon)u_\epsilon = e^{\lambda_\epsilon(T_\epsilon - S_\epsilon)}u_\epsilon.$$

It then follows from (5.2.36) and (5.2.37) that

$$\lambda_\epsilon = \frac{\ln \|U_{a_\epsilon}(T_\epsilon, S_\epsilon)u_\epsilon\|}{T_\epsilon - S_\epsilon} \leq \lambda_{\min}(a) + 2\epsilon. \tag{5.2.38}$$

By [39, Appendix C, Theorem 6], there is a sequence of C^∞ functions $a_\epsilon^{(n)}(t, x)$ which are periodic in t with period $T_\epsilon - S_\epsilon$ such that

$$a_\epsilon^{(n)} \to a_\epsilon \quad \text{as} \quad n \to \infty \quad \text{in} \quad L_2([0, T_\epsilon - S_\epsilon] \times D_0)$$

for any compact subset D_0 of D. Let $D_n \subset D$ be a sequence of C^∞ subdomains of D such that $D_n \to D$ as $n \to \infty$ (see Lemma 4.5.2). Let $\lambda_\epsilon^{(n)} := \lambda_{\mathrm{princ}}(a_\epsilon^{(n)}, D_n)$. By Lemma 4.5.1, $\lambda_\epsilon^{(n)} \to \lambda_\epsilon$ as $n \to \infty$. Therefore, for a given $\epsilon > 0$, there is $n_1 = n_1(\epsilon) > 0$ such that

$$\lambda_\epsilon^{(n)} \leq \lambda_{\min}(a) + 3\epsilon \quad \text{for} \quad n \geq n_1. \tag{5.2.39}$$

Define

$$\hat{a}_\epsilon^{(n)}(x) := \frac{1}{T_\epsilon - S_\epsilon} \int_{S_\epsilon}^{T_\epsilon} a_\epsilon^{(n)}(t, x)\, dt, \quad x \in \bar{D}.$$

Let $\hat{\lambda}_\epsilon^{(n)} := \lambda_{\mathrm{princ}}(\hat{a}_\epsilon^{(n)}, D_n)$. By Theorem 5.2.2,

$$\hat{\lambda}_\epsilon^{(n)} \leq \lambda_\epsilon^{(n)}. \tag{5.2.40}$$

Note that for any compact subset D_0 of D,

$$\int_{D_0} \|\hat{a}_\epsilon^{(n)}(x) - \hat{a}_\epsilon(x)\|^2\, dx = \frac{1}{(T_\epsilon - S_\epsilon)^2} \int_{D_0} \left\| \int_{S_\epsilon}^{T_\epsilon} (a_\epsilon^{(n)}(t, x) - a_\epsilon(t, x))\, dt \right\|^2 dx$$

$$\leq \frac{1}{T_\epsilon - S_\epsilon} \int_{D_0} \int_{S_\epsilon}^{T_\epsilon} \|a_\epsilon^{(n)}(t, x) - a_\epsilon(t, x)\|^2\, dt\, dx$$

$$\to 0 \quad \text{as} \quad n \to \infty$$

(in the above display, $\|\cdot\|$ stands for the standard norm in $\mathbb{R}^{N^2+2N+1}$). Then by Lemma 4.5.1 again, $\hat{\lambda}_\epsilon^{(n)} \to \hat{\lambda}_\epsilon$ as $n \to \infty$. Hence there is $n_2 \geq n_1$ such that

$$\hat{\lambda}_\epsilon \leq \hat{\lambda}_\epsilon^{(n)} + \epsilon \quad \text{for} \quad n \geq n_2. \tag{5.2.41}$$

It then follows that

$$\hat{\lambda}_\epsilon \leq \lambda_{\min}(a) + 4\epsilon. \tag{5.2.42}$$

Finally, take a sequence $\epsilon_k \to 0$ such that

$$\hat{c}_0(x) := \lim_{k \to \infty} \frac{1}{T_{\epsilon_k} - S_{\epsilon_k}} \int_{S_{\epsilon_k}}^{T_{\epsilon_k}} c_0(t, x)\, dt$$

exists for a.e. $x \in D$. By Lemma 4.5.1, we have $\hat{\lambda}_{\epsilon_k} \to \lambda_{\text{princ}}(\hat{a})$, where $\hat{a} = (a_{ij}, a_i, b_i, \hat{c}_0, 0) \in \hat{Y}(a)$. This, together with (5.2.42), implies that $\lambda_{\min}(a) \geq \lambda_{\text{princ}}(\hat{a})$, which proves (1).

(2) Take any $S_n < T_n$ with $T_n - S_n \to \infty$ such that

$$\hat{c}_0(x) := \lim_{n \to \infty} \frac{1}{T_n - S_n} \int_{S_n}^{T_n} c_0(t, x)\, dt$$

exists for a.e. $x \in D$. Put $\hat{a} = (a_{ij}, a_i, b_i, \hat{c}_0, 0)$. We claim that $\lambda_{\max}(a) \geq \lambda_{\text{princ}}(\hat{a})$. In fact, without loss of generality, we may assume that $\lambda = \lim_{n \to \infty} \frac{1}{T_n - S_n} \ln \|U_a(T_n, S_n)w(a \cdot S_n)\|$ exists. It then follows from arguments as above that $\lambda_{\max}(a) \geq \lambda \geq \lambda_{\text{princ}}(\hat{a})$. $\qquad\Box$

PROOF (Proof of Theorem 5.2.4 in the Neumann and Robin boundary conditions cases)

(1) First, $\hat{w}$ denotes $\hat{w}(\cdot; S, T, a)$ unless specified explicitly. Also, we use the summation convention. Let $\eta(t; S) := \|U_a(t, S)w(a \cdot S)\|$. For any fixed $T > S$ and any $v \in V \cap L_\infty(D)$, we have

$$-\frac{1}{T - S} \int_S^T \langle w(a \cdot t), \partial_t(v\hat{w}/w(a \cdot t)) \rangle\, dt$$

$$= -\frac{1}{T - S} \int_S^T \langle w(a \cdot t)\eta(t; S), \partial_t(\frac{v\hat{w}}{\eta(t; S)w(a \cdot t)}) \rangle\, dt$$

$$- \frac{1}{T - S} \int_S^T \frac{\partial_t \eta(t; S)}{\eta(t; S)} \langle v, \hat{w} \rangle\, dt.$$

By Lemma 5.2.5, $\frac{\langle v, \hat{w} \rangle}{\eta w} \in W(S, T; V, V^*)$. Hence by Proposition 2.1.3 and Eq. (2.1.4) or Eq. (2.1.5) we have (recall that the boundary ∂D of the domain D is assumed to be Lipschitz, so the Hausdorff measure on ∂D reduces to the ordinary surface measure)

$$- \frac{1}{T-S} \int_S^T \left\langle \eta(t;S)w(a \cdot t), \partial_t \frac{v\hat{w}}{\eta(t;S)w(a \cdot t)} \right\rangle dt$$

$$= -\frac{1}{T-S} \int_S^T B_a\left(t, \eta(t;S)w(a \cdot t), \frac{v\hat{w}}{\eta(t;S)w(a \cdot t)}\right) dt$$

$$= -\frac{1}{T-S} \int_S^T \int_D a_{ij} \, \partial_{x_j}(\eta(t;S)w(a \cdot t))\partial_{x_i}\left(\frac{v\hat{w}}{\eta(t;S)w(a \cdot t)}\right) dx \, dt$$

$$- \frac{1}{T-S} \int_S^T \int_D a_i \eta(t;S)w(a \cdot t) \, \partial_{x_i}\left(\frac{v\hat{w}}{\eta(t;S)w(a \cdot t)}\right) dx \, dt$$

$$+ \frac{1}{T-S} \int_S^T \int_D b_i \, \partial_{x_i}(\eta(t;S)w(a \cdot t)) \frac{v\hat{w}}{\eta(t;S)w(a \cdot t)} dx \, dt$$

$$+ \frac{1}{T-S} \int_S^T \int_D c_0 v\hat{w} \, dx \, dt - \frac{1}{T-S} \int_S^T \int_{\partial D} d_0 v\hat{w} \, dx \, dt$$

$$= \frac{1}{T-S} \int_D \int_S^T a_{ij}\left(\frac{\partial_{x_j}w}{w}\frac{\partial_{x_i}w}{w}v\hat{w} - \frac{\partial_{x_j}w}{w}\partial_{x_i}v \, \hat{w} - \frac{\partial_{x_j}w}{w}v \, \partial_{x_i}\hat{w}\right) dt \, dx$$

$$+ \frac{1}{T-S} \int_D \int_S^T a_i\left(\frac{\partial_{x_i}w}{w}v\hat{w} - \partial_{x_i}v \, \hat{w} - v \, \partial_{x_i}\hat{w}\right) dt \, dx$$

$$+ \frac{1}{T-S} \int_D \left(\int_S^T b_i \frac{\partial_{x_i}w}{w}v\hat{w} \, dt + \int_S^T c_0 v\hat{w} \, dt\right) dx$$

$$- \frac{1}{T-S} \int_{\partial D} \int_S^T d_0 v\hat{w} \, dt \, dx.$$

By Lemma 5.1.1, for $v \in V \cap L_\infty(D)^+$

$$\frac{1}{T-S} \int_D \int_S^T a_{ij} \frac{\partial_{x_j}w}{w} \frac{\partial_{x_i}w}{w} v\hat{w} \, dt \, dx$$

$$\geq \int_D a_{ij} v\hat{w}\left(\frac{1}{T-S} \int_S^T \frac{\partial_{x_j}w}{w} \, dt\right)\left(\frac{1}{T-S} \int_S^T \frac{\partial_{x_i}w}{w} \, dt\right) dx.$$

Observe that

$$\frac{\partial_{x_i}\hat{w}}{\hat{w}} = \frac{1}{T-S} \int_S^T \frac{\partial_{x_i}w}{w} \, dt.$$

It then follows from $v \in V \cap L_\infty(D)^+$, Proposition 2.2.11 and (A5-N2) (which allows us to change the order of integration) that there holds

$$
-\frac{1}{T-S}\int_S^T \left\langle w(a\cdot t), \partial_t\left(\frac{v\hat{w}}{w(a\cdot t)}\right)\right\rangle dt
$$

$$
= \frac{1}{T-S}\int_D \left(\int_S^T \frac{\partial_t w(a\cdot t)}{w(a\cdot t)}\,dt\right) v\hat{w}\,dx
$$

$$
\geq \int_D a_{ij}\frac{\partial_{x_j}\hat{w}}{\hat{w}}\frac{\partial_{x_i}\hat{w}}{\hat{w}}v\hat{w}\,dx - \int_D a_{ij}\frac{\partial_{x_j}\hat{w}}{\hat{w}}(\partial_{x_i}v)\hat{w} - \int_D a_{ij}\frac{\partial_{x_j}\hat{w}}{\hat{w}}v(\partial_{x_i}\hat{w})\,dx
$$

$$
+ \int_D a_i(\partial_{x_i}\hat{w}\,v - \partial_{x_i}v\,\hat{w} - \partial_{x_i}\hat{w}\,v)\,dx + \int_D b_i\frac{\partial_{x_i}\hat{w}}{\hat{w}}v\hat{w}\,dx
$$

$$
+ \frac{1}{T-S}\left(\int_D\left(\int_S^T\left(c_0 - \frac{\partial_t\eta(t;S)}{\eta(t;S)}\right)dt\right)v\hat{w}\,dx - \int_{\partial D}\int_S^T d_0 v\hat{w}\,dt\,dx\right)
$$

$$
= \int_D\left(-a_{ij}\,\partial_{x_j}\hat{w}\,\partial_{x_i}v - a_i\,\hat{w}\,\partial_{x_i}v + b_i\,\partial_{x_i}\hat{w}\,v\right)dx
$$

$$
+ \frac{1}{T-S}\left(\int_D\left(\int_S^T\left(c_0 - \frac{\partial_t\eta(t;S)}{\eta(t;S)}\right)dt\right)\hat{w}v\,dx\right.
$$

$$
\left. - \int_{\partial D}\left(\int_S^T d_0 v\hat{w}\,dt\right)dx\right). \tag{5.2.43}
$$

Note that there are $S_n < T_n$ with $T_n - S_n \to \infty$ such that

$$
\frac{1}{T_n - S_n}\int_{S_n}^{T_n}\frac{\partial_t\eta(t;S_n)}{\eta(t;S_n)}\,dt \to \lambda_{\min}(a) \quad \text{as} \quad n \to \infty. \tag{5.2.44}
$$

Without loss of generality, we may assume that

$$
\frac{1}{T_n - S_n}\int_{S_n}^{T_n} c_0(t,x)\,dt \to \hat{c}_0(x) \quad \text{for a.e.} \quad x \in D
$$

and

$$
\frac{1}{T_n - S_n}\int_{S_n}^{T_n} d_0(t,x)\,dt \to \hat{d}_0(x) \quad \text{for a.e.} \quad x \in \partial D
$$

as $n \to \infty$. Let $\hat{a} := (a_{ij}, a_i, b_i, \hat{c}_0, \hat{d}_0)$.

We prove that $\lambda_{\min}(a) \geq \lambda(\hat{a})$. In fact, for any $\epsilon > 0$ there is $n_0 = n_0(\epsilon) \in \mathbb{N}$ such that for any $v \in V \cap L_\infty(D)^+$ with $\|v\| \leq 1$ and any $n \geq n_0$ we have

$$
\frac{1}{T_n - S_n}\int_{S_n}^{T_n}\frac{\partial_t\eta(t;S_n)}{\eta(t;S_n)}\,dt \leq \lambda_{\min}(a) + \epsilon,
$$

$$
\int_D\left(\frac{1}{T_n - S_n}\int_{S_n}^{T_n}(c_0(t,x) - \hat{c}_0(x))\,dt\right)\hat{w}(x;S_n,T_n)v\,dx \geq -\epsilon,
$$

$$
\int_{\partial D}\left(\frac{1}{T_n - S_n}\int_{S_n}^{T_n}(d_0(t,x) - \hat{d}_0(x))\,dt\right)\hat{w}(x;S_n,T_n)v\,dx \leq \epsilon,
$$

and

$$\int_D \left(\frac{1}{T_n - S_n} \int_{S_n}^{T_n} \frac{\partial_t w(a \cdot t)}{w(a \cdot t)} \, dt \right) v(x) \hat{w}(x; S_n, T_n) \, dx \le \epsilon$$

(for the last display, see Lemma 5.2.3). It then follows from (5.2.43) that

$$\int_D \left(-a_{ij} \, \partial_{x_j} \hat{w} \, \partial_{x_i} v - a_i \, \hat{w} \, \partial_{x_i} v + b_i \, \partial_{x_i} \hat{w} \, v + (\hat{c}_0 - \lambda_{\mathrm{princ}}(\hat{a})) \hat{w} v \right) dx$$

$$- \int_{\partial D} \hat{d}_0 \hat{w} v \, dx \ \le (\lambda_{\min}(a) - \lambda_{\mathrm{princ}}(\hat{a}) + \epsilon) \int_D \hat{w} v \, dx + 3\epsilon$$

for any $v \in V \cap L_\infty(D)^+$ with $\|v\| \le 1$ and any $n \ge n_0$, where $\hat{w}(x) = \hat{w}(x; S_n, T_n, a)$. We specialize $v(\cdot)$ to be the principal eigenfunction of the adjoint equation associated to (5.2.3) with $\hat{c}_0$ being replaced by $\hat{c}_0 - \lambda_{\mathrm{princ}}(\hat{a})$. Then

$$0 = \int_D \left(-a_{ij} \, \partial_{x_j} \hat{w} \, \partial_{x_i} v - a_i \, \hat{w} \, \partial_{x_i} v + b_i \, \partial_{x_i} \hat{w} \, v + (\hat{c}_0 - \lambda_{\mathrm{princ}}(\hat{a})) \hat{w} v \right) dx$$

$$- \int_{\partial D} \hat{d}_0 \hat{w} v \, dx.$$

Further, with the help of (5.2.35) and taking into account that v satisfy a similar estimate, we see that there is $m_1 > 0$ such that

$$\int_D \hat{w} v \, dx \ge m_1$$

for all $n \ge n_0$, where $\hat{w}(x) = \hat{w}(x; S_n, T_n, a)$. Suppose to the contrary that $\lambda_{\min}(a) < \lambda_{\mathrm{princ}}(\hat{a})$. Let $\epsilon > 0$ be so small that

$$(\lambda_{\min}(a) - \lambda_{\mathrm{princ}}(\hat{a}) + \epsilon) m_1 + 3\epsilon < 0.$$

Then

$$0 = \int_D \left(-a_{ij} \, \partial_{x_j} \hat{w} \, \partial_{x_i} v - a_i \hat{w} \, \partial_{x_i} v + b_i \, \partial_{x_i} \hat{w} \, v + (\hat{c}_0 - \lambda_{\mathrm{princ}}(\hat{a})) \hat{w} v \right) dx$$

$$- \int_{\partial D} \hat{d}_0 \hat{w} v \, dx \le (\lambda_{\min}(a) - \lambda_{\mathrm{princ}}(\hat{a}) + \epsilon) \int_D \hat{w} v \, dx + 3\epsilon < 0$$

for n sufficiently large, where $\hat{w}(x) = \hat{w}(x; S_n, T_n, a)$. This is a contradiction. Therefore we must have $\lambda_{\min}(a) \ge \lambda_{\mathrm{princ}}(\hat{a})$.

(2) For any $\hat{a} = (a_{ij}, b_i, b_i, \hat{c}_0, \hat{d}_0) \in \hat{Y}(a)$, there is $S_n < T_n$ with $T_n - S_n \to \infty$ such that

$$\frac{1}{T_n - S_n} \int_{S_n}^{T_n} c_0(t, x) \, dx \to \hat{c}_0(x) \quad \text{for a.e. } x \in D$$

and

$$\frac{1}{T_n - S_n} \int_{S_n}^{T_n} d_0(t, x) \, dx \to \hat{d}_0(x) \quad \text{for a.e. } x \in \partial D.$$

Observe that for any $\epsilon > 0$ there is $n_0 = n_0(\epsilon) > 0$ such that

$$\frac{1}{T_n - S_n} \int_{S_n}^{T_n} \frac{\partial_t \eta(t; S_n)}{\eta(t; S_n)} \, dt \le \lambda_{\max}(a) + \epsilon$$

for $n > n_0$. Then by arguments similar to those in (1), we have $\lambda_{\max}(a) \ge \lambda_{\text{princ}}(\hat{a})$. $\square$

PROOF (Proof of Theorem 5.2.3)   Let $\omega \in \Omega$ be such that

$$\lambda = \lim_{T \to \infty} \frac{\ln \|U_{E_a(\omega)}(T, 0) w(E_a(\omega))\|}{T}$$

and

$$\hat{c}_0(x) = \lim_{T \to \infty} \frac{1}{T} \int_0^T c_0(\theta_t \omega, x) \, dt \quad \text{for a.e. } x \in D$$

$$\hat{d}_0(x) = \lim_{T \to \infty} \frac{1}{T} \int_0^T d_0(\theta_t \omega, x) \, dt \quad \text{for a.e. } x \in \partial D.$$

Then by arguments similar to those in the proof of Theorem 5.2.4, we have $\lambda \ge \lambda_{\text{princ}}(\hat{a}) \; (= \hat{\lambda})$. $\square$

5.3   Influence of Spatial Variation on Principal Lyapunov Exponents and Principal Spectrum

In this section, we study the influence of spatial variation of the zeroth order terms in (5.0.1) and (5.0.2) on their principal spectrum and principal Lyapunov exponents when the boundary conditions are Neumann. We assume $a_{ij}(\omega, x) = a_{ij}(\omega)$, $a_i(\omega, x) \equiv 0$, $b_i(\omega, x) \equiv 0$ in (5.0.1) and $a_{ij}(t, x) = a_{ij}(t)$, $a_i(t, x) \equiv 0$, $b_i(t, x) \equiv 0$ in (5.0.2). That is, we consider the nonautonomous equation of form

$$\begin{cases} \dfrac{\partial u}{\partial t} = \displaystyle\sum_{i=1}^{N} \dfrac{\partial}{\partial x_i} \left(\sum_{j=1}^{N} a_{ij}(t) \dfrac{\partial u}{\partial x_j} \right) + c_0(t, x) u, & x \in D, \\[3mm] \displaystyle\sum_{i=1}^{N} \left(\sum_{j=1}^{N} a_{ij}(t) \partial_{x_j} u \right) \nu_i = 0, & x \in \partial D, \end{cases} \tag{5.3.1}$$

and the random equation of form

$$
\begin{cases}
\dfrac{\partial u}{\partial t} = \displaystyle\sum_{i=1}^{N} \dfrac{\partial}{\partial x_i}\Big(\sum_{j=1}^{N} a_{ij}(\theta_t\omega)\dfrac{\partial u}{\partial x_j}\Big) + c_0(\theta_t\omega, x)u, & x \in D, \\[4mm]
\displaystyle\sum_{i=1}^{N}\Big(\sum_{j=1}^{N} a_{ij}(\theta_t\omega)\partial_{x_j} u\Big)\nu_i = 0, & x \in \partial D,
\end{cases}
\tag{5.3.2}
$$

where $((\Omega, \mathfrak{F}, \mathbb{P}), \theta)$ is an ergodic metric dynamical system.

Let $\bar c_0(t) := \frac{1}{|D|}\int_D c_0(t, x)\,dx$ in the case of (5.3.1) and $\bar c_0(\omega) := \frac{1}{|D|}\int_D c_0(\omega, x)\,dx$ in the case of (5.3.2). Let $\bar a = (a_{ij}, 0, 0, \bar c_0, 0)$. We call $\bar a$ the *space average* of a, and call the equations

$$
\begin{cases}
\dfrac{\partial u}{\partial t} = \displaystyle\sum_{i=1}^{N} \dfrac{\partial}{\partial x_i}\Big(\sum_{j=1}^{N} a_{ij}(t)\dfrac{\partial u}{\partial x_j}\Big) + \bar c_0(t)u, & x \in D, \\[4mm]
\displaystyle\sum_{i=1}^{N}\Big(\sum_{j=1}^{N} a_{ij}(t)\partial_{x_j} u\Big)\nu_i = 0, & x \in \partial D,
\end{cases}
\tag{5.3.3}
$$

and

$$
\begin{cases}
\dfrac{\partial u}{\partial t} = \displaystyle\sum_{i=1}^{N} \dfrac{\partial}{\partial x_i}\Big(\sum_{j=1}^{N} a_{ij}(\theta_t\omega)\dfrac{\partial u}{\partial x_j}\Big) + \bar c_0(\theta_t\omega)u, & x \in D, \\[4mm]
\displaystyle\sum_{i=1}^{N}\Big(\sum_{j=1}^{N} a_{ij}(\theta_t\omega)\partial_{x_j} u\Big)\nu_i = 0, & x \in \partial D,
\end{cases}
\tag{5.3.4}
$$

the *space averaged equations* of (5.3.1) and (5.3.2), respectively.

In the sequel we will speak of exponential separation, principal eigenvalues, etc., for space averaged equations. We assume

(A5-N4) (5.3.1) *satisfies* (A5-N1), (A5-N2), *and* (5.3.3) *satisfies* (A5-N1).

(A5-R4) (5.3.2) *satisfies* (A5-R1), (A5-R2), *and* (5.3.4) *satisfies* (A5-R1).

Consider the space averaged equation (5.3.3). It is straightforward to see that the function $v(t, x) := \exp\left(\int_0^t \bar c_0(\tau)\,d\tau\right)$ $(t \in \mathbb{R},\ x \in \bar D)$ is an entire positive solution of (5.3.3). Consequently, $w(\bar a \cdot t) = |D|^{-1/2}\mathbf{1}$ for each $t \in \mathbb{R}$, where $\mathbf{1}$ means the function constantly equal to one (see a remark below Assumption (A5-N1)). This implies that (5.3.3) automatically satisfies (A5-N2).

Consider the space averaged equation (5.3.4). It is also straightforward to see that, for $\mathbb{P}$-a.e. $\omega \in \Omega$, the function $v(t, x) := \exp\left(\int_0^t \bar c_0(\theta_\tau\omega)\,d\tau\right)$ $(t \in \mathbb{R}, x \in \bar D)$ is an entire positive solution of (5.3.4). Consequently, $w(E_{\bar a}(\omega) \cdot t) = |D|^{-1/2}\mathbf{1}$ for each $t \in \mathbb{R}$ (see a remark below Assumption (A5-R1)). This also implies that (5.3.4) automatically satisfies (A5-R2).

Denote by $[\lambda_{\min}(\bar{a}), \lambda_{\max}(\bar{a})]$ and by $\lambda(\bar{a})$ the principal spectrum interval and principal Lyapunov exponent of (5.3.3) and (5.3.4), respectively. Then we have

THEOREM 5.3.1
Consider (5.3.1) and assume (A5-N4).

(1) $[\lambda_{\min}(\bar{a}), \lambda_{\max}(\bar{a})] = \{\lambda : \exists S_n < T_n \text{ with } T_n - S_n \to \infty \text{ such that } \lambda = \lim_{n \to \infty} \frac{1}{T_n - S_n} \int_{S_n}^{T_n} \bar{c}_0(t)\, dt \}.$

(2) $\lambda_{\min}(a) \geq \lambda_{\min}(\bar{a})$ *and* $\lambda_{\max}(a) \geq \lambda_{\max}(\bar{a}).$

(3) *Assume the smoothness assumption (A2-5). Then the equalities in (2) hold if and only if* $c_0(t, x) = \bar{c}_0(t)$ *for all* $x \in D$ *and* $\in \mathbb{R}.$

THEOREM 5.3.2
Consider (5.3.2) and assume (A5-R4).

(1) $\lambda(\bar{a}) = \int_{\Omega} \bar{c}_0(\omega)\, d\mathbb{P}(\omega).$

(2) $\lambda(a) \geq \lambda(\bar{a}).$

(3) *Assume the smoothness assumption (A2-5). Then the equality in (2) holds if and only if* $c_0(\omega, x) = \bar{c}_0(\omega)$ *for all* $x \in D$ *and* $\mathbb{P}$-*a.e.* $\omega \in \Omega.$

COROLLARY 5.3.1
Consider (5.3.1) and assume that $a_{ij}(t) = a_{ij}$ *and (A5-N1)–(A5-N4).*

(1) $\lambda_{\max}(a) \geq \lambda_{\text{princ}}(\hat{a}) \geq \lambda_{\min}(\bar{a})$ *for all* $\hat{a} \in \hat{Y}(a).$

(2) $\lambda_{\min}(a) \geq \lambda_{\text{princ}}(\hat{a})$ *for some* $\hat{a} \in \hat{Y}(a).$

(3) $\lambda_{\text{princ}}(\hat{a}) \geq \lambda_{\max}(\bar{a})$ *for some* $\hat{a} \in \hat{Y}(a).$

(4) *Assume the smoothness assumption (A2-5). If a is minimal and uniquely ergodic, then*

$$\lambda(a) \geq \lambda_{\text{princ}}(\hat{a}) \geq \lambda(\bar{a}) = \lim_{T \to \infty} \frac{1}{T} \int_0^T \left(\frac{1}{|D|} \int_D c_0(t, x)\, dx \right) dt.$$

COROLLARY 5.3.2
Consider (5.3.2) and assume that $a_{ij}(\omega) = a_{ij}$ *and (A5-R1)–(A5-R4) hold. Then*

$$\lambda(a) \geq \lambda_{\text{princ}}(\hat{a}) \geq \lambda(\bar{a}) = \int_{\Omega} \left(\frac{1}{|D|} \int_D c_0(\omega, x)\, dx \right) d\mathbb{P}(\omega).$$

We prove the above theorems and corollaries in the following order. First we prove Theorem 5.3.1(1), (2). Next we prove Theorem 5.3.2. Then we prove Theorem 5.3.1(3). Finally we prove Corollaries 5.3.1 and 5.3.2.

PROOF (Proof of Theorem 5.3.1(1))   Since

$$U_{\bar{a}}(T,S)w(\bar{a}\cdot S) = |D|^{-1/2}e^{\int_S^T \bar{c}_0(t)\,dt}$$

for any $S < T$, there holds

$$\frac{\ln\|U_{\bar{a}}(T,S)w(\bar{a}\cdot S)\|}{T-S} = \frac{\int_S^T \bar{c}_0(t)\,dt}{T-S}.$$

(1) then follows. ∎

PROOF (Proof of Theorem 5.3.1(2))   Note that $u(t,\cdot;w(a\cdot S),a\cdot S) = U_a(t,S)w(a\cdot S)$. By Lemma 5.2.5, $\frac{1}{u(t,\cdot;w(a\cdot S),a\cdot S)} \in W(S,T;V,V^*)$. Hence by Proposition 2.1.3 and Eq. (2.1.4) we have

$$-\frac{1}{T-S}\int_S^T \left(\frac{1}{|D|}\int_D u(t,x;w(a\cdot S),a\cdot S)\,\partial_t\left(\frac{1}{u(t,x;w(a\cdot S),a\cdot S)}\right)dx\right)dt$$

$$= \frac{1}{T-S}\int_S^T\left(\frac{1}{|D|}\int_D\sum_{i,j=1}^N a_{ij}(t)\frac{\partial u}{\partial x_i}\frac{\partial u}{\partial x_j}/u^2\,dx\right)dt$$

$$+\frac{1}{T-S}\int_S^T \bar{c}_0(t)\,dt. \tag{5.3.5}$$

This implies that

$$\frac{1}{T-S}\int_S^T \bar{c}_0(t)\,dt \le \frac{1}{|D|}\frac{1}{T-S}\int_S^T\int_D \frac{\partial_t u(t,x;w(a\cdot S),a\cdot S)}{u(t,x;w(a\cdot S),a\cdot S)}\,dx\,dt$$

$$= \frac{1}{|D|}\frac{1}{T-S}\int_D\int_S^T \frac{\partial_t u(t,x;w(a\cdot S),a\cdot S)}{u(t,x;w(a\cdot S),a\cdot S)}\,dt\,dx$$

$$= \frac{1}{|D|}\frac{1}{T-S}\int_D\left(\ln u(T,x;w(a\cdot S),a\cdot S) - \ln w(a\cdot S)(x)\right)dx$$

$$= \frac{\ln\|u(T,\cdot;w(a\cdot S),a\cdot S)\|}{T-S}$$

$$+\frac{1}{|D|}\frac{1}{T-S}\int_D\left(\ln w(a\cdot T)(x) - \ln w(a\cdot S)(x)\right)dx.$$

Note that

$$\lambda_{\min}(a) = \liminf_{T-S\to\infty}\frac{\ln\|u(T,\cdot;w(a\cdot S),a\cdot S)\|}{T-S}$$

and

$$\lambda_{\max}(a) = \limsup_{T-S\to\infty} \frac{\ln\|u(T,\cdot; w(a\cdot S), a\cdot S)\|}{T-S}.$$

Therefore,

$$\lambda_{\min}(a) \geq \lambda_{\min}(\bar{a}), \quad \lambda_{\max}(a) \geq \lambda_{\max}(\bar{a}).$$

☐

PROOF (Proof of Theorem 5.3.2)

(1) It follows from Proposition 4.1.1 and Lemma 1.2.6 that

$$\lambda(\bar{a}) = \lim_{T\to\infty} \frac{1}{T} \int_0^T \bar{c}_0(\theta_t\omega)\,dt = \int_\Omega \bar{c}_0(\cdot)\,d\mathbb{P}(\cdot)$$

for $\mathbb{P}$-a.e. $\omega \in \Omega$.

(2) By Proposition 4.1.1 again, for $\mathbb{P}$-a.e. $\omega \in \Omega$,

$$\lambda(a) = \lim_{T\to\infty} \frac{1}{T} \ln\|U_{E_a(\omega)}(T,0)w(E_a(\omega))\|$$

and

$$\lambda(\bar{a}) = \lim_{T\to\infty} \frac{1}{T} \int_0^T \bar{c}_0(\theta_t\omega)\,dt.$$

It then follows from Theorem 5.3.1(2) that

$$\lambda(a) \geq \lambda(\bar{a}).$$

(3) The "if" part is straightforward. Theorem 3.5.3 together with Part (1) imply that

$$\lambda(a) = \int_D \int_\Omega \sum_{i,j=1}^N a_{ij}(\omega)\frac{\partial w(\omega)}{\partial x_i}\frac{\partial w(\omega)}{\partial x_j}\,d\mathbb{P}(\omega)\,dx + \lambda(\bar{a}).$$

It then follows that $\lambda(a) = \lambda(\bar{a})$ if and only if

$$\int_D \int_\Omega \sum_{i,j=1}^N a_{ij}(\omega)\frac{\partial w(\omega)}{\partial x_i}\frac{\partial w(\omega)}{\partial x_j}\,d\mathbb{P}(\omega)\,dx = 0.$$

Then by the ellipticity we must have

$$\int_\Omega \sum_{i,j=1}^N a_{ij}(\omega)\frac{\partial w(\omega)}{\partial x_i}\frac{\partial w(\omega)}{\partial x_j}\,d\mathbb{P}(\omega) = 0 \quad \text{for all} \quad x \in D.$$

Since $w(\omega) \in C^1(\bar{D})$ we have that for $\mathbb{P}$-a.e. $\omega \in \Omega$ there holds

$$\frac{\partial w(\omega)}{\partial x_i} \equiv 0 \quad \text{on } \bar{D},$$

which implies that

$$w(\omega) = \text{const}$$

for $\mathbb{P}$-a.e. $\omega \in \Omega$. Lemma 3.5.4 gives that

$$c_0(\theta_t \omega) = \kappa(E_a(\omega) \cdot t)$$

for $\mathbb{P}$-a.e. $\omega \in \Omega$, $t \in \mathbb{R}$ and $x \in \bar{D}$. Consequently

$$c_0(\omega, x) = \bar{c}_0(\omega)$$

for all $x \in D$ and $\mathbb{P}$-a.e. $\omega \in \Omega$. $\square$

PROOF (Proof of Theorem 5.3.1(3))   Let $\mathbb{P}$ be the unique ergodic measure on $\hat{Y}(a)$. Then by Theorem 5.3.2(3), for $\mathbb{P}$-a.e. $\tilde{a} = (\tilde{a}_{ij}, 0, 0, \tilde{c}_0, 0) \in \hat{Y}(a)$, $\tilde{c}_0$ is independent of x. By the minimality of $(\tilde{Y}(a), \sigma)$, there is $(t_n)_{n=1}^{\infty} \subset \mathbb{R}$ such that $\tilde{c}_0 \cdot t_n \to c_0$ in the open-compact topology. This implies that c_0 is independent of x, and hence $c_0(t, x) = \bar{c}_0(t)$. $\square$

PROOF (Proof of Corollary 5.3.1)
(1) First, by Theorem 5.2.4(2),

$$\lambda_{\max}(a) \geq \lambda_{\text{princ}}(\hat{a}) \quad \text{for all } \hat{a} \in \hat{Y}(a).$$

Now, for any $\hat{a} \in \hat{Y}(a)$, there are $S_n, T_n \in \mathbb{R}$ with $S_n < T_n$ such that $T_n - S_n \to \infty$ and

$$\hat{c}_0(x) = \lim_{n \to \infty} \frac{1}{T_n - S_n} \int_{S_n}^{T_n} c_0(t, x) \, dt \quad \text{for a.e. } x \in D.$$

Without loss of generality we can assume that

$$\check{c}_0 := \lim_{n \to \infty} \frac{1}{T_n - S_n} \int_{S_n}^{T_n} \bar{c}_0(t) \, dt$$

exists. Note that

$$\frac{1}{|D|} \int_D \hat{c}_0(x) \, dx = \frac{1}{|D|} \int_D \lim_{n \to \infty} \frac{1}{T_n - S_n} \int_{S_n}^{T_n} c_0(t, x) \, dt \, dx$$

$$= \lim_{n \to \infty} \frac{1}{T_n - S_n} \int_{S_n}^{T_n} \bar{c}_0(t) \, dt$$

$$= \check{c}_0.$$

This together with Theorem 5.3.1(2) implies that

$$\lambda_{\text{princ}}(\hat{a}) \geq \lambda_{\text{princ}}(\check{a}) \geq \lambda_{\min}(\bar{a})$$

where $\check{a} = (a_{ij}, 0, 0, \check{c}_0, 0)$.

(2) This is Theorem 5.2.4(2).

(3) There are $S_n, T_n \in \mathbb{R}$ such that $T_n - S_n \to \infty$ and

$$\lambda_{\max}(\bar{a}) = \lim_{n \to \infty} \frac{1}{T_n - S_n} \int_{S_n}^{T_n} \bar{c}_0(t) \, dt.$$

Without loss of generality we can assume that

$$\hat{c}_0(x) = \lim_{n \to \infty} \frac{1}{T_n - S_n} \int_{S_n}^{T_n} c_0(t, x) \, dt$$

exists for a.e. $x \in D$. Then by arguments as in (1) we have

$$\lambda_{\mathrm{princ}}(\hat{a}) \geq \lambda_{\max}(\bar{a})$$

where $\hat{a} = (a_{ij}, 0, 0, \hat{c}_0, 0)$.

(4) When a is minimal and uniquely ergodic then, by Lemma 5.2.2,

$$\lambda_{\max}(a) = \lambda_{\min}(a) := \lambda(a),$$
$$\hat{Y}(a) = \text{singleton}$$

and

$$\lambda_{\max}(\bar{a}) = \lambda_{\min}(\bar{a}) := \lambda(\bar{a}) = \lim_{T \to \infty} \frac{1}{T} \int_0^T \frac{1}{|D|} \int_D c_0(t, x) \, dx \, dt.$$

Then by (1)–(3),

$$\lambda(a) \geq \lambda_{\mathrm{princ}}(\hat{a}) \geq \lambda(\bar{a}) = \lim_{T \to \infty} \frac{1}{T} \int_0^T \frac{1}{|D|} \int_D c_0(t, x) \, dx \, dt.$$

$\square$

PROOF (Proof of Corollary 5.3.2)   By Proposition 4.1.1 and Lemma 1.2.6, there is $\Omega_0 \subset \Omega$ with $\mathbb{P}(\Omega_0) = 1$ such that

$$\lambda(a) = \lim_{T \to \infty} \frac{\ln \|u(T, \cdot; w(\omega), w)\|}{T},$$

$$\hat{c}_0(x) = \lim_{T \to \infty} \frac{1}{T} \int_0^T c_0(\theta_t \omega, x) \, dt,$$

and

$$\lambda(\bar{a}) = \lim_{T \to \infty} \frac{1}{T} \int_0^T \bar{c}_0(\theta_t \omega) \, dt = \int_\Omega \bar{c}_0(\cdot) \, d\mathbb{P}(\cdot)$$

for $\omega \in \Omega_0$. It then follows from Corollary 5.3.1 that

$$\lambda(a) \geq \lambda_{\mathrm{princ}}(\hat{a}) \geq \lambda(\bar{a}) = \int_\Omega \frac{1}{|D|} \int_D c_0(\omega, x) \, dx \, d\mathbb{P}(\omega).$$

$\square$

5.4 Faber–Krahn Inequalities

In this section we consider the influence of the shape of domain on the principal spectrum and principal Lyapunov exponent for the following nonautonomous equation with Dirichlet boundary condition

$$
\begin{cases}
\dfrac{\partial u}{\partial t} = \displaystyle\sum_{i=1}^{N} \dfrac{\partial}{\partial x_i}\left(\sum_{j=1}^{N} a_{ij}(t,x)\dfrac{\partial u}{\partial x_j}\right), & x \in D, \\
u = 0, & x \in \partial D,
\end{cases}
\tag{5.4.1}
$$

and for the following random equation with Dirichlet boundary condition

$$
\begin{cases}
\dfrac{\partial u}{\partial t} = \displaystyle\sum_{i=1}^{N} \dfrac{\partial}{\partial x_i}\left(\sum_{j=1}^{N} a_{ij}(\theta_t\omega,x)\dfrac{\partial u}{\partial x_j}\right), & x \in D, \\
u = 0, & x \in \partial D.
\end{cases}
\tag{5.4.2}
$$

The standing assumption in the present section is that (A5-N1), (A5-N2) hold when we consider (5.4.1), and that (A5-R1), (A5-R2) hold when we consider (5.4.2).

Denote by $\Sigma(a) = [\lambda_{\min}(a), \lambda_{\max}(a)]$ the principal spectrum interval of (5.4.1), and by $\lambda(a)$ the principal Lyapunov exponent of (5.4.2), respectively. Let λ_{sym} be the principal eigenvalue of

$$
\begin{cases}
\Delta u = \lambda u, & x \in D_{\text{sym}} \\
u = 0, & x \in \partial D_{\text{sym}},
\end{cases}
\tag{5.4.3}
$$

where D_{sym} is the ball in $\mathbb{R}^N$ with center 0 which has the same volume as D. We extend the so called Faber–Krahn inequalities for elliptic and periodic parabolic problems to general time dependent and random ones.

THEOREM 5.4.1
Consider (5.4.1). *Let* $\delta \colon \mathbb{R} \to \mathbb{R}$ *be a bounded continuous function with* $\delta(t) > 0$ *for all* $t \in \mathbb{R}$ *such that*

$$
\sum_{i,j=1}^{N} a_{ij}(t,x)\xi_i\xi_j \geq \delta(t)\sum_{i=1}^{N}\xi_i^2
$$

for a.e. $(t,x) \in \mathbb{R} \times D$ *and any* $(\xi_1, \xi_2, \ldots, \xi_N)^\top \in \mathbb{R}^N$. *Then*

(1) $\lambda_{\max}(a) \leq \delta_1\lambda_{\text{sym}}$, *where* $\delta_1 := \limsup_{t-s\to\infty} \dfrac{1}{t-s}\displaystyle\int_s^t \delta(\tau)\,d\tau$.

(2) *Assume moreover that $a = (a_{ij})$ is uniquely ergodic and recurrent, δ is recurrent with at least the same recurrence as a, $a_{ij} \in C^\alpha(\mathbb{R} \times \bar{D})$ for some $0 < \alpha < 1$, $a_{ij}(t, x)$ is analytic in x, and that the boundary ∂D is analytic. Then $\lambda_{\min}(a) = \lambda_{\max}(a)$ and $\lambda_{\max}(a) = \delta_1 \lambda_{\text{sym}}$ if and only if $D = D_{\text{sym}}$ up to translation, $w_{\text{sym}}(a \cdot t) = w(a \cdot t) = \varphi_{\text{sym}}$ up to phase shift and*

$$\sum_{i,j=1}^{N} \frac{\partial}{\partial x_i} \left(a_{ij}(t, \cdot) \frac{\partial w(a \cdot t)}{\partial x_j} \right) = \delta(t) \, \Delta w(a \cdot t)$$

for $t \in \mathbb{R}$, where φ_{sym} is the positive principal eigenfunction of (5.4.3) with $\|\varphi_{\text{sym}}\| = 1$ and w_{sym} is the Schwarz symmetrization of w.

THEOREM 5.4.2

Consider (5.4.2). Let $\delta : \Omega \to \mathbb{R}$ be a $\mathbb{P}$-integrable function with $\delta(\omega) > 0$ for $\mathbb{P}$-a.e. $\omega \in \Omega$ such that

$$\sum_{i,j=1}^{N} a_{ij}(\omega, x) \xi_i \xi_j \geq \delta(\omega) \sum_{i=1}^{N} \xi_i^2$$

for $\mathbb{P}$-a.e. $\omega \in \Omega$, a.e. $x \in D$ and any $(\xi_1, \xi_2, \ldots, \xi_N)^\top \in \mathbb{R}^N$. Then

(1) *$\lambda \leq \delta_2 \lambda_{\text{sym}}$, where $\delta_2 := \int_\Omega \delta(\omega) \, d\mathbb{P}(\omega)$.*

(2) *Assume moreover that $a_{ij}^\omega(\cdot, \cdot) \in C^\alpha(\mathbb{R} \times \bar{D})$ for some $0 < \alpha < 1$, with the $C^\alpha(\mathbb{R} \times \bar{D})$-norms bounded uniformly in $\omega \in \Omega$ ($a_{ij}^\omega(t, x) = a_{ij}(\theta_t \omega, x)$), $a_{ij}(\omega, x)$ is analytic in x for any $\omega \in \Omega$, and the boundary ∂D is analytic. Then $\lambda = \delta_2 \lambda_{\text{sym}}$ if and only if $D = D_{\text{sym}}$ up to translation and there is $\Omega_0 \subset \Omega$ with $\mathbb{P}(\Omega_0) = 1$ such that $w_{\text{sym}}(\omega) = w(\omega) = \varphi_{\text{sym}}$ up to phase shift and*

$$\sum_{i,j=1}^{N} \frac{\partial}{\partial x_i} \left(a_{ij}(\omega, \cdot) \frac{\partial w(\omega)}{\partial x_j} \right) = \delta(\omega) \, \Delta w(\omega)$$

for all $\omega \in \Omega_0$, where φ_{sym} is the positive principal eigenfunction of (5.4.3) with $\|\varphi_{\text{sym}}\| = 1$ and w_{sym} is the Schwarz symmetrization of w.

The above theorems have been proved in [82] for the smooth case. We shall then only prove Theorem 5.4.1(1) and Theorem 5.4.2(1) for the general case.

PROOF (Proof of Theorem 5.4.1) (1) First of all, there are $S_n < T_n$ with $T_n - S_n \to \infty$ such that

$$\lambda_{\max}(a) = \lim_{n \to \infty} \frac{1}{T_n - S_n} \ln \|U_a(T_n, S_n) w(a \cdot S_n)\|.$$

Let $\eta_n(t;a) := \|U_a(t, S_n)w(a \cdot S_n)\|$ and $v(t, x; a) := w(a \cdot t)(x)$. Then $U_a(t, S_n)w(a \cdot S_n) = \eta_n(t;a)v(t, \cdot; a)$. By Lemma 3.2.7, we have

$$\eta_n(t;a) = \exp\left(-\int_{S_n}^t B_a(\tau, w(a \cdot \tau), w(a \cdot \tau)) \, d\tau\right) \qquad \text{for any } S_n < t.$$

For each $\epsilon > 0$ let $B_\epsilon(\cdot) \colon \mathbb{R} \to \mathbb{R}$ be a C^∞ function such that

$$B_\epsilon(t) \to B_a(t, w(a \cdot t), w(a \cdot t)) \quad \text{for a.e. } t \in \mathbb{R} \quad \text{as } \epsilon \to 0$$

and

$$B_\epsilon(t) \to B(t, w(a \cdot t), w(a \cdot t)) \quad \text{in } L_{1,\mathrm{loc}}(\mathbb{R}) \quad \text{as } \epsilon \to 0.$$

Put

$$\eta_{n,\epsilon}(t) := \exp\left(-\int_{S_n}^t B_\epsilon(\tau) \, d\tau\right).$$

Then it is not difficult to prove that

$$\eta_{n,\epsilon}(t) \to \eta_n(t;a) \quad \text{uniformly on compact subsets of } \mathbb{R} \quad \text{as } \epsilon \to 0$$

and

$$\eta'_{n,\epsilon}(t) \to \eta'_n(t;a) \quad \text{for a.e. } t \in \mathbb{R} \quad \text{as } \epsilon \to 0.$$

Consequently,

$$\int_{S_n}^{T_n} \left\langle v\eta_n(t;a), \frac{\partial}{\partial t}\frac{v}{\eta_{n,\epsilon}(t)} \right\rangle dt \to \int_{S_n}^{T_n} \left\langle v\eta_n(t;a), \frac{\partial}{\partial t}\frac{v}{\eta_n(t;a)} \right\rangle dt \quad \text{as } \epsilon \to 0.$$

Note that for any fixed $\epsilon > 0$, $v/\eta_{n,\epsilon} \in W(S_n, T_n; V, V^*)$. By Proposition 2.1.3, we have

$$-\int_{S_n}^{T_n} \left\langle v\eta_n(t;a), \frac{\partial}{\partial t}\frac{v}{\eta_{n,\epsilon}(t)} \right\rangle dt = -\int_{S_n}^{T_n} B_a(t, v\eta_n, v/\eta_{n,\epsilon}) \, dt$$

$$+ \eta_n(S_n)/\eta_{n,\epsilon}(S_n) - \eta_n(T_n)/\eta_{n,\epsilon}(T_n).$$

Letting $\epsilon \to 0$, we obtain

$$-\int_{S_n}^{T_n} \left\langle v\eta_n(t;a), \frac{\partial}{\partial t}\frac{v}{\eta_n(t;a)} \right\rangle dt = -\int_{S_n}^{T_n} B_a(t, v\eta_n, v/\eta_n) \, dt.$$

Therefore

$$-\int_{S_n}^{T_n} \left\langle v, \frac{\partial v}{\partial t} \right\rangle dt = -\int_{S_n}^{T_n} \left\langle v\eta_n(t;a), \frac{\partial}{\partial t}\frac{v}{\eta_n(t;a)} \right\rangle dt - \int_{S_n}^{T_n} \frac{\partial \eta_n}{\partial t}\frac{1}{\eta_n} \, dt$$

$$= -\int_{S_n}^{T_n} B_a(t, v\eta_n, v/\eta_n) \, dt - \ln\|U_a(T_n, S_n)w(a \cdot S_n)\|$$

$$= -\int_{S_n}^{T_n} \int_D \sum a_{ij}\partial_{x_j}v \, \partial_{x_i}v \, dx \, dt - \ln\|U_a(T_n, S_n)w(a \cdot S_n)\|$$

$$\leq -\int_{S_n}^{T_n} \int_D \delta(t) \sum \partial_{x_i}v \, \partial_{x_i}v \, dx \, dt - \ln\|U_a(T_n, S_n)w(a \cdot S_n)\|.$$

By Lemma 5.1.3,
$$\|\nabla v_{\text{sym}}(t, x; a)\|^2 \leq \|\nabla v(t, x; a)\|^2$$

for a.e. $t \in \mathbb{R}$, where $v_{\text{sym}}(t, x; a)$ is the Schwarz symmetrization of $v(t, x; a)$. Then by the variational characterization of the principal eigenvalue of the Laplace operator with Dirichlet boundary conditions we have

$$-\|\nabla v_{\text{sym}}(t, \cdot; a)\|^2 \leq \lambda_{\text{sym}} \|v_{\text{sym}}(t, \cdot; a)\|^2 = \lambda_{\text{sym}}$$

for a.e. $t \in \mathbb{R}$. It then follows that

$$-\frac{1}{T_n - S_n} \int_{S_n}^{T_n} \left\langle v, \frac{\partial v}{\partial t} \right\rangle dt$$
$$\leq \frac{\lambda_{\text{sym}}}{T_n - S_n} \int_{S_n}^{T_n} \delta(t) \, dt - \frac{\ln \|U_a(T_n, S_n)w(a \cdot S_n)\|}{T_n - S_n}.$$

This implies that

$$\lambda_{\max}(a) \leq \lambda_{\text{sym}} \limsup_{T-S \to \infty} \frac{1}{T-S} \int_S^T \delta(t) \, dt = \delta_1 \lambda_{\text{sym}}.$$

(2) See Theorem B in [82]. $\qquad\qquad\qquad\qquad\qquad\qquad\qquad$ □

PROOF (Proof of Theorem 5.4.2)   (1) By Proposition 4.1.1 and Lemma 1.2.6, there is $\omega_0 \in \Omega$ such that

$$\lambda(a) = \lim_{T \to \infty} \frac{1}{T} \ln \|U_{E_a(\omega_0)}(T, 0)w(E_a(\omega_0))\|$$

and

$$\delta_2 = \int_\Omega \delta(\omega) \, d\mathbb{P}(\omega) = \lim_{T \to \infty} \frac{1}{T} \int_0^T \delta(\theta_t \omega_0) \, dt.$$

It then follows from the arguments in Theorem 5.4.1(1) with $a = a^\omega$ that

$$\lambda \leq \delta_2 \lambda_{\text{sym}}.$$

(2) See Theorem A in [82]. $\qquad\qquad\qquad\qquad\qquad\qquad\qquad\qquad\qquad$ □

5.5   Historical Remarks

Spectral theory for linear parabolic equations is a basic tool for the study of nonlinear parabolic equations. The influence of spatial and temporal variations and the shape of the domain of linear parabolic equations on their principal spectrum and principal Lyapunov exponents is of great interest for a lot of applied problems and has been investigated in many papers for the smooth case (both the domain and the coefficients are sufficiently smooth). For example, in [62], the authors studied the influence of temporal variations of periodic and almost periodic smooth parabolic equations on the principal eigenvalue and principal spectrum point and proved some results similar to those in Theorem 5.2.2. The results of [62] for periodic and almost periodic parabolic equations were extended to general nonautonomous and random smooth parabolic equations in [81] and [84]. In [13], the authors studied the influence of spatial variations of some time independent smooth parabolic equations on their principal eigenvalues. In [49], the Faber–Krahn inequality for elliptic equations was extended to periodic parabolic equations. The Faber–Krahn inequality for elliptic and periodic parabolic equations was further extended to general time dependent and random parabolic equations in [82]. The results in this chapter extend all the above mentioned works to general time dependent and random (nonsmooth) parabolic equations.

Chapter 6

Cooperative Systems of Parabolic Equations

The purpose of this chapter is to extend the theories developed in the previous chapters for scalar parabolic equations to cooperative systems of nonautonomous and random parabolic equations. To do so, we first consider cooperative systems of parabolic equations in the general setting. We then study the principal spectrum and principal Lyapunov exponent for cooperative systems of nonautonomous and random parabolic equations.

To be more precise, let $D \subset \mathbb{R}^N$ be a bounded domain and $\mathbf{Y}$ be a bounded subset of $L_\infty(\mathbb{R} \times D, \mathbb{R}^{K(N^2+2N+K)}) \times L_\infty(\mathbb{R} \times \partial D, \mathbb{R}^K)$ satisfying (A1-4) and (A1-5). Recall that for $\mathbf{a} \in \mathbf{Y}$, we write it as $\mathbf{a} = (a_{ij}^k, a_i^k, b_i^k, c_l^k, d_0^k)$, where $i, j = 1, 2, \ldots, N$ and $k, l = 1, 2, \ldots, K$.

We first consider the following cooperative systems of parabolic equations on D,

$$
\frac{\partial u_k}{\partial t} = \sum_{i=1}^N \frac{\partial}{\partial x_i} \Big(\sum_{j=1}^N a_{ij}^k(t, x) \frac{\partial u_k}{\partial x_j} + a_i^k(t, x) u_k \Big)
$$

$$
+ \sum_{i=1}^N b_i^k(t, x) \frac{\partial u_k}{\partial x_i} + \sum_{l=1}^K c_l^k(t, x) u_l, \quad t > s, \ x \in D, \tag{6.0.1}
$$

complemented with the boundary conditions

$$
\mathcal{B}_{a^k}(t) u_k = 0 \qquad t > s, \ x \in \partial D \tag{6.0.2}
$$

for all $\mathbf{a} = (a_{ij}^k, a_i^k, b_i^k, c_l^k, d_0^k) \in \mathbf{Y}$, where $\mathcal{B}_{a^k}$ is as in (2.0.3) with a being replaced by $a^k = ((a_{ij}^k)_{i,j=1}^N, (a_i^k)_{i=1}^N, (b_i^k)_{i=1}^N, 0, d_0^k)$, $k = 1, 2, \ldots, K$. We also assume that $\mathbf{Y}$ satisfies (A1-6), i.e., $d_0^k = 0$ for all $\mathbf{a} = (a_{ij}^k, a_i^k, b_i^k, c_l^k, d_0^k) \in \mathbf{Y}$ in the Dirichlet or Neumann cases and $d_0^k \geq 0$ for all $\mathbf{a} = (a_{ij}^k, a_i^k, b_i^k, c_l^k, d_0^k) \in \mathbf{Y}$ in the Robin case.

For convenience, we use the notion of mild solutions of (6.0.1)+(6.0.2) in this chapter. We introduce the concept of a mild solution of (6.0.1)+(6.0.2) and investigate the basic properties of the mild solutions in Section 6.1, which extends the theories developed in Chapter 2 for weak solutions of scalar parabolic equations in the general setting to mild solutions of cooperative systems of parabolic equations in the general setting. We introduce the concept of principal spectrum and principal Lyapunov exponent and exponential separation of (6.0.1)+(6.0.2), investigate their basic properties and show the existence of exponential separation and existence and uniqueness of entire positive solutions in Section 6.2, which extends the theories established in Chapter 3 for scalar parabolic equation in general setting to cooperative systems of parabolic equations in general setting.

We then consider the following cooperative systems of nonautonomous parabolic equations

$$
\begin{cases}
\dfrac{\partial u_k}{\partial t} = \displaystyle\sum_{i=1}^{N} \dfrac{\partial}{\partial x_i}\left(\sum_{j=1}^{N} a_{ij}^k(t,x)\dfrac{\partial u_k}{\partial x_j} + a_i^k(t,x)u_k\right) \\
\quad + \displaystyle\sum_{i=1}^{N} b_i^k(t,x)\dfrac{\partial u_k}{\partial x_i} + \sum_{l=1}^{K} c_l^k(t,x)u_l, \qquad\qquad t>0,\ x\in D \\
\mathcal{B}_{a^k}(t)u_k = 0, \qquad\qquad\qquad\qquad\qquad\qquad t>0,\ x\in\partial D,
\end{cases}
\tag{6.0.3}
$$

where $\mathcal{B}_{a^k}$ is of the same form as in (6.0.2), $k = 1, 2, \ldots, K$ and $\mathbf{a} = (a_{ij}^k, a_i^k, b_i^k, c_l^k, d_0^k)$ is a given element of $\mathbf{Y}$, and consider the following cooperative systems of random parabolic equations

$$
\begin{cases}
\dfrac{\partial u_k}{\partial t} = \displaystyle\sum_{i=1}^{N} \dfrac{\partial}{\partial x_i}\left(\sum_{j=1}^{N} a_{ij}^k(\theta_t\omega,x)\dfrac{\partial u_k}{\partial x_j} + a_i^k(\theta_t\omega,x)u_k\right) \\
\quad + \displaystyle\sum_{i=1}^{N} b_i^k(\theta_t\omega,x)\dfrac{\partial u_k}{\partial x_i} + \sum_{l=1}^{K} c_l^k(\theta_t\omega,x)u_l, \qquad t>0,\ x\in D \\
\mathcal{B}_{a^{k,\omega}}(t)u_k = 0, \qquad\qquad\qquad\qquad\qquad\qquad t>0,\ x\in\partial D,
\end{cases}
$$
$$
\tag{6.0.4}
$$

where $k = 1, 2, \ldots, K$, $\omega \in \Omega$, $((\Omega, \mathfrak{F}, \mathbb{P}), \{\theta_t\}_{t\in\mathbb{R}})$ is an ergodic metric dynamical system, and for each $\omega \in \Omega$, $\mathbf{a}^\omega(t,x) = (a_{ij}^k(\theta_t\omega,x), a_i^k(\theta_t\omega,x), b_i^k(\theta_t\omega), c_l^k(\theta_t\omega,x), d_0^k(\theta_t\omega,x)) \in \mathbf{Y}$ and $\mathcal{B}_{a^{k,\omega}}(t)$ is of the same form as in (6.0.2) with a^k being replaced by $a^{k,\omega}(t,x) = (a_{ij}^k(\theta_t\omega,x), a_i^k(\theta_t\omega,x), b_i^k(\theta_t\omega,x), 0, d_0^k(\theta_t\omega,x))$. We extend the theories developed in Chapters 4 and 5 for nonautonomous and random parabolic equations to cooperative systems of nonautonomous and random parabolic equations in Section 6.3.

This chapter is ended with some remarks in Section 6.4.

6.1   Existence and Basic Properties of Mild Solutions in the General Setting

In this section, we extend the theories developed in Chapter 2 for scalar parabolic equations in the general setting to cooperative systems of parabolic equations in the general setting. We first consider the nonsmooth case (both the coefficients and the domain are not smooth) and then consider the smooth case (both the coefficients and the domain are sufficiently smooth).

6.1.1 The Nonsmooth Case

Consider (6.0.1)+(6.0.2). Recall that for a given $\mathbf{a} = (a_{ij}^k, a_i^k, b_i^k, c_l^k, d_0^k) \in \mathbf{Y}$, $a^k = (a_{ij}^k, a_i^k, b_i^k, 0, d_0^k)$ denotes an element in Y, which is a subset of $L_\infty(\mathbb{R} \times D, \mathbb{R}^{N^2+2N+1}) \times L_\infty(\mathbb{R} \times \partial D, \mathbb{R})$ satisfying (A1-1)–(A1-3) (see Section 1.3) (hence for a^k, $1 \le k \le K$ is fixed and $i, j = 1, 2, \ldots, N$), and $\mathbf{C_a} = (c_l^k)_{l,k=1,2,\ldots,K}$. For a fixed $1 \le k \le K$, $P^k \colon \mathbf{Y} \to Y$ is defined by $P^k(\mathbf{a}) := a^k$ and $Y^k := \{ P^k(\mathbf{a}) : \mathbf{a} \in \mathbf{Y} \}$. Then for each k ($1 \le k \le K$), there corresponds the following family of scalar parabolic equations

$$
\begin{cases}
\dfrac{\partial u_k}{\partial t} = \displaystyle\sum_{i=1}^N \dfrac{\partial}{\partial x_i} \Big(\sum_{j=1}^N a_{ij}^k(t, x) \dfrac{\partial u_k}{\partial x_j} + a_i^k(t, x) u_k \Big) \\
\qquad + \displaystyle\sum_{i=1}^N b_i^k(t, x) \dfrac{\partial u_k}{\partial x_i}, \qquad\qquad\qquad t > 0, \ x \in D, \\
\mathcal{B}_{a^k}(t) u_k = 0, \qquad\qquad\qquad\qquad\qquad\quad t > 0, \ x \in \partial D,
\end{cases}
\tag{6.1.1}
$$

where $a^k = (a_{ij}^k, a_i^k, b_i^k, 0, d_0^k) \in Y^k$.

We start by introducing the following standing assumptions.

(A6-1) (a_{ij}^k) *satisfies* (A2-1) *and* D *satisfies* (A2-2).

(A6-2) *For any sequence* $(\mathbf{a}^{(n)})_{n=1}^\infty \subset \mathbf{Y}$ *convergent to* $\mathbf{a} \in \mathbf{Y}$, *where* $\mathbf{a}^{(n)} = (a_{ij}^{k,(n)}, a_i^{k,(n)}, b_i^{k,(n)}, c_l^{k,(n)}, d_0^{k,(n)})$ *and* $\mathbf{a} = (a_{ij}^k, a_i^k, b_i^k, c_l^k, d_0^k)$, *one has* $a_{ij}^{k,(n)}(t, x)$, $a_i^{k,(n)}(t, x)$, $b_i^{k,(n)}(t, x)$, *and* $c_l^{k,(n)}(t, x)$ *converge to* $a_{ij}^k(t, x)$, $a_i^k(t, x)$, $b_i^k(t, x)$, *and* $c_l^k(t, x)$, *respectively, for a.e.* $(t, x) \in \mathbb{R} \times D$, *and* $d_0^{k,(n)}(t, x)$ *converges to* $d_0^k(t, x)$ *for a.e.* $(t, x) \in \mathbb{R} \times \partial D$ ($i, j = 1, 2, \ldots, N$, $k, l = 1, 2, \ldots, K$).

(A6-3) **(Cooperativity)** *For any* $\mathbf{a} = (a_{ij}^k, a_i^k, b_i^k, c_l^k, d_0^k) \in \mathbf{Y}$, $c_l^k(t, x) \ge 0$ *for* $(t, x) \in \mathbb{R} \times D$ *and* $k, l = 1, 2, \ldots, K$ *with* $k \ne l$.

(A6-4) **(Irreducibility)** *For any* $\mathbf{a} = (a_{ij}^k, a_i^k, b_i^k, c_l^k, d_0^k) \in \mathbf{Y}$, *any* $t_0 \in \mathbb{R}$ *and any partition* $I = \{i_1, i_2, \ldots, i_k\}$ *and* $J = \{j_1, j_2, \ldots, j_l\}$ *of* $\{1, 2, \ldots, K\}$ *(i.e.,* $I \ne \emptyset$, $J \ne \emptyset$, $I \cap J = \emptyset$, $I \cup J = \{1, 2, \ldots, K\}$) *there are* $i \in I$ *and* $j \in J$ *such that* $c_i^j(t, \cdot) \ge 0$ *and* $c_i^j(t, \cdot) \not\equiv 0$ *for* t *in any sufficiently small neighborhood of* t_0.

Throughout this section, we assume (A6-1). At some places, we will also assume (A6-2) and/or (A6-3) and/or (A6-4), which will be pointed out explicitly. For a given $\mathbf{u} = (u_1, u_2, \ldots, u_K) \in \mathbb{R}^K$, $\mathbf{u}_\pm := (u_{1\pm}, u_{2\pm}, \ldots, u_{K\pm})$, where $u_{l+} := u_l$ ($u_{l-} := 0$) if $u_l > 0$ and $u_{l+} := 0$ ($u_{l-} := -u_l$) if $u_l \le 0$, $l = 1, 2, \ldots, K$. Hence $\mathbf{u} = \mathbf{u}_+ - \mathbf{u}_-$. We write $|\mathbf{u}|$ for $(|u_1|, |u_2|, \ldots, |u_K|)$.

For $1 \leq p < \infty$ we write $\|\mathbf{u}\|_p$ for $(|u_1|^p + \cdots + |u_K|^p)^{1/p}$, and we write $\|\mathbf{u}\|_\infty$ for $\sup\{|u_l| : 1 \leq l \leq K\}$. For a given $\mathbf{u} \colon D \to \mathbb{R}^K$, $\mathbf{u} \in L_p(D, \mathbb{R}^K)$ if and only if $u_l \in L_p(D)$ for all $l = 1, 2, \ldots, K$.

Recall that if $1 \leq p < \infty$ we denote the norm in $\mathbf{L}_p(D)$ by

$$\|\mathbf{u}\|_p := \left(\sum_{k=1}^K \int_D |u_k(x)|^p \, dx \right)^{1/p}.$$

We denote the norm in $\mathbf{L}_\infty(D)$ by

$$\|\mathbf{u}\|_\infty := \max_{1 \leq k \leq K} \operatorname{ess\,sup}\{|u_k(x)| : x \in D\}.$$

First of all, by Propositions 2.1.5, 2.2.1, and 2.2.2, we have

LEMMA 6.1.1
For any k $(1 \leq k \leq K)$, any $u_k^0 \in L_2(D)$, and any $s \in \mathbb{R}$, (6.1.1) with initial condition $u_k(s) = u_k^0$ has a unique global weak solution $[\,[s, \infty) \ni t \mapsto U_{a^k}(t, s) u_k^0 \in L_2(D)\,]$. Moreover, for any $s < t$ the linear operator $U_{a^k}(t, s)$ can be extended to an operator in $\mathcal{L}(L_p(D), L_p(D))$ $(1 \leq p \leq \infty)$ and there are $M > 0$ and $\gamma > 0$ such that

$$\|U_{a^k}(t, 0)\|_{p,q} \leq M t^{-\frac{N}{2}\left(\frac{1}{p} - \frac{1}{q}\right)} e^{\gamma t} \tag{6.1.2}$$

for $a^k \in Y^k$, $1 \leq p \leq q \leq \infty$, $t > 0$, and $k = 1, 2, \ldots, K$. When $1 < p < \infty$, the mapping $[\,[0, \infty) \ni t \mapsto U_{a^k}(t, 0) \in \mathcal{L}_s(L_p(D))\,]$ is continuous.

By Proposition 2.2.4, we have

LEMMA 6.1.2
For any $t_1 < t_2$, there exists $\alpha \in (0, 1)$ such that for any $\mathbf{a} \in Y$, any $1 \leq k \leq K$, any $u_k^0 \in L_p(D)$, and any compact subset $D_0 \subset D$ the function $[t_1, t_2] \times D_0 \ni (t, x) \mapsto (U_{a^k}(t, 0) u_k^0)(x)\,]$ belongs to $C^{\alpha/2, \alpha}([t_1, t_2] \times D_0)$. Moreover, for fixed t_1, t_2 and D_0, the $C^{\alpha/2, \alpha}([t_1, t_2] \times D_0)$-norm of the above restriction is bounded above by a constant depending on $\|u_k^0\|_p$ only.

LEMMA 6.1.3
Denote $M_1 := \sup\{\|\mathbf{C_a}\|_{L_\infty(\mathbb{R} \times D, \mathbb{R}^{K^2})} : \mathbf{a} \in Y\}$.

(1) *For all $\mathbf{a} \in Y$ and a.e. $t \in \mathbb{R}$ there hold*

$$\mathbf{C_a}(t)(\cdot) := (c_l^k(t, \cdot))_{k,l=1,2,\ldots,K} \in L_\infty(D, \mathbb{R}^{K^2})$$

and

$$\|\mathbf{C_a}(t)(\cdot)\|_{L_\infty(D, \mathbb{R}^{K^2})} \leq M_1.$$

(2) *If* $1 \le p \le \infty$ *then* $\mathcal{C}_{\mathbf{a}}(t, \mathbf{u})(\cdot) := \mathbf{C}_a(t)(\cdot)\mathbf{u} \in L_p(D, \mathbb{R}^K)$ *for any* $\mathbf{a} \in \mathbf{Y}$, *a.e.* $t \in \mathbb{R}$, *and any* $\mathbf{u} \in L_p(D, \mathbb{R}^K)$. *Moreover*

$$\|\mathcal{C}_{\mathbf{a}}(t, \mathbf{u})(\cdot)\|_p \le KM_1\|\mathbf{u}\|_p.$$

(3) *For any* $-\infty \le t_1 < t_2 \le \infty$, *any* $1 \le p < \infty$, *and any* $1 \le q \le \infty$, *if* $\mathbf{u}(\cdot) \in L_q((t_1, t_2), L_p(D, \mathbb{R}^K))$ *then* $[t \mapsto \mathcal{C}_{\mathbf{a}}(t, \mathbf{u})] \in L_q((t_1, t_2), L_p(D, \mathbb{R}^K))$. *Further, the linear operator* $[\mathbf{u} \mapsto \mathcal{C}_{\mathbf{a}}(\cdot, \mathbf{u}))]$ *belongs to* $\mathcal{L}(L_q((t_1, t_2), L_p(D, \mathbb{R}^K)))$ *and has norm* $\le KM_1$.

Recall that necessary facts on measurability, etc., of functions taking values in a separable Banach space are collected in Chapter 1.

PROOF (Proof of Lemma 6.1.3) (1) For a.e. $t \in \mathbb{R}$ the function $\mathbf{C}_{\mathbf{a}}(t)(\cdot)$, as the section of a function in $L_\infty(\mathbb{R} \times D, \mathbb{R}^{K^2})$, is Lebesgue measurable. By the fact that $c_l^k(\cdot, \cdot) \in L_\infty(\mathbb{R} \times D)$, for any $\epsilon > 0$ there is a measurable set $E \subset \mathbb{R} \times D$ with $|E| = 0$ such that $|c_l^k(t, x)| \le \|c_l^k\|_\infty + \epsilon$ for $(t, x) \in (\mathbb{R} \times D) \setminus E$. Let $E(t) := \{ x \in D : (t, x) \in E \}$. Then by [42, Theorem 2.36], $|E| = \int_{\mathbb{R}} |E(t)| \, dt = 0$. Therefore $|E(t)| = 0$ for a.e. $t \in \mathbb{R}$. This implies that for a.e. $t \in \mathbb{R}$, $\mathbf{C}_{\mathbf{a}}(t)(\cdot) \in L_\infty(D, \mathbb{R}^{K^2})$ and $\|\mathbf{C}_{\mathbf{a}}(t)\|_{L_\infty(D, \mathbb{R}^{K^2})} \le \|\mathbf{C}_{\mathbf{a}}(\cdot, \cdot)\|_{L_\infty(\mathbb{R} \times D, \mathbb{R}^{K^2})}$.

(2) The first statement is a consequence of (1) and the fact that for any $\mathbf{v} \in L_\infty(D, \mathbb{R}^{K^2})$ and any $\mathbf{u} \in L_p(D, \mathbb{R}^K)$ their product $\mathbf{vu}$ belongs to $L_p(D, \mathbb{R}^K)$.

To prove the second statement for $1 < p < \infty$, let $\frac{1}{p} + \frac{1}{p^*} = 1$. Then we have

$$
\begin{aligned}
\|\mathcal{C}_{\mathbf{a}}(t, \mathbf{u})(\cdot)\|_p &= \left(\sum_{k=1}^K \int_D \left| \sum_{l=1}^K c_l^k(t, x) u_l(x) \right|^p dx \right)^{\frac{1}{p}} \\
&\le \left(\sum_{k=1}^K \int_D \left(\sum_{l=1}^K |c_l^k(t, x)|^{p^*} \right)^{\frac{p}{p^*}} \cdot \left(\sum_{l=1}^K |u_l(x)|^p \right) dx \right)^{\frac{1}{p}} \\
&\le KM_1 \left(\sum_{l=1}^K \int_D |u_l(x)|^p \, dx \right)^{\frac{1}{p}} \\
&= KM_1 \|\mathbf{u}\|_p.
\end{aligned}
$$

The second statement for $p = 1$ and $p = \infty$ can be proved in a similar way.

(3) Assume that $\mathbf{u}(\cdot) \in L_q((t_1, t_2), L_p(D, \mathbb{R}^K))$. There exists a sequence $(\mathbf{u}_n)_{n=1}^\infty$ of simple functions such that $\mathbf{u}_n(t)$ converge in $L_p(D, \mathbb{R}^K)$ to $\mathbf{u}(t)$ as $n \to \infty$, for a.e. $t \in (t_1, t_2)$.

For $n = 1, 2, 3, \ldots$, let $\mathbf{u}_n = \chi_{E_1} \mathbf{u}^1 + \cdots + \chi_{E_{m_n}} \mathbf{u}^{m_n}$ be the representation of the simple function $\mathbf{u}_n$. We prove first that the function $[t \mapsto \mathcal{C}_{\mathbf{a}}(t, \mathbf{u}_n)(\cdot) \in L_p(D, \mathbb{R}^K)]$ is measurable. By the Pettis theorem (see Definition 1.2.1(iii)), it suffices to prove that the function in question is weakly measurable. Further, it reduces to showing that its restrictions to E_i, $1 \leq i \leq m_n$, are weakly measurable. Let $\mathbf{v}^* \in L_{p^*}(D, \mathbb{R}^K)$, where $\frac{1}{p} + \frac{1}{p^*} = 1$. The function $[\, E_i \ni t \mapsto \int_D \mathcal{C}_\mathbf{a}(t)(x)\mathbf{u}_i(x)\mathbf{v}^*(x)\, dx\,]$ is measurable.

It is a consequence of Part (2) that for a.e. $t \in (t_1, t_2)$, $\mathcal{C}_\mathbf{a}(t)\mathbf{u}_n(t)$ converge, as $n \to \infty$, to $\mathcal{C}_\mathbf{a}(t)\mathbf{u}(t)$, in $L_p(D, \mathbb{R}^K)$. Therefore the function $[\, t \mapsto \mathcal{C}_\mathbf{a}(t)\mathbf{u}(t) \in L_p(D, \mathbb{R}^K)\,]$, as the a.e. pointwise limit of a sequence of measurable functions, is measurable. The fact that the mapping $[\, t \mapsto \|\mathcal{C}_\mathbf{a}(t)\mathbf{u}(t)\|_{L_p(D, \mathbb{R}^K)}\,]$ belongs to $L_q((t_1, t_2), \mathbb{R})$ follows again from (2). $\square$

Let $\mathbf{U}_\mathbf{a}^0(t, s)$ be defined by

$$\mathbf{U}_\mathbf{a}^0(t, s) := (U_{a^1}(t, s), U_{a^2}(t, s), \ldots, U_{a^K}(t, s)). \tag{6.1.3}$$

Observe that

$$\mathbf{U}_\mathbf{a}^0(t, s) = \mathbf{U}_{\mathbf{a} \cdot s}^0(t - s, 0) \quad \text{for any } t \geq s \tag{6.1.4}$$

and

$$\mathbf{U}_a^0(t + s, s) \circ \mathbf{U}_\mathbf{a}^0(s, 0) = \mathbf{U}_\mathbf{a}^0(t + s, 0) \quad \text{for all } s,\ t \geq 0 \tag{6.1.5}$$

(see Propositions 2.1.6 and 2.1.7).

Recall that

$$\mathbf{L}_p(D) = (L_p(D))^K = L_p(D, \mathbb{R}^K).$$

We denote the norm in $\mathbf{L}_2(D)$ by $\|\cdot\|$, and the norm in $\mathbf{L}_p(D)$ by $\|\cdot\|_p$. The symbol $\|\cdot\|_{p,q}$ stands for the norm in $\mathcal{L}(\mathbf{L}_p(D), \mathbf{L}_q(D))$.

LEMMA 6.1.4

For any $1 \leq p \leq q \leq \infty$, $\mathbf{a} \in \mathbf{Y}$ and $s < t$ there holds

$$\|\mathbf{U}_\mathbf{a}^0(t, s)\|_{p,q} \leq M(t - s)^{-\frac{N}{2}(\frac{1}{p} - \frac{1}{q})} e^{\gamma(t-s)} \tag{6.1.6}$$

where M and γ are as in Lemma 6.1.1.

PROOF   First, let $1 \leq p \leq q < \infty$. Then for any $\mathbf{a} \in \mathbf{Y}$, $s < t$, and $\mathbf{u}^0 \in \mathbf{L}_p(D)$, by Lemma 6.1.1, $\mathbf{U}_\mathbf{a}^0(t, s)\mathbf{u}^0 \in \mathbf{L}_q(D)$ and

$$\|\mathbf{U}_\mathbf{a}^0(t, s)\mathbf{u}^0\|_q^q = \sum_{k=1}^{K} \int_D |(U_{a^k}(t, s)u_k^0)(x)|^q\, dx$$

$$\leq \sum_{k=1}^{K} \left(M(t - s)^{-\frac{N}{2}(\frac{1}{p} - \frac{1}{q})} e^{\gamma(t-s)} \right)^q \left(\int_D |u_k^0(x)|^p\, dx \right)^{q/p}$$

$$= \left(M(t - s)^{-\frac{N}{2}(\frac{1}{p} - \frac{1}{q})} e^{\gamma(t-s)} \right)^q \left(\sum_{k=1}^{K} \left(\int_D |u_k^0(x)|^p\, dx \right)^{q/p} \right)$$

$$\leq \left(M(t - s)^{-\frac{N}{2}(\frac{1}{p} - \frac{1}{q})} e^{\gamma(t-s)} \right)^q \left(\sum_{k=1}^{K} \int_D |u_k^0(x)|^p\, dx \right)^{q/p},$$

which implies

$$\|\mathbf{U}_\mathbf{a}^0(t,s)\mathbf{u}^0\|_q \leq M(t-s)^{-\frac{N}{2}(\frac{1}{p}-\frac{1}{q})}e^{\gamma(t-s)}\|\mathbf{u}^0\|_p,$$

hence (6.1.6) holds.

Now, let $1 \leq p \leq q = \infty$. For any $\mathbf{a} \in \mathbf{Y}$, $s < t$, and $\mathbf{u}^0 \in \mathbf{L}_p(D)$, by Lemma 6.1.1 again, we have

$$\|\mathbf{U}_\mathbf{a}^0(t,s)\mathbf{u}^0\|_\infty = \max_{1 \leq k \leq K}\|U_{a^k}(t,s)u_k^0\|_\infty$$

$$\leq \max_{1 \leq k \leq K} M(t-s)^{-\frac{N}{2}(\frac{1}{p}-\frac{1}{q})}e^{\gamma(t-s)}\|u_k^0\|_p$$

$$\leq M(t-s)^{-\frac{N}{2}(\frac{1}{p}-\frac{1}{q})}e^{\gamma(t-s)}\|\mathbf{u}^0\|_p.$$

Hence (6.1.6) also holds. □

DEFINITION 6.1.1 (Mild solution) *For given $1 < p < \infty$ and $\mathbf{u}^0 \in \mathbf{L}_p(D)$, a continuous function $\mathbf{u}: [s,T) \to \mathbf{L}_p(D)$ is called a mild $(\mathbf{L}_p(D)\text{-})$solution of (6.0.1)+(6.0.2) on $[s,T)$ ($s < T \leq \infty$) with $\mathbf{u}(s) = \mathbf{u}^0$ if*

$$\mathbf{u}(t) = \mathbf{U}_\mathbf{a}^0(t,s)\mathbf{u}^0 + \int_s^t \mathbf{U}_\mathbf{a}^0(t,\tau)(\mathbf{C}_\mathbf{a}(\tau)\mathbf{u}(\tau))\,d\tau \quad \textit{for any} \quad s < t < T. \quad (6.1.7)$$

THEOREM 6.1.1 (Existence of mild solution)
For any given $1 < p < \infty$, $\mathbf{a} \in \mathbf{Y}$, $\mathbf{u}^0 \in \mathbf{L}_p(D)$ and $s \in \mathbb{R}$, there is a unique mild $\mathbf{L}_p(D)$-solution $\mathbf{u}(\cdot)$ of (6.0.1)+(6.0.2) on $[s,\infty)$ with $\mathbf{u}(s) = \mathbf{u}^0$.

PROOF As in Lemma 6.1.3, let $M_1 = \sup\{\,\|\mathbf{C}_\mathbf{a}\|_{L_\infty(\mathbb{R}\times D,\mathbb{R}^{K^2})} : \mathbf{a} \in \mathbf{Y}\,\}$.

From now until the end of the proof, we consider $1 < p < \infty$ and $\mathbf{a} \in \mathbf{Y}$ fixed, so we suppress the symbols p and $\mathbf{a}$ from the notation.

First of all, for given $h > 0$ and $r > 0$, let

$$\mathbf{M}_r(h,s) := \{\, \mathbf{u} \in C([s,s+h], \mathbf{L}_p(D)) :$$
$$\|\mathbf{u}(\tau)\|_p \leq r \text{ for all } \tau \in [s,s+h]\,\} \quad (6.1.8)$$

equipped with the topology given by the supremum norm. $\mathbf{M}_r(h,s)$ is a complete metric space.

For a given $\mathbf{u}^0 \in \mathbf{L}_p(D)$ we define an operator $\mathfrak{G}_{\mathbf{u}^0}$, acting from $\mathbf{M}_r(h,s)$ into the set of functions from $[s,s+h]$ into $\mathbf{L}_p(D)$, by the formula

$$\mathfrak{G}_{\mathbf{u}^0}(\mathbf{u})(t) := \mathbf{U}^0(t,s)\mathbf{u}^0 + \int_s^t \mathbf{U}^0(t,\tau)(\mathbf{C}(\tau)\mathbf{u}(\tau))\,d\tau, \quad t \in [s,s+h]. \quad (6.1.9)$$

The integral on the right-hand side of (6.1.9) is well defined. Indeed, for any $\mathbf{u} \in \mathbf{M}_r(h,s)$ and any $s < t < s+h$ there holds $\mathbf{u}|_{(s,t)} \in L_\infty((s,t), \mathbf{L}_p(D))$.

By Lemma 6.1.3, the function $[(s,t) \ni \tau \mapsto \mathbf{C}(\tau)\mathbf{u}(\tau)]$ belongs to $L_\infty((s,t),$ $\mathbf{L}_p(D))$. Proposition 2.2.6 states that the mapping $[(\tau, \mathbf{v}) \ni (s,t) \times \mathbf{L}_p(D) \mapsto \mathbf{U}^0(t,\tau)\mathbf{v} \in \mathbf{L}_p(D)]$ is continuous, which implies, via Lemma 1.2.2, that the mapping $[(s,t) \ni \tau \mapsto \mathbf{U}^0(t,\tau)(\mathbf{C}(\tau)\mathbf{u}(\tau)) \in \mathbf{L}_p(D)]$ is measurable. Proposition 2.2.2 (L_p–L_p estimates) yields that this function is in $L_\infty((s,t), \mathbf{L}_p(D))$.

We prove now that for any $\mathbf{u} \in \mathbf{M}_r(h,s)$ the function $[[s, s+h] \ni t \mapsto \mathfrak{G}_{\mathbf{u}^0}(\mathbf{u})(t) \in \mathbf{L}_p(D)]$ is continuous. Indeed, the mapping $[[s, s+h] \ni t \mapsto \mathbf{U}^0(t,s)\mathbf{u}^0 \in \mathbf{L}_p(D)]$ is continuous by Lemma 6.1.1. Denote $\mathbf{v}(\tau) := \mathbf{C}(\tau)\mathbf{u}(\tau)$, $\tau \in [s, s+h]$. For $t \in [s, s+h]$ we define a function $[[s, s+h] \ni \tau \mapsto \tilde{\mathbf{U}}^0(t,\tau) \in \mathcal{L}(\mathbf{L}_p(D))]$ by the formula

$$\tilde{\mathbf{U}}^0(t,\tau) := \begin{cases} \mathbf{U}^0(t,\tau) & \text{for } \tau \in [s,t] \\ \mathbf{0} & \text{for } \tau \in (t, s+h]. \end{cases}$$

Let $s \le t_1 < t_2 \le s + h$. There holds

$$\int_s^{t_2} \mathbf{U}^0(t_2, \tau)\mathbf{v}(\tau)\, d\tau - \int_s^{t_1} \mathbf{U}^0(t_1, \tau)\mathbf{v}(\tau)\, d\tau$$

$$= \int_s^{s+h} \left(\tilde{\mathbf{U}}^0(t_2, \tau) - \tilde{\mathbf{U}}^0(t_1, \tau) \right) \mathbf{v}(\tau)\, d\tau.$$

We consider separately two cases: first, t_1 is fixed and t_2 is let to approach t_1 from the right, second, t_2 is fixed and t_1 is let to approach t_2 from the left. In both cases, we deduce from Lemma 6.1.1 that the integrand converges for a.e. $t \in [s, s+h]$ to zero in the $\mathbf{L}_p(D)$-norm, which gives with the help of the Dominated Convergence Theorem that the integral converges to zero. Further, it follows from Lemma 6.1.1 that $\int_s^t \mathbf{U}^0(t, \tau)\mathbf{v}(\tau)\, d\tau$ converges to zero in $\mathbf{L}_p(D)$, as $t \searrow s$.

We claim that for any $\varrho > 0$, there are $h > 0$ and $r > 0$ such that for any $\mathbf{u}^0 \in \mathbf{L}_p(D)$ with $\|\mathbf{u}^0\|_p < \varrho$ there holds $\mathfrak{G}_{\mathbf{u}^0}: \mathbf{M}_r(h,s) \to \mathbf{M}_r(h,s)$. In fact, by Lemma 6.1.4, $\|\mathbf{U}^0(t,s)\|_p \le Me^{\gamma(t-s)}$ for all $t \ge s$. An application of Lemma 6.1.3(3) with $q = \infty$ gives that if $\mathbf{u}(\cdot) \in \mathbf{M}_r(h,s)$ then $\|\mathbf{C}(\tau)\mathbf{u}(\tau)\|_p \le rKM_1$ for a.e. $\tau \in (s, s+h)$. Consequently, we have

$$\|\mathfrak{G}_{\mathbf{u}^0}(\mathbf{u})(t)\|_p \le \varrho Me^{\gamma(t-s)} + rMM_1 \int_s^t e^{\gamma(t-\tau)}\, d\tau$$

$$= \varrho Me^{\gamma(t-s)} + rMM_1(e^{\gamma(t-s)} - 1)/\gamma \qquad (6.1.10)$$

for all $t \in [s, s+h]$. Therefore there are $h > 0$ (h is such that $MM_1(e^{\gamma h} - 1)/\gamma < 1/2$) and $r > 0$ ($r = \frac{Me^{\gamma h}\varrho}{2}$) such that

$$\|\mathfrak{G}_{\mathbf{u}^0}(\mathbf{u})(t)\|_p \le r \quad \text{for all } t \in [s, s+h]. \qquad (6.1.11)$$

We claim that $\mathfrak{G}_{\mathbf{u}^0}: \mathbf{M}_r(h,s) \to \mathbf{M}_r(h,s)$ is a contraction. In fact, for any given $\mathbf{u}, \mathbf{v} \in \mathbf{M}_r(h, s)$ we have

$$
\begin{aligned}
\|\mathfrak{G}_{\mathbf{u}^0}(\mathbf{u})(t) - \mathfrak{G}_{\mathbf{u}^0}(\mathbf{v})(t)\|_p &= \left\| \int_s^t \mathbf{U}^0(t, \tau)(\mathbf{C}(\tau)(\mathbf{u}(\tau) - \mathbf{v}(\tau))) \, d\tau \right\|_p \\
&\leq \int_s^t M e^{\gamma(t-\tau)} K M_1 \|\mathbf{u}(\tau) - \mathbf{v}(\tau)\|_p \, d\tau \\
&\leq M K M_1 \|\mathbf{u} - \mathbf{v}\|_{C([s,s+h], \mathbf{L}_p(D))} \int_s^t e^{\gamma(t-\tau)} \, d\tau \\
&\leq \frac{M K M_1 (e^{\gamma(t-s)} - 1)}{\gamma} \|\mathbf{u} - \mathbf{v}\|_{C([s,s+h], \mathbf{L}_p(D))}
\end{aligned}
$$

(6.1.12)

for all $t \in [s, s+h]$. By reducing h, if necessary, we can obtain that $\mathfrak{G}_{\mathbf{u}^0}$ is a contraction.

Let $\mathbf{u} \in \mathbf{M}_r(h, s)$ be the unique fixed point of $\mathfrak{G}_{\mathbf{u}^0}$, that is,

$$
\mathbf{u}(t) = \mathbf{U}^0(t, s)\mathbf{u}^0 + \int_s^t \mathbf{U}^0(t, \tau)(\mathbf{C}(\tau)\mathbf{u}(\tau)) \, d\tau, \qquad t \in [s, s+h].
$$

It follows from Lemma 6.1.1 that the $\mathbf{U}^0(t, s)\mathbf{u}^0$ converges to $\mathbf{u}^0$ in $\mathbf{L}_p(D)$, as $t \searrow s$. Again, by the estimate (6.1.2) and Lemma 6.1.3(2), the second term on the right-hand side converges to zero in $\mathbf{L}_p(D)$, as $t \searrow s$. Therefore, $\mathbf{u}(\cdot)$ is a (necessarily unique) mild $\mathbf{L}_p(D)$-solution (6.0.1)+(6.0.2) on $[s, s+h]$.

Moreover, by the arguments of [34, Theorem 3.8], this solution can be extended to $[s, \infty)$. The theorem is thus proved. □

Theorem 6.1.1 allows us, for any $\mathbf{a} \in \mathbf{Y}$, any $1 < p < \infty$, any $s \in \mathbb{R}$ and any $\mathbf{u}^0 \in \mathbf{L}_p(D)$, to denote by $[\,[s, \infty) \ni t \mapsto \mathbf{U}_{\mathbf{a},p}(t, s)\mathbf{u}^0 \in \mathbf{L}_p(D)\,]$ the unique mild $\mathbf{L}_p(D)$-solution of (6.0.1)+(6.0.2) with $\mathbf{u}(s) = \mathbf{u}^0$.

LEMMA 6.1.5
For any $\mathbf{a} \in \mathbf{Y}$ and any $1 < p < \infty$ we have

(1) $\mathbf{U}_{\mathbf{a},p}(t, s) = \mathbf{U}_{\mathbf{a}\cdot s, p}(t - s, 0)$ *for all $t \geq s$.*

(2) $\mathbf{U}_{\mathbf{a},p}(t + s, s) \circ \mathbf{U}_{\mathbf{a},p}(s, 0) = \mathbf{U}_{\mathbf{a},p}(t + s, 0)$ *for all $s, t \geq 0$.*

PROOF Fix $1 < p < \infty$ and $\mathbf{a} \in \mathbf{Y}$. We will write $\mathbf{U}_{\mathbf{a}}$ instead of $\mathbf{U}_{\mathbf{a},p}$.

(1) For $s \in \mathbb{R}$ and $\mathbf{u}^0 \in \mathbf{L}_p(D)$ we denote $\mathbf{u}_1(t) := \mathbf{U}_{\mathbf{a}\cdot s}(t - s, 0)\mathbf{u}^0$, $t \geq s$. By (6.1.7) the function $\mathbf{u}_1(\cdot)$ satisfies

$$
\mathbf{u}_1(t) = \mathbf{U}^0_{\mathbf{a}\cdot s}(t - s, 0)\mathbf{u}^0 + \int_0^{t-s} \mathbf{U}^0_{\mathbf{a}\cdot s}(t - s, \tau)(\mathbf{C}_{\mathbf{a}\cdot s}(\tau)\mathbf{u}_1(\tau + s)) \, d\tau
$$

for all $t \geq s$. It follows from (6.1.4) and (6.1.5) that $\mathbf{U}^0_{\mathbf{a}\cdot s}(t-s, \tau) = \mathbf{U}^0_{\mathbf{a}}(t, \tau+s)$ for any $\tau \in [0, t - s]$. Further, $\mathbf{C}_{\mathbf{a}\cdot s}(\tau)\mathbf{u}_1(\tau + s) = \mathbf{C}_{\mathbf{a}}(\tau + s)\mathbf{u}_1(\tau + s)$ for a.e.

$\tau \in [0, t - s]$. We have thus

$$\mathbf{u}_1(t) = \mathbf{U}_\mathbf{a}^0(t, s)\mathbf{u}^0 + \int_s^t \mathbf{U}_\mathbf{a}^0(t, r)(\mathbf{C}_\mathbf{a}(r)\mathbf{u}_1(r)) \, dr \quad \text{for all } t > s$$

$(r = \tau + s)$. By uniqueness, $\mathbf{u}_1(t) = \mathbf{U}_\mathbf{a}(t, s)\mathbf{u}^0$ for all $t \geq s$.

(2) Fix $\mathbf{u}^0 \in \mathbf{L}_p(D)$, and put $\mathbf{u}(t) := \mathbf{U}_\mathbf{a}(t, 0)\mathbf{u}^0$, $t \geq 0$. By (6.1.4) and (6.1.5), we have, for $s \geq 0$, $t \geq 0$,

$$\mathbf{U}_\mathbf{a}(t + s, 0)\mathbf{u}^0 = \mathbf{U}_\mathbf{a}^0(t + s, 0)\mathbf{u}^0 + \int_0^{t+s} \mathbf{U}_\mathbf{a}^0(t + s, \tau)(\mathbf{C}_\mathbf{a}(\tau)\mathbf{u}(\tau)) \, d\tau$$

$$= \mathbf{U}_\mathbf{a}^0(t + s, s)\left(\mathbf{U}_\mathbf{a}^0(s, 0)\mathbf{u}^0 + \int_0^s \mathbf{U}_\mathbf{a}^0(s, \tau)(\mathbf{C}_\mathbf{a}(\tau)\mathbf{u}(\tau)) \, d\tau\right)$$

$$+ \int_s^{t+s} \mathbf{U}_\mathbf{a}^0(t + s, \tau)(\mathbf{C}_\mathbf{a}(\tau)\mathbf{u}(\tau)) \, d\tau$$

$$= \mathbf{U}_\mathbf{a}^0(t + s, s)\mathbf{u}(s) + \int_s^{t+s} \mathbf{U}_\mathbf{a}^0(t + s, \tau)(\mathbf{C}_\mathbf{a}(\tau)\mathbf{u}(\tau)) \, d\tau$$

$$= \mathbf{U}_\mathbf{a}(t + s, s)(\mathbf{U}_\mathbf{a}(s, 0)\mathbf{u}^0).$$

$\square$

Since the set $\mathbf{Y}$ is translation invariant, in view of Lemma 6.1.5(1) we will formulate (when possible) results concerning properties of the solution operator for $s = 0$ only.

THEOREM 6.1.2 (L_p–L_q estimates)

(1) *There are* $\overline{M} > 0$ *and* $\bar{\gamma}$ *such that*

$$\|\mathbf{U}_{\mathbf{a},p}(t, 0)\|_{p,p} \leq \overline{M}e^{\bar{\gamma}t} \tag{6.1.13}$$

for all $1 < p < \infty$, *all* $\mathbf{a} \in \mathbf{Y}$, *and all* $t > 0$.

(2) *For any* $1 < p \leq q \leq \infty$ *and* $\mathbf{u}^0 \in \mathbf{L}_p(D)$, $\mathbf{U}_{\mathbf{a},p}(t, 0)\mathbf{u}^0 \in \mathbf{L}_q(D)$ *for all* $t > 0$. *Moreover, for any* $1 < p < \infty$ *and any* $0 < T_0 \leq T$ *there is* $\widetilde{M} = \widetilde{M}(p, T_0, T) > 0$ *such that*

$$\|\mathbf{U}_{\mathbf{a},p}(t, 0)\|_{p,q} \leq \widetilde{M}' \tag{6.1.14}$$

for all $q \in (p, \infty]$, *all* $\mathbf{a} \in \mathbf{Y}$, *and all* $t \in [T_0, T]$.

PROOF   (1) We write $\mathbf{U}_\mathbf{a}(t, 0)$ for $\mathbf{U}_{\mathbf{a},p}(t, 0)$. Also, we denote

$$M_1 := \sup\{\, \|\mathbf{C}_\mathbf{a}\|_{L_\infty(\mathbb{R} \times D, \mathbb{R}^{K^2})} : \mathbf{a} \in \mathbf{Y}\,\}.$$

Applying Lemma 6.1.4 with $q = p$ and Lemma 6.1.3 to (6.1.7) we have that

$$\|\mathbf{U_a}(t,0)\mathbf{u}^0\|_p \leq Me^{\gamma t}\|\mathbf{u}^0\|_p + KM_1M \int_0^t e^{\gamma(t-\tau)}\|\mathbf{U_a}(\tau,0)\mathbf{u}^0\|_p \, d\tau$$

for all $\mathbf{a} \in \mathbf{Y}$ and all $t > 0$. It then follows from the regular Gronwall inequality that there are $\overline{M} > 0$ and $\bar{\gamma}$ such that

$$\|\mathbf{U_a}(t,0)\mathbf{u}^0\|_p \leq \overline{M}e^{\bar{\gamma}t}\|\mathbf{u}^0\|_p \quad \text{for } t > 0 \text{ and } \mathbf{a} \in \mathbf{Y}.$$

Hence (6.1.13) holds.

(2) We first assume that $q < \infty$. By Lemmas 6.1.4, 6.1.3(2) and Part (1), we have that

$$\|\mathbf{U_a^0}(t,0)\mathbf{u}^0\|_q \leq Mt^{-\frac{N}{2}(\frac{1}{p}-\frac{1}{q})}e^{\gamma t}\|\mathbf{u}^0\|_p \quad \text{for } t > 0 \tag{6.1.15}$$

and

$$\|\mathbf{U_a^0}(t,\tau)(\mathbf{C_a}(\tau)\mathbf{U_a}(\tau,0)\mathbf{u}^0)\|_q \tag{6.1.16}$$
$$\leq MKM_1\overline{M}(t-\tau)^{-\frac{N}{2}(\frac{1}{p}-\frac{1}{q})}e^{\gamma(t-\tau)}e^{\bar{\gamma}\tau}\|\mathbf{u}^0\|_p \quad \text{for } 0 < \tau < t.$$

Fix for the moment $t > 0$. By Lemma 6.1.3(3), the function $[(0,t) \ni \tau \mapsto \mathbf{C_a}(\tau)\mathbf{U_a}(\tau,0)\mathbf{u}^0 \in \mathbf{L}_p(D)]$ is measurable. Proposition 2.2.6 states that the mapping $[(\tau,\mathbf{v}) \ni (0,t) \times \mathbf{L}_p(D) \mapsto \mathbf{U}^0(t,\tau)\mathbf{v} \in \mathbf{L}_q(D)]$ is continuous, consequently, by Lemma 1.2.2, the mapping

$$[(0,t) \ni \tau \mapsto \mathbf{U_a^0}(t,\tau)(\mathbf{C_a}(\tau)\mathbf{U_a}(\tau,0)\mathbf{u}^0) \in \mathbf{L}_q(D)]$$

is measurable. (6.1.16) implies that the above mapping is in $L_1((0,t),\mathbf{L}_q(D))$, provided that $1 < p \leq q < \infty$ and $\frac{N}{2}(\frac{1}{p} - \frac{1}{q}) < 1$. Consequently $\mathbf{U_a}(t,0)\mathbf{u}^0 \in \mathbf{L}_q(D)$ for $t > 0$ and for such p, q.

Fix $T > 0$. An application of (6.1.15) and (6.1.16) to (6.1.7) gives

$$\|\mathbf{U_a}(t,0)\mathbf{u}^0\|_q \leq Mt^{-\frac{N}{2}(\frac{1}{p}-\frac{1}{q})}e^{\gamma t}\|\mathbf{u}^0\|_p$$
$$+ MKM_1\overline{M} \int_0^t (t-\tau)^{-\frac{N}{2}(\frac{1}{p}-\frac{1}{q})}e^{\gamma(t-\tau)}e^{\bar{\gamma}\tau}\|\mathbf{u}^0\|_p \, d\tau$$
$$\leq \widetilde{M}_0 t^{-\frac{N}{2}(\frac{1}{p}-\frac{1}{q})}\|\mathbf{u}^0\|_p$$
$$\tag{6.1.17}$$

for all $0 < t \leq T$, provided that $1 < p \leq q < \infty$ and $\frac{N}{2}(\frac{1}{p} - \frac{1}{q}) < \frac{1}{2}$, where

$$\widetilde{M}_0 := Me^{\gamma T}(1 + 2KM_1\overline{M}e^{\bar{\gamma}T}T).$$

Observe that $\widetilde{M}_0$ is independent of p and q.

Now, for any given $p > 1$ there is $J \in \mathbb{N}$ (J is independent of $q \in (p,\infty)$) such that for any $q \in (p,\infty)$ there are $q_0 = p < q_1 < q_2 < q_{J-1} < q_J = q$ with the property that

$$\frac{N}{2}\left(\frac{1}{q_{j-1}} - \frac{1}{q_j}\right) < \frac{1}{2} \quad \text{for } j = 1, 2, \dots, J.$$

By (6.1.17), there holds

$$\|\mathbf{U_a}(t,0)\|_{q_{j-1},q_j} \le \widetilde{M}_0 t^{-\frac{N}{2}(\frac{1}{q_{j-1}} - \frac{1}{q_j})}, \quad 0 < t \le T, \tag{6.1.18}$$

where $j = 1, 2, \ldots, J$.

Fix $T_0 \in (0, T]$. Let $0 < \delta_1 < \delta_2 < \cdots < \delta_J < T_0$ and $\mathbf{u}^0 \in \mathbf{L}_p(D)$. Then

$$
\begin{aligned}
\|\mathbf{U_a}(t,0)\mathbf{u}^0\|_q &= \|\mathbf{U_a}(t,\delta_J)\mathbf{U_a}(\delta_J,0)\mathbf{u}^0\|_q \\
&\le \widetilde{M}_0 (t - \delta_J)^{-\frac{N}{2}(\frac{1}{q_{J-1}} - \frac{1}{q})} \|\mathbf{U_a}(\delta_J, 0)\mathbf{u}^0\|_{q_{J-1}} \\
&\le (\widetilde{M}_0)^2 (t - \delta_J)^{-\frac{N}{2}(\frac{1}{q_{J-1}} - \frac{1}{q})} \\
&\quad \cdot (\delta_J - \delta_{J-1})^{-\frac{N}{2}(\frac{1}{q_{J-2}} - \frac{1}{q_{J-1}})} \|\mathbf{U_a}(\delta_{J-1}, 0)\mathbf{u}^0\|_{q_{J-2}} \\
&\le \cdots \\
&\le \widetilde{M} \|\mathbf{u}^0\|_p
\end{aligned}
$$

for $T_0 < t \le T$, where

$$
\begin{aligned}
\widetilde{M} := (\widetilde{M}_0)^J (T_0 - \delta_J)^{-\frac{N}{2}(\frac{1}{q_{J-1}} - \frac{1}{q})} \\
\cdot (\delta_J - \delta_{J-1})^{-\frac{N}{2}(\frac{1}{q_{J-2}} - \frac{1}{q_{J-1}})} \cdots (\delta_2 - \delta_1)^{-\frac{N}{2}(\frac{1}{q_1} - \frac{1}{q_2})} \delta_1^{-\frac{N}{2}(\frac{1}{p} - \frac{1}{q_1})}.
\end{aligned}
$$

(6.1.14) then follows, with $\widetilde{M} > 0$ depending on p, T_0, and T only, but independent of $q \in (p, \infty)$.

It remains now to prove (2) for $q = \infty$. Suppose to the contrary that there are $\mathbf{a} \in \mathbf{Y}$, $\mathbf{u}^0 \in \mathbf{L}_p(D)$ $(1 < p < \infty)$, and $t > 0$ such that $\mathbf{U_a}(t,0)\mathbf{u}^0$ is not in $\mathbf{L}_\infty(D)$. This means that for any $n \in \mathbb{N}$ there is $E_n \subset D$ with $|E_n| > 0$ and k_n $(1 \le k_n \le K)$ such that $|(\mathbf{U_a}(t,0)\mathbf{u}^0)_{k_n}(x)| \ge n$ for all $x \in E_n$. Consequently,

$$\liminf_{q \to \infty} \|\mathbf{U_a}(t,0)\mathbf{u}^0\|_q \ge n \lim_{q \to \infty} |E_n|^{1/q} = n$$

for all $n \in \mathbb{N}$, which contradicts the already proven fact that $\{\|\mathbf{U_a}(t,0)\mathbf{u}^0\|_q : q \in (p, \infty)\}$ is bounded. The satisfaction of (6.1.14) for $q = \infty$ follows from the observation that $\|\mathbf{U_a}(t,0)\mathbf{u}^0\|_\infty = \lim_{q \to \infty}\|\mathbf{U_a}(t,0)\mathbf{u}^0\|_q$ and that (6.1.14) holds for all $q \in (p, \infty)$ with $\widetilde{M}$ independent of q. □

In view of the above theorem we can (and will) legitimately speak of a *mild solution* of (6.0.1)+(6.0.2). Accordingly, we write simply $\mathbf{U_a}(t,0)$ for the solution operator of (6.0.1)+(6.0.2).

THEOREM 6.1.3 (Compactness)

For any given $0 < t_1 \le t_2$, if $\mathbf{E}$ is a bounded subset of $\mathbf{L}_p(D)$ $(1 < p < \infty)$ then $\{\mathbf{U_a}(\tau, 0)\mathbf{u}^0 : \mathbf{a} \in \mathbf{Y}, \tau \in [t_1, t_2], \mathbf{u}^0 \in \mathbf{E}\}$ is relatively compact in $\mathbf{L}_p(D)$.

PROOF Let $(\tau_n)_{n=1}^{\infty} \subset [t_1, t_2]$, $(\mathbf{a}^{(n)})_{n=1}^{\infty} \subset \mathbf{Y}$, and $(\mathbf{u}^n)_{n=1}^{\infty} \subset \mathbf{E}$. Note that

$$\mathbf{U}_{\mathbf{a}^{(n)}}(\tau_n, 0)\mathbf{u}^n = \mathbf{U}_{\mathbf{a}^{(n)}}^0(\tau_n, 0)\mathbf{u}^n + \int_0^{\tau_n} \mathbf{U}_{\mathbf{a}^{(n)}}^0(\tau_n, s)(\mathbf{C}_{\mathbf{a}^{(n)}}(s)\mathbf{U}_{\mathbf{a}^{(n)}}(s, 0)\mathbf{u}^n)\, ds.$$

Since $\{\, \mathbf{U}_{\mathbf{a}^{(n)}}^0(\tau_n, 0)\mathbf{u}^n \,:\, n \in \mathbb{N} \,\}$ is relatively compact in $\mathbf{L}_p(D)$, we only need to prove that $\left\{\, \int_0^{\tau_n} \mathbf{U}_{\mathbf{a}^{(n)}}^0(\tau_n, s)(\mathbf{C}_{\mathbf{a}^{(n)}}(s)\mathbf{U}_{\mathbf{a}^{(n)}}(s, 0)\mathbf{u}^n)\, ds \,:\, n \in \mathbb{N} \right\}$ is relatively compact in $\mathbf{L}_p(D)$.

Put $\mathbf{v}^n(t) := \mathbf{C}_{\mathbf{a}^{(n)}}(t)\mathbf{U}_{\mathbf{a}^{(n)}}(t, 0)\mathbf{u}^n$. We claim that for any fixed $m \in \mathbb{N}$ the set

$$\widetilde{\mathbf{E}}_m := \left\{\, \int_0^{\tau_n - \frac{1}{m}} \mathbf{U}_{\mathbf{a}^{(n)}}^0(\tau_n, s)\mathbf{v}^n(s)\, ds \,:\, n \in \mathbb{N} \right\}$$

$$= \left\{\, \int_0^{\tau_n - \frac{1}{m}} \mathbf{U}_{\mathbf{a}^{(n)} \cdot s}^0(\tau_n - s, 0)\mathbf{v}^n(s)\, ds \,:\, n \in \mathbb{N} \right\}$$

is relatively compact in $\mathbf{L}_p(D)$. By Theorem 6.1.2 and Lemma 6.1.3(3), there is $M_0 > 0$ such that the $\mathbf{L}_\infty((0, t_2), \mathbf{L}_p(D))$-norms of $\mathbf{v}^n$ are $\leq M_0$ for any $n \in \mathbb{N}$. Denote by $\widehat{\mathbf{E}}_m$ the closure in $\mathbf{L}_p(D)$ of the set

$$\{\, \mathbf{U}_{\tilde{\mathbf{a}}}^0(s, 0)\tilde{\mathbf{u}} \,:\, \tilde{\mathbf{a}} \in \mathbf{Y},\ s \in [\tfrac{1}{m}, t_2],\ \|\tilde{\mathbf{u}}\|_p \leq M_0 \,\}.$$

The set $\widehat{\mathbf{E}}_m$ is compact. It is straightforward that for any simple function $\mathbf{w}$ whose $\mathbf{L}_\infty((0, t_2), \mathbf{L}_p(D))$-norm is $\leq M_0$, $\int_0^{t_n - \frac{1}{m}} \mathbf{U}_{\mathbf{a}^{(n)}}^0(\tau_n, s)\mathbf{w}(s)\, ds \in t_2 \cdot \overline{\mathrm{co}}\, \widehat{\mathbf{E}}_m$ for any $n \in \mathbb{N}$, where $\overline{\mathrm{co}}$ stands for the closed (in $\mathbf{L}_p(D)$) convex hull of a set. As, by Lemma 1.2.3, any $\mathbf{v}^n$ can be approximated by such simple functions, we see that $\widetilde{\mathbf{E}}_m \subset t_2 \cdot \overline{\mathrm{co}}\, \widehat{\mathbf{E}}_m$.

Note that $t_2 \cdot \overline{\mathrm{co}}\, \widehat{\mathbf{E}}_m$ is compact, for any $m \in \mathbb{N}$. Consequently, by a diagonal process we can assume without loss of generality that, for each $m = 1, 2, \dots$, the integrals $\int_0^{\tau_n - \frac{1}{m}} \mathbf{U}_{\mathbf{a}^{(n)}}(\tau_n, s)\mathbf{v}^n(s)\, ds$ converge, as $n \to \infty$, in $\mathbf{L}_p(D)$. For any $\epsilon > 0$ there is $m_0 \in \mathbb{N}$ such that

$$\left\| \int_{\tau_n - \frac{1}{m_0}}^{\tau_n} \mathbf{U}_{\mathbf{a}^{(n)}}^0(\tau_n, s)\mathbf{v}^n(s)\, ds \right\|_p < \epsilon$$

for all $n \in \mathbb{N}$. Take $n_0 \in \mathbb{N}$ so large that

$$\left\| \int_0^{\tau_{n_1} - \frac{1}{m_0}} \mathbf{U}_{\mathbf{a}^{(n_1)}}^0(\tau_{n_1}, s)\mathbf{v}^{n_1}(s)\, ds - \int_0^{\tau_{n_2} - \frac{1}{m_0}} \mathbf{U}_{\mathbf{a}^{(n_2)}}^0(\tau_{n_2}, s)\mathbf{v}^{n_2}(s)\, ds \right\|_p < \epsilon$$

for any $n_1, n_2 \geq n_0$. Therefore

$$\left\| \int_0^{\tau_{n_1}} \mathbf{U}_{\mathbf{a}^{(n_1)}}^0(\tau_{n_1}, s)\mathbf{v}^{n_1}(s)\, ds - \int_0^{\tau_{n_2}} \mathbf{U}_{\mathbf{a}^{(n_2)}}^0(\tau_{n_2}, s)\mathbf{v}^{n_2}(s)\, ds \right\|_p < 3\epsilon$$

for any $n_1, n_2 \geq n_0$. Hence $(\int_0^{T_n} \mathbf{U}_{\mathbf{a}^{(n)}}^0(T_n, s)(\mathbf{C}_{\mathbf{a}^{(n)}}(s)\mathbf{U}_{\mathbf{a}^{(n)}}(s, 0)\mathbf{u}^n) \, ds)_{n=1}^{\infty}$ is a Cauchy sequence in $\mathbf{L}_p(D)$. The theorem then follows. $\qquad\square$

THEOREM 6.1.4 (Joint continuity)
Assume (A6-1) *and* (A6-2). *For any sequence* $\{\mathbf{a}^{(n)}\}_{n=1}^{\infty} \subset \mathbf{Y}$, *any sequence* $(t_n)_{n=1}^{\infty}$, *and any sequence* $(\mathbf{u}^n)_{n=1}^{\infty} \subset \mathbf{L}_p(D)$ ($2 \leq p < \infty$), *if* $\mathbf{a}^{(n)} \to \mathbf{a}$, $t_n \to t$, *where* $t > 0$, *and* $\mathbf{u}^n \to \mathbf{u}^0$ *in* $\mathbf{L}_p(D)$ *as* $n \to \infty$, *then* $\mathbf{U}_{\mathbf{a}^{(n)}}(t_n, 0)\mathbf{u}^n$ *converges in* $\mathbf{L}_p(D)$ *to* $\mathbf{U}_{\mathbf{a}}(t, 0)\mathbf{u}^0$.

PROOF   Take $T > 0$ such that $t_n, t \leq T$ ($n = 1, 2, \dots$). For a given $\mathbf{u}(\cdot) \in C((0, T], \mathbf{L}_p(D))$ define

$$\mathfrak{G}_n(\mathbf{u})(t) := \mathbf{U}_{\mathbf{a}^{(n)}}^0(t, 0)\mathbf{u}^n + \int_0^t \mathbf{U}_{\mathbf{a}^{(n)}}^0(t, \tau)(\mathbf{C}_{\mathbf{a}^{(n)}}(\tau)\mathbf{u}(\tau)) \, d\tau, \quad t \in (0, T],$$

and

$$\mathfrak{G}(\mathbf{u})(t) := \mathbf{U}_{\mathbf{a}}^0(t, 0)\mathbf{u} + \int_0^t \mathbf{U}_{\mathbf{a}}^0(t, \tau)(\mathbf{C}_{\mathbf{a}}(\tau)\mathbf{u}(\tau)) \, d\tau, \quad t \in (0, T].$$

By [34, Theorem 4.4], if $\sup \{ \|\mathfrak{G}_n(\mathbf{u})(t) - \mathfrak{G}(\mathbf{u})(t)\|_p : t \in [\tau, T] \} \to 0$ as $n \to \infty$ for any $0 < \tau \leq T$ and any $\mathbf{u}(\cdot) \in C((0, T], \mathbf{L}_p(D))$, then $\mathbf{U}_{\mathbf{a}^{(n)}}(t, 0)\mathbf{u}^n \to \mathbf{U}_{\mathbf{a}}(t, 0)\mathbf{u}^0$ as $n \to \infty$, uniformly for t in compact intervals of $(0, T]$. It therefore suffices to verify that $\mathfrak{G}_n(\mathbf{u})(t) \to \mathfrak{G}(\mathbf{u})(t)$ in $\mathbf{L}_p(D)$, uniformly for t in compact subsets of $(0, T]$.

By Proposition 2.2.13, we have that $\mathbf{U}_{\mathbf{a}^{(n)}}^0(t, 0)\mathbf{u}^n \to \mathbf{U}_{\mathbf{a}}^0(t, 0)\mathbf{u}^0$ as $n \to \infty$ in $\mathbf{L}_p(D)$, uniformly for t in compact subsets of $(0, T]$.

Note that for any $\mathbf{u} \in C((0, T], \mathbf{L}_p(D))$, by (A6-2) we have

$$((\mathbf{C}_{\mathbf{a}^{(n)}}(\tau) - \mathbf{C}_{\mathbf{a}}(\tau))\mathbf{u}(\tau))(x) \to 0 \quad \text{for a.e. } (\tau, x) \in (0, T) \times D.$$

Hence

$$\int_0^T \|(\mathbf{C}_{\mathbf{a}^{(n)}}(\tau) - \mathbf{C}_{\mathbf{a}}(\tau))\mathbf{u}(\tau)\|_p \, d\tau \to 0$$

and then (by the $\mathbf{L}_p$–$\mathbf{L}_p$ estimates in Lemma 6.1.4)

$$\left\| \int_0^t (\mathbf{U}_{\mathbf{a}^{(n)}}^0(t, \tau)((\mathbf{C}_{\mathbf{a}^{(n)}}(\tau) - \mathbf{C}_{\mathbf{a}}(\tau))\mathbf{u}(\tau)) \, d\tau \right\|_p \to 0, \tag{6.1.19}$$

uniformly for t in compact subsets of $(0, T]$.

For any $0 < \tau < t$ we have

$$\|(\mathbf{U}_{\mathbf{a}^{(n)}}^0(t, \tau) - \mathbf{U}_{\mathbf{a}}^0(t, \tau))(\mathbf{C}_{\mathbf{a}}(\tau)\mathbf{u}(\tau))\|_p \to 0$$

as $n \to \infty$. This, together $\mathbf{L}_p$–$\mathbf{L}_p$ estimates (Lemma 6.1.4), implies that

$$\int_0^t \|(\mathbf{U}_{\mathbf{a}^{(n)}}^0(t, \tau) - \mathbf{U}_{\mathbf{a}}^0(t, \tau))(\mathbf{C}_{\mathbf{a}}(\tau)\mathbf{u}(\tau))\|_p \, d\tau \to 0 \tag{6.1.20}$$

and then

$$\left\| \int_0^t (\mathbf{U}^0_{\mathbf{a}^{(n)}}(t,\tau) - \mathbf{U}^0_{\mathbf{a}}(t,\tau))(\mathbf{C}_{\mathbf{a}}(\tau)\mathbf{u}(\tau)) \, d\tau \right\|_p \to 0 \tag{6.1.21}$$

as $n \to \infty$, uniformly for t in compact subsets of $(0,T]$. By (6.1.19) and (6.1.21),

$$\int_0^t \mathbf{U}^0_{\mathbf{a}^{(n)}}(t,\tau)(\mathbf{C}_{\mathbf{a}^{(n)}}(\tau)\mathbf{u}(\tau)) \, d\tau - \int_0^t \mathbf{U}^0_{\mathbf{a}}(t,\tau)(\mathbf{C}_{\mathbf{a}}(\tau)\mathbf{u}(\tau)) \, d\tau$$

$$= \int_0^t \mathbf{U}^0_{\mathbf{a}^{(n)}}(t,\tau)((\mathbf{C}_{\mathbf{a}^{(n)}}(\tau) - \mathbf{C}_{\mathbf{a}}(\tau))\mathbf{u}(\tau)) \, d\tau$$

$$+ \int_0^t (\mathbf{U}^0_{\mathbf{a}^{(n)}}(t,\tau) - \mathbf{U}^0_{\mathbf{a}}(t,\tau))(\mathbf{C}_{\mathbf{a}}(\tau)\mathbf{u}(\tau)) \, d\tau$$

$$\to 0$$

in $\mathbf{L}_p(D)$ as $n \to \infty$, uniformly for t in compact subsets of $(0,T]$.

It then follows from [34, Theorem 4.4] that

$$\|\mathbf{U}_{\mathbf{a}^{(n)}}(t_n,0)\mathbf{u}^n - \mathbf{U}_{\mathbf{a}}(t,0)\mathbf{u}^0\|_p \le \|\mathbf{U}_{\mathbf{a}^{(n)}}(t_n,0)\mathbf{u}^n - \mathbf{U}_{\mathbf{a}}(t_n,0)\mathbf{u}^0\|_p$$

$$+ \|\mathbf{U}_{\mathbf{a}}(t_n,0)\mathbf{u}^0 - \mathbf{U}_{\mathbf{a}}(t,0)\mathbf{u}^0\|_p$$

$$\to 0$$

as $n \to \infty$. $\qquad\qquad\qquad\qquad\qquad\qquad\qquad\qquad\qquad\qquad\square$

For given $\mathbf{a} = (a^k_{ij}, a^k_i, b^k_i, c^k_l, d^k_0) \in \mathbf{Y}$ and $\mathbf{u}^0 \in \mathbf{L}_p(D)$ $(1 < p < \infty)$, we denote by $\mathbf{U}^D_{\mathbf{a}}(t,s)\mathbf{u}^0$ the mild solution of (6.0.1) with the Dirichlet boundary condition: $u_k(t,x) = 0$ for $x \in \partial D$ and $1 \le k \le K$, denote by $\mathbf{U}^R_{\mathbf{a}}(t,s)\mathbf{u}^0$ the mild solution of (6.0.1) with the Robin boundary condition: $\sum_{i=1}^N (\sum_{j=1}^N a^k_{ij}(t, x)\partial_{x_j} u_k + a^k_i(t,x)u_k)\nu_i + d^k_0(t,x)u_k = 0$ for $x \in \partial D$ and $1 \le k \le K$, and denote by $\mathbf{U}^N_{\mathbf{a}}(t,s)\mathbf{u}^0$ the mild solution of (6.0.1) with the Neumann boundary condition: $\sum_{i=1}^N (\sum_{j=1}^N a^k_{ij}(t,x)\partial_{x_j} u_k + a^k_i(t,x)u_k)\nu_i = 0$ for $x \in \partial D$ and $1 \le k \le K$.

THEOREM 6.1.5 (Monotonicity of mild solution)
Let $1 < p < \infty$.

(1) *Assume (A6-1) and (A6-3). If $\mathbf{u}^0 \in \mathbf{L}_p(D)^+ \setminus \{\mathbf{0}\}$ with $1 \le k_1 \le K$ such that $u^0_{k_1} > 0$ then $\mathbf{U}_{\mathbf{a}}(t,0)\mathbf{u}^0 > \mathbf{0}$ for all $\mathbf{a} \in \mathbf{Y}$ and $t > 0$, and $(\mathbf{U}_{\mathbf{a}}(t,0)\mathbf{u}^0)_{k_1}(x) > 0$ for all $\mathbf{a} \in \mathbf{Y}$, $t > 0$, and $x \in D$. It follows that if $\mathbf{u}^0 \in \mathbf{L}_p(D)^+$ then $\mathbf{U}_{\mathbf{a}}(t,0)\mathbf{u}^0 \in \mathbf{L}_p(D)^+$ for all $\mathbf{a} \in \mathbf{Y}$ and all $t > 0$.*

(2) *Assume (A6-1), (A6-3), and (A6-4). If $\mathbf{u}^0 \in \mathbf{L}_p(D)^+ \setminus \{\mathbf{0}\}$ then $(\mathbf{U}_{\mathbf{a}}(t,0)\mathbf{u}^0)_k(x) > 0$ for all $\mathbf{a} \in \mathbf{Y}$, $1 \le k \le K$, $x \in D$, and $t > 0$.*

(3) *Assume* (A6-1) *and* (A6-3). *Assume also that* $\mathbf{a}^{(1)}, \mathbf{a}^{(2)} \in \mathbf{Y}$ *satisfy* $a_{ij}^{k,(1)} = a_{ij}^{k,(2)}$, $a_i^{k,(1)} = a_i^{k,(2)}$, $b_i^{k,(1)} = b_i^{k,(2)}$ $(i, j = 1, 2, \ldots, N,\ k = 1, 2, \ldots, K)$, *and* $c_l^{k,(2)} \geq c_l^{k,(1)}$ *and* $d_0^{k,(2)} \leq d_0^{k,(1)}$ $(l, k = 1, 2, \ldots, K)$, *where the equalities and inequalities are considered a.e. on* $\mathbb{R} \times D$, *or on* $\mathbb{R} \times \partial D$. *Then* $\mathbf{U}_{\mathbf{a}^{(2)}}(t, 0)\mathbf{u}^0 \geq \mathbf{U}_{\mathbf{a}^{(1)}}(t, 0)\mathbf{u}^0$ *for any* $\mathbf{u}^0 \in \mathbf{L}_p(D)^+$ *and* $t \geq 0$.

(4) *Assume* (A6-1) *and* (A6-3). *Then*

$$\mathbf{U}_{\mathbf{a}}^{\mathrm{D}}(t, 0)\mathbf{u}^0 \leq \mathbf{U}_{\mathbf{a}}^{\mathrm{R}}(t, 0)\mathbf{u}^0 \leq \mathbf{U}_{\mathbf{a}}^{\mathrm{N}}(t, 0)\mathbf{u}^0$$

for any $\mathbf{a} \in \mathbf{Y}$, $\mathbf{u}^0 \in \mathbf{L}_p(D)^+$, *and* $t \geq 0$.

PROOF   (1) Recall that in the proof of Theorem 6.1.1, $\left(\mathbf{U}_{\mathbf{a}}(t, 0)\mathbf{u}^0\right)|_{[0,h]}$ is a fixed point of $\mathfrak{G}_{\mathbf{u}^0}$ in the space

$$\mathbf{M}_r(h, s) = \{\, \mathbf{u} \in C([0, h], \mathbf{L}_p(D)) : \|\mathbf{u}(\tau)\|_p \leq r \text{ for all } \tau \in [0, h] \,\},$$

for some $h > 0$ and $r > 0$, where $\mathfrak{G}_{\mathbf{u}^0}$ is defined by

$$\mathfrak{G}_{\mathbf{u}^0}(\mathbf{u})(t) = \mathbf{U}_{\mathbf{a}}^0(t, 0)\mathbf{u}^0 + \int_0^t \mathbf{U}_{\mathbf{a}}^0(t, \tau)(\mathbf{C}_{\mathbf{a}}(\tau)\mathbf{u}(\tau))\, d\tau, \quad t \in [0, h].$$

Assume first that $c_k^k(t, x) \geq 0$ for all $1 \leq k \leq K$ and a.e. $(t, x) \in \mathbb{R} \times D$. Then, for any $\mathbf{u}(\tau) \geq \mathbf{0}$, $\mathbf{C}_{\mathbf{a}}(\tau)\mathbf{u}(\tau) \geq \mathbf{0}$. By Proposition 2.2.9, $U_{a^k}(t, 0)u_k^0 \geq 0$ for all $1 \leq k \leq K$ and $(U_{a^{k_1}}(t, 0)u_{k_1}^0)(x) > 0$ for all $x \in D$, which implies $\mathbf{U}_{\mathbf{a}}^0(t, 0)\mathbf{u}^0 > \mathbf{0}$, for all $t > 0$. Then by Proposition 2.2.9(1), for $\mathbf{u}(\tau) \geq \mathbf{0}$ for all $\tau \in [0, t]$, $\mathbf{U}_{\mathbf{a}}^0(t, \tau)(\mathbf{C}_{\mathbf{a}}(\tau)\mathbf{u}(\tau)) \geq \mathbf{0}$ for all $\tau \in (0, h]$. This together with the arguments of Theorem 6.1.1 implies that the fixed point of $\mathfrak{G}_{\mathbf{u}^0}$ is nonnegative. Therefore $\left(\mathbf{U}_{\mathbf{a}}(t, 0)\mathbf{u}^0\right)_{k_1}(x) > 0$ $(x \in D)$ and $\mathbf{U}_{\mathbf{a}}(t, 0)\mathbf{u}^0 > \mathbf{0}$ for $t \in (0, h]$, consequently $\left(\mathbf{U}_{\mathbf{a}}(t, 0)\mathbf{u}^0\right)_{k_1}(x) > 0$ $(x \in D)$ and $\mathbf{U}_{\mathbf{a}}(t, 0)\mathbf{u}^0 > \mathbf{0}$ for all $t > 0$.

We proceed now to the general case. For $\mathbf{a} = (a_{ij}^k, a_i^k, b_i^k, c_i^k, d_0^k) \in \mathbf{Y}$ and $r_0 \in \mathbb{R}$ denote $\mathbf{a} + r_0 := (a_{ij}^k, a_i^k, b_i^k, \tilde{c}_i^k, d_0^k)$, where $\tilde{c}_l^k = c_l^k$ for $k \neq l$ and $\tilde{c}_k^k := c_k^k + r_0$. Further, for $1 \leq k \leq K$ put $(\mathbf{a} + r_0)^k := (a_{ij}^k, a_i^k, b_i^k, r_0, d_0^k)$, and denote

$$\mathbf{U}_{\mathbf{a}+r_0}^0 := (U_{(a+r_0)^1}, \ldots, U_{(a+r_0)^K}).$$

By Lemma 4.3.1(ii), $\mathbf{U}_{\mathbf{a}+r_0}^0(t, 0) = e^{r_0 t}\mathbf{U}_{\mathbf{a}}^0(t, 0)$. Notice that $\mathbf{U}_{\mathbf{a}+r_0}(t, 0)\mathbf{u}^0$ satisfies

$$\mathbf{U}_{\mathbf{a}+r_0}(t, 0)\mathbf{u}^0 = \mathbf{U}_{\mathbf{a}+r_0}^0(t, 0)\mathbf{u}^0 + \int_0^t \mathbf{U}_{\mathbf{a}+r_0}^0(t, \tau)(\mathbf{C}_{\mathbf{a}}(\tau)\mathbf{U}_{\mathbf{a}+r_0}(\tau, 0)\mathbf{u}^0)\, d\tau.$$

Therefore $\mathbf{U}_{\mathbf{a}+r_0}(t, 0) = e^{r_0 t}\mathbf{U}_{\mathbf{a}}(t, 0)$ for all $t \geq 0$. It suffices now to take $r_0 > 0$ so large that $c_k^k + r_0 \geq 0$ a.e. on $\mathbb{R} \times D$, and apply the reasoning from the above paragraph.

(2) Without loss of generality we may assume that $c_k^k(t, x) \geq 0$ for all $1 \leq k \leq K$ and a.e. $(t, x) \in \mathbb{R} \times D$. For otherwise, we may replace $\mathbf{a}$ with $\mathbf{a} + r_0$ for some sufficiently large number r_0, as in the above arguments. Let $\mathbf{u}^0 \in \mathbf{L}_p(D) \setminus \{\mathbf{0}\}$, and let $1 \leq k_1 \leq K$ be such that $u_{k_1}^0 > 0$. By Part (1), $(\mathbf{U_a}(t, 0)\mathbf{u}^0)_{k_1}(x) > 0$ for all $t > 0$ and $x \in D$.

Now, for any $t_2 > 0$, by (A6-4), there is $1 \leq k_2 \leq K$ with $k_1 \neq k_2$ such that $c_{k_1}^{k_2}(t, \cdot) \geq 0$ and $c_{k_1}^{k_2}(t, \cdot) \not\equiv 0$ for t in any sufficiently small neighborhood of t_2. Then, again by Proposition 2.2.9(2), for given $t > t_2$,

$$\left(\mathbf{U_a^0}(t, \tau)(\mathbf{C_a}(\tau)\mathbf{u}(\tau))\right)_{k_2}(x) > 0$$

for $\tau(< t)$ in a sufficiently small neighborhood of t_2 and each $x \in D$. This implies that

$$\left(\mathbf{U_a}(t, 0)\mathbf{u}^0\right)_{k_2}(x) > 0 \quad \text{for } t > t_2, \ x \in D.$$

For any $t_3 > t_2$, by (A6-4) again, there is $1 \leq k_3 \leq K$ with $k_3 \neq k_1, k_2$ such that $c_{k_1}^{k_3}(t, \cdot) \geq 0$ and $c_{k_1}^{k_3}(t, \cdot) \not\equiv 0$ for t in any sufficiently small neighborhood of t_3 or $c_{k_2}^{k_3}(t, \cdot) \geq 0$ and $c_{k_2}^{k_3}(t, \cdot) \not\equiv 0$ for t in a sufficiently small neighborhood of t_3. This implies that $(\mathbf{U_a}(t, 0)\mathbf{u}^0)_{i_3}(x) > 0$ for $t > t_3$ and $x \in D$.

By induction, there are $t_K > t_{K-1} > \cdots > t_2 > 0 = t_1$ and $k_K, k_{K-1}, \ldots, k_1$ with $k_m \neq k_n$ for $m \neq n$ such that $(\mathbf{U_a}(t, 0)\mathbf{u}^0)_{k_m}(x) > 0$ for $t > t_m$ and $x \in D$. Then by the arbitrariness of $t_2, t_3, \ldots, t_K$, we have $(\mathbf{U_a}(t, 0)\mathbf{u}^0)_k(x) > 0$ for any $t > 0$, $x \in D$ and $1 \leq k \leq K$.

(3) Again, without loss of generality, we assume $c_k^k(t, x) \geq 0$ for all $1 \leq k \leq K$ and a.e. $(t, x) \in \mathbb{R} \times D$. Let

$$\mathfrak{S}_{\mathbf{u}_0}^{(m)}(\mathbf{u}) = \mathbf{U}_{\mathbf{a}^{(m)}}^0(t, 0)\mathbf{u}^0 + \int_0^t \mathbf{U}_{\mathbf{a}^{(m)}}^0(t, \tau)(\mathbf{C}_{\mathbf{a}^{(m)}}(\tau)\mathbf{u}(\tau)) \, d\tau$$

for $m = 1, 2$, $\mathbf{u} \in \mathbf{M}_r(h, s)$, where $\mathbf{M}_r(h, s)$ is as in (6.1.8). By Proposition 2.2.10,

$$\mathbf{U}_{\mathbf{a}^{(2)}}^0(t, 0)\mathbf{u}^0 \geq \mathbf{U}_{\mathbf{a}^{(1)}}^0(t, 0)\mathbf{u}^0$$

for any $\mathbf{u}^0 \geq \mathbf{0}$ and $t \geq 0$. Hence for any $\mathbf{u}(\cdot) \in \mathbf{M}_r(h, s)$ with $\mathbf{u}(t) \geq \mathbf{0}$ for all $\tau \in [0, h]$,

$$\mathfrak{S}_{\mathbf{u}^0}^{(2)}(\mathbf{u}) \geq \mathfrak{S}_{\mathbf{u}^0}^{(1)}(\mathbf{u}).$$

This together with the arguments in (1) implies that

$$\mathbf{U}_{\mathbf{a}^{(2)}}(t, 0)\mathbf{u}^0 \geq \mathbf{U}_{\mathbf{a}^{(1)}}(t, 0)\mathbf{u}^0$$

for any $\mathbf{u}^0 \geq \mathbf{0}$ and $t \geq 0$.

(4) It can be proved by arguments similar to those in (3). □

By the above theorems, under the assumptions (A6-1), (A6-2), (6.0.1)+ (6.0.2) generate a topological linear skew-product semiflow $\mathbf{\Pi} = \{\mathbf{\Pi}_t\}_{t \geq 0}$

$$\mathbf{\Pi}_t \colon \mathbf{L}_2(D) \times \mathbf{Y} \to \mathbf{L}_2(D) \times \mathbf{Y}, \tag{6.1.22}$$

where

$$\mathbf{\Pi}_t(\mathbf{u}^0, \mathbf{a}) := (\mathbf{U_a}(t,0)\mathbf{u}^0, \sigma_t \mathbf{a}). \tag{6.1.23}$$

Under additional assumption (A6-3) it follows from Theorem 6.1.5(1) that the solution operator $\mathbf{U_a}(t,0)$ ($\mathbf{a} \in \mathbf{Y}$, $t > 0$) has the property that, for any $\mathbf{u}^1, \mathbf{u}^2 \in \mathbf{L}_2(D)$ with $\mathbf{u}^1 < \mathbf{u}^2$ there holds $\mathbf{U_a}(t,0)\mathbf{u}^1 < \mathbf{U_a}(t,0)\mathbf{u}^2$. Adjusting the terminology used for semiflows on ordered metric spaces (see, e.g., [57]) to skew-product semiflows with ordered fibers we can say that then the (topological) linear skew-product semiflow $\mathbf{\Pi}$ is *strictly monotone*.

In view of Theorem 6.1.5(2), we say that a (global) mild solution $\mathbf{u}(\cdot) = (u_1(\cdot), \ldots, u_K(\cdot))$ of (6.0.1)+(6.0.2) satisfying (A6-1), (A6-3), and (A6-4) is a *positive mild solution* on $[s, \infty) \times D$ if $u_k(t)(x) > 0$ for all $t > s$, all $x \in D$ and all $k = 1, \ldots, K$.

6.1.2   The Smooth Case

In this subsection we show that if both $\mathbf{a}(t,x)$ and ∂D are sufficiently smooth, then $\mathbf{U_a}(t,0)\mathbf{u}^0$ is also smooth. To be more precise, we make the following standing assumption.

(A6-5) ∂D *is an* $(N-1)$*-dimensional manifold of class* $C^{3+\alpha}$*. There is* $M > 0$ *such that for any* $\mathbf{a} \in \mathbf{Y}$*, the* $C^{2+\alpha,2+\alpha}(\mathbb{R} \times \bar{D})$*-norms of* a_{ij}^k*,* a_i^k *$(i,j = 1,2,\ldots,N$, $k = 1,2,\ldots,K)$, the* $C^{2+\alpha,1+\alpha}(\mathbb{R} \times \bar{D})$*-norm of* b_i^k *and* c_l^k *$(i,j = 1,2,\ldots,N$, $l,k = 1,2,\ldots,K)$, and the* $C^{2+\alpha,2+\alpha}(\mathbb{R} \times \partial D)$*-norms of* d_0^k *$(k = 1,2,\ldots,K)$ are bounded by* M*.*

First of all, similar to Lemma 2.5.1, we have

LEMMA 6.1.6
Assume (A6-5). *Let* $a^{(n)} = (a_{ij}^{k,(n)}, a_i^{k,(n)}, b_i^{k,(n)}, c_l^{k,(n)}, d_0^{k,(n)}) \in \mathbf{Y}$ *and* $a = (a_{ij}^k, a_i^k, b_i^k, c_l^k, d_0^k) \in \mathbf{Y}$. *Then* $\lim_{n \to \infty} \mathbf{a}^{(n)} = \mathbf{a}$ *if and only if* $a_{ij}^{k,(n)}$ *converge to* a_{ij}^k*,* $a_i^{k,(n)}$ *converge to* a_i^k*,* $b_i^{k,(n)}$ *converge to* b_i^k*,* $c_l^{k,(n)}$ *converge to* c_l^k*, all uniformly on compact subsets of* $\mathbb{R} \times \bar{D}$*, and (in the Robin case)* $d_0^{k,(n)}$ *converge to* d_0^k *uniformly on compact subsets of* $\mathbb{R} \times \partial D$*.*

Throughout this section, we assume (A6-1) and (A6-5). Note that (A6-5) implies (A6-2).

THEOREM 6.1.6 (Regularity up to boundary)
Let $1 < p < \infty$ *and* $\mathbf{u}^0 \in \mathbf{L}_2(D)$*. Then for any* $\alpha \in (0, 1/2)$*, any* $\mathbf{a} \in \mathbf{Y}$*, and any* $0 < t_1 < t_2$ *the restriction* $[[t_1, t_2] \ni t \mapsto \mathbf{U_a}(t,0)\mathbf{u}^0]$ *belongs to* $C^1([t_1,t_2],(C^\alpha(\bar{D}))^K) \cap C([t_1,t_2],(C^{2+\alpha}(\bar{D}))^K)$*. Moreover, there is* $C = C(t_1, t_2, \mathbf{u}_0) > 0$ *such that*

$$\|[[t_1,t_2] \ni t \mapsto \mathbf{U_a}(t,0)\mathbf{u}_0]\|_{C^1([t_1,t_2],(C^\alpha(\bar{D}))^K)} \leq C$$

and

$$\|[\,[t_1, t_2] \ni t \mapsto \mathbf{U_a}(t,0)\mathbf{u_0}\,]\|_{C([t_1,t_2],(C^{2+\alpha}(\bar{D}))^K)} \leq C$$

for all $\mathbf{a} \in \mathbf{Y}$.

PROOF For given $1 < p < \infty$ and $\mathbf{u_0} \in \mathbf{L}_p(D)$, for any $t > 0$ and $1 < q < \infty$, one has $\mathbf{U_a}(t,0)\mathbf{u_0} \in \mathbf{L}_q(D)$ (Theorem 6.1.2). The result then follows from [3, Corollary 15.3]. $\quad\Box$

Given $1 < p < \infty$, let $V_p^1(a^k)$ be as in Section 2.5, i.e.,

$$V_p^1(a^k) := \{\, u_k \in W_p^2(D) : \mathcal{B}_{a^k}(0)u_k = 0 \,\}.$$

Let

$$\mathbf{V}_p^1(\mathbf{a}) := V_p^1(a^1) \times V_p^1(a^2) \times \cdots \times V_p^1(a^K).$$

For given $0 < \beta < 1$ and $1 < p < \infty$, let

$$V_p^\beta := \begin{cases} (L_p(D), W_p^2(D))_{\beta,p} & \text{if } 2\beta \notin \mathbb{N} \\[2mm] [L_p(D), W_p^2(D)]_\beta & \text{if } 2\beta \in \mathbb{N}, \end{cases}$$

$$\mathbf{V}_p^\beta := (V_p^\beta)^K,$$

and

$$V_p^\beta(a^k) := \begin{cases} (L_p(D), V_p^1(a^k))_{\beta,p} & \text{if } 2\beta \notin \mathbb{N} \\[2mm] [L_p(D), V_p^1(a^k)]_\beta & \text{if } 2\beta \in \mathbb{N}, \end{cases}$$

$$\mathbf{V}_p^\beta(\mathbf{a}) := V_p^\beta(a^1) \times V_p^\beta(a^2) \times \cdots \times V_p^\beta(a^K),$$

where $(\cdot, \cdot)_{\beta,p}$ is a real interpolation functor and $[\cdot, \cdot]_\beta$ is a complex interpolation functor (see [15], [105] for more detail). By Lemmas 2.5.2 and 2.5.3, we have

LEMMA 6.1.7

(1) $\mathbf{V}_p^\beta = (W_p^{2\beta}(D))^K$.

(2) If $2\beta - \frac{1}{p} \neq 0, 1$ then $\mathbf{V}_p^\beta(\mathbf{a})$ is a closed subspace of $\mathbf{V}_p^\beta$.

Also, we have the following compact embeddings:

$$W_p^{j+m}(D) \hookrightarrow C^{j+\lambda}(\bar{D}) \tag{6.1.24}$$

if $mp > N > (m-1)p$ and $0 < \lambda < m - (N/p)$, and

$$(W_p^2(D))^K \hookrightarrow \mathbf{V}_p^\beta \hookrightarrow \mathbf{L}_p(D), \tag{6.1.25}$$

$$\mathbf{V}_p^1(\mathbf{a}) \hookrightarrow \mathbf{V}_p^\beta(\mathbf{a}) \hookrightarrow \mathbf{L}_p(D) \tag{6.1.26}$$

for any $0 < \beta < 1$ and $\mathbf{a} \in \mathbf{Y}$.

By [3, Lemma 6.1 and Theorem 14.5] (see also [109]), we have

THEOREM 6.1.7

For any $1 < p < \infty$ and $\mathbf{u}^0 \in \mathbf{L}_p(D)$, $\mathbf{U_a}(t,0)\mathbf{u}^0 \in \mathbf{V}_p^1(\mathbf{a} \cdot t)$ for $t > 0$. Moreover, for any fixed $T > 0$ and $1 < p < \infty$ there is $C_p > 0$ such that

$$\|\mathbf{U_a}(t,0)\|_{\mathbf{L}_p(D),(W_p^2(D))^K} \le \frac{C_p}{t}$$

for all $\mathbf{a} \in \mathbf{Y}$ and $0 < t \le T$.

COROLLARY 6.1.1

For any $1 < p < \infty$ and any $\mathbf{u}^0 \in \mathbf{L}_2(D)$, $\mathbf{U_a}(t,0)\mathbf{u}^0 \in \mathbf{V}_p^1(\mathbf{a} \cdot t)$ for $t > 0$. Moreover, for any fixed $0 < \delta < T$ and $1 < p < \infty$ there is $C_{\delta,p} = C_{\delta,p}(T) > 0$ such that

$$\|\mathbf{U_a}(t,0)\|_{\mathbf{L}_2(D),(W_p^2(D))^K} \le C_{\delta,p}$$

for all $\mathbf{a} \in \mathbf{Y}$ and $\delta \le t \le T$.

PROOF   It follows from $\mathbf{L}_2$–$\mathbf{L}_p$ estimates and Theorem 6.1.7.   ☐

By [3, Theorems 7.1 and 14.5] we have

THEOREM 6.1.8

Suppose that $2\beta - 1/p \notin \mathbb{N}$. Then for any $t \ge 0$ and $\mathbf{u}^0 \in \mathbf{V}_p^\beta(\mathbf{a})$, $\mathbf{U_a}(t,0)\mathbf{u}^0 \in \mathbf{V}_p^\beta(\mathbf{a} \cdot t)$. Moreover, for any $T > 0$ there is $C_{p,\beta} = C_{p,\beta}(T) > 0$ such that

$$\|\mathbf{U_a}(t,0)\mathbf{u}^0\|_{\mathbf{V}_p^\beta} \le C_{p,\beta}\|\mathbf{u}^0\|_{\mathbf{V}_p^\beta}$$

for any $\mathbf{a} \in \mathbf{Y}$, $0 \le t \le T$, and $\mathbf{u}^0 \in \mathbf{V}_p^\beta(\mathbf{a})$.

THEOREM 6.1.9

For any given $0 < t_1 \le t_2$, if $\mathbf{E}$ is a bounded subset of $\mathbf{L}_2(D)$ then $\{\, \mathbf{U_a}(\tau,0)\mathbf{u}^0 : \mathbf{a} \in \mathbf{Y}, \ \tau \in [t_1,t_2], \ \mathbf{u}^0 \in \mathbf{E}\,\}$ has compact closure in $C^1(\bar{D}, \mathbb{R}^K)$.

PROOF   It is a consequence of Corollary 6.1.1 and Eq. (6.1.24) for $p > N$. ☐

THEOREM 6.1.10 (Joint continuity in $\mathbf{V}_p^\beta$)

Let $0 \le \beta < 1$ and $1 < p < \infty$ with $2\beta - 1/p \notin \mathbb{N}$. For any sequence $(\mathbf{a}^{(n)})_{n=1}^\infty \subset \mathbf{Y}$, any real sequence $(t_n)_{n=1}^\infty$, and any sequence $(\mathbf{u}^n)_{n=1}^\infty \subset$

$\mathbf{L}_2(D)$, if $\lim_{n\to\infty}\mathbf{a}^{(n)}=\mathbf{a}$, $\lim_{n\to\infty}t_n=t$, where $t>0$, and $\lim_{n\to\infty}\mathbf{u}^n=\mathbf{u}^0$ in $\mathbf{L}_2(D)$, then $\mathbf{U}_{\mathbf{a}^{(n)}}(t_n,0)\mathbf{u}^n$ converges in $\mathbf{V}_p^\beta$ to $\mathbf{U}_{\mathbf{a}}(t,0)\mathbf{u}^0$.

PROOF   First of all, by Theorem 6.1.4, we have that $\mathbf{U}_{\mathbf{a}^{(n)}}(t_n,0)\mathbf{u}^n$ converges in $\mathbf{L}_2(D)$ to $\mathbf{U}_{\mathbf{a}}(t,0)\mathbf{u}^0$.

Now Corollary 6.1.1 and (6.1.25) imply that there is a subsequence $(n_k)_{k=1}^\infty$ such that $\mathbf{U}_{\mathbf{a}^{(n_k)}}(t_{n_k},0)\mathbf{u}^{n_k}$ converges in $\mathbf{V}_p^\beta$ to some $\mathbf{u}^*$. We then must have $\mathbf{u}^*=\mathbf{U}_{\mathbf{a}}(t,0)\mathbf{u}_0$ and $\mathbf{U}_{\mathbf{a}^{(n_k)}}(t_{n_k},0)\mathbf{u}^{n_k}$ converges in $\mathbf{V}_p^\beta$ to $\mathbf{U}_{\mathbf{a}}(t,0)\mathbf{u}^0$. This implies that $\mathbf{U}_{\mathbf{a}^{(n)}}(t_n,0)\mathbf{u}^n$ converges in $\mathbf{V}_p^\beta$ to $\mathbf{U}_{\mathbf{a}}(t,0)\mathbf{u}^0$.   ⬜

THEOREM 6.1.11 (Joint continuity in $C^1(\bar{D},\mathbb{R}^K)$)
For any sequence $(\mathbf{a}^{(n)})_{n=1}^\infty\subset\mathbf{Y}$, any real sequence $(t_n)_{n=1}^\infty$, and any sequence $(\mathbf{u}^n)_{n=1}^\infty\subset\mathbf{L}_2(D)$, if $\lim_{n\to\infty}\mathbf{a}^{(n)}=\mathbf{a}$, $\lim_{n\to\infty}t_n=t$, where $t>0$, and $\lim_{n\to\infty}\mathbf{u}^n=\mathbf{u}^0$ in $\mathbf{L}_2(D)$, then $\mathbf{U}_{\mathbf{a}^{(n)}}(t_n,0)\mathbf{u}^n$ converges in $C^1(\bar{D},\mathbb{R}^K)$ to $\mathbf{U}_{\mathbf{a}}(t,0)\mathbf{u}^0$.

PROOF   It follows by (6.1.24) and similar arguments as in Theorem 6.1.10.
⬜

THEOREM 6.1.12 (Norm continuity in $C^1(\bar{D},\mathbb{R}^K)$)
For any sequence $(\mathbf{a}^{(n)})_{n=1}^\infty\subset\mathbf{Y}$ and any real sequence $(t_n)_{n=1}^\infty$, if $\mathbf{a}^{(n)}\to\mathbf{a}$ and $t_n\to t$ as $n\to\infty$, where $t>0$, then $\|\mathbf{U}_{\mathbf{a}^{(n)}}(t_n,0)-\mathbf{U}_{\mathbf{a}}(t,0)\|_{C^1(\bar{D},\mathbb{R}^K)}$ converges to 0 as $n\to\infty$.

PROOF   Assume that $\|\mathbf{U}_{\mathbf{a}^{(n)}}(t_n,0)-\mathbf{U}_{\mathbf{a}}(t,0)\|_{C^1(\bar{D},\mathbb{R}^K)}$ does not converge to 0 as $n\to\infty$. Then there are $\epsilon_0>0$ and $\mathbf{u}^{n_k}\in C^1(\bar{D},\mathbb{R}^K)$ with $\|\mathbf{u}^{n_k}\|_{C^1(\bar{D},\mathbb{R}^K)}=1$ such that

$$\|\mathbf{U}_{\mathbf{a}^{(n_k)}}(t_{n_k},0)\mathbf{u}^{n_k}-\mathbf{U}_{\mathbf{a}}(t_{n_k},0)\mathbf{u}^{n_k}\|_{C^1(\bar{D},\mathbb{R}^K)}\geq\epsilon_0 \qquad (6.1.27)$$

for any n_k. Since $C^1(\bar{D},\mathbb{R}^K)\hookrightarrow\mathbf{L}_2(D)$, we may assume without loss of generality that there is $\mathbf{u}^0\in\mathbf{L}_2(D)$ such that $\|\mathbf{u}^{n_k}-\mathbf{u}^0\|\to 0$ as $k\to\infty$. Then by Theorem 6.1.11, we have

$$\|\mathbf{U}_{\mathbf{a}^{(n_k)}}(t_{n_k},0)\mathbf{u}^{n_k}-\mathbf{U}_{\mathbf{a}}(t_{n_k},0)\mathbf{u}^{n_k}\|_{C^1(\bar{D},\mathbb{R}^K)}\to 0 \qquad (6.1.28)$$

as $k\to\infty$. This contradicts (6.1.27). The theorem is thus proved.   ⬜

We now investigate the strong monotonicity property of the solution operator $\mathbf{U}_{\mathbf{a}}(t,0)$. By Theorem 6.1.5 and arguments similar to those in Proposition 2.5.6, we have

THEOREM 6.1.13 (Strong monotonicity on initial data)
Assume (A6-1), (A6-3)–(A6-5). *Let* $1 < p < \infty$. *For any* $\mathbf{u}^1, \mathbf{u}^2 \in \mathbf{L}_p(D)$, *if* $\mathbf{u}^1 < \mathbf{u}^2$ *then, for any* $1 \le k \le K$,

(i)

$$\left(\mathbf{U_a}(t,0)\mathbf{u}^1\right)_k(x) < \left(\mathbf{U_a}(t,0)\mathbf{u}^2\right)_k(x) \quad \text{for } \mathbf{a} \in \mathbf{Y}, \ t > 0, \ x \in D$$

and

$$\frac{\partial}{\partial \boldsymbol{\nu}}\left(\mathbf{U_a}(t,0)\mathbf{u}^1\right)_k(x) > \frac{\partial}{\partial \boldsymbol{\nu}}\left(\mathbf{U_a}(t,0)\mathbf{u}^2\right)_k(x) \quad \text{for } \mathbf{a} \in \mathbf{Y}, \ t > 0, \ x \in \partial D$$

in the Dirichlet case,

(ii)

$$\left(\mathbf{U_a}(t,0)\mathbf{u}^1\right)_k(x) < \left(\mathbf{U_a}(t,0)\mathbf{u}^2\right)_k(x) \quad \text{for } \mathbf{a} \in \mathbf{Y}, \ t > 0, x \in \bar{D}$$

in the Neumann or Robin case.

In view of Lemma 1.3.2 the following result is a consequence of Theorem 6.1.13.

PROPOSITION 6.1.1
Assume (A6-1) *and* (A6-3)–(A6-5). *Let* $1 < p < \infty$. *Then*

$$\mathbf{U_a}(t,0)(\mathbf{L}_p(D)^+ \setminus \{\mathbf{0}\}) \subset \mathring{C}^1(\bar{D}, \mathbb{R}^K)^{++}$$

in the Dirichlet case, or

$$\mathbf{U_a}(t,0)(\mathbf{L}_p(D)^+ \setminus \{\mathbf{0}\}) \subset C^1(\bar{D}, \mathbb{R}^K)^{++}$$

in the Neumann or Robin cases, for all $\mathbf{a} \in \mathbf{Y}$ *and* $t > 0$.

The property that, for $\mathbf{u}^1, \mathbf{u}^2 \in \mathring{C}^1(\bar{D}, \mathbb{R}^K)$ (resp. $\mathbf{u}^1, \mathbf{u}^2 \in C^1(\bar{D}, \mathbb{R}^K)$), if $\mathbf{u}^1 < \mathbf{u}^2$ then $\mathbf{U_a}(t,0)\mathbf{u}^1 \ll \mathbf{U_a}(t,0)\mathbf{u}^2$ ($\mathbf{a} \in \mathbf{Y}$, $t > 0$), can be expressed as: For each $\mathbf{a} \in \mathbf{Y}$ and $t > 0$ the linear operator $\mathbf{U_a}(t,0)\colon \mathring{C}^1(\bar{D}, \mathbb{R}^K) \to \mathring{C}^1(\bar{D}, \mathbb{R}^K)$ (resp. $\mathbf{U_a}(t,0)\colon C^1(\bar{D}, \mathbb{R}^K) \to C^1(\bar{D}, \mathbb{R}^K)$) is *strongly positive* (or *strongly monotone*), see, e.g., [57].

In the rest of this section, we consider the adjoint problem of (6.0.1)+(6.0.2),

$$-\frac{\partial u_k}{\partial t} = \sum_{i=1}^{N} \frac{\partial}{\partial x_i}\left(\sum_{j=1}^{N} a_{ji}^k(t,x)\frac{\partial u_k}{\partial x_j} - b_i^k(t,x)u_k\right)$$

$$- \sum_{i=1}^{N} a_i^k(t,x)\frac{\partial u_k}{\partial x_i} + \sum_{l=1}^{K} c_k^l(t,x)u_l, \quad t < s, \ x \in D, \qquad (6.1.29)$$

complemented with the boundary conditions

$$\mathcal{B}^*_{a^k}(t)u_k = 0 \qquad \text{on } \partial D \quad \text{for } t < s, \tag{6.1.30}$$

where $\mathcal{B}^*_{a^k} = \mathcal{B}_{(a^*)^k}$ and $(a^*)^k = ((a^k_{ji})^N_{j,i=1}, -(b^k_i)^N_{i=1}, -(a^k_i)^N_{i=1}, 0, d^k_0)$, $k = 1, 2, \ldots, K$. We study mild solutions as well as weak solutions of $(6.0.1)+(6.0.2)$ and $(6.1.29)+(6.1.30)$, and their relations.

First, similar to Definition 6.1.1 for mild solutions of $(6.0.1)+(6.0.2)$, we can define mild solutions of $(6.1.29)+(6.1.30)$. Then by similar arguments as in Theorem 6.1.1, we can prove that for any given $1 < p < \infty$, $\mathbf{u}^0 \in \mathbf{L}_p(D)$, and $s \in \mathbb{R}$, there is a unique mild solution $u(\cdot)$ of $(6.1.29)+(6.1.30)$ on $(-\infty, s]$ with $\mathbf{u}(s) = \mathbf{u}^0$. Denote $\mathbf{U}^*_{\mathbf{a},p}(t,s)$ $(t < s)$ to be the mild solution operator of $(6.1.29)+(6.1.30)$ in $\mathbf{L}_p(D)$. We write $\mathbf{U}^*_{\mathbf{a}}(t,s)$ for $\mathbf{U}^*_{\mathbf{a},2}(t,s)$ $(t < s)$.

Let

$$\mathbf{V} := (V)^K, \tag{6.1.31}$$

where V is as in $(2.1.2)$. For given $\mathbf{a} \in \mathbf{Y}$ and $\mathbf{u}, \mathbf{v} \in \mathbf{V}$, define the bilinear form $B_{\mathbf{a}} = B_{\mathbf{a}}(t, \mathbf{u}, \mathbf{v})$ by,

$$B_{\mathbf{a}}(t, \mathbf{u}, \mathbf{v}) := \int_D \left((a^k_{ij}(t,x)\partial_{x_j}u_k + a^k_i(t,x)u_k)\partial_{x_i}v_k \right) dx$$
$$- \int_D \left((b^k_i(t,x)\partial_{x_i}u_k + c^k_l(t,x)u_l)v_k \right) dx \tag{6.1.32}$$

in the Dirichlet and Neumann boundary condition cases, and

$$B_{\mathbf{a}}(t, \mathbf{u}, \mathbf{v}) := \int_D \left((a^k_{ij}(t,x)\partial_{x_j}u_k + a^k_i(t,x)u_k)\partial_{x_i}v_k \right) dx$$
$$- \int_D \left((b^k_i(t,x)\partial_{x_i}u_k + c^k_l(t,x)u_l)v_k \right) dx$$
$$+ \int_{\partial D} d^k_0(t,x)u_k v_k \, dx \tag{6.1.33}$$

in the Robin boundary condition case. Also, define $B^*_{\mathbf{a}}(t, \mathbf{u}, \mathbf{v})$ by

$$B^*_{\mathbf{a}}(t, \mathbf{u}, \mathbf{v}) := \int_D \left((a^k_{ji}(t,x)\partial_{x_j}u_k\partial_{x_i}v_k \right) dx$$
$$- \int_D \left(b_i(t,x)u_k)\partial_{x_i}v_k + (a^k_i(t,x)\partial_{x_i}u_k - c^l_k(t,x)u_l)v_k \right) dx \tag{6.1.34}$$

in the Dirichlet and Neumann boundary condition cases, and

$$B^*_{\mathbf{a}}(t, \mathbf{u}, \mathbf{v}) := \int_D \left((a^k_{ji}(t,x)\partial_{x_j}u_k\partial_{x_i}v_k \right) dx$$
$$- \int_D \left(b^k_i(t,x)u_k)\partial_{x_i}v_k + (a^k_i(t,x)\partial_{x_i}u_k - c^l_k(t,x)u_l)v_k \right) dx$$
$$+ \int_{\partial D} d^k_0(t,x)u_k v_k \, dx \tag{6.1.35}$$

in the Robin boundary condition case. (We used the summation convention in the above).

DEFINITION 6.1.2 (Weak solution)

(1) *A function* $\mathbf{u}(\cdot) \in L_2((s,t), \mathbf{V})$ *is a* weak solution *of* (6.0.1)+(6.0.2) *on* $[s,t] \times D$ *with initial condition* $\mathbf{u}(s) = \mathbf{u}^0$ ($\mathbf{u}^0 \in \mathbf{L}_2(D)$) *if*

$$-\int_s^t \langle \mathbf{u}(\tau), \mathbf{v} \rangle \dot{\psi}(\tau)\, d\tau + \int_s^t B_\mathbf{a}(\tau, \mathbf{u}(\tau), \mathbf{v})\psi(\tau)\, d\tau - \langle \mathbf{u}^0, \mathbf{v} \rangle \psi(s) = 0$$

for all $\mathbf{v} \in \mathbf{V}$ *and* $\psi(\cdot) \in \mathcal{D}([s,t))$.

(2) *A function* $\mathbf{u}(\cdot) \in L_2((s,t), \mathbf{V})$ *is a* weak solution *of* (6.1.29)+(6.1.30) *on* $[s,t] \times D$ *with final condition* $\mathbf{u}(t) = \mathbf{u}^0$ ($\mathbf{u}^0 \in \mathbf{L}_2(D)$) *if*

$$\int_s^t \langle \mathbf{u}(\tau), \mathbf{v} \rangle \dot{\psi}(\tau)\, d\tau + \int_s^t B_\mathbf{a}^*(\tau, \mathbf{u}(\tau), \mathbf{v})\psi(\tau)\, d\tau - \langle \mathbf{u}^0, \mathbf{v} \rangle \psi(t) = 0$$

for all $\mathbf{v} \in \mathbf{V}$ *and* $\psi(\cdot) \in \mathcal{D}((s,t])$.

THEOREM 6.1.14

For any given $\mathbf{a} \in \mathbf{Y}$ *and* $\mathbf{u}^0 \in \mathbf{L}_2(D)$, $\mathbf{u}(t) := \mathbf{U}_\mathbf{a}(t,0)\mathbf{u}^0$ *is a weak solution of* (6.0.1)+(6.0.2) *on* $(0,\infty)$ *with initial condition* $\mathbf{u}(0) = \mathbf{u}^0$ *and* $\mathbf{u}^*(t) := \mathbf{U}_\mathbf{a}^*(t,0)\mathbf{u}^0$ *is a weak solution of* (6.1.29)+(6.1.30) *on* $(-\infty,0)$ *with final condition* $\mathbf{u}^*(0) = \mathbf{u}^0$.

PROOF   By Theorem 6.1.6, $\mathbf{u}(t) = \mathbf{U}_\mathbf{a}(t,0)\mathbf{u}^0$ is a classical solution of (6.0.1) +(6.0.2) on $(0,\infty)$ with $\mathbf{u}(0) = \mathbf{u}^0$ and $\mathbf{u}^*(t) = \mathbf{U}_\mathbf{a}^*(t,0)\mathbf{u}^0$ is a classical solution of (6.1.29)+(6.1.30) on $(-\infty,0)$ with $\mathbf{u}^*(0) = \mathbf{u}^0$. The theorem then follows. ☐

Thanks to Theorem 6.1.14, throughout the following sections, when both the coefficients and the domain of (6.0.1)+(6.0.2) are sufficiently smooth, a solution $\mathbf{u}(\cdot)$ of (6.0.1)+(6.0.2) ((6.1.29)+(6.1.30)) means either a mild or a weak solution.

By arguments similar to those in the proof of Proposition 2.3.2, we have

THEOREM 6.1.15

If $\mathbf{u}$ *and* $\mathbf{v}$ *are solutions of* (6.0.1)+(6.0.2) *and* (6.1.29)+ (6.1.30) *on* $[s,t] \times D$, *respectively, then* $\langle \mathbf{u}(\tau), \mathbf{v}(\tau) \rangle$ *is independent of* τ *for* $\tau \in [s,t]$.

Also, by arguments similar to those in Proposition 2.3.3, we have

THEOREM 6.1.16

$$(\mathbf{U_a}(t,s))^* = \mathbf{U_a^*}(s,t) \quad \text{for any } \mathbf{a} \in \mathbf{Y} \text{ and any } s < t.$$

6.2 Principal Spectrum and Principal Lyapunov Exponents and Exponential Separation in the General Setting

In this section, we introduce the concepts of principal spectrum, principal Lyapunov, and exponential separation of (6.0.1)+(6.0.2) and investigate their basic properties and show the existence of exponential separation and existence and uniqueness of entire positive solutions, which extends the theories established in Chapter 3 for scalar parabolic equation in the general setting to cooperative systems of parabolic equations in the general setting.

6.2.1 Principal Spectrum and Principal Lyapunov Exponents

Throughout this subsection, we assume (A6-1)–(A6-3).

Let $\mathbf{U_a}(t,s) \colon \mathbf{L}_2(D) \to \mathbf{L}_2(D)$ be the mild solution operator of (6.0.1)+(6.0.2) and let $U_{a^k}(t,s) \colon L_2(D) \to L_2(D)$, $1 \le k \le K$, be the weak solution operator of (6.1.1).

Let $\mathbf{Y}_0 \subset \mathbf{Y}$ be a compact connected translation invariant subset of $\mathbf{Y}$. For given $\mathbf{u}^0 \in \mathbf{L}_2(D)$ and $\mathbf{a} \in \mathbf{Y}_0$, let

$$\Pi_t^k(u_k^0, a^k) = (U_{a^k}(t,0)u_k^0, a^k \cdot t). \tag{6.2.1}$$

Recall that

$$\mathbf{\Pi}_t(\mathbf{u}^0, \mathbf{a}) = (\mathbf{U_a}(t,0)\mathbf{u}^0, \mathbf{a} \cdot t). \tag{6.2.2}$$

Let $r_0 \ge 0$ be such that $c_k^k(t,x) \ge -r_0$ for any $a = (a_{ij}^k, a_i^k, b_i^k, c_l^k, d_0^k) \in \mathbf{Y}_0$, a.e. $t \in \mathbb{R}$, a.e. $x \in D$, and $k = 1, 2, \ldots, K$. By (A6-3) and arguments as in the proof of Theorem 6.1.5, for $\mathbf{u}^0 \ge \mathbf{0}$,

$$(\mathbf{U_a}(t,0)\mathbf{u}^0)_k \ge e^{-r_0 t} U_{a^k}(t,0)u_k^0 \quad \text{for } t > 0, \ 1 \le k \le K. \tag{6.2.3}$$

For $\mathbf{a} \in \mathbf{Y}_0$ and $t > 0$, we define

$$\|\mathbf{U_a}(t,0)\|^+ := \sup\{\|\mathbf{U_a}(t,0)\mathbf{u}^0\| : \mathbf{u}^0 \in \mathbf{L}_2(D)^+, \ \|\mathbf{u}^0\| = 1\}.$$

LEMMA 6.2.1
For any $\mathbf{a} \in \mathbf{Y}$ and $t > 0$ one has $\|\mathbf{U_a}(t,0)\|^+ = \|\mathbf{U_a}(t,0)\|$.

PROOF   It can be proved by arguments as in Lemma 3.1.1. ☐

DEFINITION 6.2.1 (Principal resolvent)   *A real number λ belongs to the principal resolvent of $\boldsymbol{\Pi}$ over $\mathbf{Y}_0$, denoted by $\rho(\mathbf{Y}_0)$, if either of the following conditions holds:*

- *There are $\epsilon > 0$ and $M \geq 1$ such that*

$$\|\mathbf{U}_\mathbf{a}(t,0)\| \leq Me^{(\lambda-\epsilon)t} \quad \text{for } t > 0 \text{ and } \mathbf{a} \in \mathbf{Y}_0$$

    (such λ is said to belong to the upper *principal resolvent, denoted by $\rho_+(\mathbf{Y}_0)$),*

- *There are $\epsilon > 0$ and $M > 0$ such that*

$$\|\mathbf{U}_\mathbf{a}(t,0)\| \geq Me^{(\lambda+\epsilon)t} \quad \text{for } t > 0 \text{ and } \mathbf{a} \in \mathbf{Y}_0$$

    (such λ is said to belong to the lower *principal resolvent, denoted by $\rho_-(\mathbf{Y}_0)$).*

In view of Lemma 6.2.1, in the above inequalities the $\|\cdot\|$-norms can be replaced with $\|\cdot\|^+$-"norms," with the same M and ϵ.

DEFINITION 6.2.2 (Principal spectrum)   *The principal spectrum of the topological linear skew-product semiflow $\boldsymbol{\Pi}$ over $\mathbf{Y}_0$, denoted by $\Sigma(\mathbf{Y}_0)$, equals the complement in $\mathbb{R}$ of the principal resolvent of $\boldsymbol{\Pi}$ over $\mathbf{Y}_0$.*

LEMMA 6.2.2

(i) *For any $t_2 > 0$ there is $K_1 = K_1(t_2) > 0$ such that $\|\mathbf{U}_\mathbf{a}(t,0)\| \leq K_1$ for all $\mathbf{a} \in \mathbf{Y}_0$ and all $t \in [0, t_2]$.*

(ii) *For any $t_2 > 0$ there is $K_2 = K(t_2) > 0$ such that $\|\mathbf{U}_\mathbf{a}(t,0)\| \geq K_2$ for all $\mathbf{a} \in \mathbf{Y}_0$ and all $t \in [0, t_2]$.*

PROOF   Part (i) is a consequence of the $\mathbf{L}_2$–$\mathbf{L}_2$ estimates (Theorem 6.1.2). To prove (ii), take $\boldsymbol{\phi} := (\phi_1, \phi_2, \ldots, \phi_K)$ with $\phi_k \equiv 1$ $(k = 1, 2, \ldots, K)$, and notice $(\mathbf{U}_\mathbf{a}(t,0)\boldsymbol{\phi})_k \geq e^{-r_0 t}U_{a_k}(t,0)\phi_k$. (ii) then follows from Lemma 3.1.2(ii). $\Box$

LEMMA 6.2.3
There exist $\delta_0 > 0$, $M_1 > 0$, and a real $\underline{\lambda}$ such that $\|\mathbf{U}_\mathbf{a}(t,0)\| \geq M_1 e^{\underline{\lambda} t}$ for all $\mathbf{a} \in \mathbf{Y}_0$ and all $t \geq \delta_0$.

PROOF   Note that for any $\mathbf{u}^0 = (u_1^0, \ldots, u_K^0) \geq \mathbf{0}$, $(\mathbf{U}_\mathbf{a}(t,0)\mathbf{u}^0)_k \geq e^{-r_0 t}U_{a_k}(t,0)u_k^0$. The lemma then follows from Lemma 3.1.4. $\Box$

THEOREM 6.2.1
The principal spectrum of $\mathbf{\Pi}$ *over* $\mathbf{Y}_0$ *is a compact nonempty interval* $[\lambda_{\min},$
$\lambda_{\max}]$.

PROOF We prove first that the upper principal resolvent $\rho_+(\mathbf{Y}_0)$ is nonempty. Indeed, by the $\mathbf{L}_2$–$\mathbf{L}_2$ estimates (see Theorem 6.1.2), there are $\bar{M} > 0$ and $\bar{\gamma} > 0$ such that $\|\mathbf{U}_{\mathbf{a}}(t, 0)\| \leq \bar{M} e^{\bar{\gamma} t}$ for all $\mathbf{a} \in \mathbf{Y}_0$ and $t > 0$, hence $\bar{\gamma} + 1 \in \rho_+(\mathbf{Y}_0)$. Further, $\rho_+(\mathbf{Y}_0)$ is a right-unbounded open interval $(\lambda_{\max}, \infty)$.

The lower principal resolvent $\rho_-(\mathbf{Y}_0)$ is nonempty, too, since it contains, by Lemma 6.2.3, the real number $\underline{\lambda} - 1$.

Consequently, as $\rho_-(\mathbf{Y}_0) \cup \rho_+(\mathbf{Y}_0) = \rho(\mathbf{Y}_0)$ and $\rho_-(\mathbf{Y}_0) \cap \rho_+(\mathbf{Y}_0) = \emptyset$, one has $\Sigma(\mathbf{Y}_0) = \mathbb{R} \setminus \rho(\mathbf{Y}_0) = [\lambda_{\min}, \lambda_{\max}]$. ☐

Similarly to Theorem 3.1.2 we have

THEOREM 6.2.2

(1) *For any sequence* $(\mathbf{a}^{(n)})_{n=1}^{\infty} \subset \mathbf{Y}_0$ *and any real sequences* $(t_n)_{n=1}^{\infty}$, $(s_n)_{n=1}^{\infty}$ *such that* $t_n - s_n \to \infty$ *as* $n \to \infty$ *there holds*

$$\lambda_{\min} \leq \liminf_{n \to \infty} \frac{\ln \|\mathbf{U}_{\mathbf{a}^{(n)}}(t_n, s_n)\|}{t_n - s_n} \leq \limsup_{n \to \infty} \frac{\ln \|\mathbf{U}_{\mathbf{a}^{(n)}}(t_n, s_n)\|}{t_n - s_n} \leq \lambda_{\max}.$$

(2A) *There exist a sequence* $(\mathbf{a}^{(n,1)})_{n=1}^{\infty} \subset \mathbf{Y}_0$ *and a sequence* $(t_{n,1})_{n=1}^{\infty} \subset (0, \infty)$ *such that* $t_{n,1} \to \infty$ *as* $n \to \infty$, *and*

$$\lim_{n \to \infty} \frac{\ln \|\mathbf{U}_{\mathbf{a}^{(n,1)}}(t_{n,1}, 0)\|}{t_{n,1}} = \lambda_{\min}.$$

(2B) *There exist a sequence* $(\mathbf{a}^{(n,2)})_{n=1}^{\infty} \subset \mathbf{Y}_0$ *and a sequence* $(t_{n,2})_{n=1}^{\infty} \subset (0, \infty)$ *such that* $t_{n,2} \to \infty$ *as* $n \to \infty$, *and*

$$\lim_{n \to \infty} \frac{\ln \|\mathbf{U}_{\mathbf{a}^{(n,2)}}(t_{n,2}, 0)\|}{t_{n,2}} = \lambda_{\max}.$$

Assume that μ is an invariant ergodic Borel probability measure for the topological flow σ on $\mathbf{Y}_0$. Similarly to Theorem 3.1.5 we have

THEOREM 6.2.3
There exist a Borel set $\mathbf{Y}_1 \subset \mathbf{Y}_0$ *with* $\mu(\mathbf{Y}_1) = 1$ *and a real number* $\lambda(\mu)$ *such that*
$$\lim_{t \to \infty} \frac{\ln \|\mathbf{U}_{\mathbf{a}}(t, 0)\|}{t} = \lambda(\mu) \quad \text{for all} \quad \mathbf{a} \in \mathbf{Y}_1.$$

DEFINITION 6.2.3 (Principal Lyapunov exponent) $\lambda(\mu)$ *as defined above is called the* principal Lyapunov exponent *of* $\boldsymbol{\Pi}$ *for the ergodic invariant measure* μ.

Similarly to Theorem 3.1.6 we have

THEOREM 6.2.4
For any ergodic invariant measure μ *for* $\sigma|_{\mathbf{Y}_0}$ *the principal Lyapunov exponent* $\lambda(\mu)$ *belongs to the principal spectrum* $[\lambda_{\min}, \lambda_{\max}]$ *of* $\boldsymbol{\Pi}$ *on* $\mathbf{Y}_0$.

In the rest of this subsection, we assume (A6-5).
We define

$$\mathbf{X} := \begin{cases} \mathring{C}^1(\bar{D}, \mathbb{R}^K) & \text{(Dirichlet)} \\ C^1(\bar{D}, \mathbb{R}^K) & \text{(Neumann or Robin)}. \end{cases}$$

By Corollary 6.1.1, there is $C_{1,p} > 0$ such that

$$\|\mathbf{U_a}(1,0)\|_{\mathbf{L}_2(D),(W_p^2(D))^K} \leq C_{1,p} \quad \text{for all} \mathbf{a} \in \mathbf{Y}.$$

This implies, via (6.1.24), that for p sufficiently large there is $\tilde{C}_p > 0$ such that

$$\|\mathbf{U_a}(1,0)\|_{\mathbf{L}_2(D),\mathbf{X}} \leq \tilde{C}_p \quad \text{for all} \quad \mathbf{a} \in \mathbf{Y}.$$

LEMMA 6.2.4
There are $M_1, M_2 > 0$ *such that*

$$\|\mathbf{U_a}(t,0)\|_{\mathbf{X}} \leq M_1 \|\mathbf{U_a}(t-1,0)\|$$

for all $\mathbf{a} \in \mathbf{Y}$ *and all* $t > 1$, *and*

$$\|\mathbf{U_a}(t,0)\| \leq M_2 \|\mathbf{U_{a \cdot 1}}(t-1,0)\|_{\mathbf{X}}$$

for all $\mathbf{a} \in \mathbf{Y}$ *and all* $t > 1$.

PROOF For $1 \leq p \leq \infty$, denote by $\hat{M}_p$ the norm of the embedding $\mathbf{X} \hookrightarrow \mathbf{L}_p(D)$.
Fix some $p \in (N, \infty)$. Then, by (6.1.24), $(W_p^2(D))^K \hookrightarrow C^1(\bar{D}, \mathbb{R}^K)$. Denote by $\check{M}$ the norm of this embedding. We estimate

$$\begin{aligned}
\|\mathbf{U_a}(t,0)\mathbf{u}^0\|_{\mathbf{X}} &\leq \check{M} \|\mathbf{U_a}(t,0)\mathbf{u}^0\|_{(W_p^2(D))^K} \\
&= \check{M} \|\mathbf{U_a}(t,t-1)U_{\mathbf{a}}(t-1,0)\mathbf{u}^0\|_{(W_p^2(D))^K} \\
&\leq \check{M} C_{1,p} \|\mathbf{U_a}(t-1,0)\mathbf{u}^0\| \\
&\leq \check{M} C_{1,p} \|\mathbf{U_a}(t-1,0)\| \cdot \|\mathbf{u}^0\| \\
&\leq \check{M} C_{1,p} \hat{M}_2 \|\mathbf{U_a}(t-1,0)\| \cdot \|\mathbf{u}^0\|_{\mathbf{X}}
\end{aligned}$$

for all $\mathbf{a} \in \mathbf{Y}$, $u_0 \in \mathbf{X}$ and all $t > 1$, which gives the first inequality.

Further, we estimate

$$\|\mathbf{U}_\mathbf{a}(t,0)\mathbf{u}^0\| \leq \hat{M}_2 \|\mathbf{U}_\mathbf{a}(t,0)\mathbf{u}^0\|_\mathbf{X} \leq \hat{M}_2 \|\mathbf{U}_\mathbf{a}(t,1)\|_\mathbf{X} \cdot \|\mathbf{U}_\mathbf{a}(1,0)\mathbf{u}^0\|_\mathbf{X}$$
$$\leq \hat{M}_2 \tilde{C}_p \|\mathbf{U}_\mathbf{a}(t,1)\|_\mathbf{X} \cdot \|\mathbf{u}^0\| = \hat{M}_2 \tilde{C}_p \|\mathbf{U}_{\mathbf{a}\cdot 1}(t-1,0)\|_\mathbf{X} \cdot \|\mathbf{u}^0\|$$

for all $\mathbf{a} \in \mathbf{Y}$, $u_0 \in \mathbf{L}_2(D)$ and all $t > 1$, which gives the second inequality. $\quad\square$

THEOREM 6.2.5

(1) λ *belongs to the upper principal resolvent of* $\mathbf{\Pi}$ *over* $\mathbf{Y}_0$ *if and only if there are* $\epsilon > 0$ *and* $\tilde{M} \geq 1$ *such that*

$$\|\mathbf{U}_\mathbf{a}(t,0)\|_\mathbf{X} \leq \tilde{M} e^{(\lambda-\epsilon)t} \quad \textit{for } t > 1 \textit{ and } \mathbf{a} \in \mathbf{Y}_0.$$

(2) λ *belongs to the lower principal resolvent of* $\mathbf{\Pi}$ *over* $\mathbf{Y}_0$ *if and only if there are* $\epsilon > 0$ *and* $\tilde{M} > 0$ *such that*

$$\|\mathbf{U}_\mathbf{a}(t,0)\|_\mathbf{X} \geq \tilde{M} e^{(\lambda+\epsilon)t} \quad \textit{for } t > 0 \textit{ and } \tilde{a} \in \mathbf{Y}_0$$

PROOF It is an application of Lemma 6.2.4 and the $\mathbf{L}_p$–$\mathbf{L}_q$ estimates to Definition 6.2.1. $\quad\square$

The next two results follow by applying Lemma 6.2.4 to Theorems 6.2.2 and 6.2.3, respectively.

THEOREM 6.2.6

(1) *For any sequence* $(\mathbf{a}^{(n)})_{n=1}^\infty \subset \mathbf{Y}_0$ *and any real sequences* $(t_n)_{n=1}^\infty$, $(s_n)_{n=1}^\infty$ *such that* $t_n - s_n \to \infty$ *as* $n \to \infty$ *there holds*

$$\lambda_{\min} \leq \liminf_{n\to\infty} \frac{\ln \|\mathbf{U}_{\mathbf{a}^{(n)}}(t_n, s_n)\|_\mathbf{X}}{t_n - s_n} \leq \limsup_{n\to\infty} \frac{\ln \|\mathbf{U}_{\mathbf{a}^{(n)}}(t_n, s_n)\|_\mathbf{X}}{t_n - s_n} \leq \lambda_{\max}.$$

(2A) *There exist a sequence* $(\mathbf{a}^{(n,1)})_{n=1}^\infty \subset \mathbf{Y}_0$ *and a sequence* $(t_{n,1})_{n=1}^\infty \subset (0,\infty)$ *such that* $t_{n,1} \to \infty$ *as* $n \to \infty$, *and*

$$\lim_{n\to\infty} \frac{\ln \|\mathbf{U}_{\mathbf{a}^{(n,1)}}(t_{n,1}, 0)\|_\mathbf{X}}{t_{n,1}} = \lambda_{\min}.$$

(2B) *There exist a sequence* $(\mathbf{a}^{(n,2)})_{n=1}^\infty \subset \mathbf{Y}_0$ *and a sequence* $(t_{n,2})_{n=1}^\infty \subset (0,\infty)$ *such that* $t_{n,2} \to \infty$ *as* $n \to \infty$, *and*

$$\lim_{n\to\infty} \frac{\ln \|\mathbf{U}_{\mathbf{a}^{(n,2)}}(t_{n,2}, 0)\|_\mathbf{X}}{t_{n,2}} = \lambda_{\max}.$$

THEOREM 6.2.7
Let μ be an ergodic invariant measure for $\sigma|_{\mathbf{Y}_0}$. There exist a Borel set $\mathbf{Y}_1 \subset \mathbf{Y}_0$ with $\mu(\mathbf{Y}_1) = 1$ such that

$$\lim_{t \to \infty} \frac{\ln \|\mathbf{U}_{\mathbf{a}}(t,0)\|_{\mathbf{X}}}{t} = \lambda(\mu) \quad \text{for all} \quad \mathbf{a} \in \mathbf{Y}_1.$$

6.2.2   Exponential Separation: Basic Properties

Throughout this subsection, we assume (A6-1)–(A6-4). Let $\mathbf{Y}_0 \subset \mathbf{Y}$ be a compact connected translation invariant subset of $\mathbf{Y}$. The concepts of one-dimensional subbundle of $\mathbf{L}_2(D) \times \mathbf{Y}_0$ and one-codimensional subbundle of $\mathbf{L}_2(D) \times \mathbf{Y}_0$ are defined in a way similar to that in Section 3.2.

DEFINITION 6.2.4 (Exponential separation)   *We say that $\mathbf{\Pi}$ admits an* exponential separation *with separating exponent $\gamma_0 > 0$ over $\mathbf{Y}_0$ if there are an invariant one-dimensional subbundle $\mathbf{X}_1$ of $\mathbf{L}_2(D) \times \mathbf{Y}_0$ with fibers $\mathbf{X}_1(\mathbf{a}) = \operatorname{span}\{\mathbf{w}(\mathbf{a})\}$, and an invariant complementary one-codimensional subbundle $\mathbf{X}_2$ of $\mathbf{L}_2(D) \times \mathbf{Y}_0$ with fibers $\mathbf{X}_2(\mathbf{a}) = \{\, \mathbf{v} \in \mathbf{L}_2(D) : \langle \mathbf{v}, \mathbf{w}^*(\mathbf{a}) \rangle = 0\,\}$, where $\mathbf{w}, \mathbf{w}^* : \mathbf{Y}_0 \to \mathbf{L}_2(D)$ are continuous with $\|\mathbf{w}(\mathbf{a})\| = \|\mathbf{w}^*(\mathbf{a})\| = 1$ for all $\mathbf{a} \in \mathbf{Y}_0$, having the following properties:*

(i) *$\mathbf{w}(\mathbf{a}) \in \mathbf{L}_2(D)^+$ for all $\mathbf{a} \in \mathbf{Y}_0$,*

(ii) *$\mathbf{X}_2(\mathbf{a}) \cap \mathbf{L}_2(D)^+ = \{\mathbf{0}\}$ for all $\mathbf{a} \in \mathbf{Y}_0$,*

(iii) *there is $M \geq 1$ such that for any $\mathbf{a} \in \mathbf{Y}_0$ and any $\mathbf{v} \in \mathbf{X}_2(\mathbf{a})$ with $\|\mathbf{v}\| = 1$,*

$$\|\mathbf{U}_{\mathbf{a}}(t,0)\mathbf{v}\| \leq M e^{-\gamma_0 t} \|\mathbf{U}_{\mathbf{a}}(t,0)\mathbf{w}(\mathbf{a})\| \quad (t > 0).$$

Similarly, we can define exponential separation for discrete time as in Definition 3.2.2.

DEFINITION 6.2.5 (Exponential separation for discrete time)
Let $\mathbf{Y}_0$ be a compact connected invariant subset of $\mathbf{Y}$, and let $T > 0$. $\mathbf{\Pi}$ is said to admit an exponential separation *with separating exponent γ_0' for the discrete time T over $\mathbf{Y}_0$ if there are a one-dimensional subbundle $\mathbf{X}_1$ of $\mathbf{L}_2(D) \times \mathbf{Y}_0$ with fibers $\mathbf{X}_1(\mathbf{a}) = \operatorname{span}\{\mathbf{w}(\mathbf{a})\}$, and a one-codimensional subbundle $\mathbf{X}_2$ of $\mathbf{L}_2(D) \times \mathbf{Y}_0$ with fibers $\mathbf{X}_2(\mathbf{a}) = \{\, \mathbf{v} \in \mathbf{L}_2(D) : \langle \mathbf{v}, \mathbf{w}^*(\mathbf{a}) \rangle = 0\,\}$, where $\mathbf{w}, \mathbf{w}^* : \mathbf{Y}_0 \to \mathbf{L}_2(D)$ are continuous with $\|\mathbf{w}(\mathbf{a})\| = \|\mathbf{w}^*(\mathbf{a})\| = 1$ for all $\mathbf{a} \in \mathbf{Y}_0$, having the following properties:*

(a) *$\mathbf{U}_{\mathbf{a}}(T,0)\mathbf{X}_1(\mathbf{a}) = \mathbf{X}_1(\mathbf{a} \cdot T)$ and $\mathbf{U}_{\mathbf{a}}(T,0)\mathbf{X}_2(\mathbf{a}) \subset \mathbf{X}_2(\mathbf{a} \cdot T)$ for all $\mathbf{a} \in \mathbf{Y}_0$.*

(b) $\mathbf{w}(\mathbf{a}) \in \mathbf{L}_2(D)^+$ *for all* $\mathbf{a} \in \mathbf{Y}_0$,

(c) $\mathbf{X}_2(\mathbf{a}) \cap \mathbf{L}_2(D)^+ = \{\mathbf{0}\}$ *for all* $\mathbf{a} \in \mathbf{Y}_0$,

(d) *there is* $M' \geq 1$ *such that for any* $\mathbf{a} \in \mathbf{Y}_0$ *and any* $\mathbf{v} \in \mathbf{X}_2(\mathbf{a})$ *with* $\|\mathbf{v}\| = 1$,

$$\|\mathbf{U}_\mathbf{a}(nT, 0)\mathbf{v}\| \leq M' e^{-\gamma'_0 n} \|\mathbf{U}_\mathbf{a}(nT, 0)\mathbf{w}(\mathbf{a})\| \quad (n = 1, 2, 3, \dots).$$

The following lemma directly follows from Theorem 6.1.5(2).

LEMMA 6.2.5
The function $\mathbf{w} = (w_1, \dots, w_K)$ *in Definitions 6.2.4 and 6.2.5 satisfies:* $w_k(\mathbf{a}) \in L_2(D)^+ \setminus \{0\}$ *for any* $\mathbf{a} \in \mathbf{Y}$ *and any* $1 \leq k \leq K$.

By arguments as in Theorem 3.2.2, we have

THEOREM 6.2.8
Assume (A6-1)–(A6-4) *as well as* (A6-5). *If* $\mathbf{\Pi}$ *admits an exponential separation with separating exponent* $\gamma'_0 > 0$ *for some discrete time* $T > 0$ *over a compact invariant* $\mathbf{Y}_0 \subset \mathbf{Y}$, *then* $\mathbf{\Pi}$ *admits an exponential separation over* $\mathbf{Y}_0$, *with separating exponent* $\gamma_0 = \gamma'_0$.

We remark that, to prove Theorem 6.2.8 by using the arguments as in Theorem 3.2.2, we need that the adjoint operator $(\mathbf{U}_\mathbf{a}(t, s))^*$ of $\mathbf{U}_\mathbf{a}(t, s)$, $s < t$, is the same as the mild solution operator $\mathbf{U}_\mathbf{a}^*(s, t)$ of (6.1.29)+(6.1.30). Under (A6-5), it is proved in Theorem 6.1.16 that $(\mathbf{U}_\mathbf{a}(t, s))^* = \mathbf{U}_\mathbf{a}^*(s, t)$. We conjecture that Theorem 6.2.8 holds under (A6-1)–(A6-4) only.

In the rest of this subsection, we assume that $\mathbf{\Pi}$ admits an exponential separation over some $\mathbf{Y}_0$.

Similarly to Lemma 3.2.5 we have

LEMMA 6.2.6
Let $\lambda \in \mathbb{R}$, $(\mathbf{a}^{(n)})_{n=1}^\infty \subset \mathbf{Y}_0$, *and* $(s_n)_{n=1}^\infty \subset \mathbb{R}$, $(t_n)_{n=1}^\infty \subset \mathbb{R}$ *with* $t_n - s_n \to \infty$. *Then the following conditions are equivalent:*

(i) $\displaystyle\lim_{n \to \infty} \frac{\ln \|\mathbf{U}_{\mathbf{a}^{(n)}}(t_n, s_n)\mathbf{w}(\mathbf{a}^{(n)} \cdot s_n)\|}{t_n - s_n} = \lambda.$

(ii) $\displaystyle\lim_{n \to \infty} \frac{\ln \|\mathbf{U}_{\mathbf{a}^{(n)}}(t_n, s_n)\mathbf{u}^0\|}{t_n - s_n} = \lambda$ *for any* $\mathbf{u}^0 \in \mathbf{L}_2(D)^+ \setminus \{0\}$.

(iii) $\displaystyle\lim_{n \to \infty} \frac{\ln \|\mathbf{U}_{\mathbf{a}^{(n)}}(t_n, s_n)\|}{t_n - s_n} = \lim_{n \to \infty} \frac{\ln \|\mathbf{U}_{\mathbf{a}^{(n)}}(t_n, s_n)\|^+}{t_n - s_n} = \lambda.$

For the topological linear skew-product flow $\mathbf{\Pi}|_{\mathbf{X}_1}$ on the one-dimensional bundle $\mathbf{X}_1$ its *dynamical spectrum* (or the *Sacker–Sell spectrum*) is defined as the complement of the set of those $\lambda \in \mathbb{R}$ for which either of the following conditions holds:

- There are $\epsilon > 0$ and $M \geq 1$ such that
$$\|\mathbf{U_a}(t,0)\mathbf{w(a)}\| \leq Me^{(\lambda-\epsilon)t} \quad \text{for } t > 0 \text{ and } \mathbf{a} \in \mathbf{Y}_0,$$

- There are $\epsilon > 0$ and $M > 0$ such that
$$\|\mathbf{U_a}(t,0)\mathbf{w(a)}\| \geq Me^{(\lambda+\epsilon)t} \quad \text{for } t > 0 \text{ and } \mathbf{a} \in \mathbf{Y}_0.$$

THEOREM 6.2.9
The Sacker–Sell spectrum of $\mathbf{\Pi}|_{\mathbf{X}_1}$ equals $\Sigma(\mathbf{Y}_0)$.

THEOREM 6.2.10
For an ergodic invariant measure μ for $\sigma|_{\mathbf{Y}_0}$ there is $\mathbf{Y}_1 \subset \mathbf{Y}_0$ with $\mu(\mathbf{Y}_1) = 1$ such that the following holds.

(a) *For any $\mathbf{a} \in \mathbf{Y}_1$ and any $\mathbf{u}^0 \in \mathbf{L}_2(D) \setminus \mathbf{X}_2(\mathbf{a})$ (in particular, for any $\mathbf{u}^0 \in \mathbf{L}_2(D)^+ \setminus \{\mathbf{0}\}$) one has*
$$\lim_{t\to\infty} \frac{\ln\|\mathbf{U_a}(t,0)\mathbf{u}^0\|}{t} = \lambda(\mu) \in \Sigma(\mathbf{Y}_0). \tag{6.2.4}$$

(b) *For any $\mathbf{a} \in \mathbf{Y}_1$ and any $\mathbf{u}^0 \in \mathbf{X}_2(\mathbf{a}) \setminus \{\mathbf{0}\}$ one has*
$$\limsup_{t\to\infty} \frac{\ln\|\mathbf{U_a}(t,0)\mathbf{u}^0\|}{t} \leq \lambda(\mu) - \gamma_0. \tag{6.2.5}$$

THEOREM 6.2.11
There exist ergodic invariant measures $\mu_{\min}$ and $\mu_{\max}$ for $\sigma|_{\mathbf{Y}_0}$ such that $\lambda_{\min} = \lambda(\mu_{\min})$ and $\lambda_{\max} = \lambda(\mu_{\max})$.

Similarly to Theorem 3.2.7, we have

THEOREM 6.2.12
If $\mathbf{Y}_0 = \mathrm{cl}\{\mathbf{a}^{(0)} \cdot t : t \in \mathbb{R}\}$ for some $\mathbf{a}^{(0)} \in \mathbf{Y}_0$, where the closure is taken in the weak-* topology, then

(1) (i) *There are sequences $(s'_n)_{n=1}^\infty, (t'_n)_{n=1}^\infty \subset \mathbb{R}$, $t'_n - s'_n \to \infty$ as $n \to \infty$, such that*
$$\lambda_{\min} = \lim_{n\to\infty} \frac{\ln\|\mathbf{U}_{\mathbf{a}^{(0)}}(t'_n, s'_n)\mathbf{w}(\mathbf{a}^{(0)} \cdot s'_n)\|}{t'_n - s'_n}$$
$$= \lim_{n\to\infty} \frac{\ln\|\mathbf{U}_{\mathbf{a}^{(0)}}(t'_n, s'_n)\mathbf{u}^0\|}{t'_n - s'_n} = \lim_{n\to\infty} \frac{\ln\|\mathbf{U}_{\mathbf{a}^{(0)}}(t'_n, s'_n)\|}{t'_n - s'_n}$$

for each $\mathbf{u}^0 \in \mathbf{L}_2(D)^+ \setminus \{\mathbf{0}\}$.

(ii) There are sequences $(s_n'')_{n=1}^\infty, (t_n'')_{n=1}^\infty \subset \mathbb{R},\ t_n'' - s_n'' \to \infty$ as $n \to \infty$, such that

$$\lambda_{\max} = \lim_{n \to \infty} \frac{\ln \|\mathbf{U}_{\mathbf{a}^{(0)}}(t_n'', s_n'')\mathbf{w}(\mathbf{a}^{(0)} \cdot s_n'')\|}{t_n'' - s_n''}$$

$$= \lim_{n \to \infty} \frac{\ln \|\mathbf{U}_{\mathbf{a}^{(0)}}(t_n'', s_n'')\mathbf{u}^0\|}{t_n'' - s_n''} = \lim_{n \to \infty} \frac{\ln \|\mathbf{U}_{\mathbf{a}^{(0)}}(t_n'', s_n'')\|}{t_n'' - s_n''}$$

for each $\mathbf{u}^0 \in \mathbf{L}_2(D)^+ \setminus \{\mathbf{0}\}$.

(2) For any $\mathbf{u}^0 \in \mathbf{L}_2(D)^+ \setminus \{\mathbf{0}\}$ there holds

$$\lambda_{\min} = \liminf_{t-s \to \infty} \frac{\ln \|\mathbf{U}_{\mathbf{a}^{(0)}}(t, s)\mathbf{w}(\mathbf{a}^{(0)} \cdot s)\|}{t - s}$$

$$= \liminf_{t-s \to \infty} \frac{\ln \|\mathbf{U}_{\mathbf{a}^{(0)}}(t, s)\mathbf{u}^0\|}{t - s} = \liminf_{t-s \to \infty} \frac{\ln \|\mathbf{U}_{\mathbf{a}^{(0)}}(t, s)\|}{t - s}$$

$$\leq \limsup_{t-s \to \infty} \frac{\ln \|\mathbf{U}_{\mathbf{a}^{(0)}}(t, s)\|}{t - s} = \limsup_{t-s \to \infty} \frac{\ln \|\mathbf{U}_{\mathbf{a}^{(0)}}(t, s)\mathbf{u}^0\|}{t - s}$$

$$= \limsup_{t-s \to \infty} \frac{\ln \|\mathbf{U}_{\mathbf{a}^{(0)}}(t, s)\mathbf{w}(\mathbf{a}^{(0)} \cdot s)\|}{t - s} = \lambda_{\max}.$$

(3) For each $\lambda \in [\lambda_{\min}, \lambda_{\max}]$ there are sequences $(k_n)_{n=1}^\infty, (l_n)_{n=1}^\infty \subset \mathbb{Z}$, $l_n - k_n \to \infty$ as $n \to \infty$, such that

$$\lambda = \lim_{n \to \infty} \frac{\ln \|\mathbf{U}_{\mathbf{a}^{(0)}}(l_n, k_n)\mathbf{w}(\mathbf{a}^{(0)} \cdot k_n)\|}{l_n - k_n}$$

$$= \lim_{n \to \infty} \frac{\ln \|\mathbf{U}_{\mathbf{a}^{(0)}}(l_n, k_n)\mathbf{u}^0\|}{l_n - k_n} = \lim_{n \to \infty} \frac{\ln \|\mathbf{U}_{\mathbf{a}^{(0)}}(l_n, k_n)\|}{l_n - k_n}$$

for each $\mathbf{u}^0 \in \mathbf{L}_2(D)^+ \setminus \{\mathbf{0}\}$.

6.2.3 Existence of Exponential Separation and Entire Positive Solutions

In this subsection, we study the existence of exponential separation and the existence and uniqueness of entire positive solutions. We restrict ourselves to the smooth case. Let (A6-6) denote the following standing assumption:

(A6-6) *For any $T > 0$ the mapping*

$$[\mathbf{Y} \ni \mathbf{a} \mapsto [[0, T] \ni t \mapsto \mathbf{U}_\mathbf{a}(t, 0)] \in B([0, T], \mathcal{L}(\mathbf{L}_2(D), \mathbf{L}_2(D)))]$$

is continuous, where $\mathcal{L}(\mathbf{L}_2(D), \mathbf{L}_2(D))$ represents the space of all bounded linear operators from $\mathbf{L}_2(D)$ into itself, endowed with the norm topology and

$B(\cdot, \cdot)$ *stands for the Banach space of bounded functions, endowed with the supremum norm.*

It should be pointed out that in [6] and [91] conditions, for some special cases (for example, the Dirichlet boundary condition case and the case with infinitely differentiable coefficients), are given that guarantee the continuous dependence of $[\,[0, T] \ni t \mapsto \mathbf{U}_{\mathbf{a}}(t, 0)\,] \in B([0, T], \mathcal{L}(\mathbf{L}_2(D), \mathbf{L}_2(D)))$ on the coefficients.

Let $\mathbf{Y}_0$ be a compact connected invariant subset of $\mathbf{Y}$.

We first study the existence of exponential separation.

THEOREM 6.2.13 (Existence of exponential separation)
Assume (A6-1)–(A6-6). *Then* $\mathbf{\Pi}$ *admits an exponential separation over* $\mathbf{Y}_0$.

To prove the above theorem, we first show the following lemma.

LEMMA 6.2.7
Assume (A6-1)–(A6-6). *Then the map* $[\,\mathbf{Y}_0 \times (0, \infty) \ni (\mathbf{a}, t) \mapsto \mathbf{U}_{\mathbf{a}}(t, 0) \in \mathcal{L}(\mathbf{L}_2(D), \mathbf{X})\,]$ *is continuous.*

PROOF   Assume that $\mathbf{a}^{(n)}$ converges to $\mathbf{a}$ in $\mathbf{Y}_0$ and that t_n converges to $t > 0$. Suppose to the contrary that

$$\|\mathbf{U}_{\mathbf{a}^{(n)}}(t_n, 0) - \mathbf{U}_{\mathbf{a}}(t, 0)\|_{\mathbf{L}_2(D), \mathbf{X}} \nrightarrow 0$$

as $n \to \infty$. Then there are $\epsilon_0 > 0$ and a sequence $(\mathbf{u}^n)_{n=1}^{\infty} \subset \mathbf{L}_2(D)$ with $\|\mathbf{u}^n\| = 1$ such that

$$\|\mathbf{U}_{\mathbf{a}^{(n)}}(t_n, 0)\mathbf{u}^n - \mathbf{U}_{\mathbf{a}}(t, 0)\mathbf{u}^n\|_{\mathbf{X}} \geq \epsilon_0$$

for all n. By Corollary 6.1.1 and Eq. (6.1.24), there are $\mathbf{u}^*, \mathbf{u}^{**} \in \mathbf{X}$ such that (after possibly extracting a subsequence)

$$\mathbf{U}_{\mathbf{a}^{(n)}}(t_n, 0)\mathbf{u}^n \to \mathbf{u}^* \qquad \text{and} \qquad \mathbf{U}_{\mathbf{a}}(t, 0)\mathbf{u}^n \to \mathbf{u}^{**}$$

in $\mathbf{X}$, as $n \to \infty$. Without loss of generality, we may also assume that there is $\tilde{\mathbf{u}}^* \in \mathbf{X}$ such that $\mathbf{U}_{\mathbf{a}}(t/2, 0)\mathbf{u}^n \to \tilde{\mathbf{u}}^*$ in $\mathbf{X}$ as $n \to \infty$.

By the property (A6-6) we have $\|\mathbf{U}_{\mathbf{a}^{(n)}}(t_n, 0)\mathbf{u}^n - \mathbf{U}_{\mathbf{a}}(t_n, 0)\mathbf{u}^n\| \to 0$ and by Theorem 6.1.4 we have

$$\|\mathbf{U}_{\mathbf{a}}(t_n, 0)\mathbf{u}^n - \mathbf{U}_{\mathbf{a}}(t, 0)\mathbf{u}^n\|$$
$$= \|\mathbf{U}_{\mathbf{a}}(t_n, t/2)\mathbf{U}_{\mathbf{a}}(t/2, 0)\mathbf{u}^n - \mathbf{U}_{\mathbf{a}}(t, t/2)\mathbf{U}_{\mathbf{a}}(t/2, 0)\mathbf{u}^n\|$$
$$\to 0$$

as $n \to \infty$. Therefore

$$\|\mathbf{U}_{\mathbf{a}^{(n)}}(t_n, 0)\mathbf{u}^n - \mathbf{U}_{\mathbf{a}}(t, 0)\mathbf{u}^n\|$$
$$\leq \|\mathbf{U}_{\mathbf{a}^{(n)}}(t_n, 0)\mathbf{u}^n - \mathbf{U}_{\mathbf{a}}(t_n, 0)\mathbf{u}^n\| + \|\mathbf{U}_{\mathbf{a}}(t_n, 0)\mathbf{u}^n - \mathbf{U}_{\mathbf{a}}(t, 0)\mathbf{u}^n\|$$
$$\to 0$$

as $n \to \infty$. Then we must have $\mathbf{u}^* = \mathbf{u}^{**}$, hence

$$\|\mathbf{U}_{\mathbf{a}^{(n)}}(t_n, 0)\mathbf{u}^n - \mathbf{U}_{\mathbf{a}}(t, 0)\mathbf{u}^n\|_{\mathbf{X}} \to 0$$

as $n \to \infty$, a contradiction. $\qquad\square$

PROOF (Proof of Theorem 6.2.13) We show that $\mathbf{\Pi}$ admits an exponential separation over Y_0 for discrete time $T = 1$, and apply Theorem 6.2.8.

We define a mapping $\tilde{\mathbf{\Pi}} : \mathbf{X} \times \mathbf{Y}_0 \to \mathbf{X} \times \mathbf{Y}_0$ by the formula

$$\tilde{\mathbf{\Pi}}(\mathbf{u}^0, \mathbf{a}) := (\mathbf{U}_{\mathbf{a}}(1, 0)\mathbf{u}^0, \mathbf{a} \cdot 1), \quad \mathbf{a} \in \mathbf{Y}_0, \ \mathbf{u}^0 \in \mathbf{X}.$$

We have that, for each $\mathbf{a} \in \mathbf{Y}_0$, $\mathbf{U}_a(1, 0)$ is a compact (completely continuous) operator in $\mathcal{L}(\mathbf{X})$ (Theorem 6.1.9), having the property that $\mathbf{U}_a(1, 0)(\mathbf{X}^+ \setminus \{\mathbf{0}\}) \subset \mathbf{X}^{++}$ (Proposition 6.1.1). Moreover, the mapping $[\mathbf{Y}_0 \ni \mathbf{a} \mapsto \mathbf{U}_a(1, 0) \in \mathcal{L}(\mathbf{X})]$ is continuous (by Lemma 6.2.7). These allow us to use the results contained in [94] to conclude that there are continuous functions $\tilde{\mathbf{w}} : \mathbf{Y}_0 \to \mathbf{X}$, $\tilde{\mathbf{w}}^* : \mathbf{Y}_0 \to \mathbf{X}^*$, $\|\tilde{\mathbf{w}}(\mathbf{a})\|_{\mathbf{X}} = \|\tilde{\mathbf{w}}^*(\mathbf{a})\|_{\mathbf{X}^*} = 1$ for all $\mathbf{a} \in \mathbf{Y}_0$, such that

(i) $\tilde{\mathbf{w}}(\mathbf{a}) \in \mathbf{X}^{++}$, for each $\mathbf{a} \in \mathbf{Y}_0$.

(ii) $(\mathbf{v}, \tilde{\mathbf{w}}^*(\mathbf{a}))_{\mathbf{X}, \mathbf{X}^*} > 0$ for each $\mathbf{a} \in \mathbf{Y}_0$ and each nonzero $\mathbf{v} \in \mathbf{X}^+$.

(iii) For each $\mathbf{a} \in \mathbf{Y}_0$ there is $d_1 = d_1(\mathbf{a}) > 0$ such that $\mathbf{U}_{\mathbf{a}}(1, 0)\tilde{\mathbf{w}}(\mathbf{a}) = d_1 \tilde{\mathbf{w}}(\mathbf{a} \cdot 1)$.

(iv) For each $\mathbf{a} \in \mathbf{Y}_0$ there is $d_2^* = d_2^*(\mathbf{a}) > 0$ such that $(\mathbf{U}_{\mathbf{a}}(1, 0))^* \tilde{\mathbf{w}}^*(\mathbf{a} \cdot 1) = d_2^* \tilde{\mathbf{w}}^*(\mathbf{a})$, where $(\mathbf{U}_{\mathbf{a}}(1, 0))^* : \mathbf{X}^* \to \mathbf{X}^*$ stands for the linear operator dual to $\mathbf{U}_{\mathbf{a}}(1, 0)$.

(v) There are constants $\tilde{C} > 0$ and $0 < \gamma < 1$ such that

$$\|\mathbf{U}_{\mathbf{a}}(n, 0)\mathbf{u}^0\|_{\mathbf{X}} \leq \tilde{C}\gamma^n \|\mathbf{U}_{\mathbf{a}}(n, 0)\tilde{\mathbf{w}}(\mathbf{a})\|_{\mathbf{X}} \qquad (6.2.6)$$

for any $\mathbf{a} \in \mathbf{Y}_0$, any $\mathbf{u}^0 \in \mathbf{X}$ with $\|\mathbf{u}^0\|_{\mathbf{X}} = 1$ satisfying $(\mathbf{u}^0, \tilde{\mathbf{w}}^*(\mathbf{a}))_{\mathbf{X}, \mathbf{X}^*} = 0$, and any $n \in \mathbb{N}$.

Define $\mathbf{w} : \mathbf{Y}_0 \to \mathbf{L}_2(D)$ by $\mathbf{w}(\mathbf{a}) := \tilde{\mathbf{w}}(\mathbf{a}) / \|\tilde{\mathbf{w}}(\mathbf{a})\|$. Since $\mathbf{X} \hookrightarrow \mathbf{L}_2(D)$, the function $\mathbf{w}$ is well defined and continuous. Further, we deduce from Lemma 6.2.7 that the mapping $[\mathbf{Y}_0 \ni \mathbf{a} \mapsto (\mathbf{U}_{\mathbf{a}}(1, 0))^* \in \mathcal{L}(\mathbf{X}^*, \mathbf{L}_2(D))]$ is continuous. This together with (iv) gives that the function $\mathbf{w}^* : \mathbf{Y}_0 \to \mathbf{L}_2(D)$, $\mathbf{w}^*(\mathbf{a}) := \tilde{\mathbf{w}}^*(\mathbf{a})/\|\tilde{\mathbf{w}}^*(\mathbf{a})\|$, is well defined and continuous.

For $\mathbf{a} \in \mathbf{Y}_0$ put $\mathbf{X}_1(\mathbf{a}) := \text{span}\{\mathbf{w}(\mathbf{a})\}$, $\mathbf{X}_2(\mathbf{a}) := \{\mathbf{v} \in \mathbf{L}_2(D) : \langle \mathbf{v}, \mathbf{w}^*(\mathbf{a}) \rangle = 0\}$. We see that the properties (a), (b), and (c) in Definition 6.2.5 are satisfied.

Let M_1 denote the norm of the embedding $\mathbf{X} \hookrightarrow \mathbf{L}_2(D)$. Also, by the continuity of $\tilde{\mathbf{w}}$ on the compact space $\mathbf{Y}_0$ there is $M_2 > 0$ such that $\|\tilde{\mathbf{w}}(\mathbf{a})\|_\mathbf{X} \leq M_2\|\mathbf{w}(\mathbf{a})\|$ for all $\mathbf{a} \in \mathbf{Y}_0$. Further, put

$$D_1 := \sup\{\|\mathbf{U}_\mathbf{a}(1,0)\|_{\mathbf{L}_2(D),\mathbf{X}} : \mathbf{a} \in \mathbf{Y}_0\} \; (< \infty),$$
$$D_2 := \inf\{\|\mathbf{U}_\mathbf{a}(1,0)\tilde{\mathbf{w}}(\mathbf{a})\|_\mathbf{X} : \mathbf{a} \in \mathbf{Y}_0\} \; (> 0).$$

Take $\mathbf{a} \in \mathbf{Y}_0$ and $\mathbf{u}^0 \in \mathbf{X}_2$ with $\|\mathbf{u}^0\| = 1$. It follows from (iv) that $\mathbf{U}_\mathbf{a}(1,0)\mathbf{u}^0 \in \mathbf{X} \cap \mathbf{X}_2(\mathbf{a} \cdot 1)$. For $n = 2, 3, 4, \ldots$, we estimate

$$\|\mathbf{U}_\mathbf{a}(n,0)\mathbf{u}^0\| \leq M_1 \|\mathbf{U}_\mathbf{a}(n,1)(\mathbf{U}_\mathbf{a}(1,0)\mathbf{u}^0)\|_\mathbf{X}$$
$$\leq M_1 \tilde{C} \gamma^{n-1} \|\mathbf{U}_\mathbf{a}(n,1)\tilde{\mathbf{w}}(\mathbf{a} \cdot 1)\|_\mathbf{X} \, \|\mathbf{U}_\mathbf{a}(1,0)\mathbf{u}^0\|_\mathbf{X} \quad \text{(by (v))}$$
$$= \frac{M_1 \tilde{C}}{\gamma} \gamma^n \frac{\|\mathbf{U}_\mathbf{a}(n,0)\tilde{\mathbf{w}}(\mathbf{a})\|_\mathbf{X}}{\|\mathbf{U}_\mathbf{a}(1,0)\tilde{\mathbf{w}}(\mathbf{a})\|_\mathbf{X}} \|\mathbf{U}_\mathbf{a}(1,0)\mathbf{u}^0\|_\mathbf{X}$$
$$\leq \frac{M_1 M_2 D_1 \tilde{C}}{D_2 \gamma} \gamma^n \|\mathbf{U}_\mathbf{a}(n,0)\mathbf{w}(\mathbf{a})\|.$$

Clearly, $\|\mathbf{U}_\mathbf{a}(1,0)\mathbf{u}^0\| \leq \frac{M_1 M_2 D_1}{D_2} \|\mathbf{U}_\mathbf{a}(1,0)\mathbf{w}(\mathbf{a})\|$ for all $\mathbf{a} \in \mathbf{Y}_0$ and all $\mathbf{u}^0 \in \mathbf{L}_2(D)$ with $\|\mathbf{u}^0\| = 1$. Therefore (d) in Definition 6.2.5 is satisfied. ☐

We note that exponential separation and entire positive solutions defined in the following are strongly related.

DEFINITION 6.2.6 (Entire positive solution)   *A function $[\mathbb{R} \ni t \mapsto \mathbf{u}(t) \in \mathbf{L}_2(D)]$ is called an* entire positive solution *of (6.0.1)+(6.0.2) if*

$$\mathbf{u}(t) = \mathbf{U}_\mathbf{a}^0(t,s)\mathbf{u}(s) + \int_s^t \mathbf{U}_\mathbf{a}^0(t,\tau)(\mathbf{C}_\mathbf{a}(\tau)\mathbf{u}(\tau)) \, d\tau$$

for any $s < t$, and $u_k(t)(x) > 0$ for all $t \in \mathbb{R}$, $x \in D$, and $k = 1, 2, \ldots, K$.

The following lemma follows from Theorems 3.3.1 and 3.3.3.

LEMMA 6.2.8

(1) *For each k $(1 \leq k \leq K)$ and $a^k \in Y_0^k$, (6.1.1) has a unique (up to multiplication by positive scalars) entire positive solution.*

(2) *For each k $(1 \leq k \leq K)$, Π^k admits an exponential separation over Y_0^k (with the one-dimensional invariant subbundle X_1^k and the one-codimensional invariant subbundle X_2^k).*

For $1 \leq k \leq K$, let $w_k, w_k^* \colon a^k \to L_2(D)$ be continuous functions such that $\|w_k(a^k)\| = \|w_k^*(a^k)\| = 1$, $X_1^k(a^k) = \mathrm{span}\,\{w_k(a^k)\}$, $X_2^k(a^k) = \{v \in L_2(D) : \langle v, w_k^*(a^k)\rangle = 0\}$ $(a^k \in Y_0^k)$.

Recall that for any $a^k \in Y_0^k$ the value at time $t = 0$ of the unique normalized entire positive solution of $(6.1.1)$ equals $w_k(a^k)$.

THEOREM 6.2.14 (Existence of entire positive solution)
Let (A6-1) *through* (A6-4) *be satisfied. Then for each* $\mathbf{a} \in \mathbf{Y}_0$, $(6.0.1)+(6.0.2)$ *has an entire positive solution.*

PROOF   We apply the same idea as in the proof of Theorem 3.3.1.

Fix $\mathbf{a} \in \mathbf{Y}_0$ and $\mathbf{u}^0 \in \mathbf{L}_2(D)^+$ with $\|\mathbf{u}^0\| = 1$. Define a sequence $(\mathbf{u}^n)_{n=1}^\infty$ by

$$\mathbf{u}^n := \frac{\mathbf{U_a}(0, -n)\mathbf{u}^0}{\|\mathbf{U_a}(-1, -n)\mathbf{u}^0\|} = \mathbf{U_a}(0, -1)\frac{\mathbf{U_a}(-1, -n)\mathbf{u}^0}{\|\mathbf{U_a}(-1, -n)\mathbf{u}^0\|}.$$

By Theorem 6.1.2, the set $\{\,\|\mathbf{u}^n\| : n = 1, 2, \dots\,\}$ is bounded above. Moreover, from Theorem 6.1.3 we deduce that there is a sequence $(n_k)_{k=1}^\infty$ such that $\lim_{k\to\infty} n_k = \infty$ and $\lim_{k\to\infty} \mathbf{u}^{n_k} = \tilde{\mathbf{u}}^0$ in $\mathbf{L}_2(D)$. By $(6.2.3)$ and Lemma 3.3.4, the set $\{\,\|\mathbf{u}^n\| : n = 1, 2, \dots\,\}$ is bounded away from zero, consequently $\tilde{\mathbf{u}}^0 \in \mathbf{L}_2(D)^+ \setminus \{\mathbf{0}\}$.

By arguments as in Theorem 3.3.1, we have that there is an entire positive weak solution $\hat{\mathbf{u}}$ of $(6.0.1)+(6.0.2)$ such that $\hat{\mathbf{u}}(0) = \tilde{\mathbf{u}}_0$.   □

THEOREM 6.2.15 (Uniqueness of entire positive solution)
Let (A6-1) *through* (A6-5) *hold. Then for each* $\mathbf{a} \in \mathbf{Y}_0$, *an entire positive solution of* $(6.0.1)+(6.0.2)$ *is unique in the following sense: If* $\mathbf{u}^1$ *and* $\mathbf{u}^2$ *are entire positive solutions then there is* $\tilde{\beta} > 0$ *such that* $\mathbf{u}^2(t) = \tilde{\beta}\mathbf{u}^1(t)$ *for all* $t \in \mathbb{R}$.

Theorem 6.2.15 can be proved by arguments similar to those in [92] for scalar parabolic equations. For convenience, we provide a proof here. To do so, we first show some lemmas. We assume until the end of the subsection that (A6-1)–(A6-5) are satisfied.

Choose a $\boldsymbol{\phi}^* \in \mathbf{X}^{++}$. For any $\mathbf{u} \in \mathbf{L}_2(D)$, define $\|\mathbf{u}\|_{\boldsymbol{\phi}^*} := \langle |\mathbf{u}|, \boldsymbol{\phi}^*\rangle$.

LEMMA 6.2.9
Given $\tau > \delta > 0$, *there exists a constant* $C = C(\tau, \delta) > 0$ *such that for any* $\mathbf{a} \in \mathbf{Y}$ *and any* $\mathbf{u} \in \mathbf{L}_2(D)$ *there holds*

$$\|\mathbf{U_a}(t, s)\mathbf{u}\|_\mathbf{X} \leq C\|\mathbf{u}\|_{\boldsymbol{\phi}^*} \quad \text{for} \quad t,\, s \in \mathbb{R},\ \tau \geq t - s \geq \delta,$$

$$\|\mathbf{U_a^*}(t, s)\mathbf{u}\|_\mathbf{X} \leq C\|\mathbf{u}\|_{\boldsymbol{\phi}^*} \quad \text{for} \quad t,\, s \in \mathbb{R},\ \tau \geq s - t \geq \delta.$$

PROOF   First, by Theorem 6.1.11, for given $\tau > \delta > 0$ there is $\tilde{C} = \tilde{C}(\tau, \delta) > 0$ such that

$$\|\mathbf{U_a}(t,s)\mathbf{u}\|_\mathbf{X} \leq \tilde{C}\|\mathbf{U_a}(t-\delta/2,s)\mathbf{u}\|$$

for all $\mathbf{a} \in \mathbf{Y_0}$ and $\tau \geq t - s \geq \delta$. Note that

$$\begin{aligned}
\|\mathbf{U_a}(t-\delta/2,s)\mathbf{u}\| &= \sup\left\{\,\langle\mathbf{U_a}(t-\delta/2,s)\mathbf{u},\mathbf{v}\rangle : \mathbf{v} \in \mathbf{L_2}(D),\ \|\mathbf{v}\| = 1\,\right\}\\
&= \sup\left\{\,\langle\mathbf{u},\mathbf{U_a^*}(s,t-\delta/2)\mathbf{v}\rangle : \mathbf{v} \in \mathbf{L_2}(D),\ \|\mathbf{v}\| = 1\,\right\}\\
&\leq \sup\left\{\,\langle|\mathbf{u}|,|\mathbf{U_a^*}(s,t-\delta/2)\mathbf{v}|\rangle : \mathbf{v} \in \mathbf{L_2}(D),\ \|\mathbf{v}\| = 1\,\right\}.
\end{aligned}$$

We claim that there is $\hat{C} = \hat{C}(\tau,\delta) > 0$ such that

$$|\mathbf{U_a^*}(s,t-\delta/2)\mathbf{v}| \leq \hat{C}\boldsymbol{\phi}^*$$

for all $\mathbf{a} \in \mathbf{Y_0}$, $\tau \geq t - s \geq \delta$ and $\mathbf{v} \in \mathbf{L_2}(D)$ with $\|\mathbf{v}\| = 1$. Since $|\mathbf{U_a^*}(s,t-\delta/2)\mathbf{v}| \leq \mathbf{U_a^*}(s,t-\delta/2)|\mathbf{v}|$ and $\|\mathbf{v}\| = \|\,|\mathbf{v}|\,\|$, it suffices to show the inequality for $\mathbf{v} \in \mathbf{L_2}(D)^+$. And this is a consequence of Theorems 6.1.11 and 6.1.13.

It then follows that

$$\|\mathbf{U_a}(t,s)\mathbf{u}\|_\mathbf{X} \leq C\|\mathbf{u}\|_{\boldsymbol{\phi}^*},$$

with $C = \tilde{C}\hat{C}$.

The inequality for the adjoint equation is proved in an analogous way.   ∎

LEMMA 6.2.10

(1) *There are constants M, ϱ_0 such that the following holds: If $\mathbf{v}$ is a positive solution of (6.1.29)+(6.1.30) on the interval $(-\infty, \tau)$, then*

$$\frac{\|\mathbf{v}(t)\|_{\boldsymbol{\phi}^*}}{\|\mathbf{v}(s)\|_{\boldsymbol{\phi}^*}} \leq Me^{\varrho_0|t-s|} \quad \text{for } t,\ s < \tau.$$

(2) *For each positive solution $\mathbf{v}$ of (6.1.29)+(6.1.30) on $(-\infty, \tau)$, there is a constant $\eta_0 > 0$ such that*

$$\frac{\mathbf{v}(t)}{\|\mathbf{v}(t)\|_{\boldsymbol{\phi}^*}} \geq \eta_0\boldsymbol{\phi}^* \quad \text{for } t \leq \tau - 1.$$

Analogous results hold for positive solutions of (6.0.1)+(6.0.2) on intervals (τ, ∞), $\tau \in \mathbb{R}$.

PROOF   (1) For a given $\mathbf{a} \in \mathbf{Y_0}$, let $\phi_k(\mathbf{a}) := w^k(P^k(\mathbf{a}))$. It follows from Lemma 6.2.8 that $\phi_k: \mathbf{Y_0} \to C^1(\bar{D})$ is continuous. Put $\boldsymbol{\phi}(\mathbf{a}) :=$

$(\phi_1(\mathbf{a}), \ldots, \phi_K(\mathbf{a}))$, $\mathbf{a} \in \mathbf{Y}_0$. Proposition 2.5.7(2) implies the existence of $0 < q_1 < q_2$ such that

$$q_1 \boldsymbol{\phi}(\mathbf{a}) \le \boldsymbol{\phi}^* \le q_2 \boldsymbol{\phi}(\mathbf{a})$$

for any $\mathbf{a} \in \mathbf{Y}_0$.

Further, let

$$\kappa_k(\mathbf{a}) := B_{\mathbf{a}^k}(0, \phi_k(\mathbf{a}), \phi_k(\mathbf{a})),$$

where $B_{\mathbf{a}^k}(\cdot, \cdot, \cdot)$ is as in (2.1.4) with a being replaced by a^k in the Dirichlet and Neumann boundary condition cases, and is as in (2.1.5) with a being replaced by a^k in the Robin boundary condition case. The functions $\kappa_k \colon \mathbf{Y}_0 \to \mathbb{R}$ are continuous. Moreover, by Lemma 3.5.4, $u_k(t) := \phi_k(\mathbf{a} \cdot t)$ satisfies

$$\begin{cases} \dfrac{\partial u_k}{\partial t} = \displaystyle\sum_{i=1}^{N} \dfrac{\partial}{\partial x_i} \left(\sum_{j=1}^{N} a_{ij}^k(t, x) \dfrac{\partial u_k}{\partial x_j} + a_i^k(t, x) u_k \right) \\ \qquad + \displaystyle\sum_{i=1}^{N} b_i^k(t, x) \dfrac{\partial u_k}{\partial x_i} - \kappa_k(\mathbf{a} \cdot t) u_k, \qquad t > 0, \ x \in D \\ \mathcal{B}_{\mathbf{a}^k}(t) u_k = 0, \qquad\qquad\qquad\qquad\qquad\qquad t > 0, \ x \in \partial D. \end{cases} \tag{6.2.7}$$

Choose $\varrho_0 > 0$ such that

$$|\kappa_k(\mathbf{a})| + K \|\mathbf{C}_\mathbf{a}(\cdot, \cdot)\|_{L_\infty(\mathbb{R} \times D)} \le \varrho_0 \quad \text{for all} \quad \mathbf{a} \in \mathbf{Y}_0.$$

Observe that, by Proposition 2.1.3,

$$\left| \frac{d}{dt} \langle \mathbf{v}(t), \boldsymbol{\phi}(\mathbf{a} \cdot t) \rangle \right| = \left| -\sum_{k=1}^{K} \int_D v_k(t) \kappa_k(\mathbf{a} \cdot t) \phi_k(\mathbf{a} \cdot t) \, dx + \langle \mathbf{C}_\mathbf{a}^*(t) v(t), \boldsymbol{\phi}(\mathbf{a} \cdot t) \rangle \right|$$

$$\le \varrho_0 \langle \mathbf{v}(t), \boldsymbol{\phi}(\mathbf{a} \cdot t) \rangle$$

where $\mathbf{C}_\mathbf{a}^*(t, x) = (c_k^l(t, x))$ (recall that $\mathbf{C}_\mathbf{a}(t, x) = (c_l^k(t, x))$. It then follows that

$$\langle \mathbf{v}(t), \boldsymbol{\phi}(\mathbf{a} \cdot t) \rangle \le \langle \mathbf{v}(s), \boldsymbol{\phi}(\mathbf{a} \cdot s) \rangle e^{\varrho_0 |t-s|}.$$

Therefore there is $M > 0$ such that

$$\langle \mathbf{v}(t), \boldsymbol{\phi}^* \rangle \le M \langle \mathbf{v}(s), \boldsymbol{\phi}^* \rangle e^{\varrho_0 |t-s|}.$$

(1) then follows.

(2) Suppose that the statement is not true. Then there are $t_n \le \tau - 1$ $(n = 1, 2, \ldots)$ such that

$$\eta_n := \sup \left\{ \eta_0 \ge 0 : \frac{\mathbf{v}(t_n)}{\|\mathbf{v}(t_n)\|_{\boldsymbol{\phi}^*}} \ge \eta_0 \boldsymbol{\phi}^* \right\} \to 0$$

as $n \to \infty$. By $\mathbf{v}(t) \in \mathbf{X}^{++}$, we must have $t_n \to -\infty$. Let

$$\mathbf{v}^n(t) := \frac{\mathbf{v}(t_n + t)}{\|\mathbf{v}(t_n)\|_{\boldsymbol{\phi}^*}} \quad \text{for } t < \tau - t_n.$$

Then $\mathbf{v}^n(t)$ is solution of (6.1.29)+(6.1.30) with $\mathbf{a}$ replaced by $\mathbf{a} \cdot t_n$ and $\mathbf{v}^n(0) = \frac{\mathbf{v}(t_n)}{\|\mathbf{v}(t_n)\|_{\boldsymbol{\phi}^*}}$ (hence $\|\mathbf{v}^{(n)}(0)\|_{\boldsymbol{\phi}^*} = 1$).

Without loss of generality, we may assume that $\mathbf{a} \cdot t_n \to \mathbf{a}^* \in \mathbf{Y}_0$. By (1), Lemma 6.2.9, and the embedding $\mathbf{X} \hookrightarrow \mathbf{L}_2(D)$, $\|\mathbf{v}^n(2)\|$ is bounded in n. Then by Theorem 6.1.3, we may assume that $\mathbf{v}^n(3/2) \to \mathbf{v}^*$ in $\mathbf{L}_2(D)$. By Theorem 6.1.11, we have that $\mathbf{v}^n(t) \to \mathbf{v}^*(t)$ in $\mathbf{X}$ for $t \in [-1, 1]$, where $\mathbf{v}^*(t)$ is solution of (6.1.29)+(6.1.30) with $\mathbf{a}$ replaced by $\mathbf{a}^*$ and $\mathbf{v}^*(3/2) = \mathbf{v}^*$. We therefore have $\|\mathbf{v}^*(0)\|_{\boldsymbol{\phi}^*} = 1$ and $\mathbf{v}^*(0) \geq \eta_0^* \boldsymbol{\phi}^*$ for some $\eta_0^* > 0$. This implies that

$$\frac{\mathbf{v}(t_n)}{\|\mathbf{v}(t_n)\|_{\boldsymbol{\phi}^*}} \geq \frac{\eta_0^*}{2} \boldsymbol{\phi}^* \quad \text{for } n \text{ sufficiently large,}$$

which contradicts $\eta_n \to 0$ as $n \to \infty$. (2) is thus proved. $\qquad \Box$

A solution $\mathbf{u} \colon J \to \mathbf{L}_2(D)$ of (6.0.1)+(6.0.2) is *nontrivial* if, for each $t \in J$, there holds $\mathbf{u}(t) \neq \mathbf{0}$.

LEMMA 6.2.11

(1) If $\mathbf{u}$ is a nontrivial solution of (6.0.1)+(6.0.2) on J and $\mathbf{v}$ is a positive solution of (6.1.29)+(6.1.30) on the same interval J such that $\langle \mathbf{u}(t), \mathbf{v}(t) \rangle = 0$ for some (hence every) $t \in J$, then $\xi(t) := \langle |\mathbf{u}(t)|, \mathbf{v}(t) \rangle$ is a decreasing function on J.

(2) If $\mathbf{u}$ is a nontrivial solution of (6.0.1)+(6.0.2) on $(-\infty, t_0]$, $\mathbf{v}$ is a positive solution of (6.1.29)+(6.1.30) on $(-\infty, t_0]$, and $\langle \mathbf{u}(t_0), \mathbf{v}(t_0) \rangle = 0$, then there is some $0 < \varrho_0 < 1$ such that

$$\frac{\xi(t+1)}{\xi(t)} \leq \varrho_0 \quad \text{for } t \leq t_0 - 1.$$

PROOF (1) Note that $\mathbf{u}(t) = \mathbf{u}_+(t) - \mathbf{u}_-(t)$ and

$$\mathbf{u}(t) = \mathbf{U}_\mathbf{a}(t, s)\mathbf{u}_+(s) - \mathbf{U}_\mathbf{a}(t, s)\mathbf{u}_-(s)$$

for $t > s$. We then have

$$\mathbf{u}_+(t) \leq \mathbf{U}_\mathbf{a}(t, s)\mathbf{u}_+(s), \quad \mathbf{u}_-(t) \leq \mathbf{U}_\mathbf{a}(t, s)\mathbf{u}_-(s)$$

for $t > s$. It follows that $|\mathbf{u}(t)| \leq \mathbf{U}_\mathbf{a}(t, s)|\mathbf{u}(s)|$. Since $\mathbf{v}(t) > 0$ for $t \in J$ and $\mathbf{u}$ is a nontrivial solution, $\langle \mathbf{u}(t), \mathbf{v}(t) \rangle = 0$ implies that $\mathbf{u}(t)$ changes sign. We therefore must have $|\mathbf{u}(t)| < \mathbf{U}_\mathbf{a}(t, s)|\mathbf{u}(s)|$. Consequently,

$$\xi(t) < \langle \mathbf{U}_\mathbf{a}(t, s)|\mathbf{u}(s)|, \mathbf{v}(t) \rangle = \langle |\mathbf{u}(s)|, \mathbf{v}(s) \rangle = \xi(s).$$

Therefore ξ is a decreasing function on J.

(2) Suppose that the statement is false. Then there are $t_n \to -\infty$ such that

$$1 \geq \frac{\xi(t_n + t)}{\xi(t_n)} \geq \frac{\xi(t_n + 1)}{\xi(t_n)} \to 1 \quad (t \in [0, 1]) \tag{6.2.8}$$

as $n \to \infty$. Let

$$\mathbf{u}^n(t) := \frac{\mathbf{u}(t_n + t)}{\|\mathbf{u}(t_n)\|_{\phi^*}} \quad (t < t_0 - t_n)$$

and

$$\mathbf{v}^n(t) := \frac{\mathbf{v}(t_n + t)}{\|\mathbf{v}(t_n)\|_{\phi^*}} \quad (t < t_0 - t_n).$$

Then $\mathbf{u}^n$ is a solution of (6.0.1)+(6.0.2) on $(-\infty, t_0 - t_n)$ with $\mathbf{a}$ replaced by $\mathbf{a} \cdot t_n$ and $\mathbf{u}^n(0) = \frac{\mathbf{u}(t_n)}{\|\mathbf{u}(t_n)\|_{\phi^*}}$, and $\mathbf{v}^n$ is a solution of (6.1.29)+(6.1.30) on $(-\infty, t_0 - t_n)$ with $\mathbf{a}$ replaced by $\mathbf{a} \cdot t_n$ and $\mathbf{v}^n(0) = \frac{\mathbf{v}(t_n)}{\|\mathbf{v}(t_n)\|_{\phi^*}}$. Moreover, $\|\mathbf{u}^n(0)\|_{\phi^*} = \|\mathbf{v}^n(0)\|_{\phi^*} = 1$, $\langle \mathbf{u}^n(t), \mathbf{v}^n(t) \rangle = 0$ for $t < t_0 - t_n$, and $\mathbf{v}^n(t) \geq \mathbf{0}$ for $t < t_0 - t_n$.

Without loss of generality we may assume that $\mathbf{a} \cdot t_n \to \mathbf{a}^* \in \mathbf{Y}$.

By Lemma 6.2.10(1), $\|\mathbf{v}^n(2)\|_{\phi^*}$ is bounded in n, and then by Lemma 6.2.9, $\|\mathbf{v}^n(3/2)\|$ is bounded in n. Hence, by Theorems 6.1.3 and 6.1.11, we may assume that $\mathbf{v}^n(t) \to \mathbf{v}^*(t)$ in $\mathbf{X}$ for $t \in [0, 1]$, where $\mathbf{v}^*$ is a solution of (6.1.29)+(6.1.30) on $[0, 1]$ with $\mathbf{a}$ replaced by $\mathbf{a}^*$.

By Lemma 6.2.9, $\|\mathbf{u}^n(\delta)\|$ is bounded in n, for any $0 < \delta < 1$. Then, by Theorems 6.1.3 and 6.1.11 again, we may assume that $\mathbf{u}^n(t) \to \mathbf{u}^*(t)$ in $\mathbf{X}$ for $0 < t \leq 1$, where $\mathbf{u}^*$ is a solution of (6.0.1)+(6.0.2) on $(0, 1]$ with $\mathbf{a}$ replaced by $\mathbf{a}^*$, and $\langle \mathbf{u}^*(t), \mathbf{v}^*(t) \rangle = 0$ for $0 < t \leq 1$.

We claim that $\mathbf{u}^*$ and $\mathbf{v}^*$ are nontrivial solutions. In fact, by (6.2.8), for any $s \in (0, 1)$ and sufficiently large n, $\xi(t_n + s) \geq \xi(t_n)/2$, and hence

$$\langle |\mathbf{u}^*(s)|, \mathbf{v}^*(s) \rangle = \lim_{n \to \infty} \langle |\mathbf{u}(t_n + s)|/\|\mathbf{u}(t_n)\|_{\phi^*}, \mathbf{v}(t_n + s)/\|\mathbf{v}(t_n)\|_{\phi^*} \rangle$$

$$\geq \limsup_{n \to \infty} \frac{1}{2} \langle |\mathbf{u}(t_n)|/\|\mathbf{u}(t_n)\|_{\phi^*}, \mathbf{v}(t_n)/\|\mathbf{v}(t_n)\|_{\phi^*} \rangle.$$

By Lemma 6.2.10(2), there holds

$$\langle |\mathbf{u}(t_n)|, \mathbf{v}(t_n)/\|\mathbf{v}(t_n)\|_{\phi^*} \rangle \geq \eta_0 \langle |\mathbf{u}(t_n)|, \phi^* \rangle = \eta_0 \|\mathbf{u}(t_n)\|_{\phi^*}$$

for some $\eta_0 > 0$. It then follows that

$$\langle |\mathbf{u}^*(s)|, \mathbf{v}^*(s) \rangle \geq \frac{\eta_0}{2},$$

hence $\mathbf{u}^*$ and $\mathbf{v}^*$ are nontrivial solutions.

Now, for any $s, \tau \in (0, 1)$ with $s < \tau$, by Part (1)

$$\frac{\langle |\mathbf{u}^*(s)|, \mathbf{v}^*(s) \rangle}{\langle |\mathbf{u}^*(\tau)|, \mathbf{v}^*(\tau) \rangle} > 1.$$

By (6.2.8), we have

$$\frac{\langle |\mathbf{u}^*(s)|, \mathbf{v}^*(s)\rangle}{\langle |\mathbf{u}^*(\tau)|, \mathbf{v}^*(\tau)\rangle} = \lim_{n\to\infty} \frac{\xi(t_n + s)}{\xi(t_n + \tau)} = 1,$$

which is a contradiction. This proves (2).   □

PROOF (Proof of Theorem 6.2.15)   Suppose that $\mathbf{u}_1$ and $\mathbf{u}_2$ are two entire positive solutions of (6.0.1)+(6.0.2). Choose a nonzero $\mathbf{v}_0 \in \mathbf{X}^+$, and let $\mathbf{v}(t) := \mathbf{U}_\mathbf{a}^*(t, t_0)\mathbf{v}_0$. Clearly, there is a constant q such that

$$\langle \mathbf{u}_1(t_0) - q\mathbf{u}_2(t_0), \mathbf{v}_0 \rangle = 0.$$

Let $\mathbf{u}(t) := \mathbf{u}_1(t) - q\mathbf{u}_2(t)$. We then have $\mathbf{u}(t) = 0$ for all $t \in \mathbb{R}$ or $\mathbf{u}(t) \neq 0$ for all $t \leq t_1$ and some $t_1 \leq t_0$. Then by Lemma 6.2.11(2), there are $C^* > 0$ and $\gamma^* > 0$ such that

$$\langle |\mathbf{u}(t)|, \mathbf{v}(t)\rangle \leq C^* e^{-\gamma^*(t-s)} \langle |\mathbf{u}(s)|, \mathbf{v}(s)\rangle \quad t_1 > t > s.$$

On the other hand, we have

$$\begin{aligned}
\langle |\mathbf{u}(t)|, \mathbf{v}(t)\rangle &= \langle |\mathbf{u}_1(t) - q\mathbf{u}_2(t)|, \mathbf{v}(t)\rangle \\
&\leq \langle \mathbf{u}_1(t), \mathbf{v}(t)\rangle + |q|\langle \mathbf{u}_2(t), \mathbf{v}(t)\rangle \\
&= \text{const} \quad \text{for} \quad t \leq t_1.
\end{aligned}$$

It then follows that we must have $\mathbf{u}(t) = \mathbf{0}$ and then $\mathbf{u}_1(t) = q\mathbf{u}_2(t)$, for all $t \in \mathbb{R}$. Therefore an entire positive solution of (6.0.1)+(6.0.2) is unique up to a constant positive multiple.   □

6.3   Principal Spectrum and Principal Lyapunov Exponents in Nonautonomous and Random Cases

In this section, we study the principal spectrum and principal Lyapunov exponents of (6.0.3) and (6.0.4) and extend the theories developed in Chapters 4 and 5 for scalar parabolic equations to cooperative systems of parabolic equations.

6.3.1   The Random Case

In this subsection, we consider (6.0.4), i.e.,

$$
\begin{cases}
\dfrac{\partial u_k}{\partial t} = \displaystyle\sum_{i=1}^{N} \dfrac{\partial}{\partial x_i}\left(\sum_{j=1}^{N} a_{ij}^k(\theta_t\omega, x)\dfrac{\partial u_k}{\partial x_j} + a_i^k(\theta_t\omega, x)u_k\right) \\[3mm]
\quad + \displaystyle\sum_{i=1}^{N} b_i^k(\theta_t\omega, x)\dfrac{\partial u_k}{\partial x_i} + \sum_{l=1}^{K} c_l^k(\theta_t\omega, x)u_l, \qquad t > 0,\ x \in D \\[3mm]
\mathcal{B}_{a^{k,\omega}}(t)u_k = 0, \hspace{5.3cm} t > 0,\ x \in \partial D,
\end{cases}
\tag{6.3.1}
$$

where $k = 1, 2, \ldots, K$, $\omega \in \Omega$, $((\Omega, \mathfrak{F}, \mathbb{P}), \{\theta_t\}_{t\in\mathbb{R}})$ is an ergodic metric dynamical system, and for each $\omega \in \Omega$, $\mathbf{a}^\omega(t, x) := (a_{ij}^k(\theta_t\omega, x), a_i^k(\theta_t\omega, x), b_i^k(\theta_t\omega, x)$, $c_l^k(\theta_t\omega, x), d_0^k(\theta_t\omega, x)) \in \mathbf{Y}$ and $\mathcal{B}_{a^{k,\omega}}(t)$ is of the same form as in (6.0.2) with a^k being replaced by $a^{k,\omega}(t, x) = (a_{ij}^k(\theta_t\omega, x), a_i^k(\theta_t\omega, x), b_i^k(\theta_t\omega, x), 0, d_0^k(\theta_t\omega, x))$.

For a given $\omega \in \Omega$, let

$$
E_{\mathbf{a}}(\omega)(t, x) := \mathbf{a}^\omega(t, x) = \mathbf{a}(\theta_t\omega, x). \tag{6.3.2}
$$

Throughout this subsection, we assume

(A6-7) $\{E_{\mathbf{a}}(\omega) : \omega \in \Omega\} \subset \mathbf{Y}$ *and* $E_{\mathbf{a}} : \Omega \to \tilde{\mathbf{Y}}(\mathbf{a}) := \mathrm{cl}\,\{E_{\mathbf{a}}(\omega) : \omega \in \Omega\}$ *are measurable, where the closure is taken in the weak-* topology. Moreover, (A6-1)–(A6-4) are satisfied with* $\mathbf{Y}$ *replaced by* $\tilde{\mathbf{Y}}(\mathbf{a})$.

We say $\mathbf{a}$ is $\mathbf{Y}$-*admissible* if $\mathbf{a}$ satisfies (A6-7).

Let $\tilde{\mathbb{P}}$ be the image of the measure $\mathbb{P}$ under $E_{\mathbf{a}} : \forall A \in \mathfrak{B}(\tilde{\mathbf{Y}}(\mathbf{a}))$, $\tilde{\mathbb{P}}(A) := \mathbb{P}(E_{\mathbf{a}}^{-1}(A))$. Then $(\tilde{\mathbf{Y}}(\mathbf{a}), \{\sigma_t\}_{t\in\mathbb{R}})$ is a topological dynamical system with an ergodic invariant measure $\tilde{\mathbb{P}}$. Put

$$
\tilde{\mathbf{Y}}_0(\mathbf{a}) := \mathrm{supp}\,\tilde{\mathbb{P}}.
$$

Then $\tilde{\mathbf{Y}}_0(\mathbf{a})$ is a closed (hence compact) and $\{\sigma_t\}$–invariant subset of $\tilde{\mathbf{Y}}(\mathbf{a})$, with $\tilde{\mathbb{P}}(\tilde{\mathbf{Y}}_0(\mathbf{a})) = 1$. Moreover, $\tilde{\mathbf{Y}}_0(a)$ is connected. (See Chapter 4 for the reasonings.)

Similarly to Lemma 4.1.2, we have

LEMMA 6.3.1
There exists $\Omega_0 \subset \Omega$ with $\mathbb{P}(\Omega_0) = 1$ such that

$$
\tilde{\mathbf{Y}}_0(\mathbf{a}) = \mathrm{cl}\,\{E_{\mathbf{a}}(\theta_t\omega) : t \in \mathbb{R}\}
$$

for any $\omega \in \Omega_0$, where the closure is taken in the weak-* topology.

Denote by $\mathbf{\Pi}(\mathbf{a}) = \{\mathbf{\Pi}_t(\mathbf{a})\}_{t\geq 0}$ the topological skew-product semiflow generated by (6.3.1),

$$
\mathbf{\Pi}_t(\mathbf{a})(\mathbf{u}^0, \omega) := (\mathbf{U}_{\tilde{\mathbf{a}}}(t, 0)\mathbf{u}^0, \sigma_t\tilde{\mathbf{a}}) \quad \text{for } \tilde{\mathbf{a}} \in \tilde{\mathbf{Y}}_0(\mathbf{a}),\ \mathbf{u}^0 \in \mathbf{L}_2(D),\ t > 0.
$$

Instead of $\mathbf{U}_{E_{\mathbf{a}}(\omega)}(t, s)$ we will write $\mathbf{U}_\omega(t, s)$.

DEFINITION 6.3.1 (Principal spectrum and principal Lyapunov exponent)

(1) *The* principal spectrum *of (6.3.1), denoted by* $\Sigma(\mathbf{a}) = [\lambda_{\min}(\mathbf{a}), \lambda_{\max}(\mathbf{a})]$, *is defined to be the principal spectrum of* $\mathbf{\Pi}$ *over* $\tilde{\mathbf{Y}}_0(\mathbf{a})$.

(2) *The* principal Lyapunov exponent *of (6.3.1), denoted by* $\lambda(\mathbf{a})$, *is defined to be the principal Lyapunov exponent of* $\mathbf{\Pi}$ *over* $\tilde{\mathbf{Y}}_0(\mathbf{a})$ *for the ergodic invariant measure* $\tilde{\mathbb{P}}$.

THEOREM 6.3.1

Let Ω_0 be as in Lemma 6.3.1.

(1) *There is $\Omega_1 \subset \Omega_0$ with $\mathbb{P}(\Omega_1) = 1$ such that for any $\omega \in \Omega_1$ one has*

$$\lim_{t \to \infty} \frac{\ln \|\mathbf{U}_\omega(t, 0)\|}{t} = \lambda(\mathbf{a}). \tag{6.3.3}$$

(2) *For any sequence $(\omega^{(n)})_{n=1}^\infty \subset \Omega_0$ and any real sequences $(s_n)_{n=1}^\infty$, $(t_n)_{n=1}^\infty$ such that $t_n - s_n \to \infty$ one has*

$$\lambda_{\min}(\mathbf{a}) \leq \liminf_{n \to \infty} \frac{\ln \|\mathbf{U}_{\omega^{(n)}}(t_n, s_n)\|}{t_n - s_n}$$

$$\leq \limsup_{n \to \infty} \frac{\ln \|\mathbf{U}_{\omega^{(n)}}(t_n, s_n)\|}{t_n - s_n} \leq \lambda_{\max}(\mathbf{a}).$$

(3A) *There exist a sequence $(\omega^{(n)})_{n=1}^\infty \subset \Omega_0$ and a sequence $(t_{n,1})_{n=1}^\infty \subset (0, \infty)$ such that $t_{n,1} \to \infty$ as $n \to \infty$, and*

$$\lim_{n \to \infty} \frac{\ln \|\mathbf{U}_{\omega^n}(t_{n,1}, 0)\|}{t_{n,1}} = \lambda_{\min}(\mathbf{a}).$$

(3B) *There exist a sequence $(\omega^{(n)})_{n=1}^\infty \subset \Omega_0$ and a sequence $(t_{n,2})_{n=1}^\infty \subset (0, \infty)$ such that $t_{n,2} \to \infty$ as $n \to \infty$, and*

$$\lim_{n \to \infty} \frac{\ln \|\mathbf{U}_{\omega^{(n)}}(t_{n,2}, 0)\|}{t_{n,2}} = \lambda_{\max}(\mathbf{a}).$$

PROOF   (1) By Theorem 6.2.3, there is $\tilde{\mathbf{Y}}_1 \subset \tilde{\mathbf{Y}}_0$ with $\tilde{\mathbb{P}}(\tilde{\mathbf{Y}}_1) = 1$ such that

$$\lim_{t \to \infty} \frac{\ln \|\mathbf{U}_\omega(t, 0)\|}{t} = \lambda(\mathbf{a})$$

for any $\omega \in \Omega$ with $E_{\mathbf{a}}(\omega) \in \tilde{\mathbf{Y}}_1$. (1) then follows with $\Omega_1 = \Omega_0 \cap E_{\mathbf{a}}^{-1}(\tilde{\mathbf{Y}}_1)$.

(2) follows from Theorem 6.2.2(1).

(3) follows from Theorem 6.2.2(2A) and (2B). □

Let $\mathbf{a}^{(1)}$, $\mathbf{a}^{(2)}$ be $\mathbf{Y}$-admissible. Assume there is $\tilde{\Omega} \subset \Omega$ with $\mathbb{P}(\tilde{\Omega}) = 1$ such that for each $\omega \in \tilde{\Omega}$, $\mathbf{a}^{\omega,(m)}(t,x) = \mathbf{a}^{(m)}(\theta_t\omega,x) = (a_{ij}^{k,\omega,(m)}(t,x)$, $a_i^{k,\omega,(m)}(t,x)$, $b_i^{k,\omega,(m)}(t,x)$, $c_l^{k,\omega,(m)}(t,x)$, $d_0^{k,\omega,(m)}(t,x))$ $(m=1,2)$ satisfies

- $a_{ij}^{k,\omega,(1)}(t,x) = a_{ij}^{k,\omega,(2)}(t,x)$, $a_i^{k,\omega,(1)}(t,x) = a_i^{k,\omega,(2)}(t,x)$, $b_i^{k,\omega,(1)}(t,x) = b_i^{k,\omega,(2)}(t,x)$, for a.e. $(t,x) \in \mathbb{R} \times D$,

- $c_l^{k,\omega,(1)}(t,x) \le c_l^{k,\omega,(2)}(t,x)$ for a.e. $(t,x) \in \mathbb{R} \times D$,

- $d_0^{k,\omega,(1)}(t,x) \ge d_0^{k,\omega,(2)}(t,x)$ for a.e. $(t,x) \in \mathbb{R} \times \partial D$.

THEOREM 6.3.2 (Monotonicity with respect to zero order terms)

(1) $\lambda(\mathbf{a}^{(1)}) \le \lambda(\mathbf{a}^{(2)})$.

(2) $\lambda_{\min}(\mathbf{a}^{(1)}) \le \lambda_{\min}(\mathbf{a}^{(2)})$ *and* $\lambda_{\max}(\mathbf{a}^{(1)}) \le \lambda_{\max}(\mathbf{a}^{(2)})$.

PROOF It follows from Theorem 6.1.5 and Theorem 6.3.1. □

For a given $\mathbf{Y}$-admissible $\mathbf{a}$, denote by $\lambda^{\mathrm{D}}(\mathbf{a})$, $\lambda^{\mathrm{R}}(\mathbf{a})$, and $\lambda^{\mathrm{N}}(\mathbf{a})$ the Lyapunov exponents of (6.3.1) with Dirichlet, Robin, and Neumann boundary conditions, respectively. Denote by $[\lambda_{\min}^{\mathrm{D}}(\mathbf{a}), \lambda_{\max}^{\mathrm{D}}(\mathbf{a})]$, $[\lambda_{\min}^{\mathrm{R}}(\mathbf{a}), \lambda_{\max}^{\mathrm{R}}(\mathbf{a})]$, and $[\lambda_{\min}^{\mathrm{N}}(\mathbf{a}), \lambda_{\max}^{\mathrm{N}}(\mathbf{a})]$ the principal spectrum intervals of (6.3.1) with Dirichlet, Robin, and Neumann boundary conditions, respectively.

THEOREM 6.3.3 (Monotonicity with respect to boundary conditions)

(1) $\lambda^{\mathrm{D}}(\mathbf{a}) \le \lambda^{\mathrm{R}}(\mathbf{a}) \le \lambda^{\mathrm{N}}(\mathbf{a})$.

(2) $\lambda_{\min}^{\mathrm{D}}(\mathbf{a}) \le \lambda_{\min}^{\mathrm{R}}(\mathbf{a}) \le \lambda_{\min}^{\mathrm{N}}(\mathbf{a})$ *and* $\lambda_{\max}^{\mathrm{D}}(\mathbf{a}) \le \lambda_{\max}^{\mathrm{R}}(\mathbf{a}) \le \lambda_{\max}^{\mathrm{N}}(\mathbf{a})$.

PROOF It also follows from Theorem 6.1.5 and Theorem 6.3.1. □

In the rest of this subsection, we assume that (A6-1)–(A6-6) are satisfied. Then $\mathbf{\Pi}$ admits an exponential separation on $\tilde{\mathbf{Y}}_0(\mathbf{a})$.

THEOREM 6.3.4

Let Ω_0 be as in Lemma 6.3.1.

(1) There is $\Omega_1 \subset \Omega_0$ with $\mathbb{P}(\Omega_1) = 1$ such that for any $\omega \in \Omega_1$ and any $\mathbf{u}^0 \in \mathbf{L}_2^+(D) \setminus \{\mathbf{0}\}$ one has

$$\lim_{t \to \infty} \frac{\ln \| \mathbf{U}_\omega(t, 0) \mathbf{u}^0 \|}{t} = \lambda(\mathbf{a}). \qquad (6.3.4)$$

(2) For any sequence $(\omega^{(n)})_{n=1}^\infty \subset \Omega_0$ and any real sequences $(s_n)_{n=1}^\infty$, $(t_n)_{n=1}^\infty$ such that $t_n - s_n \to \infty$ one has

$$\lambda_{\min}(\mathbf{a}) \le \liminf_{n \to \infty} \frac{\ln \| \mathbf{U}_{\omega^{(n)}}(t_n, s_n) \mathbf{w}(E(\omega^{(n)}) \cdot s_n) \|}{t_n - s_n}$$

$$= \liminf_{n \to \infty} \frac{\ln \| \mathbf{U}_{\omega^{(n)}}(t_n, s_n) \mathbf{u}^0 \|}{t_n - s_n}$$

$$\le \limsup_{n \to \infty} \frac{\ln \| \mathbf{U}_{\omega^{(n)}}(t_n, s_n) \mathbf{w}(E(\omega^{(n)}) \cdot s_n) \|}{t_n - s_n}$$

$$= \limsup_{n \to \infty} \frac{\ln \| \mathbf{U}_{\omega^{(n)}}(t_n, s_n) \mathbf{u}^0 \|}{t_n - s_n} \le \lambda_{\max}(\mathbf{a})$$

for each $\mathbf{u}^0 \in \mathbf{L}_2(D)^+ \setminus \{\mathbf{0}\}$.

(3) For each $\omega \in \Omega_0$, there are sequences $(s_n')_{n=1}^\infty, (t_n')_{n=1}^\infty \subset \mathbb{R}$, $t_n' - s_n' \to \infty$ as $n \to \infty$, such that

$$\lambda_{\min}(\mathbf{a}) = \lim_{n \to \infty} \frac{\ln \| \mathbf{U}_\omega(t_n', s_n') \mathbf{w}(E(\omega) \cdot s_n') \|}{t_n' - s_n'} = \lim_{n \to \infty} \frac{\ln \| \mathbf{U}_\omega(t_n', s_n') \mathbf{u}^0 \|}{t_n' - s_n'}$$

for each $\mathbf{u}^0 \in \mathbf{L}_2(D)^+ \setminus \{\mathbf{0}\}$.

(4) For each $\omega \in \Omega_0$, there are sequences $(s_n'')_{n=1}^\infty, (t_n'')_{n=1}^\infty \subset \mathbb{R}$, $t_n'' - s_n'' \to \infty$ as $n \to \infty$, such that

$$\lambda_{\max}(\mathbf{a}) = \lim_{n \to \infty} \frac{\ln \| \mathbf{U}_\omega(t_n'', s_n'') \mathbf{w}(E(\omega) \cdot s_n'') \|}{t_n'' - s_n''} = \lim_{n \to \infty} \frac{\ln \| \mathbf{U}_\omega(t_n'', s_n'') \mathbf{u}^0 \|}{t_n'' - s_n''}$$

for each $\mathbf{u}^0 \in \mathbf{L}_2(D)^+ \setminus \{\mathbf{0}\}$.

PROOF   (1) It follows from Lemma 6.2.6 and Theorem 6.3.1(1).
(2) It follows from Lemma 6.2.6 and Theorem 6.3.1(2).
(3) It follows from Lemma 6.2.6 and Theorem 6.2.12. $\square$

In the theorem below a $\mathbf{Y}$-admissible $\mathbf{a}^{(0)}$ is fixed. $\|\cdot\|_\infty$ stands for the norm in $L_\infty(\mathbb{R} \times D, \mathbb{R}^{K(N^2+2N+K)}) \times L_\infty(\mathbb{R} \times \partial D, \mathbb{R}^K)$.

THEOREM 6.3.5 (Continuous dependence on coefficients)

(1) *For each $\epsilon > 0$ there is $\delta > 0$ such that for any* **Y**-*admissible* **a**, *if* $\|E_{\mathbf{a}}(\omega) - E_{\mathbf{a}^{(0)}}(\omega)\|_\infty < \delta$ *for* $\mathbb{P}$-*a.e.* $\omega \in \Omega$ *then*

$$|\lambda(\mathbf{a}) - \lambda(\mathbf{a}^{(0)})| < \epsilon.$$

(2) *For each $\epsilon > 0$ there is $\delta > 0$ such that for any* **Y**-*admissible* **a**, *if* $\|E_{\mathbf{a}}(\omega) - E_{\mathbf{a}^{(0)}}(\omega)\|_\infty < \delta$ *for* $\mathbb{P}$-*a.e.* $\omega \in \Omega$ *then*

$$|\lambda_{\min}(\mathbf{a}) - \lambda_{\min}(\mathbf{a}^{(0)})| < \epsilon \quad and \quad |\lambda_{\max}(\mathbf{a}) - \lambda_{\max}(\mathbf{a}^{(0)})| < \epsilon.$$

PROOF It follows along the lines of the proofs of Theorems 4.4.1 and 4.4.2. $\quad\square$

The above theorems extend the theories developed in Chapter 4 for scalar random parabolic equations to cooperative systems of random parabolic equations.

6.3.2 The Nonautonomous Case

In this subsection, we consider (6.0.3), i.e.,

$$\begin{cases} \dfrac{\partial u_k}{\partial t} = \displaystyle\sum_{i=1}^{N} \dfrac{\partial}{\partial x_i}\left(\sum_{j=1}^{N} a_{ij}^k(t,x)\dfrac{\partial u_k}{\partial x_j} + a_i^k(t,x)u_k \right) \\ \qquad + \displaystyle\sum_{i=1}^{N} b_i^k(t,x)\dfrac{\partial u_k}{\partial x_i} + \sum_{l=1}^{K} c_l^k(t,x)u_l, \qquad t > 0,\ x \in D, \\ \mathcal{B}_{a^k}(t)u_k = 0 \qquad\qquad\qquad\qquad\qquad\qquad t > 0,\ x \in \partial D, \end{cases} \tag{6.3.5}$$

where $\mathcal{B}_{a^k}(t)$ is of the same form as in (6.0.2), $k = 1, 2, \dots, K$ and $\mathbf{a} = (a_{ij}^k, a_i^k, b_i^k, c_l^k, d_0^k)$ is a given element in **Y**. Throughout this subsection, we assume

(A6-8) $\tilde{\mathbf{Y}}(\mathbf{a}) := \mathrm{cl}\,\{\,\mathbf{a} \cdot t : t \in \mathbb{R}\,\}$, *where the closure is taken in the weak-*$*$ topology, satisfies (A6-1)–(A6-4) with* **Y** *replaced by* $\tilde{\mathbf{Y}}(\mathbf{a})$.

We say **a** is **Y**-*admissible* if it satisfies (A6-8).

Let $\boldsymbol{\Pi} = \{\boldsymbol{\Pi}_t(\mathbf{a})\}_{t\geq 0}$ be the topological linear skew-product semiflow generated by (6.3.5),

$$\boldsymbol{\Pi}_t(\mathbf{a})(\mathbf{u}_0, \tilde{\mathbf{a}}) := (\mathbf{U}_{\tilde{\mathbf{a}}}(t, 0)\mathbf{u}_0, \sigma_t\tilde{\mathbf{a}})$$

where $\mathbf{u}_0 \in \mathbf{L}_2(D)$ and $\tilde{\mathbf{a}} \in \tilde{\mathbf{Y}}(\mathbf{a})$.

DEFINITION 6.3.2 (Principal spectrum) *The principal spectrum of* (6.3.5), *denoted by* $\Sigma(\mathbf{a}) = [\lambda_{\min}(\mathbf{a}), \lambda_{\max}(\mathbf{a})]$, *is defined to be the principal spectrum of* $\boldsymbol{\Pi}$ *over* $\tilde{\mathbf{Y}}(\mathbf{a})$.

THEOREM 6.3.6

(1) *For any sequence* $(\tilde{\mathbf{a}}^{(n)})_{n=1}^\infty \subset \tilde{\mathbf{Y}}(\mathbf{a})$, *and any real sequences* $(s_n)_{n=1}^\infty$, $(t_n)_{n=1}^\infty$ *such that* $t_n - s_n \to \infty$ *one has*

$$\lambda_{\min}(\mathbf{a}) \le \liminf_{n \to \infty} \frac{\ln \|\mathbf{U}_{\tilde{\mathbf{a}}^{(n)}}(t_n, s_n)\|}{t_n - s_n}$$

$$\le \limsup_{n \to \infty} \frac{\ln \|\mathbf{U}_{\tilde{\mathbf{a}}^{(n)}}(t_n, s_n)\|}{t_n - s_n} \le \lambda_{\max}(\mathbf{a}).$$

(2A) *There exist a sequence* $(\tilde{\mathbf{a}}^{(n)})_{n=1}^\infty \subset \tilde{\mathbf{Y}}(\mathbf{a})$ *and a sequence* $(t_{n,1})_{n=1}^\infty \subset (0, \infty)$ *such that* $t_{n,1} \to \infty$ *as* $n \to \infty$, *and*

$$\lim_{n \to \infty} \frac{\ln \|\mathbf{U}_{\tilde{\mathbf{a}}^n}(t_{n,1}, 0)\|}{t_{n,1}} = \lambda_{\min}(\mathbf{a}).$$

(2B) *There exist a sequence* $(\tilde{\mathbf{a}}^{(n)})_{n=1}^\infty \subset \tilde{\mathbf{Y}}(\mathbf{a})$ *and a sequence* $(t_{n,2})_{n=1}^\infty \subset (0, \infty)$ *such that* $t_{n,2} \to \infty$ *as* $n \to \infty$, *and*

$$\lim_{n \to \infty} \frac{\ln \|\mathbf{U}_{\tilde{\mathbf{a}}^n}(t_{n,2}, 0)\|}{t_{n,2}} = \lambda_{\max}(\mathbf{a}).$$

PROOF (1) follows from Theorem 6.2.2 (1).
(2) follows from Theorem 6.2.2 (2A) and (2B). ∎

Let **Y**-admissible $\mathbf{a}^{(1)}$, $\mathbf{a}^{(2)}$ satisfy the following:

- $a_{ij}^{k,(1)}(t, x) = a_{ij}^{k,(2)}(t, x)$, $a_i^{k,(1)}(t, x) = a_i^{k,(2)}(t, x)$, $b_i^{k,(1)}(t, x) = b_i^{k,(2)}(t, x)$, for a.e. $(t, x) \in \mathbb{R} \times D$,

- $c_l^{k,(1)}(t, x) \le c_l^{k,(2)}(t, x)$ for a.e. $(t, x) \in \mathbb{R} \times D$,

- $d_0^{k,(1)}(t, x) \ge d_0^{k,(2)}(t, x)$ for a.e. $(t, x) \in \mathbb{R} \times \partial D$.

THEOREM 6.3.7 (Monotonicity with respect to zero order terms)

$\lambda_{\min}(\mathbf{a}^{(1)}) \le \lambda_{\min}(\mathbf{a}^{(2)})$ *and* $\lambda_{\max}(\mathbf{a}^{(1)}) \le \lambda_{\max}(\mathbf{a}^{(2)})$.

PROOF It follows from Theorem 6.1.5 and Theorem 6.3.6. ∎

For a given **Y**-admissible $\mathbf{a}$, denote by $[\lambda_{\min}^{\mathrm{D}}(\mathbf{a}), \lambda_{\max}^{\mathrm{D}}(\mathbf{a})]$, $[\lambda_{\min}^{\mathrm{R}}(\mathbf{a}), \lambda_{\max}^{\mathrm{R}}(\mathbf{a})]$, and $[\lambda_{\min}^{\mathrm{N}}(\mathbf{a}), \lambda_{\max}^{\mathrm{N}}(\mathbf{a})]$ the principal spectrum intervals of (6.3.5) with Dirichlet, Robin, and Neumann boundary conditions, respectively.

THEOREM 6.3.8 (Monotonicity with respect to boundary conditions)
$\lambda_{\min}^{\mathrm{D}}(\mathbf{a}) \leq \lambda_{\min}^{\mathrm{R}}(\mathbf{a}) \leq \lambda_{\min}^{\mathrm{N}}(\mathbf{a})$ *and* $\lambda_{\max}^{\mathrm{D}}(\mathbf{a}) \leq \lambda_{\max}^{\mathrm{R}}(\mathbf{a}) \leq \lambda_{\max}^{\mathrm{N}}(\mathbf{a})$.

PROOF   It follows from Theorem 6.1.5 and Theorem 6.3.6.   □

In the rest of this subsection, we assume that (A6-1)–(A6-6) are satisfied. Hence $\mathbf{\Pi}$ admits an exponential separation over $\tilde{\mathbf{Y}}(\mathbf{a})$.

THEOREM 6.3.9

(1) *For any* $\mathbf{u}^0 \in \mathbf{L}_2(D)^+ \setminus \{\mathbf{0}\}$ *and any real sequences* $(s_n)_{n=1}^\infty$, $(t_n)_{n=1}^\infty$ *such that* $t_n - s_n \to \infty$ *one has*

$$\lambda_{\min}(\mathbf{a}) \leq \liminf_{n\to\infty} \frac{\ln \|\mathbf{U}_{\mathbf{a}}(t_n, s_n)\mathbf{w}(\mathbf{a} \cdot s_n)\|}{t_n - s_n}$$
$$= \liminf_{n\to\infty} \frac{\ln \|\mathbf{U}_{\mathbf{a}}(t_n, s_n)\mathbf{u}^0\|}{t_n - s_n}$$
$$\leq \limsup_{n\to\infty} \frac{\ln \|\mathbf{U}_{\mathbf{a}}(t_n, s_n)\mathbf{w}(\mathbf{a} \cdot s_n)\|}{t_n - s_n}$$
$$= \limsup_{n\to\infty} \frac{\ln \|\mathbf{U}_{\mathbf{a}}(t_n, s_n)\mathbf{u}^0\|}{t_n - s_n} \leq \lambda_{\max}(\mathbf{a}).$$

(2) *There are sequences* $(s_n')_{n=1}^\infty, (t_n')_{n=1}^\infty \subset \mathbb{R}$, $t_n' - s_n' \to \infty$ *as* $n \to \infty$, *such that*

$$\lambda_{\min}(\mathbf{a}) = \lim_{n\to\infty} \frac{\ln \|\mathbf{U}_{\mathbf{a}}(t_n', s_n')\mathbf{w}(\mathbf{a} \cdot s_n')\|}{t_n' - s_n'} = \lim_{n\to\infty} \frac{\ln \|\mathbf{U}_{\mathbf{a}}(t_n', s_n')\mathbf{u}^0\|}{t_n' - s_n'}$$

for each $\mathbf{u}^0 \in \mathbf{L}_2(D)^+ \setminus \{\mathbf{0}\}$.

(3) *There are sequences* $(s_n'')_{n=1}^\infty, (t_n'')_{n=1}^\infty \subset \mathbb{R}$, $t_n'' - s_n'' \to \infty$ *as* $n \to \infty$, *such that*

$$\lambda_{\max}(\mathbf{a}) = \lim_{n\to\infty} \frac{\ln \|\mathbf{U}_{\mathbf{a}}(t_n'', s_n'')\mathbf{w}(\mathbf{a} \cdot s_n'')\|}{t_n'' - s_n''} = \lim_{n\to\infty} \frac{\ln \|\mathbf{U}_{\mathbf{a}}(t_n'', s_n'')\mathbf{u}^0\|}{t_n'' - s_n''}$$

for each $\mathbf{u}^0 \in \mathbf{L}_2(D)^+ \setminus \{\mathbf{0}\}$.

PROOF   (1) It follows from Lemma 6.2.6 and Theorem 6.3.6.
(2) It follows from Lemma 6.2.6 and Theorem 6.2.12.   □

In the theorem below a $\mathbf{Y}$-admissible $\mathbf{a}^{(0)}$ is fixed. $\|\cdot\|_\infty$ stands for the norm in $L_\infty(\mathbb{R} \times D, \mathbb{R}^{K(N^2+2N+K)}) \times L_\infty(\mathbb{R} \times \partial D, \mathbb{R}^K)$.

THEOREM 6.3.10 (Continuous dependence on coefficients)
For each $\epsilon > 0$ there is $\delta > 0$ such that for any $\mathbf{Y}$-admissible $\mathbf{a}$, if $\|\mathbf{a} - \mathbf{a}^{(0)}\|_\infty < \delta$ then

$$|\lambda_{\min}(\mathbf{a}) - \lambda_{\min}(\mathbf{a}^{(0)})| < \epsilon \quad and \quad |\lambda_{\max}(\mathbf{a}) - \lambda_{\max}(\mathbf{a}^{(0)})| < \epsilon.$$

PROOF   It follows by arguments similar to those in the proof of Theorem 4.4.3.   ▯

The above theorems extend the theories developed in Chapter 4 for scalar nonautonomous parabolic equations to cooperative systems of nonautonomous parabolic equations.

6.3.3   Influence of Time and Space Variations

In this subsection we study the influence of time and space variations of the zero-order terms on principal spectrum and principal Lyapunov exponent. We assume that a_{ij}^k, a_i^k, b_i^k, c_l^k ($l \neq k$), and d_0^k are independent of t, i.e., we consider

$$\begin{cases} \dfrac{\partial u_k}{\partial t} = \displaystyle\sum_{i=1}^N \dfrac{\partial}{\partial x_i}\left(\sum_{j=1}^N a_{ij}^k(x)\dfrac{\partial u_k}{\partial x_j} + a_i^k(x)u_k\right) \\[2mm] \quad + \displaystyle\sum_{i=1}^N b_i^k(x)\dfrac{\partial u_k}{\partial x_i} + \sum_{l \neq k} c_l^k(x)u_l + c_k^k(t,x)u_k, \quad t > 0,\ x \in D, \\[2mm] \mathcal{B}_{a^k}u_k = 0, \hspace{6.5cm} t > 0,\ x \in \partial D, \end{cases} \tag{6.3.6}$$

and

$$\begin{cases} \dfrac{\partial u_k}{\partial t} = \displaystyle\sum_{i=1}^N \dfrac{\partial}{\partial x_i}\left(\sum_{j=1}^N a_{ij}^k(x)\dfrac{\partial u_k}{\partial x_j} + a_i^k(x)u_k\right) \\[2mm] \quad + \displaystyle\sum_{i=1}^N b_i^k(x)\dfrac{\partial u_k}{\partial x_i} + \sum_{l \neq k} c_l^k(x)u_l + c_k^k(\theta_t\omega,x)u_k, \quad t > 0,\ x \in D, \\[2mm] \mathcal{B}_{a^k}u_k = 0, \hspace{6.5cm} t > 0,\ x \in \partial D, \end{cases}$$
$$\tag{6.3.7}$$

where $\mathcal{B}_{a^k} \equiv \mathcal{B}_{a^k}(t)$ and $\mathcal{B}_{a^k}(t)$ is as in (2.0.3) with $a = a^k = (a_{ij}^k(\cdot),\, a_i^k(\cdot),\, b_i^k(\cdot),\, 0,\, d_0^k(\cdot))$, $k = 1, 2, \ldots, K$.

Throughout this subsection, we make the following assumption.

(A6-9) $\tilde{\mathbf{Y}}(\mathbf{a})$ *induced by* (6.3.6) *(or by* (6.3.7)*) satisfies* (A6-1)–(A6-6).

In the case of (6.3.6), a function $\hat{\mathbf{c}}(x) = (\hat{c}_k^k(x))_{k=1}^K$ is called a *time averaged function* of $\mathbf{c}(t,x) = (c_k^k(t,x))_{k=1}^K$ if

$$\hat{c}_k^k(x) = \lim_{t_n - s_n \to \infty} \frac{1}{t_n - s_n} \int_{s_n}^{t_n} c_k^k(t,x)\, dt$$

for some sequence $t_n - s_n \to \infty$, uniformly for $x \in \bar{D}$.

In the case of (6.3.7), $\hat{\mathbf{c}}(x) = (\hat{c}_k^k(x))_{k=1}^K$ is called the *time averaged function* of $\mathbf{c}(\theta_t \omega, x) = (c_k^k(\theta_t \omega, x))_{k=1}^K$ if

$$\hat{c}_k^k(x) = \int_\Omega c_k^k(\omega, x) \, d\mathbb{P}(\omega)$$

for $x \in \bar{D}$.

We call the following cooperative system of parabolic equations,

$$
\begin{cases}
\dfrac{\partial u_k}{\partial t} = \displaystyle\sum_{i=1}^N \dfrac{\partial}{\partial x_i} \left(\sum_{j=1}^N a_{ij}^k(x) \dfrac{\partial u_k}{\partial x_j} + a_i^k(x) u_k \right) \\
\qquad + \displaystyle\sum_{i=1}^N b_i^k(x) \dfrac{\partial u_k}{\partial x_i} + \sum_{l \neq k} c_l^k(x) u_l + \hat{c}_k^k(x) u_k, \quad t > 0, \ x \in D, \\
\mathcal{B}_{a^k} u_k = 0, \qquad\qquad\qquad\qquad\qquad\qquad\qquad\quad t > 0, \ x \in \partial D
\end{cases}
\tag{6.3.8}
$$

a *time averaged equation* of (6.3.6) (*the time averaged equation* of (6.3.7)) if $\hat{\mathbf{c}} = (\hat{c}_k^k(x))$ is an averaged function of $\mathbf{c}(t, x) = (c_k^k(t, x))$ (the averaged function of $\mathbf{c}(\omega, x) = (c_k^k(\omega, x))$).

Let $[\lambda_{\min}(\mathbf{a}), \lambda_{\max}(\mathbf{a})]$ be the principal spectrum of (6.3.6), and let $\lambda(\mathbf{a})$ be the principal Lyapunov exponent of (6.3.7). Further, let $\lambda_{\mathrm{princ}}(\mathbf{a}, \hat{\mathbf{c}})$ be the principal eigenvalue of (6.3.8). Then we have

THEOREM 6.3.11 (Influence of temporal variation in the nonautonomous case)

Consider (6.3.6).

(1) *There is a time averaged function $\hat{\mathbf{c}}(x)$ of $\mathbf{c}(t, x)$ such that $\lambda_{\min}(\mathbf{a}) \geq \lambda_{\mathrm{princ}}(\mathbf{a}, \hat{\mathbf{c}})$.*

(2) *$\lambda_{\max}(\mathbf{a}) \geq \lambda_{\mathrm{princ}}(\mathbf{a}, \hat{\mathbf{c}})$ for any time averaged function $\hat{\mathbf{c}}(x)$ of $\mathbf{c}(t, x)$.*

THEOREM 6.3.12 (Influence of time variations in the random case)

Consider (6.3.7). *Then $\lambda(\mathbf{a}) \geq \lambda_{\mathrm{princ}}(\mathbf{a}, \hat{\mathbf{c}})$.*

To prove the above theorems, we first show a few lemmas.

Let $\mathcal{A}\mathbf{u}$ be defined by

$$
(\mathcal{A}\mathbf{u})_k := \sum_{i=1}^N \dfrac{\partial}{\partial x_i} \left(\sum_{j=1}^N a_{ij}^k(x) \dfrac{\partial u_k}{\partial x_j} + a_i^k(x) u_k \right)
$$
$$
+ \sum_{i=1}^N b_i^k(x) \dfrac{\partial u_k}{\partial x_i} + \sum_{l \neq k} c_l^k(x) u_l
\tag{6.3.9}
$$

together with boundary conditions $\mathcal{B}_{a^k} u_k = 0$, $1 \leq k \leq K$.

We denote

$$\tilde{\mathbf{X}} := \begin{cases} \mathring{C}(\bar{D}, \mathbb{R}^K) & \text{(Dirichlet)} \\ C(\bar{D}, \mathbb{R}^K) & \text{(Neumann or Robin)}. \end{cases}$$

LEMMA 6.3.2
$\mathcal{A}$ together with boundary conditions $\mathcal{B}_{a^k} u_k = 0$, $1 \leq k \leq K$, generates an analytic semigroup $\{e^{\mathcal{A}t}\}_{t \geq 0}$ on $\tilde{\mathbf{X}}$. Moreover, $e^{\mathcal{A}t} \mathbf{u}^0 \geq \mathbf{0}$ for any $\mathbf{u}^0 \geq \mathbf{0}$, $\mathbf{u}^0 \in D(\mathcal{A})$, and any $t > 0$.

PROOF For the first statement, see [87]. The second statement follows along the lines of Theorem 6.1.5(1). □

Denote by $D(\mathcal{A})$ the domain of $\mathcal{A}$.

LEMMA 6.3.3
Assume $\mathbf{u}^0 \in D(\mathcal{A})$ and $\mathbf{u}^0 \geq \mathbf{0}$. If $u_k^0(x^*) = 0$ for some $x^* \in D$ and $1 \leq k \leq K$, then $(\mathcal{A}\mathbf{u}^0)_k(x^*) \geq 0$.

PROOF Since $\mathbf{u}^0 \in D(\mathcal{A})$, we have

$$\lim_{t \to 0+} \frac{(e^{\mathcal{A}t}\mathbf{u}^0)(x) - \mathbf{u}^0(x)}{t} = (\mathcal{A}\mathbf{u}^0)(x)$$

for any $x \in \bar{D}$. But $e^{\mathcal{A}t}\mathbf{u}^0 \geq \mathbf{0}$. It then follows that $(\mathcal{A}\mathbf{u}^0)_k(x^*) \geq 0$. □

For given $S < T$, let $\eta(t; S) := \|\mathbf{U}_\mathbf{a}(t, S)\mathbf{w}(\mathbf{a} \cdot S)\|$, $\mathbf{v}(t, x; S) := (\mathbf{U}_\mathbf{a}(t, S)\mathbf{w}(\mathbf{a} \cdot S))(x)/\eta(t; S)$, and $\hat{\mathbf{w}}(x; S, T) := (\hat{w}_1(x; S, T), \hat{w}_2(x; S, T), \ldots, \hat{w}_K(x; S, T))$, where

$$\hat{w}_l(x; S, T) := \exp\left(\frac{1}{T - S} \int_S^T \ln v_l(t, x; S)\, dt\right) \tag{6.3.10}$$

for $x \in D$ and $\hat{w}_l(x; S, T) = 0$ for $x \in \partial D$ in the Dirichlet case, and

$$\hat{w}_l(x; S, T) := \exp\left(\frac{1}{T - S} \int_S^T \ln v_l(t, x; S)\, dt\right) \tag{6.3.11}$$

for $x \in \bar{D}$ in the Neumann and Robin cases ($l = 1, 2, \ldots, K$).

For $S < T$, $\mathbf{v}(T, \cdot; S) = \mathbf{w}(\mathbf{a} \cdot T) \in D(\mathcal{A})$. Also, it follows from Theorem 6.1.6 that $\hat{\mathbf{w}}(\cdot; S, T) \in D(\mathcal{A})$, for $S < T$.

We have

LEMMA 6.3.4

$$\frac{1}{T-S}\int_S^T \frac{\mathcal{A}_l(\mathbf{v}(t,\cdot;S))(x)}{v_l(t,x;S)}\,dt \geq \frac{\mathcal{A}_l(\hat{\mathbf{w}}(\cdot;S,T))(x)}{\hat{w}_l(x;S,T)} \quad \text{for all} \quad x \in D,$$

(6.3.12)

where $\mathcal{A}_l\mathbf{u} = (\mathcal{A}\mathbf{u})_l$, $l = 1, 2, \ldots, K$.

PROOF It follows from the arguments of [100, Proposition 2.2]. For convenience, we provide a proof here.

First of all, recall the Jensen inequality

$$\frac{1}{T-S}\int_S^T f(t)\,dt \geq \exp\left(\frac{1}{T-S}\int_S^T \ln(f(t))\,dt\right)$$

(6.3.13)

for any positive continuous function defined on $[S, T]$, with the equality if and only if f is a constant function. This implies that

$$\frac{1}{T-S}\int_S^T \frac{\mathbf{v}(t,x;S)}{v_l(t,x^*;S)}\,dt \geq \frac{\hat{\mathbf{w}}(x;S,T)}{\hat{w}_l(x^*;S,T)}$$

for any $x, x^* \in D$ and $1 \leq l \leq K$, where the inequality $\geq$ is to be understood coordinatewise.

Let

$$\mathbf{v}^l(x,x^*) := \frac{1}{T-S}\int_S^T \frac{\mathbf{v}(t,x;S)}{v_l(t,x^*;S)}\,dt - \frac{\hat{\mathbf{w}}(x;S,T)}{\hat{w}_l(x^*;S,T)}.$$

Then $\mathbf{v}^l(\cdot,x^*) \geq \mathbf{0}$ and $v_l^l(x^*,x^*) = 0$. Observe that $\mathbf{v}^l(\cdot,x^*) \in D(\mathcal{A})$. Then by Lemma 6.3.3, (6.3.12) holds at x^*. Since $x^* \in D$ is arbitrary, we have that (6.3.12) holds for any $x \in D$. □

PROOF (Proof of Theorem 6.3.11)

(1) Let $\eta(t;S)$, $\mathbf{v}(t,x;S)$, and $\hat{\mathbf{w}}(x;S,T)$ be as above. Then we have

$$\frac{\partial v_k}{\partial t}\eta + v_k\frac{\partial \eta}{\partial t} = \eta(\mathcal{A}\mathbf{v})_k + c_k^k(t,x)\eta v_k, \quad x \in D,$$

(6.3.14)

and

$$\mathcal{B}_{a^k}v_k = 0, \quad x \in \partial D.$$

(6.3.15)

By (6.3.14), we have

$$\frac{1}{T-S}\int_S^T \frac{\partial v_k}{\partial t}\frac{1}{v_k}\,dt + \frac{1}{T-S}\ln\eta(T;S)$$

$$= \frac{1}{T-S}\int_S^T \frac{(\mathcal{A}\mathbf{v})_k}{v_k}\,dt + \frac{1}{T-S}\int_S^T c_k^k(t,x)\,dt, \quad x \in D. \quad (6.3.16)$$

By Lemma 6.3.4,

$$\frac{1}{T-S}\int_S^T \frac{\partial v_k}{\partial t}\frac{1}{v_k}\,dt + \frac{1}{T-S}\ln\eta(T;S)$$
$$\geq \frac{\mathcal{A}(\hat{\mathbf{w}}(\cdot;S,T)))_k}{\hat{\mathbf{w}}(\cdot;S,T)_k} + \frac{1}{T-S}\int_S^T c_k^k(t,x)\,dt, \quad x \in D. \qquad (6.3.17)$$

Let $T_n - S_n \to \infty$ be such that

$$\lim_{n\to\infty}\frac{1}{T_n - S_n}\ln\eta(T_n;S_n) = \lambda_{\min}(\mathbf{a}).$$

Without loss of generality, assume that

$$\hat{c}_k(x) = \lim_{n\to\infty}\frac{1}{T_n - S_n}\int_{S_n}^{T_n} c_k^k(t,x)\,dt$$

exists for all $x \in D$ and $k = 1, 2, \ldots, K$. We may also assume that there is $\mathbf{w}^* = \mathbf{w}^*(x)$ such that

$$\hat{\mathbf{w}}(x;S_n,T_n) \to \mathbf{w}^*(x)$$

uniformly for $x \in \bar{D}$,

$$\frac{\partial\hat{\mathbf{w}}}{\partial x_i}(x;S_n,T_n) \to \frac{\partial\mathbf{w}^*}{\partial x_i}(x)$$

uniformly for x in compact subsets $D_0 \subset D$ (this limit is also uniform for x in $\bar{D}$ in the Neumann and Robin cases), and

$$\frac{\partial^2\hat{\mathbf{w}}}{\partial x_i\partial x_j}(x;S_n,T_n) \to \frac{\partial^2\mathbf{w}^*}{\partial x_i\partial x_j}(x)$$

uniformly for x in compact subsets $D_0 \subset D$ (this is possible by Theorem 6.1.6).
    Proceeding as in the proof of Theorem 5.2.1 we see that

$$\lim_{n\to\infty}\frac{1}{T_n - S_n}\int_{S_n}^{T_n}\frac{\partial v_k}{\partial t}(t,x;S_n)\frac{1}{v_k(t,x;S_n)}\,dt = 0$$

for all $x \in D$.
    We claim that $\lambda_{\min}(\mathbf{a}) \geq \lambda_{\text{princ}}(\mathbf{a},\hat{\mathbf{c}})$. In fact, by the above arguments,

$$\lambda_{\min}(\mathbf{a}) \geq (A_k\mathbf{w}^*)(x)/w_k^*(x) + \hat{c}_k(x), \quad x \in D, \qquad (6.3.18)$$

and

$$\mathcal{B}_{a^k}w_k^*(x) = 0 \quad \text{for } x \in \partial D, \qquad (6.3.19)$$

$1 \leq k \leq K$. This implies that $\mathbf{w}(t,x) = \mathbf{w}^*(x)$ is a supersolution of

$$\begin{cases} \mathbf{w}_t = \mathcal{A}\mathbf{w} + (\hat{\mathbf{c}} - \lambda_{\min}(\mathbf{a}))\mathbf{w}, & t > 0,\ x \in D, \\ \mathcal{B}_{a^k}w_k = 0, & t > 0,\ x \in \partial D,\ 1 \leq k \leq K. \end{cases} \qquad (6.3.20)$$

By the fact that $\mathbf{w}^* \geq \mathbf{0}$, we have that the principal eigenvalue of (6.3.20) (i.e., $\lambda_{\mathrm{princ}}(\mathbf{a}, \hat{\mathbf{c}}) - \lambda_{\min}(\mathbf{a})$) is less than or equal to zero. Hence

$$\lambda_{\min}(\mathbf{a}) \geq \lambda_{\mathrm{princ}}(\mathbf{a}, \hat{\mathbf{c}}).$$

(2) For any averaged function $\hat{\mathbf{c}}(x)$ of $\mathbf{c}(t, x)$, there is $T_n - S_n \to \infty$ such that

$$\hat{c}_k(x) = \lim_{n \to \infty} \frac{1}{T_n - S_n} \int_{S_n}^{T_n} c_k^k(t, x)\, dt$$

for $k = 1, 2, \ldots, K$ and $x \in \bar{D}$. Let $\eta(t; S)$, $\mathbf{v}(t, x; S)$, and $\hat{\mathbf{w}}(x; S, T)$ be as in (1). Without loss of generality we may assume that $\frac{1}{T_n - S_n} \ln \eta(T_n; S_n)$ converges as $n \to \infty$ exists. Note that

$$\lambda_{\max}(\mathbf{a}) \geq \lim_{n \to \infty} \frac{1}{T_n - S_n} \ln \eta(T_n; S_n).$$

It then follows from arguments as in (1) that

$$\lambda_{\max}(\mathbf{a}) \geq \lambda_{\mathrm{princ}}(\mathbf{a}, \hat{\mathbf{c}})$$

for any averaged function $\hat{\mathbf{c}}(x)$ of $\mathbf{c}(t, x)$. ⬚

PROOF (Proof of Theorem 6.3.12) Let

$$\eta(t; \omega) := \| \mathbf{U}_{E_{\mathbf{a}}(\omega)}(t, 0) \mathbf{w}(E_{\mathbf{a}}(\omega)) \|.$$

By Theorem 6.3.4(1), for $\mathbb{P}$-a.e. $\omega \in \Omega$ there holds

$$\lambda(\mathbf{a}) = \lim_{T \to \infty} \frac{1}{T} \ln \eta(T; \omega)$$

and

$$\hat{\mathbf{c}}(x) = \lim_{T \to \infty} \frac{1}{T} \int_0^T \mathbf{c}(\theta_t \omega, x)\, dt.$$

It then follows from arguments as in Theorem 6.3.11(1) that

$$\lambda(\mathbf{a}) \geq \lambda_{\mathrm{princ}}(\mathbf{a}, \hat{\mathbf{c}}).$$

The theorem is thus proved. ⬚

6.4   Remarks

In this chapter, principal spectrum, principal Lyapunov, and exponential separation for cooperative systems of nonautonomous and random parabolic equations are investigated. The notion of mild solution is adopted for convenience. Many results on principal spectrum for single nonautonomous and random parabolic equations are extended to cooperative systems of nonautonomous and random parabolic equations.

In the smooth case (both the coefficients and the domain of the systems are sufficiently smooth), it is proved that mild solutions are also weak solutions (in fact, they are classical solutions). Almost all of the principal spectrum theories for single parabolic equations established in Chapters 3, 4, and 5 are extended to cooperative systems of parabolic equations.

In the nonsmooth case, it can also be proved that mild solutions are weak solutions in the Dirichlet boundary condition case and in the Neumann and Robin boundary conditions cases with sufficiently smooth domain (see the arguments in [33, Proposition 4.2]). We do not go into detail about this issue in the monograph. The existence of exponential separation and existence and uniqueness of entire positive solution for general single parabolic equations are proved in Chapter 3 provided that their positive solutions satisfy certain Harnack inequalities (see (A3-1) and (A3-2)). It is expected that these properties for general single parabolic equations can be extended to general cooperative systems of parabolic equations under proper conditions. We do not go into detail either about this issue in the monograph.

Chapter 7

Applications to Kolmogorov Systems of Parabolic Equations

Spectral theory for linear parabolic problems is a basic tool for the study of nonlinear parabolic problems. In this chapter, we discuss some applications of the principal spectral theory developed in previous chapters to uniform persistence of systems of random and nonautonomous nonlinear equations of Kolmogorov type. We first consider applications to random and nonautonomous nonlinear equations of Kolmogorov type, and then consider applications to systems of such equations.

To be more precise, let $D \subset \mathbb{R}^N$ be a sufficiently smooth domain and $\mathcal{B}$ be either the Dirichlet or Neumann boundary operator, i.e.,

$$\mathcal{B} := \begin{cases} \mathrm{Id} & \text{(Dirichlet)} \\ \dfrac{\partial}{\partial \boldsymbol{\nu}} & \text{(Neumann).} \end{cases} \tag{7.0.1}$$

Let $((\Omega, \mathfrak{F}, \mathbb{P}), (\theta_t)_{t \in \mathbb{R}})$ be an ergodic metric dynamical system.

We first study the following random and nonautonomous equations of Kolmogorov type,

$$\begin{cases} \dfrac{\partial u}{\partial t} = \Delta u + f(\theta_t \omega, x, u)u, & x \in D, \\ \mathcal{B}u = 0, & x \in \partial D, \end{cases} \tag{7.0.2}$$

where $f \colon \Omega \times \bar{D} \times [0, \infty) \mapsto \mathbb{R}$,

$$\begin{cases} \dfrac{\partial u}{\partial t} = \Delta u + f(t, x, u)u, & x \in D, \\ \mathcal{B}u = 0, & x \in \partial D, \end{cases} \tag{7.0.3}$$

where $f \colon \mathbb{R} \times \bar{D} \times [0, \infty) \mapsto \mathbb{R}$. In particular, we utilize the theories developed in the previous chapters to study the uniform persistence of (7.0.2) and (7.0.3).

Among other problems, (7.0.2) and (7.0.3) are used to model population growth problem. Due to the biological reason, we are only interested in the nonnegative solutions of (7.0.2) and (7.0.3). Note that in the nonautonomous case, we are interested in the solutions with initial conditions at any $t_0 \in \mathbb{R}$. Both random and nonautonomous cases take certain temporal variations of the underline systems into account and are of great interest in practice.

As (7.0.2) and (7.0.3) can be embedded into proper families of nonlinear equations, to study their uniform persistence we start in Section 7.1 by investigating a general family of nonlinear equations of Kolmogorov type:

$$\begin{cases} \dfrac{\partial u}{\partial t} = \Delta u + g(t, x, u)u, & x \in D, \\ \mathcal{B}u = 0, & x \in \partial D, \end{cases} \tag{7.0.4}$$

were g belongs to a set Z of functions satisfying certain conditions (see (A7-1)–(A7-3) in Section 7.1) and is considered as a parameter. We collect the existence, uniqueness, and basic properties of solutions of (7.0.4) in Subsection 7.1.1. Based on the spectral theory developed in previous chapters, the linear theory of the linearization of (7.0.4) at the trivial solution (i.e., $u \equiv 0$) is presented in Subsection 7.1.2. The global attractor and uniform persistence of (7.0.4) is explored in Subsection 7.1.3 in terms of the linear theory established in 7.1.2.

We then in Section 7.2 introduce the definitions of uniform persistence for (7.0.2) and (7.0.3), and establish a uniform persistence theorem for each case based on the uniform persistence for general families of nonlinear parabolic equations of Kolmogorov type.

As for the scalar equations case, to study uniform persistence for competitive Kolmogorov systems of random and nonautonomous parabolic equations, we start in Section 7.3 by considering a family of competitive Kolmogorov systems of parabolic equations:

$$\begin{cases} \dfrac{\partial u_1}{\partial t} = \Delta u_1 + g_1(t, x, u_1, u_2)u_1, & x \in D, \\ \dfrac{\partial u_2}{\partial t} = \Delta u_2 + g_2(t, x, u_1, u_2)u_2, & x \in D, \\ \mathcal{B}u_1 = 0, & x \in \partial D, \\ \mathcal{B}u_2 = 0, & x \in \partial D, \end{cases} \tag{7.0.5}$$

where $\mathbf{g} = (g_1, g_2)$ belongs to a set $\mathbf{Z}$ of functions satisfying certain conditions (see (A7-5)–(A7-8) in Section 7.3) and is considered as a parameter. We collect the existence, uniqueness, and basic properties of solutions (7.0.5) in Subsection 7.3.1. The linear theory of the linearization of (7.0.5) at trivial and semitrivial solutions (i.e., solutions $(u_1(t), u_2(t))$ satisfying $u_1(t) \equiv 0$ or $u_2(t) \equiv 0$) is investigated in Subsection 7.3.2 based on the general spectral theory developed in previous chapters. Global attractor and uniform persistence for (7.0.5) is studied in Subsection 7.3.3 in terms of the linear theory established in 7.3.2.

We then consider in Section 7.4 competitive Kolmogorov systems of random and nonautonomous parabolic equations. We introduce the definition of uniform persistence and establish a uniform persistence theorem for either case.

This chapter ends up with some remarks on the existing works about uniform persistence and global dynamics in Section 7.5.

Throughout this chapter, we assume the following smoothness of the domain D.

(A7-D) (Boundary smoothness) ∂D *is an* $(N-1)$*-dimensional manifold of class* $C^{3+\alpha}$, *for some* $\alpha > 0$.

7.1 Semilinear Equations of Kolmogorov Type: General Theory

In this section we consider families of semilinear second order parabolic equations of Kolmogorov type

$$
\begin{cases}
\dfrac{\partial u}{\partial t} = \Delta u + g(t, x, u)u, & x \in D, \\[2mm]
\mathcal{B}u = 0, & x \in \partial D,
\end{cases}
\tag{7.1.1}
$$

where $\mathcal{B}$ is a boundary operator of either the Dirichlet or Neumann type as in (7.0.1). Here g is considered a parameter. Sometimes we write (7.1.1) as $(7.1.1)_g$.

First, we present the existence, uniqueness, and basic properties of solutions of (7.1.1) in Subsection 7.1.1. We study the linearized problem at trivial solution of (7.1.1) in Subsection 7.1.2. In Subsection 7.1.3, we establish global attractor and uniform persistence theory of (7.1.1).

7.1.1 Existence, Uniqueness, and Basic Properties of Solutions

In this subsection, we present the existence, uniqueness, and basic properties of solutions of (7.1.1) in appropriate fractional power spaces of the operator Δ (with corresponding boundary conditions) with admissible $g(\cdot, \cdot, \cdot)$s. Most properties presented in this subsection can be found in literature. For convenience, we either provide proofs or references.

First, for a continuous function $g \colon \mathbb{R} \times \bar{D} \times [0, \infty) \to \mathbb{R}$ and $t \in \mathbb{R}$ denote by $g \cdot t$ the *time-translate* of g, $g \cdot t(s, x, u) := g(s + t, x, u)$ for $s \in \mathbb{R}$, $x \in \bar{D}$, and $u \in [0, \infty)$.

Let $g^{(n)}$ $(n \in \mathbb{N})$ and g be continuous real functions defined on $\mathbb{R} \times \bar{D} \times [0, \infty)$. Recall that a sequence $g^{(n)}$ converges to g in the open-compact topology if and only if for any $M > 0$ the restrictions of $g^{(n)}$ to $[-M, M] \times \bar{D} \times [0, M]$ converge uniformly to the restriction of g to $[-M, M] \times \bar{D} \times [0, M]$.

We state the following well-known result.

LEMMA 7.1.1

If $g^{(n)}$ converge to g in the open-compact topology and t_n converge to t then $g^{(n)} \cdot t_n$ converge to $g \cdot t$ in the open-compact topology.

We shall denote by Z the set of admissible parameters of the equation (7.1.1). A generic element of Z is a (at least) continuous function $g \colon \mathbb{R} \times \bar{D} \times [0, \infty) \to \mathbb{R}$. Z is always considered with the open-compact topology.

The standing assumptions on Z are the following:

(A7-1) (1) *Z is compact in the open-compact topology.*

(2) *Z is* translation invariant*: If $g \in Z$ then $g \cdot t \in Z$, for each $t \in \mathbb{R}$.*

For $t \in \mathbb{R}$ and $g \in Z$ put $\zeta_t g := g \cdot t$. It follows from Lemma 7.1.1 that $(Z, \zeta) = (Z, \{\zeta_t\}_{t \in \mathbb{R}})$ is a compact flow.

The assumption below concerns the regularity of the functions g.

(A7-2) (Regularity) *For any $g \in Z$ and any $M > 0$ the restrictions to $\mathbb{R} \times \bar{D} \times [0, M]$ of g and its derivatives $\partial_t g$, $\partial_x g$, and $\partial_u g$ belong to $C^{1-,1-,1-}(\mathbb{R} \times \bar{D} \times [0, M])$. Moreover, for $M > 0$ fixed the $C^{1-,1-,1-}(\mathbb{R} \times \bar{D} \times [0, M])$-norms of the restrictions of those functions are bounded uniformly in Z.*

For each $g \in Z$ we denote:

$$G(t, x, u) := g(t, x, u)u, \qquad t \in \mathbb{R}, \ x \in \bar{D}, \ u \in [0, \infty).$$

LEMMA 7.1.2

Assume (A7-1)–(A7-2). For any sequence $(g^{(n)})$ converging in Z to g all the derivatives of the functions $G^{(n)}$ up to order 1 converge to the respective derivatives of G, uniformly on compact subsets of $\mathbb{R} \times \bar{D} \times [0, \infty)$.

Denote by $\mathfrak{N}$ the Nemytskiĭ (substitution) operator:

$$\mathfrak{N}(t, u, g)(x) := G(t, x, u(x)), \qquad x \in \bar{D},$$

where $t \in \mathbb{R}$, $u \colon \bar{D} \to \mathbb{R}$, and $g \in Z$.

We consider $\mathfrak{N}$ to be a mapping defined on $\mathbb{R} \times C(\bar{D})^+ \times Z$. It is straightforward to see that $\mathfrak{N}$ takes $\mathbb{R} \times C(\bar{D})^+ \times Z$ into $C(\bar{D})$.

We proceed to the issue of the differentiability of the Nemytskiĭ operator $\mathfrak{N}$ with respect to t and u.

Denote by ∂_1 the differentiation with respect to t, and denote by ∂_2 the differentiation with respect to u.

It is easy to see that

LEMMA 7.1.3

Assume (A7-1)–(A7-2).

- *The derivative $\partial_1 \mathfrak{N}$ ($\in \mathcal{L}(\mathbb{R}, C(\bar{D}))$) is defined everywhere on $\mathbb{R} \times C(\bar{D})^+ \times Z$, and is given by the formula*

$$(\partial_1 \mathfrak{N}(t, u, g)1)(x) = \frac{\partial G}{\partial t}(t, x, u(x)) \qquad x \in \bar{D},$$

  where 1 is the vector tangent at (t, u, g) to the $\mathbb{R}$-axis.

- *The derivative $\partial_2 \mathfrak{N}$ ($\in \mathcal{L}(C(\bar{D}), C(\bar{D}))$) is defined everywhere on $\mathbb{R} \times C(\bar{D})^+ \times Z$, and is given by the formula*

$$(\partial_2 \mathfrak{N}(t, u, g)v)(x) = \frac{\partial G}{\partial u}(t, x, u(x)) \cdot v(x) \qquad x \in \bar{D},$$

  where $v \in C(\bar{D})$ is a vector tangent at (t, u, g) to the $C(\bar{D})$-axis.

Regarding the differentiability with respect to u, notice that formally we need to extend $\mathfrak{N}$ to some open subset of $\mathbb{R} \times C(\bar{D})^+ \times Z$ in $\mathbb{R} \times C(\bar{D}) \times Z$. Indeed, we can do that by putting $g(t, x, u) := g(t, x, 0) + \frac{\partial g}{\partial u}(t, x, 0)u$ for $t \in \mathbb{R}$, $x \in \bar{D}$ and $u < 0$.

In view of Lemma 7.1.3 the derivatives $\partial_1 \mathfrak{N}$ and $\partial_2 \mathfrak{N}$ can (and will) be identified with the functions $\partial G / \partial t$ and $\partial G / \partial u$, respectively.

In the following two lemmas $\tilde{\mathfrak{N}}$ stand for any of the mappings $\mathfrak{N}$, $\partial_1 \mathfrak{N}$ or $\partial_2 \mathfrak{N}$.

LEMMA 7.1.4

Let (A7-1)–(A7-2) be satisfied. Then

$$\tilde{\mathfrak{N}}(t + s, u, g) = \tilde{\mathfrak{N}}(t, u, g \cdot s) \tag{7.1.2}$$

for any $t, s \in \mathbb{R}$, $u \in C(\bar{D})^+$, and $g \in Z$.

PROOF It follows immediately from the definition of the time-translate and from Lemma 7.1.3. ◻

LEMMA 7.1.5

Assume (A7-1) and (A7-2).

(i) *$[\mathbb{R} \times C(\bar{D})^+ \times Z \ni (t, u, g) \mapsto \tilde{\mathfrak{N}}(t, u, g) \in C(\bar{D})]$ is continuous.*

(ii) *For any bounded $B \subset C(\bar{D})^+$, the mapping $\tilde{\mathfrak{N}}$ satisfies the Lipschitz condition with respect to (t, u), uniformly in $(t, u, g) \in \mathbb{R} \times B \times Z$.*

(iii) *For any bounded $B \subset C(\bar{D})^+$, the image $\tilde{\mathfrak{N}}(\mathbb{R} \times B \times Z)$ is bounded in $C(\bar{D})$.*

PROOF   The proofs of (i) and (ii) are standard.

To prove (iii) observe that, by Eq. (7.1.2), $\tilde{\mathfrak{N}}(\mathbb{R} \times B \times Z) = \tilde{\mathfrak{N}}(\{0\} \times B \times Z)$, and apply Part (ii).   □

We collect now some basic properties of fractional power spaces. For proofs see [48].

For $1 < p < \infty$, let A_p stand for the realization of the operator Δ (with the corresponding boundary conditions) in $L_p(D)$. The operator $-A_p$ is sectorial. Denote by $\{e^{A_p t}\}_{t \geq 0}$ the analytic semigroup generated on $L_p(D)$ by A_p.

For $1 < p < \infty$ and $\beta \geq 0$ denote by F_p^β the fractional power space of the sectorial operator $-A_p$. We have $F_p^0 = L_p(D)$, and F_p^1 equals the domain of $-A_p$. Also

$$F_p^1 \subset W_p^2(D).$$

LEMMA 7.1.6
The following embeddings hold:

(1) $F_p^{\beta_2} \hookrightarrow F_p^{\beta_1}$ *for any* $1 < p < \infty$ *and* $0 \leq \beta_1 < \beta_2$.

(2) $F_p^\beta \hookrightarrow C^{\tilde{\beta}}(D)$, *for any* $1 < p < \infty$, $\beta \geq 0$, *and* $0 \leq \tilde{\beta} < \min\{1, 2\beta - \frac{N}{p}\}$.

(3) $F_p^\beta \hookrightarrow C^{1+\tilde{\beta}}(D)$, *for any* $1 < p < \infty$, $\beta \geq 0$, *and* $0 \leq \tilde{\beta} < \min\{1, 2\beta - \frac{N}{p} - 1\}$.

PROOF   See [48, Theorem 1.4.8] for (1) and [48, Theorem 1.6.1] for (2) and (3).   □

LEMMA 7.1.7
For any $1 < p < \infty$, $\beta \geq 0$, *and* $T > 0$ *there is* $M = M(p, \beta, T) > 0$ *with the property that*

$$\|e^{A_p t}\|_{L_p(D), F_p^\beta} \leq M t^{-\beta} \quad \text{and} \quad \|e^{A_p t}\|_{F_p^\beta} \leq M$$

for all $0 < t \leq T$.

PROOF   See [48, 1.4 and 1.5].   □

Let φ_{princ} be the unique (nonnegative) principal eigenfunction of the elliptic boundary value problem

$$\begin{cases} \Delta u = 0 & \text{on } D \\ \mathcal{B}u = 0 & \text{on } \partial D \end{cases}$$

normalized so that $\|\varphi_{\mathrm{princ}}\|_{C(\bar{D})} = 1$. It follows from the regularity theory for the Laplace operator that $\varphi_{\mathrm{princ}} \in C^2(\bar{D}) \cap F_p^1$, for any $1 < p < \infty$.

By the elliptic strong maximum principle and Hopf boundary point principle, in the Dirichlet case $\varphi_{\mathrm{princ}}(x) > 0$ for each $x \in D$ and $(\partial\varphi_{\mathrm{princ}}/\partial\boldsymbol{\nu})(x) < 0$ for each $x \in \partial D$. In the Neumann case $\varphi_{\mathrm{princ}} \equiv 1$.

Until the end of the present section we fix $1 < p < \infty$, $p > N$ and $\frac{N}{2p} + \frac{1}{2} < \beta < 1$, and put

$$X := F_p^\beta. \tag{7.1.3}$$

There holds

$$X \hookrightarrow C^1(\bar{D}).$$

Indeed, p and β are so chosen that by Lemma 7.1.6 we have

$$F_p^\beta \hookrightarrow F_p^{\beta_1} \hookrightarrow C^{1+\tilde{\beta}}(D) \hookrightarrow C^1(\bar{D}),$$

where $\frac{1}{2} + \frac{N}{2p} < \beta_1 < \beta$ and $0 < \tilde{\beta} < 2\beta_1 - \frac{N}{p} - 1$.

Recall that by X^+ we denote the nonnegative cone in X, $X^+ = \{\, u \in X : u(x) \geq 0 \text{ for all } x \in \bar{D}\,\}$.

We proceed now to the investigation of the interior X^{++} of the nonnegative cone X^+.

LEMMA 7.1.8

(1) *In the case of the Dirichlet boundary conditions X^{++} is nonempty, and is characterized by*

$$\begin{aligned} X^{++} = \{\, u \in X^+ : u(x) > 0 \quad &\text{for all} \quad x \in D \\ \text{and} \quad (\partial u/\partial\boldsymbol{\nu})(x) < 0 \quad &\text{for all} \quad x \in \partial D\,\}. \end{aligned} \tag{7.1.4}$$

(2) *In the case of the Neumann boundary conditions X^{++} is nonempty, and is characterized by*

$$X^{++} = \{\, u \in X^+ : u(x) > 0 \quad \text{for all} \quad x \in \bar{D}\,\}. \tag{7.1.5}$$

PROOF   We prove the lemma only for the Dirichlet case, the proof for the Neumann case being similar, but simpler. F_p^1 consists precisely of those elements of $W_p^2(D)$ whose trace on ∂D is zero. Since $F_p^1 \hookrightarrow C^1(\bar{D})$, any $u \in F_p^1$ is a C^1 function vanishing on ∂D. By [48, Theorem 1.4.8], the image of the embedding $F_p^1 \hookrightarrow X$ is dense. Because $X \hookrightarrow C^1(\bar{D})$, we conclude that $X \hookrightarrow \mathring{C}^1(\bar{D})$.

Denote by I the embedding $X \hookrightarrow \mathring{C}^1(\bar{D})$. It follows from Lemma 1.3.1(2) that the right-hand side of (7.1.4) equals $I^{-1}(\mathring{C}^1(\bar{D})^{++})$, where $\mathring{C}^1(\bar{D})^{++}$ is an open subset of $\mathring{C}^1(\bar{D})$. This proves the "$\supset$" inclusion. We have that $\varphi_{\mathrm{princ}} \in$

$F_p^1 \hookrightarrow X$ and that it belongs to the right-hand side of (7.1.4), consequently to X^{++}. Finally, let $u \in X^{++}$. There is $\epsilon > 0$ such that $u - \epsilon\varphi_{\mathrm{princ}} \in X^+$, therefore $u(x) \geq \epsilon\varphi_{\mathrm{princ}}(x) > 0$ for all $x \in D$, which gives further that $\frac{\partial u}{\partial \boldsymbol{\nu}}(x) \leq \epsilon\frac{\partial\varphi_{\mathrm{princ}}}{\partial \boldsymbol{\nu}}(x) < 0$ for all $x \in \partial D$. $\qquad\square$

Recall that for any $u_1, u_2 \in X$ we write

$$u_1 \ll u_2 \quad \text{if and only if} \quad u_2 - u_1 \in X^{++}.$$

The symbol $\gg$ is used in an analogous way.

We write ∂X^+ for $X^+ \setminus X^{++}$.

DEFINITION 7.1.1 (Solution)   *For $t_0 \in \mathbb{R}$, $u_0 \in X^+$, and $g \in Z$ by a solution of $(7.1.1)_g$ satisfying the initial condition $u(t_0, \cdot) = u_0$ we mean a continuous function $u \colon J \to X$, where J is a nondegenerate interval with $\inf J = t_0 \in J$, satisfying the following:*

- $u(t_0) = u_0$,

- $u(t) \in F_p^1$ *for each* $t \in J \setminus \{t_0\}$,

- u *is differentiable, as a function into* $L_p(D)$, *on* $J \setminus \{t_0\}$,

- *there holds*

$$\dot{u}(t) = A_p u(t) + \mathfrak{N}(t, u(t), g) \quad \text{for each} \quad t \in J \setminus \{t_0\}.$$

(See [86].)

A solution u of $(7.1.1)_g$ satisfying $u(t_0, \cdot) = u_0$ is nonextendible *if there is no solution u^* of $(7.1.1)_g$ defined on an interval J^* with $\sup J^* > \sup J$ such that $u^*|_J \equiv u$.*

PROPOSITION 7.1.1 (Existence and uniqueness of solution)
Let (A7-1)–(A7-2) be satisfied. Then for each $g \in Z$, each $t_0 \in \mathbb{R}$, and each $u_0 \in X^+$ there exists a unique nonextendible solution of $(7.1.1)_g$ satisfying the initial condition $u(t_0, \cdot) = u_0$, defined on an interval of the form $[t_0, \tau_{\max})$, where $\tau_{\max} = \tau_{\max}(t_0, u_0, g) > t_0$.

PROOF   See [48, Theorem 3.3.3]. $\qquad\square$

We will denote the solution of $(7.1.1)_g$ satisfying the initial condition $u(t_0, \cdot) = u_0$ by $u(\cdot; t_0, u_0, g)$.

PROPOSITION 7.1.2 (Positivity)
Given $t_0 \in \mathbb{R}$, $u_0 \in X^+$, and $g \in Z$, there holds $u(t; t_0, u_0, g) \in X^+$ for any $t \in (t_0, \tau_{\max}(t_0, u_0, g))$.

PROOF   Note that for given $t_0 \in \mathbb{R}$ and $g \in Z$, $u(t; t_0, 0, g) = 0$ for all $t \geq 0$. The proposition then follows from the comparison principle for parabolic equations.   □

In a couple of places we will make use of the fact that, in [48], a solution $u(\cdot; t_0, u_0, g)$ is defined initially as a *mild solution*. We formulate that as the following.

PROPOSITION 7.1.3 (Variation of constant formula)
Assume (A7-1)–(A7-2). *Then for any* $t_0 \in \mathbb{R}$, $u_0 \in X^+$, *and* $g \in Z$ *the unique nonextendible solution* $u(\cdot) := u(\cdot; t_0, u_0, g)$ *satisfies*

$$u(t) = e^{A_p(t-t_0)} u_0 + \int_{t_0}^{t} e^{A_p(t-s)} \mathfrak{N}(s, u(s), g) \, ds \qquad (7.1.6)$$

for $t_0 < t < \tau_{\max}(t_0, u_0, g)$.

PROPOSITION 7.1.4 (Regularity)
Let (A7-1) *and* (A7-2) *be satisfied. Then for each* $u_0 \in X^+$ *and each* $g \in Z$, $u(t, x) = u(t; t_0, u_0, g)(x)$ *is a classical solution, that is,*

- $\frac{\partial u}{\partial t}(t, \cdot) \in C(\bar{D})$ *for each* $t \in (t_0, \tau_{\max}(t_0, u_0, g))$,

- $u(t, \cdot) \in C^2(\bar{D})$ *for each* $t \in (t_0, \tau_{\max}(t_0, u_0, g))$,

- *for any* $t \in (t_0, \tau_{\max}(t_0, u_0, g))$ *and* $x \in D$ *the equation* $(7.1.1)_g$ *is satisfied pointwise,*

- *for any* $t \in (t_0, \tau_{\max}(t_0, u_0, g))$ *and* $x \in \partial D$ *the boundary condition in* (7.1.1) *is satisfied pointwise.*

PROOF   See [48, Sections 3.5 and 3.6].   □

We introduce now an assumption which guarantees, among others, that any solution is *global*, that is, defined on $[0, \infty)$:

(A7-3) *There is* $P > 0$ *such that* $g(t, x, u) < 0$ *for any* $g \in Z$, *any* $t \in \mathbb{R}$, *any* $x \in \bar{D}$, *and any* $u \geq P$.

PROPOSITION 7.1.5 (Global existence)
Assume (A7-1) *through* (A7-3). *Then for any* $t_0 \in \mathbb{R}$, *any* $u_0 \in X^+$, *and any* $g \in Z$ *the unique nonextendible solution* $u(\cdot; t_0, u_0, g)$ *is defined on* $[t_0, \infty)$.

PROOF   By Lemma 7.1.5(iii), for any $K > 0$ and any bounded $B \subset C(\bar{D})$ the image $\mathfrak{N}([-K, K] \times B \times Z)$ is bounded in $C(\bar{D})$, consequently is bounded in $L_p(D)$. In view of [48, Theorem 3.3.4] it suffices to show that for any $t_0 \in \mathbb{R}$, $u_0 \in X^+$ and $g \in Z$ the set $\{\, \|u(t; t_0, u_0, g)\|_X : t \in [t_0, \tau_{\max}(t_0, u_0, g)) \,\}$ is bounded.

For $u_0 \in X^+$, take $m := \max \{P, \|u_0\|_{C(\bar{D})}\}$. The constant function m is a supersolution of $(7.1.1)_g$ with $u_0 \leq m$, hence $(0 \leq)\, u(t; t_0, u_0, g)(x) \leq m$ for all $t \in [t_0, \tau_{\max}(t_0, u_0, g))$ and all $x \in \bar{D}$, consequently $\|u(t; t_0, u_0, g)\|_{C(\bar{D})} \leq m$ for all $t \in [t_0, \tau_{\max}(t_0, u_0, g))$.

By Lemma 7.1.7 there is $M_1 > 0$ such that

$$\|e^{A_p t}\|_X \leq M_1 \quad \text{and} \quad \|e^{A_p t}\|_{L_p(D), X} \leq M_1 t^{-\beta}$$

for all $t \in (0, \tau_{\max}(t_0, u_0, g) - t_0)$. Further, $M_2 := \sup\{\, \|\mathfrak{N}(t, u, g)\|_{C(\bar{D})} : t \in [t_0, \tau_{\max}(t_0, u_0, g)), \|u_0\|_{C(\bar{D})} \leq m,\ g \in Z \,\} < \infty$. Finally, put M_3 to be the norm of the embedding $C(\bar{D}) \hookrightarrow L_p(D)$. It follows from (7.1.6) that

$$\|u(t; t_0, u_0, g)\|_X \leq M_1 \|u_0\|_X + M_1 M_2 M_3 \int_{t_0}^{t} (t - s)^{-\beta}\, ds,$$

which is bounded for $t \in (t_0, \tau_{\max}(t_0, u_0, g))$. $\qquad\square$

LEMMA 7.1.9
Assume (A7-1)–(A7-3). Then for any $t \geq 0$, $t_0 \in \mathbb{R}$, $u_0 \in X^+$, and $g \in Z$ the following holds:

$$u(t + t_0; t_0, u_0, g) = u(t; 0, u_0, g \cdot t_0). \tag{7.1.7}$$

PROOF   Fix $t_0 \in \mathbb{R}$, $u_0 \in X^+$, and $g \in Z$. By Proposition 7.1.3 we have

$$u(t + t_0; t_0, u_0, g) = e^{A_p t} u_0 + \int_{t_0}^{t+t_0} e^{A_p(t+t_0-s)} \mathfrak{N}(s, u(s; t_0, u_0, g), g)\, ds$$

for all $t > 0$, which can be written as

$$u(t + t_0; t_0, u_0, g) = e^{A_p t} u_0 + \int_{0}^{t} e^{A_p(t-s)} \mathfrak{N}(s + t_0, u(s + t_0; t_0, u_0, g), g)\, ds$$

for all $t > 0$. Put $u_1(t) := u(t + t_0; t_0, u_0, g)$ for $t > 0$. We have thus

$$u_1(t) = e^{A_p t} u_0 + \int_{0}^{t} e^{A_p(t-s)} \mathfrak{N}(s + t_0, u_1(s), g)\, ds$$

$$= e^{A_p t} u_0 + \int_{0}^{t} e^{A_p(t-s)} \mathfrak{N}(s, u_1(s), g \cdot t_0)\, ds \qquad \text{(by Eq. (7.1.2))}$$

for all $t > 0$. Consequently, $u_1(t) = u(t; 0, u_0, g \cdot t_0)$ for all $t > 0$. $\qquad\square$

As $g \cdot t$ belongs to Z for any $g \in Z$ and any $t \in \mathbb{R}$, the above lemma allows us to restrict ourselves to considering the initial moment t_0 to be equal to 0.

LEMMA 7.1.10

Assume (A7-1)–(A7-3). Then for any $0 \leq t_0 \leq t$, $u_0 \in X^+$, and $g \in Z$ the following holds:

$$u(t; t_0, u(t_0; 0, u_0, g), g) = u(t; 0, u_0, g). \tag{7.1.8}$$

PROOF   Fix $u_0 \in X^+$ and $g \in Z$, and put $u(t) := u(t; 0, u_0, g)$. We have

$$u(t) = e^{A_p t} u_0 + \int_0^t e^{A_p(t-s)} \mathfrak{N}(s, u(s), g) \, ds \qquad t > 0,$$

which can be transformed, for $t > t_0$, into

$$
\begin{aligned}
u(t) &= e^{A_p(t-t_0)} \left(e^{A_p t} u_0 + \int_0^{t_0} e^{A_p(t_0-s)} \mathfrak{N}(s, u(s), g) \, ds \right) \\
&\quad + \int_0^t e^{A_p(t-s)} \mathfrak{N}(s, u(s), g) \, ds - \int_0^{t_0} e^{A_p(t-s)} \mathfrak{N}(s, u(s), g) \, ds \\
&= e^{A_p(t-t_0)} u(t_0) + \int_{t_0}^t e^{A_p(t-s)} \mathfrak{N}(s, u(s), g) \, ds.
\end{aligned}
$$

$\square$

We write $u(t; u_0, g)$ instead of $u(t; 0, u_0, g)$.

A consequence of Lemmas 7.1.9 and 7.1.10 is the following *cocycle property*: For all $t, s \geq 0$, $u_0 \in X^+$, and $g \in Z$ there holds

$$u(t + s; u_0, g) = u(t; u(s; u_0, g), \zeta_s g). \tag{7.1.9}$$

PROPOSITION 7.1.6 (Continuous dependence)

Let (A7-1)–(A7-3) be satisfied. Then the mapping

$$\left[[0, \infty) \times X^+ \times Z \ni (t, u_0, g) \mapsto u(t; u_0, g) \in X^+ \right]$$

is continuous.

PROOF   See [48, Theorem 3.4.1]. $\square$

PROPOSITION 7.1.7 (Monotonicity)

Assume (A7-1)–(A7-3). Let $g \in Z$.

(1) *If $u_1, u_2 \in X^+$, $u_1 \leq u_2$, then $u(t; u_1, g) \leq u(t; u_2, g)$ for each $t \geq 0$.*

(2) *If $u_1, u_2 \in X^+$, $u_1 < u_2$, then $u(t; u_1, g) \ll u(t; u_2, g)$ for each $t > 0$.*

PROOF　Fix $u_1, u_2 \in X^+$ and $g \in Z$, and denote $v(t, x) := u(t; u_2, g)(x) - u(t; u_1, g)(x)$, $t \geq 0$, $x \in \bar{D}$. The function $v = v(t, x)$ is the classical solution of the nonautonomous linear parabolic partial differential equation

$$\begin{cases} \frac{\partial v}{\partial t} = \Delta v + \tilde{G}(t, x)v, & t > 0, \ x \in D \\ \mathcal{B}v = 0, & t > 0, \ x \in \partial D \end{cases}$$

with the initial condition $v(0, x) = u_2(x) - u_1(x)$ for $x \in \bar{D}$, where

$$\tilde{G}(t, x) := \int_0^1 \frac{\partial G}{\partial u}(t, x, u_1(t; x, g) + s(u_2(t; x, g) - u_1(t; x, g))) \, ds$$

for $t \geq 0$ and $x \in \bar{D}$.

The zero-order coefficient $\tilde{G}$ is continuous on $[0, \infty) \times \bar{D}$, so the standard theory of maximum principles applies.　　　　　　　　　　　　　　□

In the existing terminology we can express Proposition 7.1.7(2) in the following way: For any $g \in Z$ and any $t > 0$ the mapping $[\, X^+ \ni u_0 \mapsto u(t; u_0, g) \in X^+\,]$ is *strongly monotone* (see, e.g., [56], or [57]).

PROPOSITION 7.1.8 (Compactness)

Assume (A7-1) through (A7-3). Then for any $\delta_0 > 0$ and any $B \subset X^+$ bounded in the $C(\bar{D})$-norm, the set $\{\, u(t; u_0, g) : t \geq \delta_0, \ u_0 \in B, \ g \in Z \,\}$ has compact closure in the X-norm.

PROOF　Take $m := \max\{P, \sup\{\, \|u_0\|_{C(\bar{D})} : u_0 \in B\,\}\}$. For any $u_0 \in B$ and $g \in Z$ the constant function m is a supersolution of $(7.1.1)_g$, hence $0 \leq u(t; u_0, g)(x) \leq m$ for all $t \in [0, \infty)$ and all $x \in \bar{D}$.

Pick $\lambda \geq 0$ larger than the supremum of the real parts of eigenvalues of the operator A_p. Then there is $\epsilon > 0$ such that for any $0 < \beta_1 < 1$ there is $M = M(\beta_1) > 0$ with the property that

$$\left\| e^{(A_p - \lambda)t} \right\|_{L_p(D), F_p^{\beta_1}} \leq M t^{-\beta_1} e^{-\epsilon t}$$

for all $t > 0$ (see [48, Section 1.5]; in fact, in the Dirichlet case we can take $\lambda = 0$, whereas in the Neumann case any $\lambda > 0$ will do). For $t \in \mathbb{R}$, $u \in C(\bar{D})^+$, and $g \in Z$ put $\mathfrak{N}_\lambda(t, u, g) := \mathfrak{N}(t, u, g) + \lambda u$. There holds

$$u(t; u_0, g) = e^{(A_p - \lambda)t} u_0 + \int_0^t e^{(A_p - \lambda)(t-s)} \mathfrak{N}_\lambda(s, u(s; u_0, g), g) \, ds, \quad t > 0.$$

It follows from Lemma 7.1.5(iii) that $M_1 := \sup\{\, \|\mathfrak{N}_\lambda(t, u, g)\|_{C(\bar{D})} : t \in \mathbb{R}, \ 0 \leq u \leq m, \ g \in Z \,\} < \infty$. Further, denote by M_2 the norm of the embedding $C(\bar{D}) \hookrightarrow L_p(D)$. Fix $\beta_1 \in (\beta, 1)$. We see that

$$\|u(t; u_0, g)\|_{F_p^{\beta_1}} \leq mMM_2 t^{-\beta_1} e^{-\epsilon t} + MM_1 M_2 \int_0^t (t-s)^{-\beta_1} e^{-\epsilon(t-s)} \, ds$$

for all $t > 0$, $u_0 \in B$, and $g \in Z$. Now it suffices to notice that the first term on the right-hand side of the above inequality is bounded (by $mMM_2 \delta_0^{-\beta_1}$) for all $t \geq \delta_0$, whereas the second term is bounded for all $t > 0$. Finally, we apply the compact embedding $F_p^{\beta_1} \hookrightarrow\hookrightarrow X$ (see Lemma 7.1.6(1)). ☐

PROPOSITION 7.1.9 (Backward uniqueness)
Assume (A7-1)–(A7-3). *For any* $u_1, u_2 \in X^+$ *and any* $g \in Z$, *if* $u_1 \neq u_2$ *then* $u(t; u_1, g) \neq u(t; u_2, g)$ *for all* $t \geq 0$.

PROOF   See [43, Chapter 6] or [44, Part II, Chapter 18]. ☐

We proceed now to the question of the differentiability of the solution operator. We will be interested in the differentiability of the first order with respect to u_0 as well as the continuous dependence of the respective derivatives.

PROPOSITION 7.1.10 (Differentiability)
Assume (A7-1)–(A7-3). *Then the following holds:*

(1) *The derivative* $\partial_2 u$ *of the mapping*

$$[\, [0, \infty) \times X^+ \times Z \ni (t, u_0, g) \mapsto u(t; u_0, g) \in X^+ \,]$$

with respect to the u_0-*variable exists and is continuous on the set* $(0, \infty) \times Z \times X^+$.

(2) *For* $u_0 \in X^+$, $g \in Z$, *and* $v_0 \in X$ *the mapping*

$$[\, (0, \infty) \ni t \mapsto \partial_2 u(t; u_0, g) v_0 \in X \,]$$

is the unique solution of the integral equation (where $v(\cdot) = \partial_2 u(\cdot; u_0, g) v_0$)

$$v(t) = e^{A_p t} v_0 + \int_0^t e^{A_p(t-s)} (\partial_2 \mathfrak{N}(s, u(s; u_0, g), g) v(s)) \, ds \qquad (7.1.10)$$

for $t > 0$. *Moreover, it is the (classical) solution of the nonautonomous linear parabolic equation:*

$$\begin{cases} \dfrac{\partial v}{\partial t} = \Delta v + \dfrac{\partial G}{\partial u}(t, x, u(t; u_0, g)) v, & t > 0, \ x \in D \\ \mathcal{B}v = 0, & t > 0, \ x \in \partial D, \end{cases} \qquad (7.1.11)$$

with the initial condition $v(0, x) = v_0(x)$ *for* $x \in D$.

PROOF   The fact that, for a fixed $g \in Z$, the statement is fulfilled, follows from [48, Theorem 3.4.4]. In particular, Part (2) holds.

The continuous dependence of $\partial_2 u$ on g is a consequence of the fact that the solution $u(\cdot; u_0, g)$ is obtained as the fixed point of the operator $\mathfrak{S}(u_0, g)$: $C([0, T], X) \to C([0, T], X))$ $(T > 0)$ defined by

$$(\mathfrak{S}(u_0, g)u)(t) := e^{A_P(t)} u_0 + \int_0^t e^{A_P(t-s)} \mathfrak{N}(s, u(s), g) \, ds$$

(see Proposition 7.1.3).   ⬚

In a manner analogous to the proof of Eq. (7.1.9) one proves the following *cocycle property*: For all $t, s \geq 0$, $u_0 \in X^+$ and $g \in Z$ there holds

$$\partial_2 u(t + s; u_0, g) = \partial_2 u(t; u(s; u_0, g), \zeta_s g) \circ \partial_2 u(s; u_0, g). \qquad (7.1.12)$$

PROPOSITION 7.1.11 (Positivity of the derivative)
Assume (A7-1)–(A7-3). Let $u_0 \in X^+$ and $g \in Z$.

(1) *If $v_0 \in X^+$ then $\partial_2 u(t; u_0, g)v_0 \in X^+$ for each $t \geq 0$.*

(2) *If $v_0 \in X^+ \setminus \{0\}$ then $\partial_2 u(t; u_0, g)v_0 \in X^{++}$ for each $t > 0$.*

PROOF   In view of Proposition 7.1.10(2) this is an application of the standard theory of maximum principles for classical solutions.   ⬚

PROPOSITION 7.1.12 (Backward uniqueness of the derivative)
Assume (A7-1)–(A7-3). For any $u_0 \in X^+$, any $v_1, v_2 \in X$, and any $g \in Z$, if $v_1 \neq v_2$ then $\partial_2 u(t; u_0, g)v_1 \neq \partial_2 u(t; u_0, g)v_2$ for all $t \geq 0$.

PROOF   See [43, Chapter 6] or [44, Part II, Chapter 18].   ⬚

It should be remarked that Lemmas 7.1.9 and 7.1.10 as well as Propositions 7.1.6 through 7.1.12 would hold in fact without Assumption (A7-3), a difference being that instead of the formulation "for all $t \in [0, \infty)$" one would have "for all $t \in [0, \tau_{\max}(0, u_0, g))$," etc.

We put

$$\Phi(t; u_0, g) = \Phi_t(u_0, g) := (u(t; u_0, g), \zeta_t g), \qquad (7.1.13)$$

where $t \geq 0$, $u_0 \in X^+$, and $g \in Z$. Proposition 7.1.6 and Eq. (7.1.9) guarantee that $\Phi = \{\Phi_t\}_{t \geq 0}$ is a topological skew-product semiflow on the product bundle $X^+ \times Z$ covering the topological flow (Z, ζ). (Notice that we have the joint continuity at $t = 0$.) It should be remarked that the property mentioned in Proposition 7.1.7(2) can be written as: The (topological) skew-product semiflow Φ is *strongly monotone* (this is an adjustment of the terminology used for semiflows on ordered metric spaces, see, e.g., [57], to skew-product semiflows with ordered fibers).

Further we put

$$\partial\Phi(t; v_0, (u_0, g)) = \partial\Phi_t(v_0, (u_0, g)) \tag{7.1.14}$$
$$:= (\partial_2 u(t; u_0, g)v_0, (u(t; u_0, g), \zeta_t g)),$$

where $t \geq 0$, $v_0 \in X$, $u_0 \in X^+$, and $g \in Z$. Proposition 7.1.10 and Eq. (7.1.12) guarantee that $\partial\Phi = \{\partial\Phi_t\}_{t\geq0}$ is a topological linear skew-product semiflow on the product Banach bundle $X \times (X^+ \times Z)$ covering the topological semiflow Φ on $X^+ \times Z$. Again, the property mentioned in Proposition 7.1.11(2) can be written as: The (topological) linear skew-product semiflow $\partial\Phi$ is *strongly monotone* (or *strongly positive*).

7.1.2 Linearization at the Trivial Solution

In the present subsection we investigate the linearization of the skew-product semiflow Φ at the trivial solution. Most results follow from the general theories developed in Chapters 2, 3, and 4. For convenience, we provide proofs for some results.

We assume throughout this subsection that Assumptions (A7-1)–(A7-3) are satisfied.

The compact set $\{0\} \times Z$ is invariant under Φ. Consider the restriction of the topological linear skew-product semiflow $\partial\Phi$ to $X \times (\{0\} \times Z)$:

$$\partial\Phi_t(v_0, (0, g)) = (\partial_2 u(t; 0, g)v_0, (0, g \cdot t)), \qquad t \geq 0, \ v_0 \in X, \ g \in Z.$$

By Proposition 7.1.10, for any $v_0 \in X$ and any $g \in Z$ the function $[(0, \infty) \ni t \mapsto \partial_2 u(t; 0, g)v_0 \in X]$ is given by the classical solution of the nonautonomous linear parabolic equation

$$\begin{cases} \frac{\partial v}{\partial t} = \Delta v + g(t, x, 0)v, & t > 0, \ x \in D \\ \mathcal{B}v = 0, & t > 0, \ x \in \partial D \end{cases} \tag{7.1.15}$$

with the initial condition $v(0, x) = v_0(x)$ for $x \in D$.

Define a mapping $\tilde{p}_0 \colon Z \to L_\infty(\mathbb{R} \times D, \mathbb{R}^{N^2+2N+1}) \times L_\infty(\mathbb{R} \times \partial D, \mathbb{R})$ by

$$\tilde{p}_0(g) := ((\delta_{ij})_{i,j=1}^N, (0)_{i=1}^N, (0)_{i=1}^N, g_0, 0),$$

and $p_0 \colon Z \to L_\infty(\mathbb{R} \times D, \mathbb{R})$ by

$$p_0(g) := g_0,$$

where δ_{ij} is the Kronecker symbol and $g_0(t, x) := g(t, x, 0)$, $t \in \mathbb{R}$, $x \in \bar{D}$.

Let $\tilde{Y}$ and Y stand for the images of Z under $\tilde{p}_0$ and p_0, respectively. We identify $\tilde{p}_0(g)$ with $p_0(g) = g_0$ and identify $\tilde{Y}$ with Y. For $g_0 \in Y$ and $t \in \mathbb{R}$ we denote $g_0 \cdot t(s, x) := g_0(t + s, x)$, $s \in \mathbb{R}$, $x \in \partial D$. We write $\sigma_t g_0$ for $g_0 \cdot t$.

Y will be always considered with the open-compact topology.

LEMMA 7.1.11

(1) *The mapping $p_0 \colon Z \to Y$ is continuous.*

(2) *Y is compact.*

(3) *$g_0 \cdot t \in Y$ for any $g_0 \in Y$ and any $t \in \mathbb{R}$.*

(4) *$p_0 \circ \zeta_t = \sigma_t \circ p_0$ for any $t \in \mathbb{R}$.*

(5) *The mapping $[\, \mathbb{R} \times Y \ni (t, g_0) \mapsto \sigma_t g_0 \in Y \,]$ is continuous.*

PROOF Part (1) follows by the definition of the open-compact topology. Part (2) is a consequence of Part (1) and the compactness of Z. Parts (3) and (4) are obvious. Part (5) is well known (compare Lemma 7.1.1). ⬚

By Lemma 7.1.11(5) (Y, σ) is a compact flow.

For $g \in Z$ (or for $g_0 \in Y$) we write (7.1.15) as (7.1.15)$_{g_0}$.

From Assumptions (A7-1) and (A7-2) it follows that (A2-1)–(A2-4) and (A3-1), (A3-2) are satisfied by both (7.1.15) and its adjoint problem. Consequently, we can apply to (7.1.15) the theories presented in Sections 2.1–2.3 and Sections 3.1–3.3.

For $t \geq 0$, $v_0 \in L_2(D)$, and $g \in Z$ (or for $g_0 \in Y$) we denote by $U_{g_0}(t, 0)v_0$ the value at time t of the (weak or, what is equivalent, classical) solution of (7.1.15)$_{g_0}$ satisfying the initial condition $v(0, \cdot) = v_0$.

Denote by $\Pi = \{\Pi_t\}_{t \geq 0}$ the topological linear skew-product semiflow defined on $L_2(D) \times Y$ by (7.1.15):

$$\Pi(t; v_0, g_0) = \Pi_t(v_0, g_0) := (U_{g_0}(t, 0)v_0, \sigma_t g_0), \quad t \geq 0, \ v_0 \in L_2(D), \ g_0 \in Y.$$

From now on, we assume that Z_0 is a nonempty compact connected invariant subset of Z. By Lemma 7.1.11, $Y_0 := p_0(Z_0)$ is a nonempty compact connected invariant subset of Y.

First of all, we have the following result stating that Π admits an exponential separation over Y_0.

THEOREM 7.1.1
There exist

- *an invariant (under* Π*) one-dimensional subbundle* X_1 *of* $L_2(D) \times Y_0$ *with fibers* $X_1(g_0) = \operatorname{span}\{w(g_0)\}$*, where* $w \colon Y_0 \to L_2(D)$ *is continuous, with* $\|w(g_0)\| = 1$ *for all* $g_0 \in Y_0$*, and*

- *an invariant (under* Π*) complementary one-codimensional subbundle* X_2 *of* $L_2(D) \times Y_0$ *with fibers* $X_2(g_0) = \{\, v \in L_2(D) : \langle v, w^*(g_0) \rangle = 0 \,\}$*, where* $w^* \colon Y_0 \to L_2(D)$ *is continuous, with* $\|w^*(g_0)\| = 1$ *for all* $g_0 \in Y_0$*,*

having the following properties:

(i) $w(g_0) \in L_2(D)^+$ *for all* $g_0 \in Y_0$*,*

(ii) $X_2(g_0) \cap L_2(D)^+ = \{0\}$ *for all* $g_0 \in Y_0$*,*

(iii) *there are* $M \geq 1$ *and* $\gamma_0 > 0$ *such that for any* $g_0 \in Y_0$ *and any* $u_0 \in X_2(g_0)$ *with* $\|u_0\| = 1$*,*

$$\|U_{g_0}(t,0)u_0\| \leq M e^{-\gamma_0 t}\|U_{g_0}(t,0)w(g_0)\| \quad \text{for} \quad t > 0.$$

PROOF See Theorem 3.3.3. ⬜

Observe that by the uniqueness of solutions we have the following equality:

$$\Pi(t; v_0, g_0) = \partial\Phi(t; v_0, (0, g)), \qquad t \geq 0, \ v_0 \in X, \ g \in Z.$$

Recall that $\partial\Phi$ is continuous as a function from $(0, \infty) \times X \times Z$ into $X \times Z$.

Next we show that we have a stronger exponential separation property. To do so, we first show some lemmas.

LEMMA 7.1.12
For any $T > 0$ *there is* $C = C(T) > 0$ *such that*

$$\|U_{g_0}(t,0)v_0\|_X \leq C\|v_0\|_X$$

for all $g_0 \in Y$*,* $0 \leq t \leq T$*, and* $v_0 \in X$*.*

PROOF By Proposition 7.1.10(2), for $g \in Z$ and $v_0 \in X$ the function $[\, [0, \infty) \ni t \mapsto U_{g_0}(t,0)v_0 \in X \,]$ satisfies

$$U_{g_0}(t,0)v_0 = e^{A_p t}v_0 + \int_0^t e^{A_p(t-s)}(\partial_2\mathfrak{N}(s,0,g)(U_{g_0}(s,0)v_0))\,ds \qquad (7.1.16)$$

for all $t > 0$.

Fix $T > 0$. By Lemma 7.1.7 there is $M > 0$ such that $\|e^{A_p t}\|_{L_p(D),X} \leq M t^{-\beta}$ and $\|e^{A_p t}\|_X \leq M$, for all $t \in (0, T]$. It follows from Lemma 7.1.5(2) that $M_1 := \sup\{\, \|\partial_2\mathfrak{N}(t,0,g)\|_{\mathcal{L}(C(\bar{D}))} : t \in [0, T], \ g \in Z \,\}$ is $< \infty$. Finally, put M_2 to be the norm of the embedding $X \hookrightarrow C(\bar{D})$ and M_3 to be the norm of the embedding $C(\bar{D}) \hookrightarrow L_p(D)$. We estimate

$$\|U_{g_0}(t,0)v_0\|_X \leq M\|v_0\|_X + MM_1M_2M_3 \int_0^t (t-s)^{-\beta} \|U_{g_0}(s,0)v_0\|_X \, ds$$

for all $t \in (0,T]$. An application of the singular Gronwall lemma (see [48, 1.2.1]) gives the existence of $C > 0$ such that the desired inequality is satisfied.
∎

LEMMA 7.1.13
For any $v_0 \in L_2(D)$, $g_0 \in Y$, and $t > 0$, $U_{g_0}(t,0)v_0 \in X$. Moreover, for any $0 < T_1 < T_2$, there is $C(T_1,T_2) > 0$ such that

$$\|U_{g_0}(t,0)v_0\|_X \leq C(T_1,T_2)\|v_0\|$$

for all $g_0 \in Y$, $T_1 \leq t \leq T_2$, and $v_0 \in L_2(D)$.

PROOF Let p and β with $1 < p < \infty$ and $\frac{N}{2p} + \frac{1}{2} < \beta < 1$ be such that $X = F_p^\beta$. By the L_p–L_q estimates (Proposition 2.2.2), there is $C_1 > 0$ such that

$$\|U_{g_0}(T_1/2,0)v_0\|_p \leq C_1\|v_0\| \quad \text{for any} \quad g_0 \in Y, v_0 \in L_2(D).$$

For a given $g_0 \in Y$, let $v_1 := U_{g_0}(T_1/2,0)v_0$ and $g_1 := g_0 \cdot (T_1/2)$. Then

$$U_{g_0}(t,0)v_0 = U_{g_1}(t - T_1/2,0)v_1 \quad \text{for} \quad t \geq T_1/2.$$

Note that $U_{g_1}(t,0)v_1 \in F_p^1 \hookrightarrow X$ for $t > 0$ (see [48, Theorem 3.3.3]). By the arbitrariness of $T_1 > 0$, we have $U_{g_0}(t,0)v_0 \in X$ for $t > 0$.
By the density of X in $L_p(D)$ and (7.1.16), we have

$$U_{g_1}(t,0)v_1 = e^{A_p t}v_1 + \int_0^t e^{A_p(t-s)}(\partial_2\mathfrak{N}(s,0,g_1)(U_{g_1}(s,0)v_1)) \, ds \quad (7.1.17)$$

for all $t > 0$. Then by arguments similar to those in the proof of Lemma 7.1.12, we have

$$\|U_{g_1}(t,0)v_1\|_X \leq Mt^{-\beta}\|v_1\|_{L_p(D)} + MM_1M_2M_3 \int_0^t (t-s)^{-\beta}\|U_{g_1}(s,0)v_1\|_X \, ds.$$

This together with the singular Gronwall lemma (see [48, 1.2.1]) implies that there is $C_2 > 0$ such that $\|U_{g_1}(t,0)v_1\|_X \leq C_2\|v_1\|_p$ for $t \in [T_1/2, T_2 - T_1/2]$, $v_1 \in L_p(D)$, and $g_1 \in Y$.
It then follows that

$$\|U_{g_0}(t,0)v_0\|_X \leq C(T_1,T_2)\|v_0\|$$

for any $t \in [T_1, T_2]$, $v_0 \in L_2(D)$, and $g \in Y$, where $C(T_1, T_2) = C_1 C_2$.   □

LEMMA 7.1.14
For any $v_0 \in L_2(D)^+ \setminus \{0\}$, $g_0 \in Y$ and $t > 0$, $U_{g_0}(t, 0)v_0 \in X^{++}$.

PROOF   By Lemma 7.1.13, $U_{g_0}(t, 0)v_0 \in X$ for $t > 0$. Then by Proposition 2.2.7, we have $U_{g_0}(t, 0)v_0 \in X^+$ for $t > 0$. It then follows from Proposition 7.1.11 that $U_{g_0}(t, 0)v_0 \in X^{++}$.   □

LEMMA 7.1.15
For any bounded set $B \subset L_2(D)$ and $0 < T_1 < T_2$, the set $\{U_{g_0}(t, 0)v_0 : t \in [T_1, T_2],\ g_0 \in Y,\ v_0 \in B\}$ is relatively compact in X.

PROOF   First, by Lemma 7.1.13, $B_1 := \{U_{g_0}(T_1/2, 0)v_0 : g_0 \in Y,\ v_0 \in B\}$ is bounded in X. Then proceeding along the lines of the proof of Proposition 7.1.8 we obtain that $\{U_{g_0 \cdot T_1/2}(t, 0)v_1 : g_0 \in Y,\ t \in [T_1/2, T_2 - T_1/2],\ v_1 \in B_1\}$ is relatively compact in X.   □

Now we have

THEOREM 7.1.2

(i) *For each $g_0 \in Y_0$, $w(g_0) \in X^{++}$.*

(ii) *The function $[Y_0 \ni g_0 \mapsto w(g_0) \in X]$ is continuous.*

(iii) *There is $M_1 \geq 1$ such that*

$$\frac{1}{M_1}\|w(g_0)\| \leq \|w(g_0)\|_X \leq M_1\|w(g_0)\| \quad \textit{for all} \quad g_0 \in Y_0.$$

PROOF   Define a function $r\colon Y_0 \to \mathbb{R}$ as $r(g_0) := 1/\|U_{g_0 \cdot (-1)}(1, 0)w(g_0 \cdot (-1))\|$. The function r is positive and continuous, hence bounded above and bounded away from zero.

For each $g_0 \in Y_0$ there holds $w(g_0) = r(g_0)U_{g_0 \cdot (-1)}(1, 0)w(g_0 \cdot (-1))$. By Lemma 7.1.13, $U_{g_0 \cdot (-1)}(1, 0)w(g_0 \cdot (-1))$, hence $w(g_0)$, belongs to X.

Also, $w(g_0 \cdot (-1)) \in L_2(D)^+ \setminus \{0\}$, hence it follows from By Lemma 7.1.14, that $w(g_0) \in X^{++}$. This concludes the proof of Part (i).

By Lemma 7.1.15, $\{w(g_0) : g_0 \in Y_0\}$ has compact closure in X. Assume that $g_0^{(n)} \to g_0$. Then $w(g_0^{(n)}) \to w(g_0)$ in $L_2(D)$. By the above arguments, there are a subsequence $(n_k)_{k=1}^{\infty}$ and $u^* \in X$ such that $w(g_0^{(n_k)}) \to u^*$ in X.

Therefore we must have $u^* = w(g_0)$ and $w(g_0^{(n)}) \to w(g_0)$ in X. This proves Part (ii)

Part (iii) follows by Part (ii) and the compactness of Y_0. $\qquad\qquad\square$

THEOREM 7.1.3

There are $\widehat{M} \geq 1$ and $\gamma_0 > 0$ (γ_0 is the same as in Theorem 7.1.1) such that

$$\frac{\|U_{g_0}(t,0)u_0\|_X}{\|U_{g_0}(t,0)w(g_0)\|_X} \leq \widehat{M} e^{-\gamma_0 t} \frac{\|u_0\|_X}{\|w(g_0)\|_X}$$

for each $t > 0$, $g_0 \in Y_0$, and $u_0 \in X_2(g_0) \cap X$.

PROOF   It can be proved by arguments similar to those in the proof of in Theorem 3.5.2. For the reader's convenience we give a proof here.

By Lemma 7.1.12 there is $C > 0$ such that $\|U_{g_0}(t,0)u_0\|_X \leq C\|u_0\|_X$ for any $g_0 \in Y_0$, $0 \leq t \leq 2$, and $u_0 \in X$. Consequently

$$\|U_{g_0}(t,0)u_0\|_X \leq C e^{2\gamma_0} e^{-\gamma_0 t} \|u_0\|_X$$

for $g_0 \in Y_0$, $0 \leq t \leq 2$, and $u_0 \in X$. On the other hand,

$$\|U_{g_0}(t,0)w(g_0)\|_X \geq \frac{M_2}{M_1^2} \|w(g_0)\|_X$$

for $0 \leq t \leq 2$, where $M_2 := \inf \{ \|U_{g_0}(t,0)w(g_0)\| : g_0 \in Y_0,\ t \in [0,2] \} > 0$. Hence

$$\frac{\|U_{g_0}(t,0)u_0\|_X}{\|U_{g_0}(t,0)w(g_0)\|_X} \leq \frac{C M_1^2 e^{2\gamma_0}}{M_2} e^{-\gamma_0 t} \frac{\|u_0\|_X}{\|w(g_0)\|_X}$$

for $0 \leq t \leq 2$, provided that $u_0 \in X$.

Assume now $t > 2$. As a consequence of Lemma 7.1.13 there is $C_1 > 0$ such that $\|U_{g_0}(t,0)u_0\|_X \leq C_1\|U_{g_0}(t-1,0)u_0\|$ for each $g_0 \in Y_0$, $t > 2$, and $u_0 \in X$. Further, an application of Theorem 7.1.2(iii) gives

$$\frac{\|U_{g_0}(t,0)u_0\|_X}{\|U_{g_0}(t,0)w(g_0)\|_X} \leq \frac{C_1 M_1}{M_2} \frac{\|U_{g_0}(t-1,0)u_0\|}{\|U_{g_0}(t-1,0)w(g_0)\|}$$

for $t > 2$. By Theorem 7.1.1,

$$\frac{\|U_{g_0}(t-1,0)u_0\|}{\|U_{g_0}(t-1,0)w(g_0)\|} \leq M e^{-\gamma_0(t-2)} \frac{\|U_{g_0}(1,0)u_0\|}{\|U_{g_0}(1,0)w(g_0)\|}.$$

Further,

$$\frac{\|U_{g_0}(1,0)u_0\|}{\|U_{g_0}(1,0)w(g_0)\|} \leq \frac{M_1 M_3}{M_2} \frac{\|u_0\|_X}{\|w(g_0)\|_X},$$

where $M_3 := \sup \{ \|U_{g_0}(1,0)\|_{X,L_2(D)} : g_0 \in Y_0 \} < \infty$ (by $X \hookrightarrow L_2(D)$ and the L_2–L_2 estimates in Proposition 2.2.2).

As a consequence,

$$\frac{\|U_{g_0}(t,0)u_0\|_X}{\|U_{g_0}(t,0)w(g_0)\|_X} \le \widehat{M}e^{-\gamma_0 t}\frac{\|u_0\|_X}{\|w(g_0)\|_X}$$

for $t \ge 0$ and $u_0 \in X_2(g_0) \cap X$, where $\widehat{M} = \frac{e^{2\gamma_0}M_1^2}{M_2}\max\{C, \frac{C_1 M_3}{M_2}\}$. ▢

For $g_0 \in Y_0$ we put $\tilde{w}(g_0) := w(g_0)/\|w(g_0)\|_X$.

Recall that by Definitions 3.1.1, 3.1.2, and Lemma 3.2.6, the principal spectrum of Π over Y_0 equals the complement of the set of those $\lambda \in \mathbb{R}$ for which either of the following conditions holds:

- There are $\epsilon > 0$ and $M \ge 1$ such that

$$\|U_{g_0}(t,0)w(g_0)\| \le Me^{(\lambda-\epsilon)t} \quad \text{for } t > 0 \text{ and } g_0 \in Y_0$$

  (such $\lambda \in \mathbb{R}$ are members of the upper principal resolvent of Π over Y_0, denoted by $\rho_+(Y_0)$).

- There are $\epsilon > 0$ and $M \in (0,1]$ such that

$$\|U_{g_0}(t,0)w(g_0)\| \ge Me^{(\lambda+\epsilon)t} \quad \text{for } t > 0 \text{ and } g_0 \in Y_0$$

  (such $\lambda \in \mathbb{R}$ are members of the lower principal resolvent of Π over Y_0, denoted by $\rho_-(Y_0)$).

In view of Theorem 7.1.2 we have the following result.

THEOREM 7.1.4

(1) $\lambda \in \mathbb{R}$ *belongs to the upper principal resolvent of* Π *over* Y_0 *if and only if there are* $\epsilon > 0$ *and* $M \ge 1$ *such that*

$$\|U_{g_0}(t,0)\tilde{w}(g_0)\|_X \le Me^{(\lambda-\epsilon)t} \quad \text{for } t > 0 \text{ and } g_0 \in Y_0.$$

(2) $\lambda \in \mathbb{R}$ *belongs to the lower principal resolvent of* Π *over* Y_0 *if and only if there are* $\epsilon > 0$ *and* $M \in (0,1]$ *such that*

$$\|U_{g_0}(t,0)\tilde{w}(g_0)\|_X \ge Me^{(\lambda+\epsilon)t} \quad \text{for } t > 0 \text{ and } g_0 \in Y_0.$$

As a consequence of Proposition 7.1.10 we have the following, which characterizes the closeness between solutions of nonlinear equations and solutions of the linearized equation at the trivial solution.

THEOREM 7.1.5
For each $t > 0$ *there holds*

$$\frac{\|u(t; \varrho u_0, g) - U_{g_0}(t,0)(\varrho u_0)\|_X}{\varrho} \to 0 \quad \text{as } \varrho \to 0^+$$

uniformly in $g \in Z$ and $u_0 \in X^+$ with $\|u_0\|_X = 1$.

PROOF   Let $t > 0$ be fixed. The statement of the proposition is equivalent to saying that for each $\epsilon > 0$ there is $\delta > 0$ such that

$$\|u(t; u_0, g) - U_{g_0}(t, 0)u_0\|_X < \epsilon \|u_0\|_X$$

for any $g \in Z$ and any $u_0 \in X^+$ with $\|u_0\|_X < \delta$.

Let $\epsilon > 0$ be fixed. Proposition 7.1.10 implies that for each $g \in Z$ there is $\delta_1 = \delta_1(g) > 0$ such that for any $h \in Z$ and any $u_0 \in X^+$, if $d(g, h) < \delta_1$ and $\|u_0\| < \delta_1$ then

$$\|\partial_2 u(t; u_0, h) - \partial_2 u(t; 0, g)\|_{\mathcal{L}(X)} < \frac{\epsilon}{2},$$

where $d(\cdot, \cdot)$ denotes the distance in Z. Since Z is compact, there are finitely many $g^{(1)}, \ldots, g^{(n)} \in Z$ such that the open balls (in Z) with center $g^{(k)}$ and radius $\delta_1(g^{(k)})$ cover Z. Set $\delta := \min\{\delta_1(g^{(1)}), \ldots, \delta_1(g^{(n)})\}$.

For $g \in Z$ let $g^{(k)}$ be such that $d(g, g^{(k)}) \leq \delta_1(g^{(k)})$. For $u_0 \in X^+$ with $\|u_0\|_X < \delta$ we estimate

$$\|\partial_2 u(t; u_0, g) - \partial_2 u(t; 0, g)\|_{\mathcal{L}(X)}$$
$$\leq \|\partial_2 u(t; u_0, g) - \partial_2 u(t; 0, g^{(k)})\|_{\mathcal{L}(X)} + \|\partial_2 u(t; 0, g^{(k)}) - \partial_2 u(t; 0, g)\|_{\mathcal{L}(X)}$$
$$< \epsilon.$$

X^+ is convex, hence there holds

$$u(t; u_0, g) = \int_0^1 \partial_2 u(t; su_0, g)u_0 \, ds$$

for any $u_0 \in X^+$ and any $g \in Z$. Therefore

$$\|u(t; u_0, g) - U_{g_0}(t, 0)u_0\|_X$$
$$= \left\| \int_0^1 (\partial_2 u(t; su_0, g) - \partial_2 u(t; 0, g))u_0 \, ds \right\|_X < \epsilon \|u_0\|_X$$

for any $g \in Z$ and any $u_0 \in X^+$ with $\|u_0\|_X < \delta$.   ☐

The following result will be used several times, so we formulate it as a separate result.

THEOREM 7.1.6

Let $E \subset X^{++} \times Y_0$ be compact. Then there are $0 < c_1 \leq c_2 < \infty$ such that

$$c_1 \varphi_{\mathrm{princ}} \ll u_0 \ll c_2 \varphi_{\mathrm{princ}} \quad \textit{for all} \quad (u_0, g_0) \in E.$$

PROOF   Suppose to the contrary that for each $n \in \mathbb{N}$ there is $(u_n, g_0^{(n)}) \in E$ such that $\frac{1}{n}\varphi_{\mathrm{princ}} \not\ll u_n$. Without loss of generality we can assume that $(u_n, g_0^{(n)})$ converge to some $(u_0, g_0) \in E$. But then $0 \not\ll u_0$, which contradicts the assumption.

Now, suppose to the contrary that for each $n \in \mathbb{N}$ there is $(u_n, g_0^{(n)}) \in E$ such that $\frac{1}{n}u_n \not\ll \varphi_{\mathrm{princ}}$. Without loss of generality we can assume that $(u_n, g_0^{(n)})$ converge to some $(u_0, g_0) \in E$. The set E is compact, hence the set $\{\, u_n : n \in \mathbb{N} \,\}$ is bounded in the X-norm, consequently $u_0 = \lim_{n \to \infty} \frac{1}{n}u_n = 0$. But then $0 \not\ll \varphi_{\mathrm{princ}}$, which is impossible. $\square$

7.1.3   Global Attractor and Uniform Persistence

In this subsection, we study global attractor and uniform persistence for the skew-product semiflow Φ. Throughout this subsection, we assume (A7-1)–(A7-3). Some results presented in this subsection can be proved by applying the general theories for dissipative systems in [46] and general theories for persistence in [47]. We choose to provide elementary proofs for all the results here. We will apply the theories in [46] and [47] in Subsection 7.3.3 to competitive Kolmogorov systems of parabolic equations.

First we study the global attractor.

We denote

$$[0, P]_X := \{\, u \in X : 0 \le u(x) \le P \text{ for all } x \in \bar{D} \,\}.$$

The set $[0, P]_X$ is convex and closed (in X).

THEOREM 7.1.7 (Absorbing set)
Assume (A7-1)–(A7-3). Let $B \subset X^+$ be bounded in the $C(\bar{D})$-norm. Then there is $T = T(B) \ge 0$ such that $u(t; u_0, g)(x) \le P$ for all $t \ge T$, $u_0 \in B$, $g \in Z$, and $x \in \bar{D}$. Moreover, if $B \subset [0, P]_X$ then $T(B)$ can be taken to be zero.

PROOF   Define a function $\tilde{g} \colon [0, \infty) \to \mathbb{R}$ by

$$\tilde{g}(w) := \sup \{\, g(t, x, w) : g \in Z, \ t \in \mathbb{R}, \ x \in \bar{D} \,\}.$$

By Assumption (A7-2) the function $\tilde{g}$ is well defined. We claim that it is locally Lipschitz continuous. To do that, fix $W > 0$. Again by (A7-2) there is $L > 0$ such that

$$|g(t, x, w_1) - g(t, x, w_2)| \le L|w_1 - w_2|$$

for any $g \in Z$, $t \in \mathbb{R}$, $x \in \bar{D}$, and $w_1, w_2 \in [0, W]$. There holds

$$-L(w_2 - w_1) \le \tilde{g}(w_1) - \tilde{g}(w_2) \le L(w_2 - w_1)$$

for any two $0 \le w_1 < w_2 \le W$. Indeed, for any $g \in Z$, $t \in \mathbb{R}$, and $x \in \bar{D}$ we have

$$g(t, x, w_2) \le g(t, x, w_1) + L(w_2 - w_1) \le \tilde{g}(w_1) + L(w_2 - w_1),$$

from which it follows that $\tilde{g}(w_2) \le \tilde{g}(w_1) + L(w_2 - w_1)$. The inequality $\tilde{g}(w_1) \le \tilde{g}(w_2) + L(w_2 - w_1)$ is proved in a similar way.

Further, $\tilde{g}(w) < 0$ for all $w \ge P$.

Fix $B \subset X^+$ as in the hypothesis, and set $w_0 := \sup \{ \, \|u_0\|_{C(\bar{D})} : u_0 \in B \, \}$. For any $u_0 \in B$ and $g \in Z$, the unique solution $w(\cdot)$ of the ODE $\dot{w} = \tilde{g}(w)w$ satisfying the initial condition $w(0) = w_0$ is a supersolution of $(7.1.1)_g$. It suffices to observe that there is $T \ge 0$ such that $0 \le w(t) \le P$ for all $t \ge T$, and that if $w_0 \in [0, P]$ then $T = 0$. $\qquad \square$

THEOREM 7.1.8 (Global attractor in X^+)

Let (A7-1) through (A7-3) be satisfied. Then the topological skew-product semiflow Φ possesses a global attractor Γ contained in $[0, P]_X \times Z$. In addition, for any $B \subset X^+$ bounded in the $C(\bar{D})$-norm one has

(1) $\emptyset \ne \omega(B \times Z) \, (\subset \Gamma)$,

(2) Γ *attracts* $B \times Z$.

PROOF We first define the set Γ as $\omega([0, P]_X \times Z)$. Consequently, Γ is closed and invariant.

It follows from Proposition 7.1.8 that $O^+(\Phi_1([0, P]_X \times Z))$ has compact closure. Consequently, Γ is, by Lemma 1.2.5, nonempty and compact.

Theorem 7.1.7 implies that $\mathrm{cl}\, O^+([0, P]_X \times Z) \subset \mathrm{cl}([0, P]_X \times Z) = [0, P]_X \times Z$, consequently $\Gamma \subset [0, P]_X \times Z$.

Let $B \subset X^+$ be bounded in the $C(\bar{D})$-norm. Proposition 7.1.8 implies that $O^+(\Phi_1(B \times Z))$ has compact closure, therefore $\omega(B \times Z)$ is, by Lemma 1.2.5, nonempty and compact. Theorem 7.1.7 guarantees the existence of $T = T(B) \ge 0$ such that $O^+(\Phi_T(B \times Z)) \subset [0, P]_X \times Z$. Consequently,

$$\omega(B \times Z) = \omega(\Phi_T(B \times Z)) \subset \omega([0, P]_X \times Z) = \Gamma.$$

Since, by Lemma 1.2.5, $\omega(B \times Z)$ attracts $B \times Z$, Γ attracts $B \times Z$, too. In particular, the (nonempty compact invariant) set Γ attracts any $B \times Z$, where $B \subset X^+$ is bounded in the X-norm, consequently is the global attractor for the semiflow Φ. $\qquad \square$

By the invariance of Γ and Proposition 7.1.9, for any $(u_0, g) \in \Gamma$ and any $t < 0$ there is a unique $\tilde{u}(t; u_0, g) \in X^+$ such that $(\tilde{u}(t; u_0, g), \zeta_t g) \in \Gamma$ and $\Phi_{-t}(\tilde{u}(t; u_0, g), \zeta_t g) = (u_0, g)$. Define a mapping $\Phi|_\Gamma \colon \mathbb{R} \times \Gamma \to \Gamma$ by

$$\Phi|_\Gamma(t, (u_0, g)) := \begin{cases} (\tilde{u}(t; u_0, g), \zeta_t g) & \text{for } t < 0, \\ (u(t; u_0, g), \zeta_t g) & \text{for } t \ge 0. \end{cases}$$

PROPOSITION 7.1.13

Assume (A7-1) *through* (A7-3). *Then* $\Phi|_\Gamma$ *is a topological flow.*

PROOF   The satisfaction of the algebraic properties (TF2) and (TF3) follows from the definition and the fact that Φ is a semiflow.

For each $t \geq 0$ the mapping $(\Phi|_\Gamma)_t$ is a continuous bijection of a compact metric space onto itself, hence it is a homeomorphism. Further, for $(u_0, g) \in \Gamma$ fixed the mapping $[\,\mathbb{R} \ni t \mapsto \tilde{u}(t; u_0, g) \in X^+\,]$ is continuous (notice that for any $t \in \mathbb{R}$, $\tilde{u}(t; u_0, g) = u(t; t_0, \tilde{u}(t_0; u_0, g), g)$ for any $t_0 < t$, so the mapping is continuous at t).

We have thus the separate continuity of $\Phi|_\Gamma$. The (joint) continuity follows from standard results on joint continuity of actions of a Baire topological group ($\mathbb{R}$ in our case) on a compact (even locally compact) metric space, see e.g. [18].   ☐

Next we study the uniform persistence. Let Z_0 be a nonempty connected compact invariant subset of Z and $Y_0 := p_0(Z_0)$.

DEFINITION 7.1.2 (Uniform persistence)   *The skew-product semi-flow* Φ *on* $X^+ \times Z$ *is said to be* uniformly persistent *over* Z_0 *if there exists* $\eta_0 > 0$ *such that for any* $u_0 \in X^+ \setminus \{0\}$ *there is* $\tau = \tau(u_0) > 0$ *with the property that*

$$u(t; u_0, g) \geq \eta_0 \varphi_{\mathrm{princ}} \quad \textit{for all} \quad g \in Z_0, \ t \geq \tau.$$

In the following, we formulate an assumption on the principal spectrum of the linearization at the trivial solution.

(A7-4) *The principal spectrum* $[\lambda_{\min}, \lambda_{\max}]$ *for* Π *over* Y_0 *is contained in* $(0, \infty)$.

In other words, (A7-4) holds if and only if $0 \in \rho_-(Y_0)$.

LEMMA 7.1.16

Assume (A7-1)–(A7-4). *Then there exists* $T > 0$ *such that*

$$U_{g_0}(T, 0)\varphi_{\mathrm{princ}} \gg 2\varphi_{\mathrm{princ}} \quad \textit{for all} \quad g \in Z_0.$$

PROOF   By (A7-4) and Theorem 7.1.4(2) there are $\epsilon > 0$ and $0 < M \leq 1$ such that

$$\|U_{g_0}(t, 0)\tilde{w}(g_0)\|_X \geq Me^{\epsilon t} \quad \textit{for all} \quad t > 0, g \in Z_0.$$

Theorem 7.1.6 applied to the compact set $E = \{ (\tilde{w}(g_0), g_0) : g_0 \in Y_0 \}$ gives the existence of $0 < c_1 \leq c_2 < \infty$ such that

$$\frac{1}{c_2} \tilde{w}(g_0) \ll \varphi_{\text{princ}} \ll \frac{1}{c_1} \tilde{w}(g_0) \quad \text{for all} \quad g \in Z_0.$$

Take some $T > 0$ such that $M e^{\epsilon T} > 2 c_2 / c_1$. We have $\| U_{g_0}(T, 0) \tilde{w}(g_0) \|_X > 2 \frac{c_2}{c_1}$. Since $U_{g_0}(T, 0) \tilde{w}(g_0)$ belongs to $\text{span}\{ \tilde{w}(g_0 \cdot T) \}$, there holds

$$U_{g_0}(T, 0) \tilde{w}(g_0) \gg 2 \frac{c_2}{c_1} \tilde{w}(g_0 \cdot T).$$

Consequently, an application of Lemma 7.1.14 yields

$$U_{g_0}(T, 0) \varphi_{\text{princ}} \gg \frac{1}{c_2} U_{g_0}(T, 0) \tilde{w}(g_0) \gg \frac{2}{c_1} \tilde{w}(g_0 \cdot T) \gg 2 \varphi_{\text{princ}} \quad \text{for all} \quad g \in Z_0.$$

$$\square$$

THEOREM 7.1.9 (Uniform persistence)

Assume (A7-1)–(A7-4). Then there exists $r_0 > 0$ with the following properties.

(1) $u(T; r \varphi_{\text{princ}}, g) \gg 2 r \varphi_{\text{princ}}$ *for all $r \in (0, r_0]$ and all $g \in Z_0$, where $T > 0$ is as in Lemma 7.1.16.*

(2) *For each compact $E \subset (X^+ \setminus \{0\}) \times Z_0$ there is $T_0 = T_0(E) > 0$ such that $u(t; u_0, g) \gg 2 r_0 \varphi_{\text{princ}}$ for all $t \geq T_0$ and all $(u_0, g) \in E$ (hence $\{\Phi_t\}_{t \geq 0}$ is uniformly persistent over Z_0).*

PROOF (1) Lemma 7.1.15 implies that the set $\{ U_{g_0}(T, 0) \varphi_{\text{princ}} - 2 \varphi_{\text{princ}} : g_0 \in Y_0 \}$ (contained, by Lemma 7.1.16, in an open set X^{++}) is compact as a subset of X. Therefore $\epsilon_0 := \inf \{ \| (U_{g_0}(T, 0) \varphi_{\text{princ}} - 2 \varphi_{\text{princ}}) - v \|_X : g_0 \in Y_0, v \in \partial X^+ \}$ is positive. By linearity

$$\inf \{ \| (U_{g_0}(T, 0)(r \varphi_{\text{princ}}) - 2 r \varphi_{\text{princ}}) - v \|_X : g_0 \in Y_0, v \in \partial X^+ \} = r \epsilon_0$$
$$(7.1.18)$$

for any $r > 0$.

It follows from Theorem 7.1.5 that there is $r_0 > 0$ such that

$$\| u(T; r \varphi_{\text{princ}}, g) - U_{g_0}(T, 0)(r \varphi_{\text{princ}}) \|_X \leq \frac{r \epsilon_0}{2} \quad \text{for all} \quad g \in Z_0, r \in (0, r_0].$$

We estimate

$$\| (u(T; r \varphi_{\text{princ}}, g) - 2 r \varphi_{\text{princ}}) - (U_{g_0}(T, 0)(r \varphi_{\text{princ}}) - 2 r \varphi_{\text{princ}}) \|_X$$
$$= \| u(T; r \varphi_{\text{princ}}, g) - U_{g_0}(T, 0)(r \varphi_{\text{princ}}) \|_X \leq \frac{r \epsilon_0}{2}$$

for any $g \in Z_0$ and $0 < r \leq r_0$. Eq. (7.1.18) gives $u(T; r\varphi_{\text{princ}}, g) - 2r\varphi_{\text{princ}} \in X^{++}$, that is, $u(T; r\varphi_{\text{princ}}, g) \gg 2r\varphi_{\text{princ}}$. This proves Part (1).

Let a compact $E \subset (X^+ \setminus \{0\}) \times Z_0$. Denote $E_1 := \Phi([1, T+1] \times E)$. The set E_1 is compact and contained in $X^{++} \times Z_0$ (by Proposition 7.1.7(2)). An application of Theorem 7.1.6 to the compact set $(\text{Id}_X, p_0)E_1$ ($\subset X^{++} \times Y_0$) gives the existence of $\tilde{r} > 0$ such that $u(t; u_0, g) \gg \tilde{r}\varphi_{\text{princ}}$ for any $t \in [1, T+1]$ and any $(u_0, g) \in E$. If $\tilde{r} \geq r_0$ then we put $T_0 = 1$. If not then, for instance,

$$T_0 = \left(\left\lfloor \frac{\ln r_0 - \ln \tilde{r}}{\ln 2} \right\rfloor + 2 \right) T + 1$$

will do. $\qquad \Box$

Recall that it is proved in Theorem 7.1.8 that under (A7-1)-(A7-3), Φ possesses a global attractor Γ contained in $[0, P]_X \times Z$. In the following we go back to study more properties of the global attractor Γ under (A7-1)-(A7-4).

For each $r > 0$ we define

$$rW^+ := (X^+ + r\varphi_{\text{princ}}) \times Z_0 \quad \text{and} \quad rW^{++} := (X^{++} + r\varphi_{\text{princ}}) \times Z_0.$$

The following result is straightforward.

LEMMA 7.1.17

(i) *For any $r > 0$ the set rW^+ is closed (in $X^+ \times Z_0$).*

(ii) *For any $r > 0$, rW^{++} equals the relative interior of rW^+ in $X^+ \times Z_0$.*

(iii) *For any $0 < r_1 < r_2$ there holds $r_2 W^+ \subset r_1 W^{++}$.*

THEOREM 7.1.10
Assume (A7-1)–(A7-4). Then for the flow $\Phi|_{\Gamma \cap (X^+ \times Z_0)}$ the compact invariant set $\{0\} \times Z_0$ is a repeller, with its dual attractor Γ^{++} equal to $\omega(r_0 W^{++} \cap \Gamma)$ and contained in $2r_0 W^+$, where $r_0 > 0$ is as in Theorem 7.1.9.

PROOF We start by showing that the set $r_0 W^{++} \cap \Gamma$ is nonempty. Pick any $(u_0, g) \in r_0 W^{++}$. Theorem 7.1.8 yields $\emptyset \neq \omega((u_0, g)) \subset \Gamma$. By Theorem 7.1.9(2), $(u(t; u_0, g), \zeta_t g)$ belong to $2r_0 W^{++}$ for t sufficiently large, consequently $\omega((u_0, g)) \subset \text{cl} \, 2r_0 W^{++} \subset 2r_0 W^+ \subset r_0 W^{++}$.

The set $r_0 W^+ \cap \Gamma$ is compact and contained in $(X^+ \setminus \{0\}) \times Z_0$. Theorem 7.1.9(2) gives the existence of $T_0 > 0$ such that $\Phi_t(r_0 W^{++} \cap \Gamma) \subset 2r_0 W^{++} \cap \Gamma$ for all $t \geq T_0$. Thus

$$\Gamma^{++} := \omega(r_0 W^{++} \cap \Gamma) = \omega(\Phi_{T_0}(r_0 W^{++} \cap \Gamma)) \subset 2r_0 W^+ \cap \Gamma.$$

It follows from Theorem 7.1.2(ii) that $r_0 W^{++} \cap \Gamma$ is relatively open in $\Gamma \cap (X^+ \times Z_0)$, so it is a neighborhood of Γ^{++} in the relative topology of $\Gamma \cap (X^+ \times Z_0)$. We have thus proved that Γ^{++} is an attractor for the flow $\Phi|_{\Gamma \cap (X^+ \times Z_0)}$.

We prove now that the attraction basin of Γ^{++} equals $\Gamma \cap ((X^+ \setminus \{0\}) \times Z_0)$, that is, the complement of $\{0\} \times Z_0$ in $\Gamma \cap (X^+ \times Z_0)$. Take any nonzero $u_0 \in X^+$ and any $g \in Z_0$ such that $(u_0, g) \in \Gamma$. Theorem 7.1.9(2) yields the existence of $T_0 > 0$ such that $\Phi(t, u_0, g) \in 2r_0 W^{++} \subset r_0 W^+$ for $t \geq T_0$. Consequently, $\omega((u_0, g)) = \omega(\Phi(T_0; u_0, g)) \subset \omega(r_0 W^+ \cap \Gamma) = \Gamma^{++}$.   ☐

The above theorem deals only with the restriction of Π to $\Gamma \cap (X^+ \times Z_0)$. That set, as a compact subset of a bundle with fibers being modeled on a subset of an infinite-dimensional Banach space with nonempty interior, is rather small. What is more, usually we do not have any useable characterization of members of Γ. These are reasons why we are interested in finding larger sets attracted by Γ^{++}. The next result gives two families of such sets.

THEOREM 7.1.11

Let (A7-1)–(A7-4) be fulfilled. Assume that $B \subset X^+ \setminus \{0\}$ satisfies one of the following conditions:

(a) *B is compact.*

(b) *B is bounded in the $C(\bar{D})$-norm and there is $\tilde{r} > 0$ such that $u_0 \geq \tilde{r}\varphi_{\mathrm{princ}}$ for each $u_0 \in B$.*

Then Γ^{++} attracts $B \times Z_0$.

PROOF   From Proposition 7.1.8 it follows that $O^+(\Phi_1(B \times Z_0))$ has compact closure (in $X^+ \times Z_0$), consequently, by Lemma 1.2.5, $\omega(B \times Z_0)$ is compact nonempty and attracts $B \times Z_0$. The remainder of the proof is devoted to showing that $\omega(B \times Z_0) \subset \Gamma^{++}$.

Put $E_1 := \mathrm{cl}(\Phi_1(B \times Z_0))$, where the closure is taken in the $(X \times Z_0)$-topology. The set E_1 is compact (in case (a) as a consequence of the continuity of Φ_1, in case (b) by Proposition 7.1.8).

We claim that $E_1 \subset X^{++} \times Z_0$. In case (a) this is a direct consequence of Proposition 7.1.7(2) and the fact that now $E_1 = \Phi_1(B \times Z_0)$. In case (b), by Proposition 7.1.7(2) $u(1; \tilde{r}\varphi_{\mathrm{princ}}, g) \gg 0$ for each $g \in Z_0$. An application of Theorem 7.1.6 to the compact set $\{u(1; \tilde{r}\varphi_{\mathrm{princ}}, g) : g \in Z_0\} \times Y_0$ gives the existence of $\tilde{r}_1 > 0$ such that $u(1; \tilde{r}\varphi_{\mathrm{princ}}, g) \gg \tilde{r}_1\varphi_{\mathrm{princ}}$ for all $g \in Z_0$. By Proposition 7.1.7(1), $u(1; u_0, g) \geq u(1; \tilde{r}\varphi_{\mathrm{princ}}, g) \gg \tilde{r}_1\varphi_{\mathrm{princ}}$ for all $u_0 \in B$ and all $g \in Z_0$, that is, $\Phi_1(B \times Z_0) \subset \tilde{r}_1 W^{++}$, which yields $E_1 \subset \tilde{r}_1 W^+ \subset X^{++} \times Z_0$.

Theorem 7.1.9(2) guarantees the existence of $T_0 = T_0(E_1) > 0$ such that $\Phi_t(E_1) \subset 2r_0 W^{++}$ for all $t \geq T_0$. We have

$$\omega(B \times Z_0) \subset \omega(E_1) = \omega(\Phi_{T_0}(E_1)) \subset 2r_0 W^+ \subset r_0 W^{++}.$$

Since $\omega(B \times Z_0) \subset \Gamma$, it follows from Theorem 7.1.10 that

$$\omega(B \times Z_0) = \omega(\omega(B \times Z_0)) = \omega(\omega(B \times Z_0) \cap \Gamma) \subset \omega(r_0 W^{++} \cap \Gamma) = \Gamma^{++}.$$

☐

THEOREM 7.1.12 (Structure of Γ^{++})
Assume (A7-1)–(A7-4). *Assume also that $\partial_u g(t, x, u) < 0$ for any $g \in Z$, $t \in \mathbb{R}$, $x \in \bar{D}$, and $u \geq 0$. Then there is a continuous $\xi \colon Z_0 \to X^{++}$ such that*

$$\Gamma^{++} = \{\, (\xi(g), g) : g \in Z_0 \,\}.$$

Moreover, for any $u_0 \in X^+ \setminus \{0\}$ and $g \in Z_0$, $u(t; u_0, g) - \xi(g \cdot t) \to 0$ in X as $t \to \infty$.

PROOF   By Theorem 7.1.6, there are $0 < c_1 < c_2$ such that $c_1 \varphi_{\mathrm{princ}} \ll u_0 \ll c_2 \varphi_{\mathrm{princ}}$ for each $(u_0, g) \in \Gamma^{++}$. Put $u^* := 2c_2 \varphi_{\mathrm{princ}}$. The set $\{\, u_0 \in X : 0 \leq u_0 \ll c_2 \varphi_{\mathrm{princ}} \,\} \times Z_0$ is an open neighborhood of Γ^{++} in the relative topology of $X^+ \times Z_0$, consequently we deduce from Theorem 7.1.11 that there is $T > 0$ such that $u(T; u^*, g) \leq u^*$ for all $t \geq T$ and all $g \in Z_0$. For $n \in \mathbb{N}$ and $g \in Z_0$ define $\xi^{(n)}(g) := u(nT; u^*, g \cdot (-nT))$. For a fixed $g \in Z_0$ the sequence $(\xi^{(n)}(g))_{n=1}^{\infty}$ is monotone (decreasing), and bounded in $L_2(D)$, consequently it has a limit in $L_2(D)$ (denoted by $\xi^+(g)$). This sequence is, by Proposition 7.1.8, relatively compact in the X-norm. As a consequence, we have that $\|\xi^{(n)}(g) - \xi^+(g)\|_X \to 0$ as $n \to \infty$, for each $g \in Z_0$. Clearly $(\xi^+(g), g) \in \omega(\{\, (u^*, g \cdot (-nT_0)) : n \in \mathbb{N} \,\}) \subset \Gamma^{++}$.

Fix for the moment $(u_0, g) \in \Gamma^{++}$. We have $(u(-nT; u_0, g), g \cdot (-nT)) \in \Gamma^{++}$ for all $n \in \mathbb{N}$, hence $u(-nT; u_0, g) \leq u^*$ for all $n \in \mathbb{N}$. Consequently

$$u_0 = u(nT; u(-nT; u_0, g), g \cdot (-nT)) \leq u(nT; u^*, g \cdot (-nT)) \quad \text{for } n = 1, 2, 3, \ldots,$$

therefore $u_0 \leq \xi^+(g)$.

In a similar way we prove, for each $g \in Z_0$, the existence of $\xi^-(g)$ such that $(\xi^-(g), g) \in \Gamma^{++}$ and $\xi^-(g) \leq u_0$ for any $(u_0, g) \in \Gamma^{++}$.

We claim that $u(t; \xi^+(g), g) = \xi^+(g \cdot t)$ and $u(t; \xi^-(g), g) = \xi^-(g \cdot t)$ for any $g \in Z_0$ and $t \in \mathbb{R}$. Suppose first that for some $g \in Z_0$ and $t < 0$ we have $u(t; \xi^+(g), g) < \xi^+(g \cdot t)$. Then $\xi^+(g) = u(-t; u(t; \xi^+(g), g), g \cdot t) \ll u(-t; \xi^+(g \cdot t), g \cdot t)$, which contradicts the characterization of $\xi^+(g)$. It remains now to observe that, by construction, $u(nT; \xi^+(g), g) = \xi^+(g \cdot nT)$ and $u(nT; \xi^-(g), g) = \xi^-(g \cdot nT)$ for all $n \in \mathbb{N}$ and apply the previous reasoning.

Next we show that $\xi^+(g) = \xi^-(g)$ for any $g \in Z_0$. To do so we introduce the so called *part metric* $\rho(\cdot, \cdot)$ on X^{++} defined by

$$\rho(u, v) := \inf \{\, \ln \alpha : \alpha > 1, \tfrac{1}{\alpha} v \leq u \leq \alpha v \,\}, \qquad u, v \in X^{++}. \tag{7.1.19}$$

By arguments as in [83, Lemma 3.2],

$$\rho(u(t; u_0, g), u(t; v_0, g)) < \rho(u(s; u_0, g), u(s; v_0, g)) \qquad (7.1.20)$$

for any $u_0, v_0 \in X^{++}$ with $u_0 \neq v_0$, $0 < s < t$, and $g \in Z_0$. This implies that if $\tilde{g} \in Z_0$ is such that $\xi^+(\tilde{g}) \neq \xi^-(\tilde{g})$, then

$$\rho(\xi^+(\tilde{g} \cdot t), \xi^-(\tilde{g} \cdot t)) < \rho(\xi^+(\tilde{g} \cdot s), \xi^-(\tilde{g} \cdot s))$$

for any $s < t$. Put $\rho_{-\infty} := \lim_{s \to -\infty} \rho(\xi^+(\tilde{g} \cdot s), \xi^-(\tilde{g} \cdot s)) \ (\neq 0)$. Let $s_n \to -\infty$. Without loss of generality we may assume that $\tilde{g} \cdot s_n \to g_*$, $\xi^+(\tilde{g} \cdot s_n) \to u_*^+$ (in X) and $\xi^-(\tilde{g} \cdot s_n) \to u_*^-$ (in X), as $n \to \infty$. Then $u_*^+ \neq u_*^-$ and $\rho(u(t; u_*^+, g_*), u(t; u_*^-, g_*)) = \rho_{-\infty}$ for all $t \geq 0$, which contradicts (7.1.20). Therefore, $\xi^+(g) = \xi^-(g) =: \xi(g)$ for any $g \in Z_0$. Since the compact set Γ^{++} equals the graph of the function ξ (with compact domain Z_0), the continuity of ϕ follows.

The last property in the statement of the theorem follows from the fact that Γ^{++} attracts any $(u_0, g) \in (X^+ \setminus \{0\}) \times Z_0$. $\quad\Box$

Finally, we provide some sufficient conditions which guarantee (A7-4) holds.

A function $\hat{g}_0 \in C(\bar{D})$ is called a *time averaged function* of $g_0 \in Y_0$ if there are $s_n < t_n$ with $t_n - s_n \to \infty$ as $n \to \infty$ such that

$$\hat{g}_0(x) = \lim_{n \to \infty} \frac{1}{t_n - s_n} \int_{s_n}^{t_n} g_0(t, x) \, dt$$

uniformly for $x \in \bar{D}$. Let $\hat{Y}_0$ be defined as follows:

$$\hat{Y}_0 := \{\, \hat{g}_0 : \hat{g}_0 \text{ is an averaged function of some } g_0 \in Y_0 \,\}$$

For a given $\hat{g}_0 \in \hat{Y}_0$, let $\lambda(\hat{g}_0)$ be the principal eigenvalue of

$$\begin{cases} \frac{\partial u}{\partial t} = \Delta u + \hat{g}_0(x)u, & x \in D \\ \mathcal{B}u = 0, & x \in \partial D \end{cases} \qquad (7.1.21)$$

where $\mathcal{B}u$ is as in (7.1.1).

THEOREM 7.1.13
If $\lambda(\hat{g}_0) > 0$ for any $\hat{g}_0 \in \hat{Y}_0$, then (A7-4) holds, and hence Theorems 7.1.9, 7.1.10, and 7.1.11 hold.

PROOF   First of all, by Theorem 3.2.5, there are g_0^- such that

$$\lambda_{\min} = \lim_{t \to \infty} \frac{1}{t} \ln \|U_{g_0^-}(t, 0)w(g_0^-)\|.$$

Then by Theorem 5.2.2, there is $\hat{g}_0^- \in \hat{Y}_0$ such that $\lambda_{\min} \geq \lambda(\hat{g}_0^-) > 0$. This implies that (A7-4) holds. The theorem thus follows. $\quad\Box$

7.2   Semilinear Equations of Kolmogorov Type: Examples

In this section, we discuss applications of the general theory established in the previous section to some random and nonautonomous semilinear equations of Kolmogorov type.

7.2.1   The Random Case

Assume that $((\Omega, \mathfrak{F}, \mathbb{P}), \{\theta_t\}_{t\in\mathbb{R}})$ is an ergodic metric dynamical system. Consider the following random parabolic equation of Kolmogorov type:

$$\begin{cases} \dfrac{\partial u}{\partial t} = \Delta u + f(\theta_t\omega, x, u)u, & t > 0, \ x \in D, \\[2mm] \mathcal{B}u = 0, & t > 0, \ x \in \partial D, \end{cases} \qquad (7.2.1)$$

where $f\colon \Omega \times \bar{D} \times [0, \infty) \mapsto \mathbb{R}$.

The first assumption in the present subsection concerns the measurability of the function f (recall that for a metric space S the symbol $\mathfrak{B}(S)$ stands for the countably additive algebra of Borel sets):

(A7-R1) (Measurability) *The function f is $(\mathfrak{F} \times \mathfrak{B}(D) \times \mathfrak{B}([0, \infty)), \mathfrak{B}(\mathbb{R}))$-measurable.*

For each $\omega \in \Omega$, let $f^\omega(t, x, u) := f(\theta_t\omega, x, u)$.
The function

$$[\Omega \times \mathbb{R} \times D \times [0, \infty) \ni (\omega, t, x, u) \mapsto f^\omega(t, x, u) \in \mathbb{R}]$$

is $(\mathfrak{F} \times \mathfrak{B}(\mathbb{R}) \times \mathfrak{B}(D) \times \mathfrak{B}([0, \infty)), \mathfrak{B}(\mathbb{R}))$-measurable (as a composite of Borel measurable functions).

As a section of a Borel measurable function, the function f^ω, is $(\mathfrak{B}(\mathbb{R}) \times \mathfrak{B}(D) \times \mathfrak{B}([0, \infty)), \mathfrak{B}(\mathbb{R}))$-measurable, for any fixed $\omega \in \Omega$.

The next assumption regards regularity of the function f:

(A7-R2) (Regularity) *For each $\omega \in \Omega$ and any $M > 0$ the restrictions to $\mathbb{R} \times \bar{D} \times [0, M]$ of f^ω and all the derivatives of the functions f^ω up to order 1 belong to $C^{1-,1-,1-}(\mathbb{R} \times \bar{D} \times [0, M])$. Moreover, for $M > 0$ fixed the $C^{1-,1-,1-}(\mathbb{R} \times \bar{D} \times [0, M])$-norms of the restrictions of f^ω and those derivatives are bounded uniformly in $\omega \in \Omega$.*

The last assumption in the present subsection concerns the negativity of the function f for large values of u.

(A7-R3) *There are $P > 0$ and a function $m\colon [P, \infty) \to (0, \infty)$ such that $f(\omega, x, u) \le -m(u)$ for any $\omega \in \Omega$, any $x \in \bar{D}$, and any $u \ge P$.*

From now on, until the end of the present subsection, assume that (A7-R1) through (A7-R3) are satisfied.

Define the mapping E from Ω into the set of continuous real functions defined on $\mathbb{R} \times \bar{D} \times [0, \infty)$ as

$$E(\omega) := f^\omega.$$

Put

$$Z := \mathrm{cl}\,\{\, E(\omega) : \omega \in \Omega \,\} \tag{7.2.2}$$

with the open-compact topology, where the closure is taken in the open-compact topology. It is a consequence of (A7-R2) via the Ascoli–Arzelà theorem that the set Z is a compact metrizable space.

The following result follows immediately from the measurability properties of f (Assumption (A7-R1)).

LEMMA 7.2.1
The mapping E is $(\mathfrak{F}, \mathfrak{B}(Z))$-measurable.

An important property of the mapping E is the following

$$\zeta_t \circ E = E \circ \theta_t \qquad \text{for each } t \in \mathbb{R}. \tag{7.2.3}$$

It follows that if $g \in E(\Omega)$ then $g \cdot t \equiv \zeta_t g \in E(\Omega)$ for all $t \in \mathbb{R}$. Further, we deduce from Lemma 7.1.1 that if $g \in Z$ then $g \cdot t \in Z$ for all $t \in \mathbb{R}$.

Hence $(Z, \{\zeta_t\}_{t\in\mathbb{R}})$ is a compact flow.

The mapping E is a homomorphism of the measurable flow $((\Omega, \mathfrak{F}), \{\theta_t\}_{t\in\mathbb{R}})$ into the measurable flow $((Z, \mathfrak{B}(Z)), \{\zeta_t\}_{t\in\mathbb{R}})$. Denote by $\tilde{\mathbb{P}}$ the image of the measure $\mathbb{P}$ under E: for any $A \in \mathfrak{B}(Z)$, $\tilde{\mathbb{P}}(A) := \mathbb{P}(E^{-1}(A))$. $\tilde{\mathbb{P}}$ is a $\{\zeta_t\}$-invariant ergodic Borel measure on Z. So, E is a homomorphism of the metric flow $((\Omega, \mathfrak{F}, \mathbb{P}), \{\theta_t\}_{t\in\mathbb{R}})$ into the metric flow $((Z, \mathfrak{B}(Z), \tilde{\mathbb{P}}), \{\zeta_t\}_{t\in\mathbb{R}})$.

We will consider a family of Eqs. (7.1.1) parameterized by $g \in Z$. We claim that Assumptions (A7-1) through (A7-3) are fulfilled. Indeed, (A7-1) is clearly satisfied. It follows from (A7-R2) through the Ascoli–Arzelà theorem that (A7-2) is satisfied. The satisfaction of (A7-3) follows from (A7-R3).

We denote by $\Phi = \{\Phi_t\}_{t\geq 0}$ the *topological skew-product semiflow generated by* (7.2.1) on the product Banach bundle $X^+ \times Z$:

$$\Phi(t; u_0, g) = \Phi_t(u_0, g) := (u(t; u_0, g), \zeta_t g), \quad t \geq 0, \; g \in Z, \; u_0 \in X^+, \tag{7.2.4}$$

where $u(t; u_0, g)$ stands for the solution of (7.1.1) with initial condition $u(0; u_0, g)(x) = u_0(x)$.

Moreover, define

$$\tilde{\Phi}(t; u_0, \omega) := (u(t; u_0, E(\omega)), \theta_t\omega), \quad t \geq 0, \; \omega \in \Omega, \; u_0 \in X^+. \tag{7.2.5}$$

We have

LEMMA 7.2.2
$\tilde{\Phi}$ *is a continuous random skew-product semiflow on the measurable bundle* $X^+ \times \Omega$, *covering the metric flow* $((\Omega, \mathfrak{F}, \mathbb{P}), \{\theta_t\}_{t \in \mathbb{R}})$.

PROOF   It follows from (7.2.3) and the definitions of Φ and $\tilde{\Phi}$ that for each $t \geq 0$ the diagram

$$
\begin{array}{ccc}
X^+ \times \Omega & \xrightarrow{\tilde{\Phi}_t} & X^+ \times \Omega \\
(\mathrm{Id}_{X^+}, E) \downarrow & & \downarrow (\mathrm{Id}_{X^+}, E) \\
X^+ \times Z & \xrightarrow{\Phi_t} & X^+ \times Z,
\end{array}
$$

commutes. Consequently, the properties (RSP1) and (RSP2) of the random skew-product semiflow are satisfied.

As regards the measurability of the mapping $\tilde{\Phi}$, its second coordinate $[\, [0, \infty) \times \Omega \ni (t, \omega) \mapsto \theta_t \omega \in \Omega \,]$ is $(\mathfrak{B}([0, \infty)) \times \mathfrak{F}, \mathfrak{F})$-measurable. The mapping

$$
[\, [0, \infty) \times X^+ \times \Omega \ni (t, u_0, \omega) \mapsto u(t; u_0, E(\omega)) \in X^+ \,]
$$

is the composition of the mapping $(\mathrm{Id}_{[0,\infty)}, \mathrm{Id}_{X^+}, E)$ (which is $(\mathfrak{B}([0, \infty)) \times \mathfrak{B}(X^+) \times \mathfrak{F}, \mathfrak{B}([0, \infty)) \times \mathfrak{B}(X^+) \times \mathfrak{B}(Z))$-measurable) and the continuous mapping $[\, [0, \infty) \times X^+ \times Z \ni (t, u_0, g) \mapsto u(t; u_0, g) \in X^+ \,]$. $\quad\square$

For $t \geq 0$, $u_0 \in X^+$, and $\omega \in \Omega$ we will write $u(t; u_0, \omega)$ instead of $u(t; u_0, E(\omega))$. Similarly, for $t_0 \in \mathbb{R}$, $t \geq t_0$, $u_0 \in X^+$, and $\omega \in \Omega$ we will write $u(t; t_0, u_0, \omega)$ instead of $u(t - t_0; u_0, E(\theta_{t_0}\omega))$.

DEFINITION 7.2.1 (Uniform persistence)   (7.2.1) *is said to be uniformly persistent if there is* $\eta_0 > 0$ *such that for any* $u_0 \in X^+ \setminus \{0\}$ *there is* $\tau = \tau(u_0) > 0$ *with the property that*

$$
u(t; t_0, u_0, \omega) \geq \eta_0 \varphi_{\mathrm{princ}} \quad \textit{for } \mathbb{P}\text{-a.e. } \omega \in \Omega, \textit{ all } t_0 \in \mathbb{R}, \textit{ and all } t \geq t_0 + \tau.
$$

It should be noted that the adjective "uniform" in the above definition means that $\eta_0 > 0$ is independent of $u_0 \in X^+ \setminus \{0\}$ (sometimes such a property is referred to as *permanence*). However, in Definition 7.2.1 we have also uniformity with respect to the initial time $t_0 \in \mathbb{R}$. It follows that our uniform persistence is both in the pullback as well as in the forward sense (compare, e.g., [71]).

Put

$$
Z_0 := \mathrm{supp}\, \tilde{\mathbb{P}} \tag{7.2.6}
$$

($g \in Z_0$ if and only if for any neighborhood U of g in Z one has $\tilde{\mathbb{P}}(U) > 0$). Z_0 is a closed (hence compact) and $\{\zeta_t\}$-invariant subset of Z, with $\tilde{\mathbb{P}}(Z_0) = 1$.

Also, Z_0 is connected, since otherwise there would exist two open sets $U_1, U_2 \subset Z$ such that $Z_0 \cap U_1$ and $Z_0 \cap U_2$ are nonempty, compact and disjoint, and their union equals Z_0. The sets $Z_0 \cap U_1$ and $Z_0 \cap U_2$ are invariant, and, by the definition of support, each of them has $\tilde{\mathbb{P}}$-measure positive, which contradicts the ergodicity of $\tilde{\mathbb{P}}$.

LEMMA 7.2.3
There exists $\Omega_0 \subset \Omega$ with $\mathbb{P}(\Omega_0) = 1$ such that $Z_0 = \mathrm{cl}\,\{E(\theta_t \omega) : t \in \mathbb{R}\}$ for any $\omega \in \Omega_0$, where the closure is taken in the open-compact topology on Z.

PROOF   A proof is a copy of the proof of Lemma 4.1.2.   ⬜

Note that $\{0\} \times Z_0$ is invariant under Φ. Consider the linearization of Φ at $\{0\} \times Z_0$. Let $Y_0 := p_0(Z_0)$. Let

$$\Pi(t; v_0, g_0) := (U_{g_0}(t, 0)v_0, \sigma_t g_0), \quad t \geq 0, \quad v_0 \in X, \quad g_0 \in Z_0,$$

where $U_{g_0}(t, 0)v_0 = \partial_2 u(t; 0, g)v_0$.

Denote the principal spectrum of Π over Y_0 by $[\lambda_{\min}(f), \lambda_{\max}(f)]$. Let Ω_0 be as in Lemma 7.2.3.

THEOREM 7.2.1 (Uniform persistence)
Assume that $\lambda_{\min}(f) > 0$. Then there is $\eta_0 > 0$ such that for each nonzero $u_0 \in X^+$ there exists $\tau = \tau(u_0) > 0$ with the property that $u(t; t_0, u_0, \omega)(x) \geq \eta_0 \varphi_{\mathrm{princ}}(x)$ for all $t_0 \in \mathbb{R}$, $t \geq t_0 + \tau$, $\omega \in \Omega_0$, and $x \in \bar{D}$.

PROOF   An application of Theorem 7.1.9 to the compact set $\{u_0\} \times Z_0 \subset (X^+ \setminus \{0\}) \times Z_0$ gives the existence of $T > 0$ such that $u(t; t_0, u_0, \omega) \geq \eta_0 \varphi_{\mathrm{princ}}$ for any $t_0 \in \mathbb{R}$, any $t \geq t_0 + T$ and any $\omega \in \Omega_0$.   ⬜

7.2.2   The Nonautonomous Case

Consider the following nonautonomous parabolic equation of Kolmogorov type:

$$\begin{cases} \dfrac{\partial u}{\partial t} = \Delta u + f(t, x, u)u, & t > 0,\ x \in D, \\[2mm] \mathcal{B}u = 0, & t > 0,\ x \in \partial D, \end{cases} \tag{7.2.7}$$

where $f \colon \mathbb{R} \times \bar{D} \times [0, \infty) \to \mathbb{R}$.

We assume

(A7-N1) (Regularity) *For any $M > 0$ the restrictions to $\mathbb{R} \times \bar{D} \times [0, M]$ of f and all the derivatives of the function f up to order 1 belong to $C^{1-,1-,1-}(\mathbb{R} \times \bar{D} \times [0, M])$.*

(A7-N2) *There are $P > 0$ and a function $m: [P, \infty) \to (0, \infty)$ such that $f(t, x, u) \leq -m(u)$ for any $t \in \mathbb{R}$, any $x \in \bar{D}$ and any $u \geq P$.*

From now on, until the end of the present subsection, assume that (A7-N1) and (A7-N2) are satisfied.

Put
$$Z := \mathrm{cl}\,\{\, f \cdot t : t \in \mathbb{R}\,\} \tag{7.2.8}$$

with the open-compact topology, where the closure is taken in the open-compact topology. It is a consequence of (A7-N1) via the Ascoli–Arzelà theorem that the set Z is a compact metrizable space.

We deduce from Lemma 7.1.1 that if $g \in Z$ then $g \cdot t \in Z$ for all $t \in \mathbb{R}$. Hence $(Z, \{\zeta_t\}_{t \in \mathbb{R}})$ is a compact flow.

We will consider a family of Eqs. (7.1.1) parameterized by $g \in Z$. We claim that Assumptions (A7-1) through (A7-3) are fulfilled. Indeed, (A7-1) is clearly satisfied. It follows from (A7-N1) through the Ascoli–Arzelà theorem that (A7-2) is satisfied. The satisfaction of (A7-3) follows from (A7-N2).

We denote by $\Phi = \{\Phi_t\}_{t \geq 0}$ the *topological skew-product semiflow generated by* (7.2.7) on the product Banach bundle $X^+ \times Z$:

$$\Phi(t; u_0, g) = \Phi_t(u_0, g) := (u(t; u_0, g), \zeta_t g), \quad t \geq 0, \ g \in Z, \ u_0 \in X^+, \tag{7.2.9}$$

where $u(t; u_0, g)$ stands for the solution of (7.1.1) with initial condition $u(0; u_0, g)(x) = u_0(x)$.

DEFINITION 7.2.2 (Uniform persistence)   (7.2.7) *is said to be uniformly persistent if there exists $\eta_0 > 0$ such that for any $u_0 \in X^+ \setminus \{0\}$ there is $\tau = \tau(u_0) > 0$ with the property that*

$$u(t; t_0, u_0, f) \geq \eta_0 \varphi_{\mathrm{princ}} \quad \text{for all} \quad t_0 \in \mathbb{R} \quad \text{and all} \quad t \geq t_0 + \tau(u_0).$$

Note that Z is connected and $\{0\} \times Z$ is invariant under Φ. Consider the linearization of Φ at $\{0\} \times Z$. Let $Y := p_0(Z)$. Let

$$\Pi(t; v_0, g_0) := (U_{g_0}(t, 0)v_0, \sigma_t g_0), \quad t \geq 0, \quad v_0 \in X, \quad g_0 \in Y,$$

where $U_{g_0}(t, 0)v_0 = \partial_2 u(t; 0, g)v_0$.
Denote the principal spectrum of Π over Y by $[\lambda_{\min}(f), \lambda_{\max}(f)]$.

THEOREM 7.2.2 (Uniform persistence)
Assume that $\lambda_{\min}(f) > 0$. Then there is $\eta_0 > 0$ such that for each nonzero $u_0 \in X^+$ there exists $\tau = \tau(u_0) > 0$ with the property that $u(t; t_0, u_0, f)(x) \geq \eta_0 \varphi_{\mathrm{princ}}(x)$ for all $t_0 \in \mathbb{R}$, $t \geq t_0 + \tau$, and $x \in \bar{D}$.

PROOF   We apply Theorem 7.1.9(2).   ▯

COROLLARY 7.2.1

If for any time averaged function $\hat{f}_0$ of $f_0(t,x) := f(t,x,0)$ there holds $\lambda(\hat{f}_0) > 0$, then there exists $\eta_0 > 0$ such that for each nonzero $u_0 \in X^+$ there is $\tau = \tau(u_0) > 0$ with the property that $u(t;t_0,u_0,f)(x) \geq \eta_0\varphi_{\mathrm{princ}}(x)$ for all $t_0 \in \mathbb{R}$, $t \geq t_0 + \tau$ and $x \in \bar{D}$.

PROOF　It follows from Theorems 5.2.2 and 7.2.2.　　　　□

7.3　Competitive Kolmogorov Systems of Semilinear Equations: General Theory

In the present section we consider families of competitive Kolmogorov systems of semilinear second order parabolic equations:

$$\begin{cases} \dfrac{\partial u_1}{\partial t} = \Delta u_1 + g_1(t,x,u_1,u_2)u_1, & x \in D, \\[2mm] \dfrac{\partial u_2}{\partial t} = \Delta u_2 + g_2(t,x,u_1,u_2)u_2, & x \in D, \\[2mm] \mathcal{B}u_1 = 0, & x \in \partial D, \\[2mm] \mathcal{B}u_2 = 0, & x \in \partial D, \end{cases} \tag{7.3.1}$$

where $\mathcal{B}$ is a boundary operator of either the Dirichlet or Neumann type as in (7.0.1), and $\mathbf{g} = (g_1, g_2)$ belongs to a certain set of functions. Sometimes we write (7.3.1) as $(7.3.1)_{\mathbf{g}}$.

We consider the existence, uniqueness, and basic properties of solutions in Subsection 7.3.1. In Subsection 7.3.2 we study the linearizations of (7.3.1) at trivial and semitrivial solutions. Global attractor and uniform persistence of (7.3.1) are investigated in Subsection 7.3.3.

7.3.1　Existence, Uniqueness, and Basic Properties of Solutions

In this subsection, we present the existence, uniqueness, and basic properties of solutions of (7.3.1) in appropriate fractional power spaces of the operator $\Delta \times \Delta$ (with corresponding boundary conditions) with admissible $\mathbf{g}(\cdot,\cdot,\cdot) = (g_1(\cdot,\cdot,\cdot), g_2(\cdot,\cdot,\cdot))$. Those properties which are similar to the scalar equations case will be stated without proofs.

As in Subsection 7.1.1, for a continuous function $\mathbf{g}\colon \mathbb{R} \times \bar{D} \times [0,\infty) \to \mathbb{R}^2$ and $t \in \mathbb{R}$ denote by $\mathbf{g} \cdot t$ the *time-translate* of $\mathbf{g}$, $\mathbf{g} \cdot t(s,x,\mathbf{u}) := \mathbf{g}(s+t,x,\mathbf{u})$ for $s \in \mathbb{R}$, $x \in \bar{D}$, and $\mathbf{u} \in [0,\infty) \times [0,\infty)$.

Let $\mathbf{g}^{(n)}$ $(n \in \mathbb{N})$ and $\mathbf{g}$ be continuous functions from $\mathbb{R} \times \bar{D} \times [0,\infty) \times [0,\infty)$ to $\mathbb{R}^2$. Recall that a sequence $\mathbf{g}^{(n)}$ converges to $\mathbf{g}$ in the open-compact topology if and only if for any $M > 0$ the restrictions of $\mathbf{g}^{(n)}$ to $[-M, M] \times \bar{D} \times [0, M] \times [0, M]$ converge uniformly to the restriction of $\mathbf{g}$ to $[-M, M] \times \bar{D} \times [0, M] \times [0, M]$.

Denote by $\mathbf{Z}$ the set of admissible functions $\mathbf{g} = (g_1, g_2)$ for (7.3.1). We assume that $\mathbf{Z}$ satisfies

(A7-5) (1) $\mathbf{Z}$ *is compact in the open-compact topology.*
(2) $\mathbf{Z}$ *is* translation invariant: *If* $\mathbf{g} \in \mathbf{Z}$ *then* $\mathbf{g} \cdot t \in \mathbf{Z}$, *for each* $t \in \mathbb{R}$.

By (A7-5), $(\mathbf{Z}, \{\zeta_t\}_{t \in \mathbb{R}})$ is a compact flow, where $\zeta_t \mathbf{g} = \mathbf{g} \cdot t$ for $\mathbf{g} \in \mathbf{Z}$ and $t \in \mathbb{R}$.

(A7-6) (Regularity) *For any* $\mathbf{g} = (g_1, g_2) \in \mathbf{Z}$ *and any* $M > 0$ *the restrictions to* $\mathbb{R} \times \bar{D} \times [0, M] \times [0, M]$ *of* g_1, g_2, *and all the derivatives of* g_1 *and* g_2 *up to order 1 belong to* $C^{1-,1-,1-,1-}(\mathbb{R} \times \bar{D} \times [0, M] \times [0, M])$. *Moreover, for* $M > 0$ *fixed the* $C^{1-,1-,1-,1-}(\mathbb{R} \times \bar{D} \times [0, M])$-*norms of the restrictions of* g_1, g_2, *and those derivatives are bounded uniformly in* $\mathbf{Z}$.

For each $\mathbf{g} = (g_1, g_2) \in \mathbf{Z}$ we denote:

$$\mathbf{G}(t, x, \mathbf{u}) = (G_1(t, x, u_1, u_2), G_2(t, x, u_1, u_2))$$
$$:= (g_1(t, x, u_1, u_2)u_1, g_2(t, x, u_1, u_2)u_2)$$

for $t \in \mathbb{R}$, $x \in \bar{D}$, $\mathbf{u} = (u_1, u_2) \in [0, \infty) \times [0, \infty)$.

As in Subsection 7.1.1, we denote by $\mathfrak{N}$ the Nemytskiĭ (substitution) operator:

$$\mathfrak{N}(t, \mathbf{u}, \mathbf{g})(x) = (\mathfrak{N}_1(t, \mathbf{u}, \mathbf{g})(x), \mathfrak{N}_2(t, \mathbf{u}, \mathbf{g})(x)) := \mathbf{G}(t, x, \mathbf{u}(x)), \quad x \in \bar{D},$$

where $t \in \mathbb{R}$, $\mathbf{u} \colon \bar{D} \to \mathbb{R} \times \mathbb{R}$, and $\mathbf{g} \in \mathbf{Z}$.

We consider $\mathfrak{N}$ to be a mapping defined on $\mathbb{R} \times (C(\bar{D})^+ \times C(\bar{D})^+) \times \mathbf{Z}$. It is straightforward to see that $\mathfrak{N}$ takes $\mathbb{R} \times (C(\bar{D})^+ \times C(\bar{D})^+) \times \mathbf{Z}$ into $C(\bar{D}) \times C(\bar{D})$.

Let A_p stand the realization of the operator Δ (with corresponding boundary conditions) in $L_p(D)$. For $1 < p < \infty$ and $\beta \geq 0$ denote by F_p^β the fractional power space of the sectorial operator $-A_p$.

Until the end of the present section we fix $1 < p < \infty$, $p > N$ and $\frac{1}{2} + \frac{N}{2p} < \beta < 1$, and put

$$X := F_p^\beta. \tag{7.3.2}$$

There holds

$$X \hookrightarrow C^1(\bar{D}).$$

Let

$$\mathbf{X} = X \times X, \tag{7.3.3}$$
$$\mathbf{X}^+ = X^+ \times X^+,$$

and

$$\mathbf{X}^{++} = X^{++} \times X^{++}.$$

Note that $\mathbf{X}^{++}$ is the interior of $\mathbf{X}^+$.

Recall that for $\mathbf{u} = (u_1, u_2)$, $\mathbf{v} = (v_1, , v_2) \in \mathbf{X}$,

$$\mathbf{u} \leq \mathbf{v} \quad \text{if} \quad (v_1 - u_1, v_2 - u_2) \in \mathbf{X}^+,$$

$$\mathbf{u} < \mathbf{v} \quad \text{if} \quad (v_1 - u_1, v_2 - u_2) \in \mathbf{X}^+ \setminus \{(0, 0)\},$$

and

$$\mathbf{u} \ll \mathbf{v} \quad \text{if} \quad (u_1 - v_1, u_2 - v_2) \in \mathbf{X}^{++}.$$

Let $\leq_2$ be the order in $\mathbf{X}$ defined as follows: for $\mathbf{u} = (u_1, u_2)$, $\mathbf{v} = (v_1, , v_2) \in \mathbf{X}$,

$$\mathbf{u} \leq_2 \mathbf{v} \quad \text{if} \quad (v_1 - u_1, u_2 - v_2) \in \mathbf{X}^+,$$

$$\mathbf{u} <_2 \mathbf{v} \quad \text{if} \quad (v_1 - u_1, u_2 - v_2) \in \mathbf{X}^+ \setminus \{(0, 0)\},$$

and

$$\mathbf{u} \ll_2 \mathbf{v} \quad \text{if} \quad (v_1 - u_1, u_2 - v_2) \in \mathbf{X}^{++}.$$

DEFINITION 7.3.1 *For $t_0 \in \mathbb{R}$, $\mathbf{u}_0 = (u_{10}, u_{20}) \in \mathbf{X}^+$, and $\mathbf{g} = (g_1, g_2) \in \mathbf{Z}$ by a solution of $(7.3.1)_{\mathbf{g}}$ satisfying the initial condition $\mathbf{u}(t_0) = (u_1(t_0, \cdot), u_2(t_0, \cdot)) = \mathbf{u}_0$ we mean a continuous function $\mathbf{u} = (u_1, u_2) \colon J \to \mathbf{X}$, where J is a nondegenerate interval with $\inf J = t_0 \in J$, satisfying the following:*

- $\mathbf{u}(t_0) = \mathbf{u}_0$,

- $\mathbf{u}(t) \in F_p^1 \times F_p^1$ *for each $t \in J \setminus \{t_0\}$*,

- $\mathbf{u}(\cdot)$ *is differentiable, as a function into $L_p(D) \times L_p(D)$, on $J \setminus \{t_0\}$*,

- *there holds*

$$\begin{cases} \dot{u}_1(t) = A_p u_1(t) + \mathfrak{N}_1(t, u_1(t), u_2(t), \mathbf{g}) \\ \dot{u}_2(t) = A_p u_2(t) + \mathfrak{N}_2(t, u_1(t), u_2(t), \mathbf{g}) \end{cases}$$

for each $t \in J \setminus \{t_0\}$.

A solution $\mathbf{u} = (u_1, u_2)$ of $(7.3.1)_{\mathbf{g}}$ satisfying $\mathbf{u}(t_0) = \mathbf{u}_0$ is nonextendible *if there is no solution $\mathbf{u}^*$ of $(7.3.1)_{\mathbf{g}}$ defined on an interval J^* with $\sup J^* > \sup J$ such that $\mathbf{u}^*|_J \equiv \mathbf{u}$.*

Similarly to Propositions 7.1.1 and 7.1.3, we have

PROPOSITION 7.3.1 (Existence and uniqueness of solution)
Let (A7-5)–(A7-6) be satisfied. Then for each $\mathbf{g} \in \mathbf{Z}$, each $t_0 \in \mathbb{R}$, and each $\mathbf{u}_0 = (u_{10}, u_{20}) \in \mathbf{X}^+$ there exists a unique nonextendible solution of $(7.3.1)_{\mathbf{g}}$ satisfying the initial condition $\mathbf{u}(t_0, \cdot) = \mathbf{u}_0$, defined on an interval of the form

$[t_0, \tau_{\max})$, *where* $\tau_{\max} = \tau_{\max}(t_0, \mathbf{u}_0, g) > t_0$. *Moreover, for any* $1 < p < \infty$ *the unique nonextendible solution* $\mathbf{u}(\cdot) := \mathbf{u}(\cdot; t_0, \mathbf{u}_0, g)$ *satisfies*

$$\mathbf{u}(t) = e^{(A_p \times A_p)(t-t_0)} \mathbf{u}_0 + \int_{t_0}^t e^{(A_p \times A_p)(t-s)} \mathfrak{N}(s, \mathbf{u}(s), \mathbf{g}) \, ds,$$

for $t \in (t_0, \tau_{\max}(t_0, \mathbf{u}_0, \mathbf{g}))$.

Now we formulate several analogs of results from Subsection 7.1.1. We give (indications of) proofs only when they differ from the proofs of the corresponding results for scalar equations.

Similarly to Proposition 7.1.2, we have

PROPOSITION 7.3.2 (Positivity)
Assume (A7-5)–(A7-6). *Given* $t_0 \in \mathbb{R}$, $\mathbf{u}_0 \in \mathbf{X}^+$, *and* $\mathbf{g} \in \mathbf{Z}$, *there holds* $\mathbf{u}(t; t_0, \mathbf{u}_0, \mathbf{g}) \in \mathbf{X}^+$ *for any* $t \in [t_0, \tau_{\max}(t_0, \mathbf{u}_0, \mathbf{g}))$.

By arguments as in Lemmas 7.1.9 and 7.1.10, we can prove

$$\mathbf{u}(t + t_0; t_0, \mathbf{u}_0, \mathbf{g}) = \mathbf{u}(t; 0, \mathbf{u}_0, \mathbf{g} \cdot t_0) \tag{7.3.4}$$

for $t \in (0, \tau_{\max}(t_0, \mathbf{u}_0, \mathbf{g}) - t_0)$ and

$$\mathbf{u}(t; t_0, \mathbf{u}(t_0; 0, \mathbf{u}_0, \mathbf{g}), \mathbf{g}) = \mathbf{u}(t; 0, \mathbf{u}_0, g) \tag{7.3.5}$$

for $t_0, t \in (0, \tau_{\max}(0, \mathbf{u}_0, \mathbf{g}))$ with $t_0 \le t$.

In the following we write $\mathbf{u}(t; \mathbf{u}_0, \mathbf{g})$ for $\mathbf{u}(t; 0, \mathbf{u}_0, \mathbf{g})$. By (7.3.4) and (7.3.5), for any $\mathbf{u}_0 \in \mathbf{X}^+$ and $\mathbf{g} \in \mathbf{Z}$ there holds

$$\mathbf{u}(t + s; \mathbf{u}_0, \mathbf{g}) = \mathbf{u}(t; \mathbf{u}(s; \mathbf{u}_0, \mathbf{g}), \zeta_s \mathbf{g}) \tag{7.3.6}$$

for $s \in [0, \tau_{\max}(0, \mathbf{u}_0, \mathbf{g}))$ and $t \in [0, \tau_{\max}(s, \mathbf{u}(s; \mathbf{u}_0, \mathbf{g}), \zeta_s \mathbf{g}))$. We write $\tau_{\max}(\mathbf{u}_0, \mathbf{g})$ for $\tau_{\max}(0, \mathbf{u}_0, \mathbf{g})$.

Similarly to Proposition 7.1.4, we have

PROPOSITION 7.3.3 (Regularity)
Let (A7-5) *and* (A7-6) *be satisfied. Then for each* $\mathbf{u}_0 \in \mathbf{X}^+$ *and each* $\mathbf{g} \in \mathbf{Z}$, $\mathbf{u}(t, x) = \mathbf{u}(t; \mathbf{u}_0, \mathbf{g})(x)$ *is a classical solution, that is,*

- $\frac{\partial \mathbf{u}}{\partial t}(t, \cdot) \in C(\bar{D}) \times C(\bar{D})$ *for each* $t \in (0, \tau_{\max}(\mathbf{u}_0, \mathbf{g}))$,

- $\mathbf{u}(t, \cdot) \in C^2(\bar{D}) \times C^2(\bar{D})$ *for each* $t \in (0, \tau_{\max}(\mathbf{u}_0, \mathbf{g}))$,

- *for any* $t \in (0, \tau_{\max}(\mathbf{u}_0, \mathbf{g}))$ *and* $x \in D$ *the equation in* (7.3.1) *is satisfied pointwise,*

- *for any* $t \in (0, \tau_{\max}(\mathbf{u}_0, \mathbf{g}))$ *and* $x \in \partial D$ *the boundary condition in* (7.3.1) *is satisfied pointwise.*

Similarly to Proposition 7.1.6 we have

PROPOSITION 7.3.4 (Continuous dependence)
Let (A7-5)–(A7-6) be satisfied. Then the mapping

$$[\, [0, \tau_{\max}(\mathbf{u}_0, \mathbf{g})) \times \mathbf{X}^+ \times \mathbf{Z} \ni (t, \mathbf{u}_0, \mathbf{g}) \mapsto u(t; \mathbf{u}_0, \mathbf{g}) \in \mathbf{X}^+\,]$$

is continuous.

Similarly to Proposition 7.1.10, we have

PROPOSITION 7.3.5 (Differentiability)
Assume (A7-5)–(A7-6). Then the derivative $\partial_2\mathbf{u}$ of the mapping

$$[\, [0, \tau_{\max}(\mathbf{u}_0, \mathbf{g})) \times \mathbf{X}^+ \times \mathbf{Z} \ni (t, \mathbf{u}_0, \mathbf{g}) \mapsto u(t; \mathbf{u}_0, \mathbf{g}) \in \mathbf{X}^+\,]$$

with respect to the $\mathbf{u}_0$-variable exists and is continuous on the set $(0, \tau_{\max}(\mathbf{u}_0, \mathbf{g})) \times \mathbf{X}^+ \times \mathbf{Z}$. For $\mathbf{u}_0 = (u_{01}, u_{02}) \in \mathbf{X}^+$ and $\mathbf{v}_0 = (v_{01}, v_{02}) \in \mathbf{X}$ the function $[t \mapsto v(t)]$, where $\mathbf{v}(t) = (v_1(t), v_2(t)) := \partial_2\mathbf{u}(t; \mathbf{u}_0, \mathbf{g})v_0$, is a classical solution of the system of parabolic equations

$$\begin{cases} \frac{\partial v_1}{\partial t} = \Delta v_1 + \frac{\partial G_1}{\partial u_1}(t, x, \mathbf{u}(t; \mathbf{u}_0, \mathbf{g})(x))v_1 \\ \qquad + \frac{\partial G_1}{\partial u_2}(t, x, \mathbf{u}(t; \mathbf{u}_0, \mathbf{g})(x))v_2, & t > 0, \ x \in D, \\ \frac{\partial v_2}{\partial t} = \Delta v_2 + \frac{\partial G_2}{\partial u_1}(t, x, \mathbf{u}(t; \mathbf{u}_0, \mathbf{g})(x))v_1 \\ \qquad + \frac{\partial G_2}{\partial u_2}(t, x, \mathbf{u}(t; \mathbf{u}_0, \mathbf{g})(x))v_2, & t > 0, \ x \in D, \\ \mathcal{B}v_1 = 0, & t > 0, \ x \in \partial D, \\ \mathcal{B}v_2 = 0, & t > 0, \ x \in \partial D, \end{cases} \tag{7.3.7}$$

with initial conditions $v_1(0) = v_{01}$, $v_2(0) = v_{02}$.

The following is a competition assumption.

(A7-7) (Strong competitiveness)
$(\partial g_1/\partial u_2)(t, x, u_1, u_2) < 0$ and $(\partial g_2/\partial u_1)(t, x, u_1, u_2) < 0$ *for all* $\mathbf{g} \in \mathbf{Z}$, $t \in \mathbb{R}$, $x \in \bar{D}$, *and* $(u_1, u_2) \in [0, \infty) \times [0, \infty)$.

PROPOSITION 7.3.6 (Order preserving)
Assume (A7-5)–(A7-7). Let $\mathbf{g} \in \mathbf{Z}$.

(1) *If* $(u_1, u_2), (v_1, v_2) \in \mathbf{X}^+$, $(u_1, u_2) \leq_2 (v_1, v_2)$, *then*

$$\mathbf{u}(t; (u_1, u_2), \mathbf{g}) \leq_2 \mathbf{u}(t; (v_1, v_2), \mathbf{g})$$

for each $t \in [0, \tau_{\max}((u_1, u_2), \mathbf{g})) \cap [0, \tau_{\max}((v_1, v_2), \mathbf{g}))$.

(2) *If (u_1, u_2), $(v_1, v_2) \in X^+$, $(u_1, u_2) <_2 (v_1, v_2)$ and $v_1 > 0$ or $u_2 > 0$, then*

$$\mathbf{u}(t; (u_1, u_2), \mathbf{g}) \ll_2 \mathbf{u}(t; (v_1, v_2), \mathbf{g})$$

for each $t \in (0, \tau_{\max}((u_1, u_2), \mathbf{g})) \cap (0, \tau_{\max}((v_1, v_2), \mathbf{g}))$.

PROOF   It follows from Proposition 7.3.2 and comparison principle for parabolic equations.   ⬜

Here is the assumption which guarantees the global existence of solutions.

(A7-8) *There is $P > 0$ such that $g_1(t, x, u_1, u_2) < 0$ for any $\mathbf{g} \in \mathbf{Z}$, any $t \in \mathbb{R}$, any $x \in \bar{D}$, and any $u_1 \geq P$, any $u_2 \in [0, \infty)$; and $g_2(t, x, u_1, u_2) < 0$ for any $\mathbf{g} \in \mathbf{Z}$, any $t \in \mathbb{R}$, any $x \in \bar{D}$, and any $u_1 \in [0, \infty)$, $u_2 \geq P$.*

The following proposition follows from Proposition 7.1.5.

PROPOSITION 7.3.7 (Semitrivial solutions)
Assume (A7-5)–(A7-8).

(1) *For any $\mathbf{u}_0 \in \mathbf{X}^+ \times \{0\}$, $\mathbf{u}(t; \mathbf{u}_0, \mathbf{g})$ exists and $\mathbf{u}(t; \mathbf{u}_0, \mathbf{g}) \in \mathbf{X}^+ \times \{0\}$ for all $t \geq 0$ and $\mathbf{g} \in \mathbf{Z}$ (such $\mathbf{u}(t; \mathbf{u}_0, \mathbf{g})$ is called a* semitrivial solution*).*

(2) *For any $\mathbf{u}_0 \in \{0\} \times X^+$, $\mathbf{u}(t; \mathbf{u}_0, \mathbf{g})$ exists and $\mathbf{u}(t; \mathbf{u}_0, \mathbf{g}) \in \{0\} \times X^+$ for all $t \geq 0$ and $\mathbf{g} \in \mathbf{Z}$ (such $\mathbf{u}(t; \mathbf{u}_0, \mathbf{g})$ is also called a* semitrivial *solution*).*

PROPOSITION 7.3.8 (Global existence)
Assume (A7-5) *through* (A7-8)*. Then for any $\mathbf{u}_0 \in \mathbf{X}^+$ and any $\mathbf{g} \in \mathbf{Z}$ the unique nonextendible solution $\mathbf{u}(\cdot; \mathbf{u}_0, \mathbf{g})$ is defined on $[0, \infty)$.*

PROOF   For any $\mathbf{u}_0 = (u_{01}, u_{02}) \in \mathbf{X}^+$ and $\mathbf{g} \in \mathbf{Z}$, by Proposition 7.3.6, we have

$$\mathbf{u}(t; (0, u_{02}), \mathbf{g}) \leq_2 \mathbf{u}(t; (u_{01}, u_{02}), \mathbf{g}) \leq_2 \mathbf{u}(t; (u_{01}, 0), \mathbf{g})$$

for all $t \in [0, \tau_{\max}(\mathbf{u}_0, \mathbf{g}))$. An application of Theorem 7.1.7 to both the first coordinate of $\mathbf{u}(\cdot; (u_{01}, 0), \mathbf{g})$ and the second coordinate of $\mathbf{u}(\cdot; (0, u_{02}), \mathbf{g})$ gives together with the above inequality that the set $\{ \|\mathbf{u}(t; \mathbf{u}_0, \mathbf{g})\|_{C(\bar{D}) \times C(\bar{D})} : t \in [0, \tau_{\max}(\mathbf{u}_0, \mathbf{g})) \}$ is bounded. The remainder of the proof goes along the lines of the proof of Proposition 7.1.5.   ⬜

Similarly to Proposition 7.1.8, we have

PROPOSITION 7.3.9 (Compactness)

Assume (A7-5) through (A7-8). Then for any $\delta_0 > 0$ and any $B \subset \mathbf{X}^+$ bounded in the $C(\bar{D}) \times C(\bar{D})$-norm, the set $\{\,\mathbf{u}(t; \mathbf{u}_0, \mathbf{g}) : t \geq \delta_0,\ \mathbf{u}_0 \in B,\ \mathbf{g} \in \mathbf{Z}\,\}$ has compact closure in the $\mathbf{X}$-norm.

In the rest of this subsection, we assume (A7-5)-(A7-8). Let

$$\Phi(t; \mathbf{u}_0, \mathbf{g}) = \Phi_t(\mathbf{u}_0, \mathbf{g}) := (\mathbf{u}(t; \mathbf{u}_0, \mathbf{g}), \zeta_t \mathbf{g}) \tag{7.3.8}$$

for $t \geq 0$, $\mathbf{u}_0 \in \mathbf{X}^+$, and $\mathbf{g} \in \mathbf{Z}$. Then $\Phi = \{\Phi_t\}_{t \geq 0}$ is a topological skew-product semiflow on the product bundle $\mathbf{X}^+ \times \mathbf{Z}$ covering the topological flow $(\mathbf{Z}, \zeta)$.

In the sequel an important role will be played by the set

$$\widetilde{\partial} \mathbf{X}^+ := (X^+ \times \{0\}) \cup (\{0\} \times X^+).$$

Notice that $\widetilde{\partial} \mathbf{X}^+$ is a proper subset of $\mathbf{X}^+ \setminus \mathbf{X}^{++}$ ($=$ the boundary of $\mathbf{X}^+$ in $\mathbf{X}$). Indeed, let a nonzero $u_0 \in X^+ \setminus X^{++}$. Then $(u_0, u_0) \in \mathbf{X}^+ \setminus \mathbf{X}^{++}$ but $(u_0, u_0) \notin \widetilde{\partial} \mathbf{X}^+$.

By Propositions 7.3.7 and 7.3.6(2), the sets $\widetilde{\partial} \mathbf{X}^+ \times \mathbf{Z}$ and $(\mathbf{X}^+ \setminus \widetilde{\partial} \mathbf{X}^+) \times \mathbf{Z}$ are forward invariant. The property in Proposition 7.3.6(2) can be written as: The restriction $\Phi|_{(\mathbf{X}^+ \setminus \widetilde{\partial} \mathbf{X}^+) \times \mathbf{Z}}$ is *strongly monotone* with respect to the order relation $\leq_2$ (cp. [57]).

Further, let

$$\begin{aligned}
\partial \Phi(t; \mathbf{v}_0, (\mathbf{u}_0, \mathbf{g})) &= \partial \Phi_t(\mathbf{v}_0, (\mathbf{u}_0, \mathbf{g})) \\
&:= (\partial_2 \mathbf{u}(t; \mathbf{u}_0, \mathbf{g}) \mathbf{v}_0, (\mathbf{u}(t; \mathbf{u}_0, \mathbf{g}), \zeta_t \mathbf{g})),
\end{aligned} \tag{7.3.9}$$

where $t \geq 0$, $\mathbf{v}_0 \in \mathbf{X}$, $\mathbf{u}_0 \in \mathbf{X}^+$, and $\mathbf{g} \in \mathbf{Z}$. Proposition 7.3.5 guarantees that $\partial \Phi = \{\partial \Phi_t\}_{t \geq 0}$ is a topological linear skew-product semiflow on the product Banach bundle $\mathbf{X} \times (\mathbf{X}^+ \times \mathbf{Z})$ covering the topological semiflow Φ on $\mathbf{X}^+ \times \mathbf{Z}$.

7.3.2 Linearization at Trivial and Semitrivial Solutions

In this subsection, we consider the linearization of (7.3.1) at trivial and semitrivial solutions. Throughout this subsection we assume (A7-5)-(A7-8). Most results in this subsection follow from the general theories developed in Chapters 2, 3, and 4.

We start by considering the linearization at trivial solution.

Note that $\{0\} \times \mathbf{Z}$ is invariant under Φ. We introduce $\mathbf{p}_0 \colon \mathbf{Z} \to L_\infty(\mathbb{R} \times D, \mathbb{R})$ by

$$\mathbf{p}_0(\mathbf{g}) := \mathbf{g}_0,$$

where $\mathbf{g}_0(t, x) := \mathbf{g}(t, x, 0, 0)$ for $\mathbf{g} \in \mathbf{Z}$. We further introduce $p_{01} \colon \mathbf{Z} \to L_\infty(\mathbb{R} \times D, \mathbb{R})$ by

$$p_{01}(\mathbf{g}) := g_{01},$$

where $g_{01}(t, x) := g_1(t, x, 0, 0)$ for $\mathbf{g} = (g_1, g_2) \in \mathbf{Z}$, and $p_{02} \colon \mathbf{Z} \to L_\infty(\mathbb{R} \times D, \mathbb{R})$ by

$$p_{02}(\mathbf{g}) := g_{02},$$

where $g_{02}(t, x) := g_2(t, x, 0, 0)$ for $\mathbf{g} = (g_1, g_2) \in \mathbf{Z}$. Denote by $\mathbf{Y}^0$, Y^{01}, and Y^{02} the images of $\mathbf{p}_0$, p_{01}, and p_{02}, respectively.

Consider the restriction of the topological linear skew-product semiflow $\partial\mathbf{\Phi}$ to $\mathbf{X} \times (\{\mathbf{0}\} \times \mathbf{Z})$:

$$\partial\mathbf{\Phi}_t(\mathbf{v}_0, (\mathbf{0}, \mathbf{g})) = (\partial_2\mathbf{u}(t; \mathbf{0}, \mathbf{g})\mathbf{v}_0, (\mathbf{0}, \mathbf{g} \cdot t)), \qquad t \geq 0, \ \mathbf{v}_0 \in \mathbf{X}, \ \mathbf{g} \in \mathbf{Z}.$$

It follows from Proposition 7.1.10 that for any $\mathbf{v}_0 = (v_{01}, v_{02}) \in \mathbf{X}$ and any $\mathbf{g} = (g_1, g_2) \in \mathbf{Z}$ the function $[\, (0, \infty) \ni t \mapsto \partial_2\mathbf{u}(t; \mathbf{0}, \mathbf{g})\mathbf{v}_0 \in \mathbf{X} \,]$ is given by the classical solution $\mathbf{v}(\cdot) = (v_1(\cdot), v_2(\cdot))$ of

$$\begin{cases} \frac{\partial v_i}{\partial t} = \Delta v_i + g_{0i}(t, x)v_i, & t > 0, \ x \in D, \\ \mathcal{B}v_i = 0, & t > 0, \ x \in \partial D, \end{cases} \qquad (7.3.10)$$

with initial condition $v_i(0) = v_{01}$, $i = 1, 2$. We sometimes write (7.3.10) as $(7.3.10)_i$, $i = 1, 2$.

Let $U^{0i}_{g_{0i}}(t, 0) \colon L_2(D) \to L_2(D)$, $i = 1, 2$, be the weak solution operator of $(7.3.10)_i$. We have

$$\partial_2\mathbf{u}(t; \mathbf{0}, \mathbf{g})\mathbf{v}_0 = (U^{01}_{g_{01}}(t, 0)v_{01}, U^{02}_{g_{02}}(t, 0)v_{02}) \qquad (7.3.11)$$

for any $\mathbf{v}_0 = (v_{01}, v_{02}) \in \mathbf{X}$ and $t \geq 0$.

We may write $U^{01}_{\mathbf{g}_0}(t, 0)$ and $U^{02}_{\mathbf{g}_0}(t, 0)$ for $U^{01}_{g_{01}}(t, 0)$ and $U^{02}_{g_{02}}(t, 0)$, respectively, if no confusion occurs, where $\mathbf{g}_0 = (g_{01}, g_{02}) = \mathbf{p}_0(\mathbf{g})$.

Let $\mathbf{Z}_0$ be a nonempty connected compact translation invariant subset of $\mathbf{Z}$. Denote $Y^0_{01} := p_{01}(\mathbf{Z}_0)$ and $Y^0_{02} := p_{02}(\mathbf{Z}_0)$.

Let

$$\Pi^0_i(t; v_{0i}, g_{0i}) := (U^{0i}_{g_{0i}}(t, 0)v_{0i}, g_{0i} \cdot t)$$

for $t \geq 0$, $v_{0i} \in L_2(D)$, and $g_{0i} \in Y^0_{0i}$, $i = 1, 2$.

Similarly to Theorem 7.1.1, following from Theorem 3.3.3 we have

THEOREM 7.3.1

Let $i = 1$ or 2. There exist

- *an invariant (under Π^0_i) one-dimensional subbundle $X^0_{i,1}$ of $L_2(D) \times Y^0_{0i}$ with fibers $X^0_{i,1}(g_{0i}) = \mathrm{span}\{w^0_i(g_{0i})\}$, where $w^0_i \colon Y^0_{0i} \to L_2(D)$ is continuous, with $\|w^0_i(g_{0i})\| = 1$ for all $g_{0i} \in Y^0_{0i}$, and*

- *an invariant (under Π^0_i) complementary one-codimensional subbundle $X^0_{i,2}$ of $L_2(D) \times Y^0_{0i}$ with fibers $X^0_{i,2}(g_{0i}) = \{\, v \in L_2(D) : \langle v, w^{0,*}_i(g_{0i})\rangle = 0 \,\}$, where $w^{0,*}_i \colon Y^0_{0i} \to L_2(D)$ is continuous, and for all $g_{0i} \in Y^0_{0i}$, $\|w^{0,*}_i(g_{0i})\| = 1$,* *having the following properties:*

(i) $w_i^0(g_{0i}) \in L_2(D)^+$ *for all* $g_{0i} \in Y_{0i}^0$,

(ii) $X_{i,2}^0(g_{01}) \cap L_2(D)^+ = \{0\}$ *for all* $g_{0i} \in Y_{0i}^0$,

(iii) *there are* $M_i^0 \geq 1$ *and* $\gamma_i^0 > 0$ *such that for any* $g_{0i} \in Y_{0i}^0$ *and any* $v_{0i} \in X_{i,2}^0(g_{0i})$ *with* $\|v_{0i}\| = 1$,

$$\|U_{g_{0i}}^{0i}(t,0)v_{0i}\| \leq M_i^0 e^{-\gamma_i^0 t}\|U_{g_{0i}}^{0i}(t,0)w_i^0(g_{0i})\| \quad \text{for} \quad t > 0.$$

For $i = 1, 2$, denote by $[\lambda_{i,\min}^0, \lambda_{i,\max}^0]$ the principal spectrum interval of $(7.3.10)_i$ over Y_{0i}^0. $[\lambda_{i,\min}^0, \lambda_{i,\max}^0]$ are referred to as the *principal spectrum intervals* of $\mathbf{\Pi}$ on $Y_0^0 := \mathbf{p}_0(\mathbf{Z}_0)$, or the principal spectrum intervals associated to $\{\mathbf{0}\} \times \mathbf{Z}_0$.

THEOREM 7.3.2

Assume (A7-5)–(A7-8). *For any* $\epsilon > 0$, *there is* $\delta > 0$ *such that for any continuous* $(u_1^*, u_2^*) : \mathbb{R} \times \bar{D} \to \mathbb{R}^2$ *satisfying* $|u_1^*(t,x)|, |u_2^*(t,x)| \leq \delta$ *for all sufficiently large* t *and all* $x \in \bar{D}$ *there holds*

$$\liminf_{t \to \infty} \frac{1}{t} \ln \|\tilde{u}_1(t; t_0, u_{01}, g_1^*)\| \geq \lambda_{1,\min}^0 - \epsilon$$

and

$$\liminf_{t \to \infty} \frac{1}{t} \ln \|\tilde{u}_2(t; t_0, u_{02}, g_2^*)\| \geq \lambda_{2,\min}^0 - \epsilon$$

for any $t_0 \in \mathbb{R}$ *and any* $u_{01}, u_{02} \in X^+ \setminus \{0\}$, *where* $\tilde{u}_i(\cdot; t_0, u_{0i}, g_i^*)$, $i = 1, 2$, *denotes the solution of* $(7.3.10)_i$ *with* $g_{0i}(t,x)$ *replaced by* $g_i^*(t,x) = g_i(t,x, u_1^*(t,x), u_2^*(t,x))$ *and the initial condition* $\tilde{u}_i(t_0; t_0, u_{0i}, g_i^*) = u_{0i}$.

PROOF Fix $\mathbf{g} = (g_1, g_2) \in \mathbf{Z}_0$. Since, by (A7-6), the derivatives $\partial g_i / \partial u_j$, $i, j = 1, 2$, are bounded uniformly on sets of the form $\mathbb{R} \times \bar{D} \times [0, M] \times [0, M]$, for any $\epsilon > 0$ there is $\delta > 0$ such that if $|u_1^*(t,x)|, |u_2^*(t,x)| \leq \delta$ for $x \in \bar{D}$ and t sufficiently large, then

$$g_i^*(t,x) \geq g_i(t,x,0,0) - \epsilon$$

for $i = 1, 2$ and t sufficiently large. Without loss of generality we may assume that $g_i^*(t,x) \geq g_i(t,x,0,0) - \epsilon$ for all $t \geq t_0$ and $x \in \bar{D}$. Then by the comparison principle for parabolic equations, we have

$$\tilde{u}_1(t; t_0, u_{01}, g_1^*) \geq e^{-\epsilon(t-t_0)} U_{\mathbf{g}_0}^{01}(t, t_0) u_{01}$$

and

$$\tilde{u}_2(t; t_0, u_{02}, g_2^*) \geq e^{-\epsilon(t-t_0)} U_{\mathbf{g}_0}^{02}(t, t_0) u_{02}$$

for $t > t_0$. By Theorem 3.1.2 and Lemma 3.2.5,

$$\liminf_{t \to \infty} \frac{1}{t} \ln \| U^{01}_{\mathbf{g}_0}(t, t_0) u_{01} \| \geq \lambda^0_{1,\min}$$

and

$$\liminf_{t \to \infty} \frac{1}{t} \ln \| U^{02}_{\mathbf{g}_0}(t, t_0) u_{02} \| \geq \lambda^0_{2,\min}.$$

The theorem thus follows.   ☐

The following assumption is about the repelling property of the trivial solution $(u_1(t), u_2(t)) \equiv (0, 0)$.

(A7-9) $\lambda^0_{i,\min} > 0$, $i = 1, 2$.

By Theorem 7.1.9, we have

THEOREM 7.3.3
Assume (A7-5)–(A7-9). *There are nonempty compact invariant sets* $\mathbf{\Gamma}_1 \subset (X^{++} \times \{0\}) \times \mathbf{Z}_0$ *and* $\mathbf{\Gamma}_2 \subset (\{0\} \times X^{++}) \times \mathbf{Z}_0$ *such that* $\mathbf{\Gamma}_1$ *attracts any point in* $((X^+ \setminus \{0\}) \times \{0\}) \times \mathbf{Z}_0$ *and* $\mathbf{\Gamma}_2$ *attracts any point in* $(\{0\} \times (X^+ \setminus \{0\})) \times \mathbf{Z}_0$.

In the rest of this subsection we assume that (A7-5)–(A7-9) hold.
We also make the following assumption.

(A7-10) $(\partial g_1 / \partial u_1)(t, x, u_1, u_2) < 0$ *and* $(\partial g_2 / \partial u_2)(t, x, u_1, u_2) < 0$ *for all* $\mathbf{g} \in \mathbf{Z}_0$, $t \in \mathbb{R}$, $x \in \bar{D}$, *and* $(u_1, u_2) \in [0, \infty) \times [0, \infty)$.

In the literature, systems satisfying both (A7-7) and (A7-10) are called *totally competitive*.
We will investigate now the linearization of $\mathbf{\Phi}$ at semitrivial solutions in $\mathbf{\Gamma}_1$. By Theorem 7.1.12, there exists a continuous function $\boldsymbol{\xi}_1 \colon \mathbf{Z}_0 \to X^{++} \times \{0\}$ such that $\mathbf{\Gamma}_1 = \{ (\boldsymbol{\xi}_1(\mathbf{g}), \mathbf{g}) : \mathbf{g} \in \mathbf{Z}_0 \}$. For $\mathbf{g} \in \mathbf{Z}_0$ denote

$$g_{12}(t, x) := g_2(t, x, u_1(t; \boldsymbol{\xi}_1(\mathbf{g}), \mathbf{g})(x), 0) \qquad t \in \mathbb{R}, \ x \in \bar{D}.$$

Further, define a mapping $p_{12} \colon \mathbf{\Gamma}_1 \to L_\infty(\mathbb{R} \times D, \mathbb{R})$ as $p_{12}(\boldsymbol{\xi}_1(\mathbf{g}), \mathbf{g}) := g_{12}(t, x)$. Put $Y_1 := p_{12}(\mathbf{\Gamma}_1)$.
Observe that at any $(\boldsymbol{\xi}_1(\mathbf{g}), \mathbf{g}) \in \mathbf{\Gamma}_1$ the second coordinate of the linearized equation (7.3.7) takes the form

$$\begin{cases} \frac{\partial v_2}{\partial t} = \Delta v_2 + g_{12}(t, x) v_2, & t > 0, \ x \in D, \\ \mathcal{B} v_2 = 0, & t > 0, \ x \in \partial D. \end{cases} \tag{7.3.12}$$

For $g_{12} \in Y_1$ denote by $U^{12}_{g_{12}}(t, 0) v_{02}$ the weak solution operator of (7.3.12). Let

$$\Pi^1_2(t; v_{02}, g_{12}) := (U^{12}_{g_{12}}(t, 0) v_{02}, g_{12} \cdot t)$$

for $t \geq 0$, $v_{02} \in L_2(D)$, and $g_{12} \in Y_1$.

By the theory developed in Chapter 3, (7.3.12) or Π_2^1 admits an exponential separation over Y_1. Namely, we have

THEOREM 7.3.4

There exist

- *an invariant (under Π_2^1) one-dimensional subbundle X_1^1 of $L_2(D) \times Y_1$ with fibers $X_1^1(g_{12}) = \text{span}\{w^1(g_{12})\}$, where $w^1 \colon Y_1 \to L_2(D)$ is continuous, with $\|w^1(g_{12})\| = 1$ for all $g_{12} \in Y_1$, and*

- *an invariant (under Π_2^1) complementary one-codimensional subbundle X_2^1 of $L_2(D) \times Y_1$ with fibers $X_2^1(g_{12}) = \{v \in L_2(D) : \langle v, w^{1,*}(g_{12})\rangle = 0\}$, where $w^{1,*} \colon Y_1 \to L_2(D)$ is continuous, with $\|w^{1,*}(g_{12})\| = 1$ for all $g_{12} \in Y_1$,*

having the following properties:

(i) $w^1(g_{12}) \in L_2(D)^+$ *for all $g_{12} \in Y_1$,*

(ii) $X_2^1(g_{12}) \cap L_2(D)^+ = \{0\}$ *for all $g_{12} \in Y_1$,*

(iii) *there are $M^1 \geq 1$ and $\gamma^1 > 0$ such that for any $g_{12} \in Y_1$ and any $v_0 \in X_2^1(g_{12})$ with $\|v_0\| = 1$,*

$$\|U_{g_{12}}^{12}(t,0)v_0\| \leq M^1 e^{-\gamma^1 t}\|U_{g_{12}}^{12}(t,0)w^1(g_{12})\| \quad \text{for} \quad t > 0.$$

We denote by $[\lambda_{\min}^1, \lambda_{\max}^1]$ the principal spectrum interval of (7.3.12) over Y_1. $[\lambda_{\min}^1, \lambda_{\max}^1]$ is also referred to as the principal spectrum interval *associated* to Γ_1.

THEOREM 7.3.5

For any $\epsilon > 0$, if a continuous $g_2^* \colon \mathbb{R} \times \bar{D} \to \mathbb{R}$ satisfies

$$|g_2^*(t,x) - g_{12}(t,x)| < \epsilon$$

for some $\mathbf{g} \in \mathbf{Z}_0$, all $x \in \bar{D}$, and t sufficiently large, then there holds

$$\liminf_{t \to \infty} \frac{1}{t} \ln \|\tilde{u}_2(t;t_0, u_{02}, g_2^*)\| > \lambda_{\min}^1 - \epsilon,$$

where $u_{02} \in X^+ \setminus \{0\}$ and $\tilde{u}_2(t;t_0, u_{02}, g_2^*)$ is the solution of (7.3.12) with g_{12} replaced by g_2^* and $\tilde{u}_2(t_0;t_0, u_{02}, g_2^*) = u_{02}$.

PROOF   If g_2^* satisfies $|g_2^*(t,x) - g_{12}(t,x)| < \epsilon$ for some $\mathbf{g} \in \mathbf{Z}_0$, all $x \in \bar{D}$, and sufficiently large t, then

$$g_2^*(t,x) \geq g_{12}(t,x) - \epsilon$$

for $x \in \bar{D}$ and large t. Without loss of generality, we may assume that $g_2^*(t, x) \geq g_{12}(t, x) - \epsilon$ for all $x \in \bar{D}$ and all $t \geq t_0$. Then we have

$$\tilde{u}_2(t; t_0, u_{02}, g_2^*) \geq e^{-\epsilon(t-t_0)} U_{g_{12}}^{12}(t, t_0) u_{02}$$

for any $u_{01} \in X^+$ and $t \geq t_0$. By Theorem 3.1.2 and Lemma 3.2.5,

$$\liminf_{t \to \infty} \frac{1}{t} \ln \|U_{g_{12}}^{12}(t, t_0) u_{02}\| \geq \lambda_{\min}^1.$$

The theorem thus follows.   □

Results analogous to those presented above hold for semi-trivial solutions in $\mathbf{\Gamma}_2$. We will not write them down. For reference, we denote by $[\lambda_{\min}^2, \lambda_{\max}^2]$ the principal spectrum interval associated to $\mathbf{\Gamma}_2$.

The next assumption will be used in the investigation of uniform persistence.

(A7-11) $\lambda_{\min}^i > 0$, $i = 1, 2$.

7.3.3   Global Attractor and Uniform Persistence

In this subsection, we study the global attractor and uniform persistence for (7.3.1).

First we study global attractor.

We denote

$$[0, P]_X \times [0, P]_X := \{\, u = (u_1, u_2) \in \mathbf{X} :$$
$$0 \leq u_i(x) \leq P \text{ for all } x \in \bar{D}, \ i = 1, 2 \,\}.$$

The set $[0, P]_X \times [0, P]_X$ is convex and closed (in $\mathbf{X}$).

THEOREM 7.3.6 (Absorbing set)
Assume (A7-5)–(A7-8). Suppose that $\mathbf{B} \subset \mathbf{X}^+$ is bounded in the $C(\bar{D}) \times C(\bar{D})$-norm. Then there is $T = T(\mathbf{B}) \geq 0$ such that $u_i(t; \mathbf{u_0}, \mathbf{g})(x) \leq P$ for all $t \geq T$, $\mathbf{u_0} \in \mathbf{B}$, $\mathbf{g} \in \mathbf{Z}$, and $x \in \bar{D}$, $i = 1, 2$. Moreover, if $\mathbf{B} \subset [0, P]_X \times [0, P]_X$ then $T(\mathbf{B})$ can be taken to be zero.

PROOF   By Theorem 7.1.7, there is $T(\mathbf{B}) > 0$ such that for any $\mathbf{u_0} = (u_{01}, u_{02}) \in \mathbf{B}$ and $\mathbf{g} \in \mathbf{Z}$, $\mathbf{\Phi}_t((u_{01}, 0), \mathbf{g}) \in [0, P]_X \times \{0\}$ and $\mathbf{\Phi}_t((0, u_{02}), \mathbf{g}) \in \{0\} \times [0, P]_X$ for any $t \geq T(\mathbf{B})$ and $\mathbf{g} \in \mathbf{Z}$. Now by Proposition 7.3.6,

$$\mathbf{\Phi}_t((0, u_{02}), \mathbf{g}) \leq_2 \mathbf{\Phi}_t((u_{01}, u_{02}), \mathbf{g}) \leq_2 \mathbf{\Phi}_t((u_{01}, 0), \mathbf{g})$$

for all $t > 0$. It then follows that $\mathbf{\Phi}_t(\mathbf{u_0}, \mathbf{g}) \in [0, P]_X \times [0, P]_X$ for any $t \geq T(\mathbf{B})$, $\mathbf{u_0} \in \mathbf{B}$ and $\mathbf{g} \in \mathbf{Z}$.   □

THEOREM 7.3.7 (Global attractor)

Assume (A7-5)–(A7-8). *Then the topological skew-product semiflow* $\mathbf{\Phi}$ *possesses a global attractor* $\mathbf{\Gamma}$ *contained in* $([0, P]_X \times [0, P]_X) \times \mathbf{Z}$. *In addition, for any* $\mathbf{B} \subset \mathbf{X}^+$ *bounded in the* $C(\bar{D}) \times C(\bar{D})$*-norm one has*

(1) $\emptyset \neq \omega(\mathbf{B} \times \mathbf{Z}) \ (\subset \mathbf{\Gamma})$,

(2) $\mathbf{\Gamma}$ *attracts* $\mathbf{B} \times \mathbf{Z}$.

PROOF　　A proof can be obtained by rewriting the proof of Theorem 7.1.10 word for word, only with Proposition 7.1.8 replaced by Proposition 7.3.9 and Theorem 7.1.7 replaced by Theorem 7.3.6.　　　　　　　　　　　　　　□

Recall that $\widetilde{\partial}\mathbf{X}^+ = (X^+ \times \{0\}) \cup (\{0\} \times X^+)$. Put

$$\mathbf{\Gamma}_{\widetilde{\partial}} := \mathbf{\Gamma} \cap (\widetilde{\partial}\mathbf{X}^+ \times \mathbf{Z}).$$

It follows immediately from Theorem 7.3.7 that $\mathbf{\Gamma}_{\widetilde{\partial}}$ is the global attractor for the restriction of the skew-product semiflow $\mathbf{\Phi}$ to the forward invariant set $\widetilde{\partial}\mathbf{X}^+ \times \mathbf{Z}$. By Proposition 7.1.13, $\mathbf{\Phi}|_{\mathbf{\Gamma}_{\widetilde{\partial}}}$ is a topological flow.

Let $\mathbf{Z}_0$ be a nonempty connected compact invariant subset of $\mathbf{Z}$. If (A7-9) is fulfilled then $\mathbf{\Gamma}_1, \mathbf{\Gamma}_2 \subset \mathbf{\Gamma} \cap (\mathbf{X}^+ \times \mathbf{Z}_0)$.

We proceed now to study uniform persistence.

DEFINITION 7.3.2 (Uniform persistence)

(7.3.1) *is said to be uniformly persistent over* $\mathbf{Z}_0$ *if there is* $\eta_0 > 0$ *such that for any* $\mathbf{u}_0 \in \mathbf{X}^+ \setminus ((X^+ \times \{0\}) \cup (\{0\} \times X^+))$ *there is* $\tau(\mathbf{u}_0) > 0$ *with the property that*

$$u_i(t; \mathbf{u}_0, \mathbf{g}) \geq \eta_0 \varphi_{\mathrm{princ}} \quad for \quad i = 1, 2, \quad all \quad t \geq \tau(\mathbf{u}_0), \ \mathbf{g} \in \mathbf{Z}_0.$$

THEOREM 7.3.8 (Uniform persistence)

Let (A7-5) *through* (A7-11) *be satisfied. Then* (7.3.1) *is uniformly persistent over* $\mathbf{Z}_0$.

To prove the theorem, we first prove some lemmas.
Define $\mathbf{\Gamma}_0 := \{(\mathbf{0}, \mathbf{g}) : \mathbf{g} \in \mathbf{Z}_0\}$.
Note that $\mathbf{\Gamma}_0$, $\mathbf{\Gamma}_1$ and $\mathbf{\Gamma}_2$ are compact invariant subsets of $\mathbf{\Gamma}_{\widetilde{\partial}} \cap (\mathbf{X}^+ \times \mathbf{Z}_0)$.

LEMMA 7.3.1

Assume that the conditions in Theorem 7.3.8 hold. Then there is $\delta_0 > 0$ *such that if for some* $\mathbf{u}_0 \in \mathbf{X}^+$ *and* $\mathbf{g} \in \mathbf{Z}_0$ *there holds* $\|\mathbf{u}(t; \mathbf{u}_0, \mathbf{g})\|_X < \delta_0$ *for all* $t \geq 0$, *then* $\mathbf{u}_0 = \mathbf{0}$. *In particular,* $\mathbf{\Gamma}_0$ *is an isolated invariant set for* $\mathbf{\Phi}|_{\mathbf{X}^+ \times \mathbf{Z}_0}$.

PROOF   Take $\epsilon_0 > 0$ be such that $\lambda_{i,\min}^0 - \epsilon_0 > 0$ for $i = 1, 2$. Let $\delta_0 = \epsilon_0$. Suppose to the contrary that there are $\mathbf{u}_0 = (u_{01}, u_{02}) \in \mathbf{X}^+ \setminus \{\mathbf{0}\}$ with $\|\mathbf{u}_0\|_\mathbf{X} < \delta_0$ and $\mathbf{g} \in \mathbf{Z}_0$ such that $\|\mathbf{u}(t; \mathbf{u}_0, \mathbf{g})\|_\mathbf{X} < \delta_0$ for all $t \geq 0$. Without loss of generality, assume $u_{02} \neq 0$. Then by Theorem 7.3.2, we have

$$\liminf_{t \to \infty} \frac{1}{t} \ln \|u_2(t; \mathbf{u}_0, \mathbf{g})\| > 0.$$

This contradicts the fact that the set $\{ \|u_2(t; \mathbf{u}_0, \mathbf{g})\|_\mathbf{X} : t \geq 0 \}$ is bounded. ∎

LEMMA 7.3.2
Suppose that the conditions in Theorem 7.3.8 hold. Then, for each $i = 1, 2$,

(i) *there is $\delta_i > 0$ such that the situation is impossible that $d(\mathbf{\Pi}(t, \mathbf{u}_0, \mathbf{g}), \mathbf{\Gamma}_i) < \delta_i$ for all $t \geq 0$ but $\mathbf{u}_0 \notin \widetilde{\partial \mathbf{X}}^+$,*

(ii) *$\mathbf{\Gamma}_i$ is an isolated invariant set for $\mathbf{\Phi}|_{\mathbf{X}^+ \times \mathbf{Z}_0}$.*

PROOF   We prove the lemma for $\mathbf{\Gamma}_1$. Similarly, we can prove that the corresponding results hold for $\mathbf{\Gamma}_2$.
   (i) Take $\epsilon_1 > 0$ such that $\lambda_{\min}^1 - \epsilon_1 > 0$. Let $\delta_1 > 0$ be such that

$$|g_2(t, x, u_1, u_2) - g_2(t, x, u_1, 0)| < \frac{\epsilon_1}{2}$$

for all $\mathbf{g} \in \mathbf{Z}_0$, $t \in \mathbb{R}$, $u_1 \in [0, P]$, and $u_2 \in [0, \delta_1]$ (the existence of such a δ_1 follows by (A7-6)). Suppose to the contrary that there are $\mathbf{u}_0 = (u_{01}, u_{02}) \in \mathbf{X}^+$ with $u_{02} \neq 0$ and $\mathbf{g} \in \mathbf{Z}_0$ such that

$$d((\mathbf{u}(t; \mathbf{u}_0, \mathbf{g}), \zeta_t \mathbf{g}), \mathbf{\Gamma}_1) < \delta_1 \quad \text{for all} \quad t \geq 0.$$

Let $\mathbf{u}_0^* := (u_{01}, 0)$. Then $\mathbf{u}_0 \leq_2 \mathbf{u}_0^*$. By Proposition 7.3.6, we have

$$\mathbf{u}(t; \mathbf{u}_0, \mathbf{g}) \leq_2 \mathbf{u}(t; \mathbf{u}_0^*, \mathbf{g}) \quad \text{for all} \quad t > 0,$$

hence

$$u_1(t; \mathbf{u}_0, \mathbf{g}) \leq u_1(t; \mathbf{u}_0^*, \mathbf{g}) \quad \text{for all} \quad t > 0.$$

There holds

$$g_2(t, x, u_1(t; \mathbf{u}_0, \mathbf{g})(x), u_2(t; \mathbf{u}_0, \mathbf{g})(x)) \tag{7.3.13}$$
$$\geq g_2(t, x, u_1(t; \mathbf{u}_0^*, \mathbf{g})(x), u_2(t; \mathbf{u}_0, \mathbf{g})(x))$$

for all $t > 0$ and all $x \in \bar{D}$. It follows from Theorem 7.3.6 that $u_1(t; \mathbf{u}_0^*, \mathbf{g})(x) \leq P$ for sufficiently large $t > 0$ and all $x \in \bar{D}$. Since $0 < u_2(t; \mathbf{u}_0, \mathbf{g})(x) < \delta_1$ for all $t > 0$, we have

$$g_2(t, x, u_1(t; \mathbf{u}_0^*, \mathbf{g})(x), u_2(t; \mathbf{u}_0, \mathbf{g})(x)) > g_2(t, x, u_1(t; \mathbf{u}_0^*, \mathbf{g})(x), 0) - \frac{\epsilon_1}{2}$$
$$\tag{7.3.14}$$

for sufficiently large $t > 0$ and all $x \in \bar{D}$. Further, by Theorem 7.1.12, $\|u_1(t; \mathbf{u}_0^*, \mathbf{g}) - \xi(\zeta_t g_1)\|_X \to 0$ as $t \to \infty$. Therefore

$$g_2(t, x, u_1(t; \mathbf{u}_0^*, \mathbf{g})(x), 0) = g_2(t, x, \mathbf{u}(t; \mathbf{u}_0^*, \mathbf{g})(x))$$
$$> g_2(t, x, \mathbf{u}(t; \xi(\zeta_t g_1), 0)(x)) - \frac{\epsilon_1}{2} \qquad (7.3.15)$$

for sufficiently large $t > 0$ and all $x \in \bar{D}$. Combining Eqs. (7.3.13)–(7.3.15) we obtain that

$$g_2(t, x, \mathbf{u}(t; \mathbf{u}_0, \mathbf{g})(x)) > g_{12}(t, x) - \epsilon_1 \qquad (7.3.16)$$

for sufficiently large $t > 0$ and all $x \in \bar{D}$. Then by Theorem 7.3.5,

$$\liminf_{t \to \infty} \frac{1}{t} \ln \|u_2(t; \mathbf{u}_0, \mathbf{g})\| > 0,$$

which is a contradiction. This proves (i).

(ii) To prove (ii), observe that, by (i), it suffices to show that $\boldsymbol{\Gamma}_1$ is an isolated invariant set in $(X^+ \times \{0\}) \times \mathbf{Z}_0$. This is so, since $\boldsymbol{\Gamma}_1$ attracts any point in $(X^{++} \times \{0\}) \times \mathbf{Z}_0$. ▯

LEMMA 7.3.3
Assume (A7-5)–(A7-9). Then $\{\boldsymbol{\Gamma}_0, \boldsymbol{\Gamma}_1, \boldsymbol{\Gamma}_2\}$ is a Morse decomposition of $\boldsymbol{\Gamma}_{\widetilde{\partial}} \cap (X^+ \times \mathbf{Z}_0)$.

PROOF Take $(\mathbf{u}_0, \mathbf{g}) \in \boldsymbol{\Gamma}_{\widetilde{\partial}} \cap (X^+ \times \mathbf{Z}_0) \setminus (\boldsymbol{\Gamma}_0 \cup \boldsymbol{\Gamma}_1 \cup \boldsymbol{\Gamma}_2)$. Then either $(\mathbf{u}_0, \mathbf{g}) \in \boldsymbol{\Gamma} \cap (((X^+ \setminus \{0\}) \times \{0\}) \times \mathbf{Z}_0)$, in which case $\omega((\mathbf{u}_0, \mathbf{g})) \subset \boldsymbol{\Gamma}_1$, or $(\mathbf{u}_0, \mathbf{g}) \in \boldsymbol{\Gamma} \cap ((\{0\} \times (X^+ \setminus \{0\})) \times \mathbf{Z}_0)$, in which case $\omega((\mathbf{u}_0, \mathbf{g})) \subset \boldsymbol{\Gamma}_2$. In both cases $\alpha((\mathbf{u}_0, \mathbf{g})) \subset \boldsymbol{\Gamma}_0$. ▯

From now on, let $d(\cdot, \cdot)$ stand for the distance between a point in and a subset of $X^+ \times \mathbf{Z}$.

LEMMA 7.3.4
Suppose the conditions in Theorem 7.3.8 hold. If there is some $\eta_1 > 0$ such that

$$\limsup_{t \to \infty} d(\boldsymbol{\Phi}(t, \mathbf{u}_0, \mathbf{g}), \boldsymbol{\Gamma}_i) \geq \eta_1 \qquad (7.3.17)$$

for all $\mathbf{u}_0 \in X^+ \setminus \widetilde{\partial} X^+$, $\mathbf{g} \in \mathbf{Z}_0$, and $i = 0, 1, 2$, then there is $\eta_2 > 0$ such that

$$\liminf_{t \to \infty} d(\boldsymbol{\Phi}(t, \mathbf{u}_0, \mathbf{g}), \widetilde{\partial} X^+) \geq \eta_2 \qquad (7.3.18)$$

for all $\mathbf{u}_0 \in X^+ \setminus \widetilde{\partial} X^+$ and $\mathbf{g} \in \mathbf{Z}_0$.

PROOF It follows from Lemmas 7.3.1, 7.3.2, 7.3.3, and [58, Theorem 4.3] (see also [47, Theorem 4.1]). ▯

PROOF (Proof of Theorem 7.3.8)   First, Lemma 7.3.2(i) implies
(7.3.17). Next, by Lemma 7.3.4, we have

$$\liminf_{t \to \infty} d(\mathbf{\Phi}(t, \mathbf{u}_0, \mathbf{g}), \widetilde{\partial \mathbf{X}^+}) \geq \eta_2$$

for all $\mathbf{u}_0 \in \mathbf{X}^+ \setminus \widetilde{\partial \mathbf{X}^+}$ and $\mathbf{g} \in \mathbf{Z}_0$. This together with Theorem 7.3.6 yields
that for each $\mathbf{u}_0 \in \mathbf{X}^+ \setminus \widetilde{\partial \mathbf{X}^+}$ and each $\mathbf{g} \in \mathbf{Z}_0$ there is $\tau_1 = \tau_1(\mathbf{u}_0, \mathbf{g}) > 0$
such that $\mathbf{u}(t; \mathbf{u}_0, \mathbf{g}) \in [0, P]_X \times [0, P]_X$ and $d(\mathbf{\Phi}(t, \mathbf{u}_0, \mathbf{g}), \widetilde{\partial \mathbf{X}^+}) \geq \frac{\eta_2}{2}$ for all
$t \geq \tau_1$. Applying ideas in the proof of [47, Theorem 3.2] to the restriction
$\mathbf{\Phi}|_{(\mathbf{X}^+ \setminus \widetilde{\partial \mathbf{X}^+}) \times \mathbf{Z}_0}$ we obtain the existence of a nonempty compact invariant set
$\mathbf{\Gamma}^{++} \subset (\mathbf{X}^+ \setminus \widetilde{\partial \mathbf{X}^+}) \times \mathbf{Z}_0$ attracting any compact $\mathbf{B} \times \mathbf{Z}_0$ with $\mathbf{B} \subset \mathbf{X}^+ \setminus \widetilde{\partial \mathbf{X}^+}$.

We claim that $\mathbf{\Gamma}^{++} \subset \mathbf{X}^{++} \times \mathbf{Z}_0$. By invariance, for any $(\mathbf{u}_0, \mathbf{g}) \in \mathbf{\Gamma}^{++}$,
there is $(\mathbf{u}_{-1}, \mathbf{g} \cdot (-1)) \in \mathbf{\Gamma}^{++}$ such that $\mathbf{u}(t; \mathbf{u}_{-1}, \mathbf{g} \cdot (-1)) = \mathbf{u}_0$. We write
$\mathbf{u}_0 = (u_{01}, u_{02})$ and $\mathbf{u}_{-1} = (u_{-1,1}, u_{-1,2})$. Since $\mathbf{\Gamma}^{++} \subset (\mathbf{X}^+ \setminus \widetilde{\partial \mathbf{X}^+}) \times \mathbf{Z}_0$,
$u_{-1,1}, u_{-1,2} \in X^+ \setminus \{0\}$. Consequently, there holds $(0, u_{-1,2}) <_2 \mathbf{u}_{-1} <_2$
$(u_{-1,1}, 0)$. It then follows from Proposition 7.3.6(2) that

$$\mathbf{u}(1; (0, u_{-1,2}), \mathbf{g} \cdot (-1)) \ll_2 \mathbf{u}_0 \ll_2 \mathbf{u}(1; (u_{-1,1}, 0), \mathbf{g} \cdot (-1)).$$

As the first and the third term are semitrivial solutions, Proposition 7.1.7(2)
gives that $u_{01} \in X^{++}$ and $u_{02} \in X^{++}$, that is, $\mathbf{u}_0 \in \mathbf{X}^{++}$.

We define $p \colon \mathbf{X}^+ \times \mathbf{Z}_0 \to [0, \infty)$ by

$$p(\mathbf{u}, \mathbf{g}) := \sup \{ \delta \geq 0 : u_i \geq \delta \varphi_{\mathrm{princ}}, i = 1, 2 \}.$$

Clearly, $p(\mathbf{u}, \mathbf{g}) > 0$ if and only if $\mathbf{u} \in \mathbf{X}^{++}$. Further, it follows from the
openness of $\mathbf{X}^{++}$ that p is lower semicontinuous. Since $p(\mathbf{u}, \mathbf{g}) > 0$ for any
$(\mathbf{u}, \mathbf{g}) \in \mathbf{\Gamma}^{++}$, the compactness of $\mathbf{\Gamma}^{++}$ and the lower semicontinuity of p imply
the existence of an open (in the relative topology of $\mathbf{X}^{++} \times \mathbf{Z}_0$) neighborhood
$\mathbf{V}$ of $\mathbf{\Gamma}^{++}$ and of a positive number η_0 such that

$$p(\mathbf{u}, \mathbf{g}) \geq \eta_0 \quad \text{for all } (\mathbf{u}, \mathbf{g}) \in \mathbf{V}.$$

We conclude the proof by noting that for a given $\mathbf{u}_0 \in \mathbf{X}^+ \setminus \widetilde{\partial \mathbf{X}^+}$ there is
$\tau > 0$ such that $\mathbf{\Pi}_t(\{\mathbf{u}_0\} \times \mathbf{Z}_0) \in \mathbf{V}$ for all $t \geq \tau$.   ⬜

7.4   Competitive Kolmogorov Systems of Semilinear Equations: Examples

In this section, we discuss applications of the general theory established in
Section 7.3 to some competitive Kolmogorov systems of random and nonautonomous semilinear equations.

7.4.1   The Random Case

Assume that $((\Omega, \mathfrak{F}, \mathbb{P}), \{\theta_t\}_{t\in\mathbb{R}})$ is an ergodic metric dynamical system.

Consider the following competitive Kolmogorov systems of random partial differential equations:

$$\begin{cases} \dfrac{\partial u_1}{\partial t} = \Delta u_1 + f_1(\theta_t\omega, x, u_1, u_2)u_1, & x \in D, \\[2mm] \dfrac{\partial u_2}{\partial t} = \Delta u_2 + f_2(\theta_t\omega, x, u_1, u_2)u_2, & x \in D, \\[2mm] \mathcal{B}u_1 = 0, & x \in \partial D, \\[1mm] \mathcal{B}u_2 = 0, & x \in \partial D, \end{cases} \qquad (7.4.1)$$

where $\mathbf{f} = (f_1, f_2)\colon \Omega \times \bar{D} \times [0, \infty) \times [0, \infty) \to \mathbb{R}^2$.

We assume

(A7-R4) (Measurability) *The function* $\mathbf{f}$ *is* $(\mathfrak{F} \times \mathfrak{B}(D) \times \mathfrak{B}([0, \infty) \times [0, \infty))$, $\mathfrak{B}(\mathbb{R}^2))$-*measurable.*

For each $\omega \in \Omega$, let $\mathbf{f}^\omega(t, x, u_1, u_2) = (f_1^\omega(t, x, u_1, u_2), f_2^\omega(t, x, u_1, u_2)) := \mathbf{f}(\theta_t\omega, x, u_1, u_2)$.

(A7-R5) (Regularity) *For each* $\omega \in \Omega$ *and any* $M > 0$ *the restrictions to* $\mathbb{R} \times \bar{D} \times [0, M] \times [0, M]$ *of* f_1^ω, f_2^ω, *and all the derivatives of the functions* f_1^ω, f_2^ω, *up to order 1 belong to* $C^{1-,1-,1-,1-}(\mathbb{R} \times \bar{D} \times [0, M] \times [0, M])$. *Moreover, for* $M > 0$ *fixed the* $C^{1-,1-,1-,1-}(\mathbb{R} \times \bar{D} \times [0, M] \times [0, M])$-*norms of the restrictions of* f_1^ω, f_2^ω, *and those derivatives are bounded uniformly in* $\omega \in \Omega$.

(A7-R6) *There are* $P > 0$ *and a function* $m\colon [P, \infty) \to (0, \infty)$ *such that* $f_1(\omega, x, u_1, u_2) \le -m(u_1)$ *for any* $\omega \in \Omega$, *any* $x \in \bar{D}$, *and any* $u_1 \ge P$, $u_2 \ge 0$, *and* $f_2(\omega, x, u_1, u_2) \le -m(u_2)$ *for any* $\omega \in \Omega$, *any* $x \in \bar{D}$, *and any* $u_1 \ge 0$, $u_2 \ge P$.

(A7-R7) (Total competitiveness) *There is a function* $\tilde{m}\colon [0, \infty) \to (0, \infty)$ *such that* $\partial_{u_1} f_1(\omega, x, u_1, u_2) \le -\tilde{m}(u_1)$, $\partial_{u_2} f_1(\omega, x, u_1, u_2) \le -\tilde{m}(u_2)$ *and* $\partial_{u_1} f_2(\omega, x, u_1, u_2) \le -\tilde{m}(u_1)$, $\partial_{u_2} f_2(\omega, x, u_1, u_2) \le -\tilde{m}(u_2)$ *for any* $\omega \in \Omega$, *any* $x \in \bar{D}$, *and any* $(u_1, u_2) \in [0, \infty) \times [0, \infty)$.

From now on, until the end of the present subsection, assume that (A7-R4)–(A7-R7) are satisfied.

Define the mapping E from Ω into the set of continuous real functions defined on $\mathbb{R} \times \bar{D} \times [0, \infty) \times [0, \infty)$ as

$$E(\omega) := \mathbf{f}^\omega.$$

Put

$$\mathbf{Z} := \mathrm{cl}\,\{\, E(\omega) : \omega \in \Omega \,\} \qquad (7.4.2)$$

with the open-compact topology, where the closure is taken in the open-compact topology. It is a consequence of (A7-R5) via the Ascoli–Arzelà theorem that the set $\mathbf{Z}$ is a compact metrizable space. By arguments similar to those in Subsection 7.2.1, $(\mathbf{Z}, \{\zeta_t\}_{t\in\mathbb{R}})$ is a compact flow, where $\zeta_t \mathbf{g}(\tau, x, u_1, u_2) = \mathbf{g} \cdot t(\tau, x, u_1, u_2) = \mathbf{g}(\tau + t, x, u_1, u_2)$.

The mapping E is a homomorphism of the measurable flow $((\Omega, \mathfrak{F}), \{\theta_t\}_{t\in\mathbb{R}})$ into the measurable flow $((\mathbf{Z}, \mathfrak{B}(\mathbf{Z})), \{\zeta_t\}_{t\in\mathbb{R}})$. Denote by $\tilde{\mathbb{P}}$ the image of the measure $\mathbb{P}$ under E: for any $A \in \mathfrak{B}(\mathbf{Z})$, $\tilde{\mathbb{P}}(A) := \mathbb{P}(E^{-1}(A))$. $\tilde{\mathbb{P}}$ is a $\{\zeta_t\}$-invariant ergodic Borel measure on $\mathbf{Z}$. So, E is a homomorphism of the metric flow $((\Omega, \mathfrak{F}, \mathbb{P}), \{\theta_t\}_{t\in\mathbb{R}})$ into the metric flow $((\mathbf{Z}, \mathfrak{B}(\mathbf{Z}), \tilde{\mathbb{P}}), \{\zeta_t\}_{t\in\mathbb{R}})$.

We will consider a family of Eqs. (7.3.1) parameterized by $\mathbf{g} \in \mathbf{Z}$. By (A7-R4)–(A7-R7), (A7-5) through (A7-8) as well as (A7-10) are fulfilled.

We denote by $\boldsymbol{\Phi} = \{\boldsymbol{\Phi}_t\}_{t\geq 0}$ the *topological skew-product semiflow generated by* (7.4.1) *on the product Banach bundle* $\mathbf{X}^+ \times \mathbf{Z}$:

$$\boldsymbol{\Phi}(t; \mathbf{u}_0, \mathbf{g}) = \boldsymbol{\Phi}_t(\mathbf{u}_0, \mathbf{g}) := (\mathbf{u}(t; \mathbf{u}_0, \mathbf{g}), \zeta_t \mathbf{g}) \tag{7.4.3}$$

for $t \geq 0$, $\mathbf{g} \in \mathbf{Z}$, $\mathbf{u}_0 \in \mathbf{X}^+$, where $\mathbf{u}(t; \mathbf{u}_0, \mathbf{g})$ stands for the solution of $(7.3.1)_{\mathbf{g}}$ with the initial condition $u(0; \mathbf{u}_0, \mathbf{g})(x) = \mathbf{u}_0(x)$ for $x \in \bar{D}$.

Moreover, define

$$\tilde{\boldsymbol{\Phi}}(t; \mathbf{u}_0, \omega) := (\mathbf{u}(t; \mathbf{u}_0, E(\omega)), \theta_t \omega), \quad t \geq 0, \ \omega \in \Omega, \ \mathbf{u}_0 \in \mathbf{X}^+. \tag{7.4.4}$$

We have

LEMMA 7.4.1

$\tilde{\boldsymbol{\Phi}}$ *is a continuous random skew-product semiflow on the measurable bundle* $\mathbf{X}^+ \times \Omega$, *covering the metric flow* $((\Omega, \mathfrak{F}, \mathbb{P}), \{\theta_t\}_{t\in\mathbb{R}})$.

For $t \geq 0$, $\mathbf{u}_0 \in \mathbf{X}^+$, and $\omega \in \Omega$ we will write $\mathbf{u}(t; \mathbf{u}_0, \omega)$ instead of $\mathbf{u}(t; \mathbf{u}_0, E(\omega))$. Similarly, for $t_0 \in \mathbb{R}$, $t \geq t_0$, $\mathbf{u}_0 \in \mathbf{X}^+$, and $\omega \in \Omega$ we will write $\mathbf{u}(t; t_0, \mathbf{u}_0, \omega)$ instead of $\mathbf{u}(t - t_0; \mathbf{u}_0, E(\theta_{t_0}\omega))$.

DEFINITION 7.4.1 (Uniform persistence)   (7.4.1) *is said to be* uniformly persistent *if there is* $\eta_0 > 0$ *such that for any* $\mathbf{u}_0 \in (X^+ \setminus \{0\}) \times (X^+ \setminus \{0\})$ *there is* $\tau(\mathbf{u}_0) > 0$ *with the property that*

$$u_i(t; t_0, \mathbf{u}_0, \omega) \geq \eta_0 \varphi_{\mathrm{princ}}$$

for $i = 1, 2$, $\mathbb{P}$-*a.e.* $\omega \in \Omega$, *any* $t_0 \in \mathbb{R}$, *and any* $t \geq t_0 + \tau(\mathbf{u}_0)$.

Put

$$\mathbf{Z}_0 := \operatorname{supp} \tilde{\mathbb{P}} \tag{7.4.5}$$

($\mathbf{g} \in \mathbf{Z}_0$ if and only if for any neighborhood V of $\mathbf{g}$ in $\mathbf{Z}$ one has $\tilde{\mathbb{P}}(V) > 0$). $\mathbf{Z}_0$ is a closed (hence compact) and $\{\zeta_t\}$-invariant subset of $\mathbf{Z}$, with $\tilde{\mathbb{P}}(\mathbf{Z}_0) = 1$. Also, $\mathbf{Z}_0$ is connected.

Similarly to Lemma 7.2.3 we have

LEMMA 7.4.2

There exists $\Omega_0 \subset \Omega$ with $\mathbb{P}(\Omega_0) = 1$ such that $\mathbf{Z}_0 = \mathrm{cl}\{\, E(\theta_t \omega) : t \in \mathbb{R} \,\}$ for any $\omega \in \Omega_0$, where the closure is taken in the open-compact topology on $\mathbf{Z}$.

Observe that the set $\{\mathbf{0}\} \times \mathbf{Z}_0$ is invariant under the semiflow $\mathbf{\Phi}$. Consider the linearization of $\mathbf{\Phi}$ at $\{\mathbf{0}\} \times \mathbf{Z}_0$,

$$\partial \mathbf{\Phi}_t(\mathbf{v}_0, (\mathbf{0}, \mathbf{g})) = (\partial_2 \mathbf{u}(t; \mathbf{0}, \mathbf{g})\mathbf{v}_0, (\mathbf{0}, \mathbf{g} \cdot t)), \quad t \geq 0, \ \mathbf{v}_0 \in \mathbf{X}, \ \mathbf{g} \in \mathbf{Z}_0, \quad (7.4.6)$$

where $\partial_2 \mathbf{u}(t; \mathbf{0}, \mathbf{g})\mathbf{v}_0 = (U_{\mathbf{g}}^{01}(t, 0)v_{01}, U_{\mathbf{g}}^{02}(t, 0)v_{02})$, and $U_{\mathbf{g}}^{0i}(t, 0)$, $i = 1, 2$, is the solution operator of $(7.3.10)_i$.

Let $[\lambda_{i,\min}^0, \lambda_{i,\max}^0]$, $i = 1, 2$, be the principal spectrum intervals of $(7.3.10)_i$ over $p_{0i}(\mathbf{Z}_0)$.

THEOREM 7.4.1

Suppose that $\lambda_{i,\min}^0 > 0$ for $i = 1, 2$. Then there are nonempty compact invariant sets $\mathbf{\Gamma}_1 \subset (X^{++} \times \{0\}) \times \mathbf{Z}_0$ and $\mathbf{\Gamma}_2 \subset (\{0\} \times X^{++}) \times \mathbf{Z}_0$ such that $\mathbf{\Gamma}_1$ attracts any point in $((X^+ \setminus \{0\}) \times \{0\}) \times \mathbf{Z}_0$ and $\mathbf{\Gamma}_2$ attracts any point in $(\{0\} \times (X^+ \setminus \{0\})) \times \mathbf{Z}_0$.

PROOF   It follows from Theorem 7.3.3.   ▯

Let $[\lambda_{\min}^1, \lambda_{\max}^1]$ be the principal spectrum interval associated to $\mathbf{\Gamma}_1$ and $[\lambda_{\min}^2, \lambda_{\max}^2]$ be the principal spectrum interval associated to $\mathbf{\Gamma}_2$.

THEOREM 7.4.2 (Uniform persistence)

Suppose that $\lambda_{i,\min}^0 > 0$ and $\lambda_{\min}^i > 0$ for $i = 1, 2$. Then $(7.4.1)$ is uniformly persistent.

PROOF   It follows from Theorem 7.3.8.   ▯

7.4.2 The Nonautonomous Case

Consider the following competitive Kolmogorov systems of nonautonomous partial differential equations:

$$\begin{cases} \dfrac{\partial u_1}{\partial t} = \Delta u_1 + f_1(t, x, u_1, u_2)u_1, & x \in D, \\[2mm] \dfrac{\partial u_2}{\partial t} = \Delta u_2 + f_2(t, x, u_1, u_2)u_2, & x \in D, \\[2mm] \mathcal{B}u_1 = 0, & x \in \partial D, \\[2mm] \mathcal{B}u_2 = 0, & x \in \partial D, \end{cases} \tag{7.4.7}$$

where $\mathbf{f} = (f_1, f_2) \colon \mathbb{R} \times \bar{D} \times [0, \infty) \times [0, \infty) \to \mathbb{R}^2$.

We assume

(A7-N3) (Regularity) *For any $M > 0$ the restrictions to $\mathbb{R} \times \bar{D} \times [0, M] \times [0, M]$ of f_1, f_2, and all the derivatives of the functions f_1, f_2 up to order 1 belong to $C^{1-,1-,1-,1-}(\mathbb{R} \times \bar{D} \times [0, M] \times [0, M])$.*

(A7-N4) *There are $P > 0$ and a function $m \colon [P, \infty) \to (0, \infty)$ such that $f_1(t, x, u_1, u_2) \leq -m(u_1)$ for any $t \in \mathbb{R}$, any $x \in \bar{D}$, and any $u_1 \geq P$, $u_2 \geq 0$ and $f_2(t, x, u_1, u_2) \leq -m(u_2)$ for any $t \in \mathbb{R}$, any $x \in \bar{D}$, and any $u_1 \geq 0$, $u_2 \geq P$.*

(A7-N5) (Total competitiveness) *There is a function $\tilde{m} \colon [0, \infty) \to (0, \infty)$ such that $\partial_{u_1} f_1(t, x, u_1, u_2) \leq -\tilde{m}(u_1)$, $\partial_{u_2} f_1(t, x, u_1, u_2) \leq -\tilde{m}(u_2)$, and $\partial_{u_1} f_2(t, x, u_1, u_2) \leq -\tilde{m}(u_1)$, $\partial_{u_2} f_2(t, x, u_1, u_2) \leq -\tilde{m}(u_2)$ for all $t \in \mathbb{R}$, $x \in \bar{D}$, and $u_1, u_2 \in [0, \infty)$.*

Throughout this subsection, we assume (A7-N3)–(A7-N5).

Put

$$\mathbf{Z} := \mathrm{cl}\{\mathbf{f} \cdot t : t \in \mathbb{R}\} \tag{7.4.8}$$

with the open-compact topology, where the closure is taken in the open-compact topology. It is a consequence of (A7-N3) via the Ascoli–Arzelà theorem that the set $\mathbf{Z}$ is a compact metrizable space.

We deduce from Lemma 7.1.1 that if $\mathbf{g} \in \mathbf{Z}$ then $\mathbf{g} \cdot t \in \mathbf{Z}$ for all $t \in \mathbb{R}$.

Hence $(\mathbf{Z}, \{\zeta_t\}_{t\in\mathbb{R}})$ is a compact flow, where $\zeta_t \mathbf{g} = \mathbf{g} \cdot t$.

We will consider a family of Eqs. (7.3.1) parametcrized by $\mathbf{g} \in \mathbf{Z}$. By (A7-N3)–(A7-N5), (A7-5) through (A7-8) as well as (A7-10) hold.

We denote by $\mathbf{\Phi} = \{\mathbf{\Phi}_t\}_{t\geq 0}$ the *topological skew-product semiflow generated by* (7.4.7) on the product Banach bundle $\mathbf{X}^+ \times \mathbf{Z}$:

$$\mathbf{\Phi}(t; \mathbf{u}_0, \mathbf{g}) = \mathbf{\Phi}_t(\mathbf{u}_0, \mathbf{g}) := (\mathbf{u}(t; \mathbf{u}_0, \mathbf{g}), \zeta_t \mathbf{g}) \tag{7.4.9}$$

for $t \geq 0$, $\mathbf{g} \in \mathbf{Z}$, $\mathbf{u}_0 \in \mathbf{X}^+$, where $\mathbf{u}(t; \mathbf{u}_0, \mathbf{g})$ represents the solution of $(7.3.1)_{\mathbf{g}}$ with initial condition $\mathbf{u}(0; \mathbf{u}_0, \mathbf{g})(x) = \mathbf{u}_0(x)$ for $x \in D$.

For $t_0 \in \mathbb{R}$, $t \geq t_0$, and $\mathbf{u}_0 \in \mathbf{X}^+$ we write $\mathbf{u}(t; t_0, \mathbf{u}_0, \mathbf{f})$ instead of $\mathbf{u}(t - t_0; \mathbf{u}_0, \mathbf{f} \cdot t_0)$.

DEFINITION 7.4.2 (Uniform persistence)   (7.4.7) *is said to be uniformly persistent if there is* $\eta_0 > 0$ *such that for any* $\mathbf{u}_0 \in (X^+ \setminus \{0\}) \times (X^+ \setminus \{0\})$ *there is* $\tau(\mathbf{u}_0) > 0$ *with the property that*

$$u_i(t; t_0, \mathbf{u}_0, \mathbf{f}) \geq \eta_0 \varphi_{\text{princ}}$$

for $i = 1, 2$, *all* $t_0 \in \mathbb{R}$, *and* $t \geq t_0 + \tau(\mathbf{u}_0)$.

Note that $\mathbf{Z}$ is connected and $\Gamma_0 = \{0\} \times \mathbf{Z}$ is invariant under $\mathbf{\Phi}$.
Let $[\lambda^0_{i,\min}, \lambda^0_{i,\max}]$, $i = 1, 2$, be the principal spectrum intervals of $(7.3.10)_i$ over $p_{0i}(\mathbf{Z})$.

THEOREM 7.4.3
Suppose that $\lambda^0_{i,\min} > 0$ *for* $i = 1, 2$. *Then there are nonempty compact invariant sets* $\Gamma_1 \subset (X^{++} \times \{0\}) \times \mathbf{Z}$ *and* $\Gamma_2 \subset (\{0\} \times X^{++}) \times \mathbf{Z}$ *such that* Γ_1 *attracts any point in* $((X^+ \setminus \{0\}) \times \{0\}) \times \mathbf{Z}$ *and* Γ_2 *attracts any point in* $(\{0\} \times (X^+ \setminus \{0\})) \times \mathbf{Z}$.

PROOF   It follows from Theorem 7.3.3.   ▯

Let $[\lambda^1_{\min}, \lambda^1_{\max}]$ be the principal spectrum interval associated to Γ_1 and $[\lambda^2_{\min}, \lambda^2_{\max}]$ be the principal spectrum interval associated to Γ_2.

THEOREM 7.4.4 (Uniform persistence)
Suppose that $\lambda^0_{i,\min} > 0$ *and* $\lambda^i_{\min} > 0$ *for* $i = 1, 2$. *Then* (7.4.7) *is uniformly persistent.*

PROOF   It follows from Theorem 7.3.8.   ▯

7.5   Remarks

As mentioned in the introduction of this monograph, principal spectral theory for linear parabolic equations under various special conditions has been studied in a lot of papers (see [29], [30], [31], [32], [28], [35], [49], [50], [59], [60], [61], [62], [79], [81], [82], [83], [84], [92], [93], [94], etc.) and has found many applications (see [51], [64], [83], [85], [93], etc.). In the previous chapters of this monograph, we developed the principal spectral theory for general random and nonautonomous parabolic equations. This theory will certainly also find great applications to lots of nonlinear problems. In the present chapter, we considered its application to the uniform persistence problem in random and nonautonomous semilinear parabolic equations of Kolmogorov type and two dimensional competitive systems of such equations.

It should be pointed out that uniform persistence as well as many other dynamical aspects in semilinear parabolic equations of Kolmogorov type and competitive Kolmogorov systems of parabolic equations have been widely studied. See for example, [9], [23], [25], [41], [52], [54], [101], [106], [111], etc., for scalar parabolic equations of Kolmogorov type, and [47], [53], [58], [80], [104], [108], [110], etc., for two species competitive Kolmogorov systems of parabolic equations. However, most equations in the literature are not as general as those in the present chapter, except [85], which in fact utilized principal spectral theory to consider the uniform persistence in quite general n-dimensional nonautonomous and random parabolic Kolmogorov systems.

References

[1] R. A. Adams, "Sobolev Spaces," Pure and Applied Mathematics, Vol. **65**, Academic Press, New York–London, 1975. **MR 56 #9247**

[2] E. Akin, "The General Topology of Dynamical Systems," Grad. Stud. Math., **1**, American Mathematical Socety, Providence, RI, 1993. **MR 94f:58041**

[3] H. Amann, Existence and regularity for semilinear parabolic evolution equations, *Ann. Scuola Norm. Sup. Pisa Cl. Sci. (4)* **11** (1984), no. 4, 593–676. **MR 87h:34088**

[4] H. Amann, Dynamic theory of quasilinear parabolic equations. II. Reaction–diffusion systems, *Differential Integral Equations* **3** (1990), no. 1, 13–75. **MR 90i:35124**

[5] H. Amann, Semigroups and nonlinear evolution equations, in: Proceedings of the Symposium on Operator Theory (Athens, 1985), *Linear Algebra Appl.* **84** (1986), 3–32. **MR 88b:35103**

[6] H. Amann, Quasilinear evolution equations and parabolic systems *Trans. Amer. Math. Soc.* **292** (1986), no. 1, 191–227. **MR 87d:35070**

[7] L. Arnold, "Random Dynamical Systems," Springer Monogr. Math., Springer, Berlin, 1998. **MR 2000m:37087**

[8] L. Arnold and I. Chueshov, Cooperative random and stochastic differential equations, *Discrete Contin. Dynam. Systems* **7** (2001), no. 1, 1–33. **MR 2002b:37086**

[9] L. Arnold and I. Chueshov, A limit set trichotomy for order-preserving random systems, *Positivity* **5** (2001), no. 2, 95–114. **MR 2002i:37072**

[10] D. G. Aronson, Non-negative solutions of linear parabolic equations, *Ann. Scuola Norm. Sup. Pisa (3)* **22** (1968), 607–694. **MR 55 #8533**

[11] J. M. Arrieta, Elliptic equations, principal eigenvalue and dependence on the domain, *Comm. Partial Differential Equations* **21** (1996), no. 5-6, 971–991. **MR 97h:35037**

[12] C. Bandle, "Isoperimetric Inequalities and Applications," Monogr. Stud. Math., **7**, Pitman (Advanced Publishing Program), Boston, Mass.–London, 1980. **MR 81e:35095**

[13] H. Berestycki, F. Hamel and L. Roques, Analysis of the periodically fragmented environment model. I. Species persistence, *J. Math. Biol.* **51** (2005), no. 1, 75–113. **MR 2007d:35151**

[14] H. Berestycki, L. Nirenberg and S. R. S. Varadhan, The principal eigenvalue and maximum principle for second-order elliptic operators in general domains, *Comm. Pure Appl. Math.* **47** (1994), no. 1, 47–92. **MR 95h:35053**

[15] J. Bergh and J. Löfström, "Interpolation Spaces. An Introduction," Grundlehren Math. Wiss., No. **223**, Springer, Berlin–New York, 1976. **MR 58 #2349**

[16] R. S. Cantrell and C. Cosner, "Spatial Ecology via Reaction–Diffusion Equations," Wiley Ser. Math. Comput. Biol., Wiley, Chichester, 2003. **MR 2007a:92069**

[17] T. Caraballo, J. A. Langa and J. C. Robinson, Stability and random attractors for a reaction–diffusion equation with multiplicative noise, *Discrete Contin. Dynam. Syst.* **6** (2000), no. 4, 875–892. **MR 2001j:37093**

[18] P. Chernoff and J. Marsden, On continuity and smoothness of group actions, *Bull. Amer. Math. Soc.* **76** (1970), 1044–1049. **MR 42 #419**

[19] C. Chicone and Y. Latushkin, "Evolution Semigroups in Dynamical Systems and Differential Equations," Math. Surveys Monogr., **70**, Amer. Math. Soc., Providence, RI, 1999. **MR 2001e:47068**

[20] S.-N. Chow and H. Leiva, Existence and roughness of the exponential dichotomy for skew-product semiflow in Banach spaces, *J. Differential Equations* **120** (1995), no. 2, 429–477. **MR 97a:34121**

[21] S.-N. Chow, K. Lu and J. Mallet-Paret, Floquet theory for parabolic differential equations, *J. Differential Equations* **109** (1994), no. 1, 147–200. **MR 95c:35116**

[22] S.-N. Chow, K. Lu and J. Mallet-Paret, Floquet bundles for scalar parabolic equations, *Arch. Rational Mech. Anal.* **129** (1995), no. 3, 245–304. **MR 96c:35070**

[23] I. D. Chueshov, "Monotone Random Systems — Theory and Applications," Lecture Notes in Math., **1779**, Springer, Berlin, 2002. **MR 2003d:37072**

[24] I. D. Chueshov and B. Schmalfuß, Averaging of attractors and inertial manifolds for parabolic PDE with random coefficients, *Adv. Nonlinear Stud.* **5** (2005), no. 4, 461–492. **MR 2006i:37163**

[25] I. D. Chueshov and P.-A. Vuillermot, Long-time behavior of solutions to a class of quasilinear parabolic equations with random coefficients,

Ann. Inst. H. Poincaré Anal. Non Linéaire **15** (1998), no. 2, 191–232. **MR 99b:35225**

[26] H. Crauel, A. Debussche and F. Flandoli, Random attractors, *J. Dynam. Differential Equations* **9** (1997), no. 2, 307–341. **MR 98c:60066**

[27] H. Crauel and F. Flandoli, Attractors for random dynamical systems, *Probab. Theory Related Fields* **100** (1994), no. 3, 365–393. **MR 95k:58092**

[28] E. N. Dancer and D. Daners, Domain perturbation for elliptic equations subject to Robin boundary conditions, *J. Differential Equations* **138** (1997), no. 1, 86–132. **98e:35017**

[29] D. Daners, Existence and perturbation of principal eigenvalues for a periodic-parabolic problem, in: Proceedings of the Conference on Nonlinear Differential Equations (Coral Gables, FL, 1999), Electron. J. Differ. Equ. Conf., **5**, Southwest Texas State Univ., San Marcos, TX, 2000, 51–67. **MR 2001j:35125**

[30] D. Daners, Heat kernel estimates for operators with boundary conditions, *Math. Nachr.* **217** (2000), 13–41. **MR 2002f:35109**

[31] D. Daners, Robin boundary value problems on arbitrary domains, *Trans. Amer. Math. Soc.* **352** (2000), no. 9, 4207–4236. **MR 2000m:35048**

[32] D. Daners, Dirichlet problems on varying domains, *J. Differential Equations* **188** (2003), no. 2, 591–624. **MR 2004a:35042**

[33] D. Daners, Domain perturbation for linear and nonlinear parabolic equations, *J. Differential Equations* **129** (1996), no. 2, 358–402. **MR 97h:35071**

[34] D. Daners, Perturbation of semi-linear evolution equations under weak assumptions at initial time, *J. Differential Equations* **210** (2005), no. 2, 352–382. **MR 2005j:34080**

[35] D. Daners and P. Koch Medina, "Abstract Evolution Equations, Periodic Problems and Applications," Pitman Res. Notes Math. Ser., **279**, Longman/Wiley, Harlow/New York, 1992. **MR 94b:34002**

[36] R. Dautray and J.-L. Lions, "Mathematical Analysis and Numerical Methods for Science and Technology. Vol. 5: Evolution Problems. I," with the collaboration of M. Artola, M. Cessenat and H. Lanchon, translated from the French by A. Craig, Springer, Berlin, 1992. **MR 92k:00006**

[37] J. Duan, K. Lu and B. Schmalfuß, Invariant manifolds for stochastic partial differential equations, *Ann. Probab.* **31** (2003), no. 4, 2109–2135. **MR 2004m:60136**

[38] J. Duan, K. Lu and B. Schmalfuß, Smooth stable and unstable manifolds for stochastic evolutionary equations, *J. Dynam. Differential Equations* **16** (2004), no. 4, 949–972. **MR 2005j:60124**

[39] L. C. Evans, "Partial Differential Equations," Grad. Stud. Math., **19**, American Mathematical Society, Providence, RI, 1998. **MR 99e:35001**

[40] E. B. Fabes, M. V. Safonov and Y. Yuan, Behavior near the boundary of positive solutions of second order parabolic equations II. *Trans. Amer. Math. Soc.* **351** (1999), no. 12, 4947–4961. **MR 2000c:35085**

[41] R. A. Fisher, Gene frequencies in a cline determined by selection and diffusion, *Biometrics* **6** (1950), 353–361.

[42] G. B. Folland, "Real Analysis. Modern Techniques and Their Applications," second edition, Pure Appl. Math. (N. Y.), Wiley, New York, 1999. **MR 2000c:00001**

[43] A. Friedman, "Partial Differential Equations of Parabolic Type," Prentice–Hall, Englewood Cliffs, N.J., 1964. **MR 31 #6062**

[44] A. Friedman, "Partial Differential Equations," Holt, Rinehart and Winston, New York–Montreal, Que.–London, 1969. **MR 56 #3433**

[45] V. Girault and P.-A. Raviart, "Finite Element Methods for Navier–Stokes Equations. Theory and Algorithms," Springer Ser. Comput. Math., **5**, Springer, Berlin, 1986. **MR 88b:65129**

[46] J.K. Hale, "Asymptotic Behavior of Dissipative Systems," Math. Surveys Monogr, **25**, American Mathematical Society, Providence, RI, 1988. **MR 89g:58059**

[47] J.K. Hale and P. Waltman, Persistence in infinite-dimensional systems, *SIAM J. Math. Anal.* **20** (1989), no. 2, 388–395. **MR 90b:58156**

[48] D. Henry, "Geometric Theory of Semilinear Parabolic Equations," Lecture Notes in Math., **840**, Springer, Berlin–New York, 1981. **MR 83j:35084**

[49] P. Hess, An isoperimetric inequality for the principal eigenvalue of a periodic-parabolic problem, *Math. Z.* **194** (1987), no. 1, 121–125. **MR 88d:35091**

[50] P. Hess, "Periodic-Parabolic Boundary Value Problems and Positivity," Pitman Res. Notes Math. Ser., **247**, Longman/Wiley, Harlow/New York, 1991. **MR 92h:35001**

[51] P. Hess and P. Poláčik, Boundedness of prime periods of stable cycles and convergence to fixed points in discrete monotone dynamical systems, *SIAM J. Math. Anal.* **24** (1993), no. 5, 1312–1330. **MR 94i:47087**

[52] P. Hess and H. Weinberger, Convergence to spatial-temporal clines in the Fisher equation with time-periodic fitness, *J. Math. Biol.* **28** (1990), no. 1, 83–98. **MR 90k:35125**

[53] G. Hetzer and W. Shen, Uniform persistence, coexistence, and extinction in almost periodic/nonautonomous competition diffusion systems, *SIAM J. Math. Anal.* **34** (2002), no. 1, 204–227. **MR 2003m:37143**

[54] G. Hetzer, W. Shen and S. Zhu, Asymptotic behavior of positive solutions of random and stochastic parabolic equations of Fisher and Kolmogorov types, *J. Dynam. Differential Equations* **14** (2002), no. 1, 139–188. **MR 2003e:60140**

[55] M. W. Hirsch, "Differential Topology," corrected reprint of the 1976 original, Grad. Texts in Math., **33**, Springer, New York, 1994. **MR 96c:57001**

[56] M. W. Hirsch, Stability and convergence in strongly monotone dynamical systems, *J. Reine Angew. Math.* **383** (1988), 1–53. **MR 89c:58108**

[57] M. W. Hirsch and H. L. Smith, Monotone maps: a review, *J. Difference Equ. Appl.* **11** (2005), no. 4–5, 379–398. **MR 2006b:37001**

[58] M. W. Hirsch, H. L. Smith and X.-Q. Zhao, Chain transitivity, attractivity, and strong repellors for semidynamical systems, *J. Dynam. Differential Equations* **13** (2001), no. 1, 107–131. **MR 2002a: 37014**

[59] J. Húska, Harnack inequality and exponential separation for oblique derivative problems on Lipschitz domains, *J. Differential Equations* **226** (2006), no. 2, 541–557. **MR 2007h:35144**

[60] J. Húska and P. Poláčik, The principal Floquet bundle and exponential separation for linear parabolic equations, *J. Dynam. Differential Equations* **16** (2004), no. 2, 347–375. **MR 2006e:35147**

[61] J. Húska, P. Poláčik and M. V. Safonov, Harnack inequality, exponential separation, and perturbations of principal Floquet bundles for linear parabolic equations, *Ann. Inst. H. Poincaré Anal. Non Linéaire* **24** (2007), no. 5, 711–739. **MR 2348049**

[62] V. Hutson, W. Shen and G. T. Vickers, Estimates for the principal spectrum point for certain time-dependent parabolic operators, *Proc. Amer. Math. Soc.* **129** (2000), no. 6, 1669–1679. **MR 2001m:35243**

[63] A. V. Ivanov, The Harnack inequality for generalized solutions of quasilinear parabolic equations of second order, *Dokl. Akad. Nauk SSSR* **173** (1967), 752–754; English translation: *Soviet Math. Dokl.* **8** (1967), 463–466. **MR 35 #4598**

[64] J.-F. Jiang and X.-Q. Zhao, Convergence in monotone and uniformly stable skew-product semiflows with applications, *J. Reine Angew. Math.* **589** (2005), 21–55. **MR 2006k:37031**

[65] R. A. Johnson, K. J. Palmer and G. R. Sell, Ergodic properties of linear dynamical systems, *SIAM J. Math. Anal.* **18** (1987), no. 1, 1–33. **MR 88a:58112**

[66] B. Kawohl, "Rearrangements and Convexity of Level Sets in PDE," Lecture Notes in Math., **1150**, Springer, Berlin, 1985. **MR 87a:35001**

[67] U. Krengel, "Ergodic Theorems," with a supplement by A. Brunel, de Gruyter Stud. Math., **6**, Walter de Gruyter, Berlin, 1985. **MR 87i:28001**

[68] S. N. Kruzhkov [S. N. Kružkov] and I. M. Kolodiĭ, A priori estimates and Harnack's inequality for generalized solutions of degenerate quasi-linear parabolic equations, *Sibirsk. Mat. Zh.* **18** (1977), no. 3, 608–628; English translation: *Siberian Math. J.* **18** (1977), no. 3, 434–449. **MR 57 #10246**

[69] N. V. Krylov and M. V. Safonov, A property of the solutions of parabolic equations with measurable coefficients (Russian), *Izv. Akad. Nauk SSSR Ser. Mat.* **44** (1980) no. 1, 161–175. **MR 83c:35059**

[70] O. A. Ladyzhenskaya [O. A. Ladyženskaja], V. A. Solonnikov and N. N. Ural'tseva [N. N. Ural'ceva], "Linear and Quasilinear Equations of Parabolic Type," translated from the Russian by S. Smith, Transl. Math. Monogr., Vol. **23**, American Mathematical Society, Providence, RI, 1967. **MR 39 #3159b**

[71] J. A. Langa, J. C. Robinson and A. Suárez, Pullback permanence in a non-autonomous competitive Lotka–Volterra model, *J. Differential Equations* **190** (2003), no. 1, 214–238. **MR 2004a:35098**

[72] Z. Lian and K. Lu, Lyapunov exponents and invariant manifolds for random dynamical systems in a Banach space, preprint.

[73] G. M. Lieberman, "Second Order Parabolic Differential Equations," World Scientific, River Edge, NJ, 1996. **MR 98k:35003**

[74] T.-W. Ma, "Banach–Hilbert Spaces, Vector Measures and Group Representations," World Scientific, River Edge, NJ, 2002. **MR 2003e:00004**

[75] R. Mañé, "Lyapounov Exponents and Stable Manifolds for Compact Transformations," Geometric Dynamics (Rio de Janeiro, 1981), Lecture Notes in Math., **1007**, Springer, Berlin, 1983, 522–577. **MR 85j:58126**

[76] J. Mierczyński, Flows on ordered bundles, unpublished manuscript, 1991.

[77] J. Mierczyński, Globally positive solutions of linear parabolic PDEs of second order with Robin boundary conditions, *J. Math. Anal. Appl.* **209** (1997), no. 1, 47–59. **MR 98c:35071**

[78] J. Mierczyński, Globally positive solutions of linear parabolic partial differential equations of second order with Dirichlet boundary conditions, *J. Math. Anal. Appl.* **226** (1998), no. 2, 326–347. **MR 99m:35096**

[79] J. Mierczyński, The principal spectrum for linear nonautonomous parabolic PDEs of second order: Basic properties, *in:* Special issue in celebration of Jack K. Hale's 70th birthday, Part 2 (Atlanta, GA/Lisbon, 1998), *J. Differential Equations* **168** (2000), no. 2, 453–476. **MR 2001m:35147**

[80] J. Mierczyński and S. J. Schreiber, Kolmogorov vector fields with robustly permanent subsystems, *J. Math. Anal. Appl.* **267** (2002), no. 1, 329–337. **MR 2003b:37041**

[81] J. Mierczyński and W. Shen, Exponential separation and principal Lyapunov exponent/spectrum for random/nonautonomous parabolic equations, *J. Differential Equations* **191** (2003), no. 1, 175–205. **MR 2004h:35232**

[82] J. Mierczyński and W. Shen, The Faber–Krahn inequality for random/nonautonomous parabolic equations, *Commun. Pure Appl. Anal.* **4** (2005), no. 1, 101–114. **MR 2006b:35358**

[83] J. Mierczyński and W. Shen, Lyapunov exponents and asymptotic dynamics in random Kolmogorov models, *J. Evol. Equ.* **4** (2004), no. 3, 371–390. **MR 2005k:37122**

[84] J. Mierczyński and W. Shen, Time averaging for nonautonomous/random linear parabolic equations, *Discrete Contin. Dyn. Syst.*, to appear.

[85] J. Mierczyński, W. Shen and X.-Q. Zhao, Uniform persistence for nonautonomous and random parabolic Kolmogorov systems, *J. Differential Equations* **204** (2004), no. 2, 471–510. **MR 2006f:37111**

[86] M. Miklavčič, "Applied Functional Analysis and Partial Differential Equations," World Scientific, River Edge, NJ, 1998. **MR 2001k:47001**

[87] X. Mora, Semilinear parabolic problems define semiflows on C^k spaces, *Trans. Amer. Math. Soc.* **278** (1983), no. 1, 21–55. **MR 84e:35077**

[88] J. Moser, A Harnack inequality for parabolic differential equations, *Comm. Pure Appl. Math.* **17** (1964), 101–134. **MR 28 #2357**

[89] V. V. Nemytskiĭ and V. V. Stepanov, "Qualitative Theory of Differential Equations," Princeton Math. Ser., No. **22** Princeton University Press, Princeton, N.J., 1960. **MR 22 #12258**

[90] J. Oxtoby, Ergodic sets, *Bull. Amer. Math. Soc.* **58** (1952), 116–136. **MR 13,850e**

[91] F. D. Penning, Continuous dependence results for parabolic problems with time-dependent coefficients, *Quaestiones Math.* **14** (1991), no. 1, 33–49. **MR 92b:35021**

[92] P. Poláčik, On uniqueness of positive entire solutions and other properties of linear parabolic equations, *Discrete Contin. Dyn. Syst.* **12** (2005), no. 1, 13–26. **MR 2005k:35170**

[93] P. Poláčik and I. Tereščák, Convergence to cycles as a typical asymptotic behavior in smooth strongly monotone discrete-time dynamical systems, *Arch. Rational Mech. Anal.* **116** (1991), no. 4, 339–360. **MR 93b:58088**

[94] P. Poláčik and I. Tereščák, Exponential separation and invariant bundles for maps in ordered Banach spaces with applications to parabolic equations, *J. Dynam. Differential Equations* **5** (1993), no. 2, 279–303; Erratum, *J. Dynam. Differential Equations* **6** (1994), no. 1, 245–246. **MR 94d:47064**

[95] D. Ruelle, Analyticity properties of the characteristic exponents of random matrix products, *Adv. in Math.* **32** (1979), no. 1, 68–80. **MR 80e:58035**

[96] D. Ruelle, Characteristic exponents and invariant manifolds in Hilbert space, *Ann. of Math. (2)* **115** (1982), no. 2, 243–290. **MR 83j:58097**

[97] R. J. Sacker and G. R. Sell, Dichotomies for linear evolutionary equations in Banach spaces, *J. Differential Equations* **113** (1994), no. 1, 17–67. **MR 96k:34136**

[98] B. Schmalfuß, Attractors for nonautonomous and random dynamical systems perturbed by impulses, *Discrete Contin. Dyn. Syst.* **9** (2003), no. 3, 727–744. **MR 2004c:37112**

[99] Š. Schwabik and G. Ye, "Topics in Banach Space Integration," Ser. Real Anal., **10**, World Scientific, Hackensack, NJ, 2005. **MR 2006g:28002**

[100] W. Shen and G. T. Vickers, Spectral theory for general nonautonomous/random dispersal evolution operators, *J. Differential Equations* **235** (2007), no. 1, 262–297. **MR 2309574**

[101] W. Shen and Y. Yi, Convergence in almost periodic Fisher and Kolmogorov models, *J. Math. Biol.* **37** (1998), no. 1, 84–102. **MR 99k:92037**

[102] W. Shen and Y. Yi, "Almost Automorphic and Almost Periodic Dynamics in Skew-product Semiflows, Part II. Skew-product Semiflows," *Mem. Amer. Math. Soc.* **136** (1998), no. 647, 23–52. **MR 99d:34088**

[103] I. Tereščák, Dynamics of C^1 smooth strongly monotone discrete-time dynamical systems, preprint.

[104] H. R. Thieme, Uniform persistence and permanence for non-autonomous semiflows in population biology, *Math. Biosci.* **166** (2000), no. 2, 173–201. **MR 2001j:34102**

[105] H. Triebel, "Interpolation Theory, Function Spaces, Differential Operators," North-Holland Math. Library, **18**, North-Holland, Amsterdam–New York, 1978. **MR 80i:46032b**

[106] P.-A. Vuillermot, Almost periodic attractors for a class of nonautonomous reaction diffusion equations on $\mathbb{R}^N$, I. Global stabilization processes, *J. Differential Equations* **94** (1991), no. 2, 228–253. **MR 93c:35076a**

[107] P. Walters, "An Introduction to Ergodic Theory," Grad. Texts in Math., **79**, Springer, New York–Berlin, 1982. **MR 84e:28017**

[108] P. Waltman, A brief survey of persistence in dynamical systems, *in:* Delay Differential Equations and Dynamical Systems (Claremont, CA, 1991), Lecture Notes in Math., **1475**, Springer, Berlin, 1991, 31–40. **MR 92j:34093**

[109] A. Yagi, On the abstract linear evolution equations in Banach spaces, *J. Math. Soc. Japan* **28** (1976), no. 2, 290–303. **MR 53 #1337**

[110] X.-Q. Zhao, Uniform persistence in processes with application to nonautonomous competitive models, *J. Math. Anal. Appl.* **258** (2001), no. 1, 87–101. **MR 2002b:92033**

[111] X.-Q. Zhao, Global attractivity in monotone and subhomogeneous almost periodic systems, *J. Differential Equations* **187** (2003), no. 2, 494–509. **MR 2004i:37157**

Index